Baruch S. Blumberg

Molecular Biology of the Hepatitis B Virus

Molecular Biology of the Hepatitis B Virus

Editor

Alan McLachlan, Ph.D.
Associate Member
Division of Biochemistry
Department of Molecular and Experimental Medicine
The Scripps Research Institute
La Jolla, California

CRC Press
Boca Raton Ann Arbor Boston London

Library of Congress Cataloging-in-Publication Data

McLachlan, Alan
Molecular biology of the hepatitis B virus / Alan McLachlan
p. cm.
Includes bibilographical references and index.
ISBN 0-8493-5516-8
1. Hepatitis B virus. 2. Biology—molecular. I. McLachlan, Alan. II. Title.
[DNLM: 1. Hepatitis B virus. QW 710 G289h]
QR749 .H64G78
616'.0149—dc20
DNLM/DLC
for Library of Congress 91-32641
CIP

Direct all inquiries to CRC Press, Inc., 2000 Corporate Blvd., N. W., Boca Raton, Florida, 33431.

International Standard Book Number 0-8493-5516-8

Library of Congress Card Number 91-32641
Printed in the United States 0 1 2 3 4 5 6 7 8 9

PREFACE

The principal aim of this volume, *Molecular Biology of the Hepatitis B Virus,* is to present a comprehensive and precise account of the current state of knowledge regarding the various molecular aspects of the life cycle of the hepatitis B virus (HBV). The areas of the molecular biology of HBV covered include the animal model systems, sequence data on the hepadnavirus genomes, the transcripts coded for by the viral genome and the sequence elements involved in regulating their expression, hepadnavirus replication, analysis of the various HBV gene products and their role in virion synthesis and assembly, a description of the consequences of long-term exposure to hepadnavirus infection and its association with hepatocellular carcinoma, the use of recombinant technologies in the generation of second generation vaccines, and the utilization of recombinant technologies to analyze an immune mediated disease. The volume, therefore, serves as a detailed source of information on the molecular aspects of hepadnavirus biology and contains only enough clinical and immunological data to place the molecular data in the appropriate context for an immunologically mediated disease.

THE EDITOR

Alan McLachlan, Ph.D., is an Associate Member in the Division of Biochemistry, Department of Molecular and Experimental Medicine at The Scripps Research Institute in La Jolla, California.

Dr. McLachlan received his B.Sc. and Ph.D. degrees in 1977 and 1980, respectively, from the Department of Biochemistry, University of Aberdeen, Aberdeen, Scotland. After doing postdoctoral work at Harvard University, Cambridge, Massachusetts, he was appointed an Assistant Member at the Research Institute of Scripps Clinic in 1986.

Dr. McLachlan is the recipient of research grants from the National Institutes of Health. He has published more than 50 papers. His current research interests are in gene regulation and the biogenesis of hepatitis B virus.

DEDICATION

This book is dedicated to my father and mother, John and Margaret McLachlan, for the sacrifices they made to guarantee that I was given the best possible educational opportunities.

ACKNOWLEDGMENTS

I am grateful to all of the authors for their contributions. Their efforts are greatly appreciated. In addition, I am especially grateful to Anneke Raney for her assistance with every stage and aspect of the generation of this book. I am indebted to Judith Preston and Lynn Keyes for their dedication in preparing the final manuscript.

CONTRIBUTORS

Ronald W. Ellis, Ph.D.
Senior Director and Head
Department of Cellular and
Molecular Biology
Merck Sharpe and Dohme Research Labs
West Point, Pennsylvania

Doris Fernholz, M.S.
Max-Planck-Institut für Biochemie
Martinsried, Germany

Markus Fischer
ZMBH University of Heidelberg
Heidelberg, Germany

Wolfram H. Gerlich, Ph.D.
Professor
Department of Medical Microbiology
Georg-August-University
Göttingen, Germany

Klaus-Hinrich Heermann, M.D.
Professor
Department of Medical Microbiology
Georg-August-University
Göttingen, Germany

Peter J. Kniskern, Ph.D.
Associate Director
Department of Cellular and
Molecular Biology
Merck Sharpe and Dohme Research Labs
West Point, Pennsylvania

Patricia L. Marion, Ph.D.
Senior Research Scientist
Department of Infectious Disease
Stanford University
Stanford, California

Kenichi Matsubara, Ph.D.
Institute for Molecular and Cellular Biology
Osaka University
Suita, Osaka, Japan

Alan McLachlan, Ph.D.
Associate Member
Department of Molecular
and Experimental Medicine
The Scripps Research Institute
La Jolla, California

David R. Milich, Ph.D.
Associate Member
Department of Molecular Biology
The Scripps Research Institute
La Jolla, California

Roger H. Miller, Ph.D.
Senior Staff Fellow
Hepatitis Viruses Section
National Institutes of Health
Bethesda, Maryland

Christine Pourcel, Ph.D.
Institut Pasteur
Paris, France

Anneke K. Raney
Research Assistant
Department of Molecular and
Experimental Medicine
The Scripps Research Institute
La Jolla, California

Heinz Schaller, Dr.rer.nat.
Professor
ZMBH University of Heidelberg
Heidelberg, Germany

Nancy Schek, Ph.D.
Professor
ZMBH University of Heidelberg
Heidelberg, Germany

H.-J. Schlicht, Ph.D.
Department of Virology
University of Ulm
Ulm, Germany

Ralf Schneider, M.S.
Max-Planck-Institut für Biochemie
Martinsried, Germany

Florian Schödel, M.D.
Max-Planck-Institut für Biochemie
Martinsried, Germany

Christoph Seeger, Ph.D.
Fox Chase Cancer Center
Philadelphia, Pennsylvania

Yosef Shaul, Ph.D.
Senior Scientist
Department of Molecular Genetics
and Virology
The Weizmann Institute of Science
Rehovot, Israel

Robert J. Shepherd, Ph.D.
Department of Plant Pathology
University of Kentucky
Lexington, Kentucky

Aleem Siddiqui, Ph.D.
Associate Professor
Department of Microbiology
University of Colorado Medical School
Denver, Colorado

Rolf Sprengel, Ph.D.
Max-Planck-Institut für Biochemie
Martinsried, Germany

David N. Standring, Ph.D.
Assistant Professor
Hormone Research Institute
University of California School of Medicine
San Francisco, California

Thomas Weimer, Ph.D.
Max-Planck-Institut für Biochemie
Martinsried, Germany

Gerhild Wildner, Ph.D.
Max-Planck-Institut für Biochemie
Martinsried, Germany

Hans Will, Ph.D.
Max-Planck-Institut für Biochemie
Martinsried, Germany

TABLE OF CONTENTS

Chapter 1

THE BIOLOGY OF HEPATITIS B VIRUS

Anneke K. Raney and Alan McLachlan

TABLE OF CONTENTS

I. INTRODUCTION

Hepatitis B virus (HBV) is the prototype member of a family of DNA viruses that primarily infect the liver and share a similar viral morphology and cellular life cycle.[1] The other identified and characterized members of this family are the woodchuck hepatitis virus (WHV),[2] the ground squirrel hepatitis virus (GSHV),[3] the duck hepatitis B virus (DHBV),[4] and the heron hepatitis B virus (HHBV).[5] In addition to these five viruses, the tree squirrel hepatitis B virus (TSHV) possibly represents an additional member of this virus family.[6] As a consequence of the unique features of this group of viruses, they have been classified as a separate family of viruses known as the hepadnaviruses.[1] The name *hepadna*virus reflects the *hepa*totropism of these *DNA* viruses.

II. THE HISTORY OF THE DISCOVERY, ISOLATION, AND CHARACTERIZATION OF HEPATITIS B VIRUS

The path that ultimately led to the discovery of one of the viruses, hepatitis B virus (HBV), responsible for parenterally transmitted hepatitis (serum or type B hepatitis), began in 1965 with the observation by Dr. Baruch Blumberg and colleagues of a precipitin reaction between the sera from an Australian aborigine and a frequently transfused hemophilia patient from New York city.[7] The lipoprotein present in the serum of the aborigine responsible for the precipitin reaction was called "Australia antigen." The subsequent observation that Australia (Au) antigen, or serum hepatitis (SH) antigen as it was also named,[8] was present at a much higher frequency in the sera of acute and chronic hepatitis patients than in control subjects led to the hypothesis that this antigen may be associated with an infectious agent responsible for "viral hepatitis."[9-15] With a view to testing this idea, electron microscopic analysis of Au antigen-positive sera revealed particles that reacted with antibodies against Au antigen.[16] These particles were predominantly spheres and filaments approximately 22 nm in diameter (Figure 1). The length of the filaments varied from less than 50 nm up to 1000 nm.[16-19] In addition to these forms of Au antigen, a larger particle, the Dane particle, which was much less abundant than the smaller particles,[20] was found subsequently in the sera of serum hepatitis patients.[17,21,22] This particle is 42 nm in diameter and comprises a 28-nm diameter inner body, the nucleocapsid or core, surrounded by a 7-nm outer coat (Figure 1). Since the 22-nm spheres and filaments aggregated with the larger particles in the presence of antibodies against Au antigen, it was suggested that all of these particles shared a common surface or envelope antigen,[17] subsequently called hepatitis B surface antigen (HBsAg).[23]

The identification of the various particulate forms in the sera of hepatitis patients did not resolve which, if any of these particles, represented the infectious agent. On the basis of morphology, it was suggested that the 42-nm Dane particle represented the agent responsible for serum hepatitis and the 22-nm spheres and filaments represented excess virus coat material.[17] Support for this suggestion came from the observation that the 22-nm spheres appeared to lack nucleic acid.[24] Further analysis of Dane particle structure was achieved by detergent treatment that released the inner body, the nucleocapsid, as a spherical 28-nm component that can form aggregates in the presence of posthepatitis but not prehepatitis sera.[25] This represented the identification of an additional antigen–antibody system specific for Au antigen-positive hepatitis. It also permitted further physical and biochemical characterization of the various viral and subviral components present in the sera of serum hepatitis patients.

The physical characterization of the 22-nm spheres demonstrated that these particles had an estimated buoyant density in cesium chloride (CsCl) of 1.18 to 1.22 g/cm^3 and a sedimentation coefficient in the range from 40 to 54.[24,26-28] From further analysis, the approximate

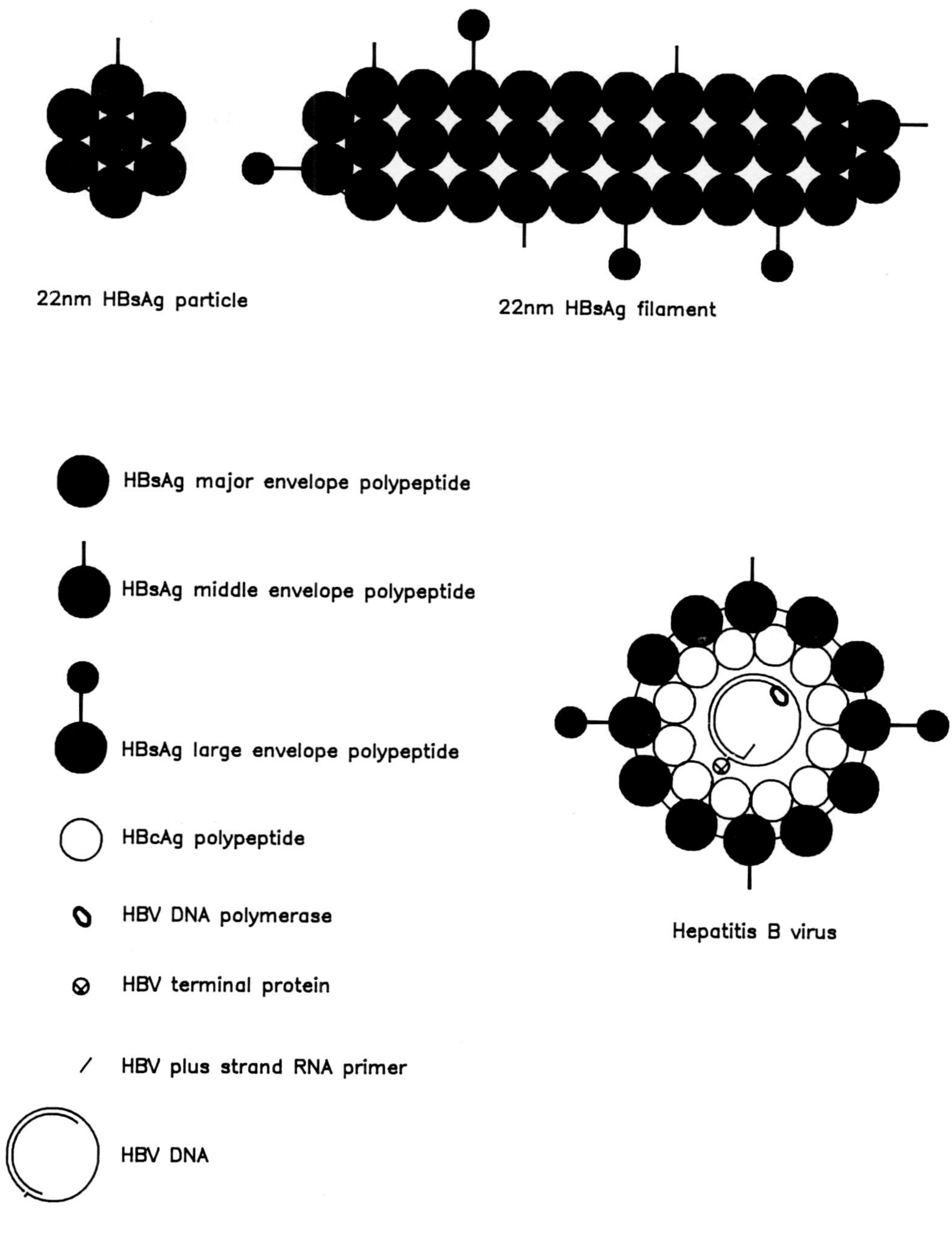

FIGURE 1. Diagrammatic representation of the structure of the 22-nm hepatitis B surface antigen (HBsAg) sphere, 22-nm HBsAg filament, and hepatitis B virus (Dane particle). HBcAg, hepatitis B core antigen.

molecular weight of 2 to 4 × 10^6 was estimated for the 22-nm sphere.[26,28] The density of the 22-nm filaments is similar to that of the 22-nm spheres.[19,28] The 28-nm nucleocapsid of the Dane particle was shown to have a density in CsCl of 1.30 to 1.36 g/cm^3 [29-33] and a sedimentation coefficient of approximately 110.[34] The density of the Dane particle, 1.24 to 1.27 g/cm^3,[31,33] was found to be intermediate between those of the 22-nm spheres and the 28-nm inner body, as might be expected based on its composition. Dane particles and nucleocapsids have been shown to exist as two subpopulations with different densities.[31] The less dense populations lack the HBV DNA and represent defective particles, whereas the more dense populations contain HBV DNA.[31]

Characterization of the physical and biochemical properties of the various viral and subviral particles demonstrated that there was a DNA polymerase activity and endogenous primer–template complex tightly associated with the nucleocapsid of the Dane particle.[34,35] The DNA polymerase activity was not associated with the 22-nm surface antigen particles.[34] The DNA template that represented the substrate for this polymerase was isolated from the nucleocapsid of the Dane particle and by electron microscopy was shown to be a circular molecule of 0.78 μm.[36] A double-stranded DNA molecule of this length has a molecular weight of 1.6×10^6, which corresponds to approximately 2450 nucleotide pairs.[36,37] This was the first evidence that the genome of HBV was very small. Restriction enzyme and electron microscopic analyses of the HBV genome before and after the Dane particle DNA polymerase reaction demonstrated that the majority of circular DNA molecules within the virion possessed a single-stranded region encompassing 15 to 50% of the length of the genome.[38-40] The endogenous viral DNA polymerase reaction is responsible for the conversion of this single-stranded region to double-stranded DNA.[38-40] Based on these studies,[38-40] the size of the HBV genome after modification by the endogenous DNA polymerase reaction was reestimated to be approximately 3200 nucleotide pairs. Therefore, the HBV genome consists of a long strand of 3200 nucleotides and a short strand of variable length. Subsequent analysis demonstrated that the long strand of the viral DNA also contained a nick or short gap.[41] In addition, it was shown that this discontinuity in the long strand and the nick or short gap remaining in the short strand after completion of the endogenous DNA polymerase reaction are located at unique positions approximately 226 nucleotides apart in the HBV DNA.[41-43] Hence, the circularity of the genome is maintained by the approximately 226-nucleotide 5′-terminal cohesive overlap between the ends of the long and short DNA strands[41] (Figure 1). Additional studies of the HBV genome demonstrated that there is a protein, the terminal protein, covalently attached to the 5′ terminus of the long strand.[44] There is probably also a short oligoribonucleotide attached to the 5′ end of the short strand of the HBV genome in the virion,[43] as observed in DHBV,[45] GSHV,[46] and WHV.[46]

The inability to infect permanent cell lines in culture or standard laboratory animals with HBV restricted the source of virus to patients' sera. This limited the amount of HBV DNA that was available for study. However, the detailed analysis of the structure of the viral DNA permitted the cloning[47,48] and subsequent sequencing of the complete HBV genome.[49-61] The HBV genome sizes varied between 3182 and 3221 nucleotides (Figure 2). As a consequence of cloning the HBV genome, large amounts of pure HBV DNA of defined sequence could be obtained, analyzed, and manipulated. In addition, the infectivity and functionality of cloned DNA was tested by inoculating the complete HBV genome into the livers of chimpanzees.[62,63] The observation of a typical, self-limited, acute hepatitis with hepatitis B surface antigenemia indicated that a nondefective HBV genome had been introduced into the livers of the chimpanzees, and therefore the cloned DNA encoded all of the essential information to complete the viral life cycle.[62,63] This critical experiment verified that the 3200-nucleotide, partially single-stranded DNA present in the Dane particle represents the HBV genome and that the Dane particle is almost certainly the infectious agent responsible for serum or type B hepatitis. Further confirmation of the biological activity of cloned HBV came from transfection experiments where the complete HBV genome was introduced into various hepatoma cell lines or transgenic mice and Dane particles were subsequently produced.[33,64-69] In two cases, the Dane particles secreted by the cell lines were shown to be infectious in chimpanzees.[70,71]

```
                  10                  30                  50
                   .         .         .         .         .         .
adw2    :  AATTCCACTGCCTTCCACCAAACTCTGCAGGATCCCAGAGTCAGGGGTCTGTATCTTCCT
adw     :                G      G
adr(1)  :    C     AA A         G      TA           G     C  A   T
adr(2)  :          AA A         G      TA           G     C  A   T
ayr     :    C     AA A         G      TA           G     C  A   T
ayw(1)  :          AA                   A           G  A  C      T C
ayw(2)  :    C     AA                   A           G  A  C      T C

                  70                  90                 110
                   .         .         .         .         .         .
adw2    :  GCTGGTGGCTCCAGTTCAGGAACAGTAAACCCTGCTCCGAATATTGCCTCTCACATCTCG
adw     :
adr(1)  :                   C                T     C  C  T   A C   A
adr(2)  :                   C                T     C  C      A C   A
ayr     :                   C                T     C  C      A C   A
ayw(1)  :                                    T  T  C  C        C T A
ayw(2)  :                                    T     C  C  T     C   A

                 130                 150                 170
                   .         .         .      Si  .        .         .
adw2    :  TCAATCTCCGCGAGGACTGGGGACCCTGTGACGAACATGGAGAACATCACATCAGGATTC
adw     :                                    T
adr(1)  :         T T                  CAC            G  CA
adr(2)  :         T T                  CAC               CA
ayr     :         T T                  CAC            G  CA
ayw(1)  :         T T      T           C CT
ayw(2)  :         T T      T           C CT

                 190                 210                 230
                   .         .         .         .         .         .
adw2    :  CTAGGACCCCTGCTCGTGTTACAGGCGGGGTTTTTCTTGTTGACAAGAATCCTCACAATA
adw     :
adr(1)  :
adr(2)  :
ayr     :
ayw(1)  :             T
ayw(2)  :

                 250                 270                 290
                   .         .         .         .         .         .
adw2    :  CCGCAGAGTCTAGACTCGTGGTGGACTTCTCTCAATTTTCTAGGGGGATCTCCCGTGTGT
adw     :                                                    A
adr(1)  :    A                                             G A   AC
adr(2)  :    A                                             G A   AC
ayr     :    A                                             G A   AC
ayw(1)  :                                                  A  A
ayw(2)  :                                                  A  A

                 310                 330                 350
                   .         .         .         .         .         .
adw2    :  CTTGGCCAAAATTCGCAGTCCCCAACCTCCAATCACTCACCAACCTCCTGTCCTCCAATT
adw     :
adr(1)  :   C                                             T
adr(2)  :   C                                             T
ayr     :   C                                             T          C
ayw(1)  :                                                 T          C
ayw(2)  :                                                            C
```

FIGURE 2. Comparison of the HBV DNA sequences of the major subtypes. The HBV genomes are subtypes *adw*$_2$,[52] *adw*,[54] *adr* (sequences 1[61] and 2[54]), *ayr*,[60] and *ayw* (sequences 1[50] and 2[56]). The sequences were aligned with the *adw*$_2$ sequence using the first A residue of the *Eco*RI site as nucleotide 1. The differences between the *adw*$_2$ sequence and the other sequences are indicated. Dots indicate the location of gaps necessary to permit maximum alignment of the nucleotide sequences. PS1^{i}a and PS1^{i}b, initiation codons for the 119- and 108-amino acid preS1 regions, respectively; PS2^i, initiation codon for preS2 region; S^i and S^t, initiation and termination codon for the major HBsAg open reading frame; PC^i, initiation codon for the precore region; C^i and C^t, initiation and termination codon for the core open reading frame; X^i and X^t, initiation and termination codon for the X gene open reading frame; P^i and P^t, initiation and termination codon for the polymerase open reading frame.

```
               370                 390                 410
                 .         .         .         .         .         .
adw2  : TGTCCTGGTTATCGCTGGATGTGTCTGCGGCGTTTTATCATATTCCTCTTCATCCTGCTG
adw   :
adr(1):
adr(2):         C
ayr   :         C
ayw(1):                                          C
ayw(2):                                          C

               430                 450                 470
                 .         .         .         .         .         .
adw2  : CTATGCCTCATCTTCTTATTGGTTCTTCTGGATTATCAAGGTATGTTGCCCGTTTGTCCT
adw   :
adr(1):                  G              C  C
adr(2):                  G              C  C
ayr   :                  G              C  C
ayw(1):                  G              C
ayw(2):                  G              C

               490                 510                 530
                 .         .         .         .         .         .
adw2  : CTAATTCCAGGATCAACAACAACCAGTACGGGACCATGCAAAACCTGCACGACTCCTGCT
adw   :
adr(1):    C        A  T    T     C              G          T
adr(2):    C        A  T    C     C     G        G          T
ayr   :    C        A  T    T     C             G           T
ayw(1):               CT          C            CGG       T    A
ayw(2):               TT    T     C             G

               550                 570                 590
                 .         .         .         .         .         .
adw2  : CAAGGCAACTCTATGTTTCCCTCATGTTGCTGTACAAAACCTACGGATGGAAATTGCACC
adw   :              A
adr(1):      A C               T                  T    C     C     T
adr(2):      A C               T                  T    C     C     T
ayr   :      A C               T                  T    C     C     T
ayw(1):      A C        A      C           C      T    C
ayw(2):      A C        A      C           C      T    C

               610                 630                 650
                 .         .         .         .         .         .
adw2  : TGTATTCCCATCCCATCGTCCTGGGCTTTCGCAAAATACCTATGGGAGTGGGCCTCAGTC
adw   :
adr(1):                  A                G  T
adr(2):                  A                G  T
ayr   :                  A                G  T
ayw(1):                  A             G     T                    C
ayw(2):                  A             G     T                    C

               670                 690                 710
                 .         .         .         .         .         .
adw2  : CGTTTCTCTTGGCTCAGTTTACTAGTGCCATTTGTTCAGTGGTTCGTAGGGCTTTCCCCC
adw   :
adr(1):         C                                     C
adr(2):         C
ayr   :         C
ayw(1):         C
ayw(2):         C
```

FIGURE 2 (continued).

```
                 730                 750                 770
                  .         .         .         .         .         .
adw2   : ACTGTTTGGCTTTCAGCTATATGGATGATGTGGTATTGGGGGCCAAGTCTGTACAGCATC
adw    :
adr(1):                  T                                      A
adr(2):                  T                                      A
ayr    :                                                        A
ayw(1):                  T
ayw(2):                  T

                 790                 810                 830
                  .         .         .         .         .     st  .
adw2   : GTGAGTCCCTTTATACCGCTGTTACCAATTTTCTTTTGTCTCTGGGTATACATTTAAACC
adw    :
adr(1):  T           T    T  A                    T             G
adr(2):  T           T    T  A                    T
ayr    : T           T    T  A                    T             G
ayw(1):  T           T                            T
ayw(2):  T           T                      C     T

                 850                 870                 890
                  .         .         .         .         .         .
adw2   : CTAACAAAACAAAAAGATGGGGTTATTCCCTAAACTTCATGGGCTACATAAATTGGAAGTT
adw    :
adr(1):      T     C   C T     C  C     T           A  TG
adr(2):      T     C   C T     C  C     T           A  TG
ayr    :     T     C   C T     C  C     T           A  TG
ayw(1):               G           C  T     T  T     T  TG C      T
ayw(2):                           C  TT  C T           TG C      T

                 910                 930                 950
                  .         .         .         .         .         .
adw2   : GGGGAACTTTGCCACAGGATCATATTGTACAAAAGATCAAACACTGTTTTAGAAAACTTC
adw    :
adr(1):      T     A        A         TT   AC    G  A      C       G
adr(2):      T     A  G     A          T   AC    G  A      C     T G
ayr    :     T     A  G     A              AC    G  A      C     T G
ayw(1):  AT  GT C        A  A  C  CA       A      G A
ayw(2):  AT  GT A        A     C  CA    G  A      G A  C

                 970                 990                1010
                  .         .         .         .         .         .
adw2   : CTGTTAACAGGCCTATTGATTGGAAAGTATGTCAAAGAATTGTGGGTCTTTTGGGCTTTG
adw    :
adr(1):      A  T  A
adr(2):      A  T  C
ayr    :     A  T  A
ayw(1):    A                                C                   T
ayw(2):                              C      C T                 T

                1030                1050                1070
                  .         .         .         .         .         .
adw2   : CTGCTCCATTTACACAATGTGGATATCCTGCCTTAATGCCTTTGTATGCATGTATACAAG
adw    :
adr(1):      C  T              C           G                        T
adr(2):      C  T           C  C           G        A               T
ayr    :     C  T              C           G        A               T
ayw(1):      C  T              T        G  G                    T   T
ayw(2):      C  T              T        T                       T  GT
```

FIGURE 2 (continued).

```
                1090                1110                1130
                   .         .         .         .         .         .
adw2    : CTAAACAGGCTTTCACTTTCTCGCCAACTTACAAGGCCTTTCTAAGTAAACAGTACATGA
adw     :
adr(1)  :     G                          T           GT  C    A   C  C
adr(2)  :     G                          T           GT       A  TC
ayr     :     G                                      GT       A  TC A
ayw(1)  :     G                                      GT       A   C
ayw(2)  :  G  G        T                             GT       A   C

                1150                1170                1190
                   .         .         .         .         .         .
adw2    : ACCTTTACCCCGTTGCTCGGCAACGGCCTGGTCTGTGCCAAGTGTTTGCTGACGCAACCC
adw     :
adr(1)  :                 C         T A     C
adr(2)  :                 C         T A     C
ayr     :                 C         T A     C
ayw(1)  :                 C           A
ayw(2)  :                 C           A

                1210                1230                1250
                   .         .         .         .         .         .
adw2    : CCACTGGCTGGGGCTTGGCCATAGGCCATCAGCGCATGCGTGGAACCTTTGTGGCTCCTC
adw     :                 A
adr(1)  :        A                      G                       T
adr(2)  :        A                      G
ayr     :     G  T                      G
ayw(1)  :                   T   G                           TC
ayw(2)  :                   T   G                           C      G

                1270                1290                1310
                   .         .         .         .         .         .
adw2    : TGCCGATCCATACTGCGGAACTCCTAGCCGCTTGTTTTGCTCGCAGCCGGTCTGGAGCAA
adw     :
adr(1)  :                             A               GA
adr(2)  :                             A                             G
ayr     :                             A                             G
ayw(1)  :                                                A
ayw(2)  :                                                A

                1330                1350                1370
                   .         .         .         .         .     Xi  .
adw2    : AGCTCATCGGAACTGACAATTCTGTCGTCCTCTCGCGGAAATATACATCGTTTCCATGGC
adw     :                                                  A
adr(1)  :  A  T     G        C  G  T        T        C  C  C  C
adr(2)  :  A  T        C     C     T        T        C  C  C
ayr     :  A  T        C     C  A  T        T        C  C  C
ayw(1)  :  CA T     G     T  C     T     A  C  C
ayw(2)  :  CA TC    G  G  T  C     T  T     C  C             A

                1390                1410                1430
                   .         .         .         .         .         .
adw2    : TGCTAGGCTGTACTGCCAACTGGATCCTTCGCGGGACGTCCTTTGTTTACGTCCCGTCGG
adw     :
adr(1)  :     C  G   G                G                 C
adr(2)  :        G   G                G                 C
ayr     :            G                G                 C
ayw(1)  :            G                G
ayw(2)  :            G                G
```

FIGURE 2 (continued).

```
               1450                1470                1490
                  .         .         .         .         .         .
adw2   : CGCTGAATCCCGCGGACGACCCCTCTCGGGGCCGCTTGGGACTCTCTCGTCCCCTTCTCC
adw    :
adr(1) :                       G           T     C    AC      T G  TT
adr(2) :                       G           T     G    AC           T
ayr    :                       G           T     C    AC           T
ayw(1) :           T           T        T
ayw(2) :                       T        T

               1510                1530                1550
                  .         .         .         .         .         .
adw2   : GTCTGCCGTTCCAGCCGACCACGGGGCGCACCTCTCTTTACGCGGTCTCCCCGTCTGTGC
adw    :
adr(1) : C
adr(2) : T           G
ayr    : A           G
ayw(1) :             GA                                A
ayw(2) :           T GA                                A

               1570                1590                1610
                  .         .         .         .         .         .
adw2   : CTTCTCATCTGCCGGTCCGTGTGCACTTCGCTTCACCTCTGCACGTTGCATGGAGACCAC
adw    :                                                      C
adr(1) :                A                              C
adr(2) :                A                              C
ayr    :                A                              A
ayw(1) :                A                              C
ayw(2) :                A                              C

               1630                1650                1670
          pt      .         .         .         .         .         .
adw2   : CGTGAACGCCCATCAGATCCTGCCCAAGGTCTTACATAAGAGGACTCTTGGACTCCCAGC
adw    :
adr(1) :         G   C   G  T        C                          T
adr(2) :             C   G  T                                   T
ayr    :             C   G  T                C                  T
ayw(1) :             C  A  AT                                   T
ayw(2) :       A     A CAT  T                                   T T T

               1690                1710                1730
                  .         .         .         .         .         .
adw2   : AATGTCAACGACCGACCTTGAGGCCTACTTCAAAGACTGTGTGTTTAAGGACTGGGAGGA
adw    :
adr(1) :          A              A               T       A
adr(2) : C                       A                       A
ayr    :                         A               T       A
ayw(1) :                         A               T       A
ayw(2) :                         A               T       A

               1750                1770                1790
                  .         .         .         .         .         .
adw2   : GCTGGGGGAGGAGATTAGGTTAAAGGTCTTTGTATTAGGAGGCTGTAGGCACAAATTGGT
adw    :  T                     T A                          T
adr(1) :  T                                C                 T
adr(2) :  T                                C                 .........
ayr    :  T                                C                 T
ayw(1) :  T                                C                 T
ayw(2) :  T                A                                 T
```

(In the original, "TGA" after "CG" in the 1630 block adw2 line is underlined, with "pt" above it.)

FIGURE 2 (continued).

```
               1810                1830                1850
                .     PCi .         .       Xt.         .         .
adw2   : CTGCGCACCAGCACCATGCAACTTTTTCACCTCTGCCTAATCATCTCTTGTACATGTCCC
adw    :
adr(1) :    TT                                          A   T       T
adr(2) : ..................T                            A   T       T
ayr    :    TT                       C                  A   T       T
ayw(1) :                                                    T       T
ayw(2) :                                                    T       T

               1870                1890                1910
                .         .         .         .  Ci     .         .
adw2   : ACTGTTCAAGCCTCCAAGCTGTGCCTTGGGTGGCTTTGGGGCATGGACATTGACCCTTAT
adw    :
adr(1) :                                                         G
adr(2) :                                                         G
ayr    :                                                         G
ayw(1) :                                                   C
ayw(2) :                                                      T

               1930                1950                1970
                .         .         .         .         .         .
adw2   : AAAGAATTTGGAGCTACTGTGGAGTTACTCTCGTTTTTGCCTTCTGACTTCTTTCCTTCC
adw    :
adr(1) :                T                T                          T
adr(2) :               AT                T                       G  T
ayr    :                T                T                          T
ayw(1) :                                                            A
ayw(2) :                                                            A

               1990                2010                2030
                .         .         .         .         .         .
adw2   : GTCAGAGATCTCCTAGACACCGCCTCAGCTCTGTATCGAGAAGCCTTAGAGTCTCCTGAG
adw    :   AC
adr(1) : A TC          C           T           G  G              G  A
adr(2) : A TC          T           T           G  G              G  A
ayr    : A TC          C           T           G  G              G  A
ayw(1) :   AC       T     T                    G
ayw(2) :   AC       T     T A                  G

               2050                2070                2090
                .         .         .         .         .         .
adw2   : CATTGCTCACCTCACCATACTGCACTCAGGCAAGCCATTCTCTGCTGGGGGGAATTGATG
adw    :
adr(1) :      T              A              T     G  T     T  G
adr(2) :      T              A              T     G  T     T  G  A
ayr    :      T              A              T     G  T     T  G
ayw(1) :      T                             A     T            C A
ayw(2) :      T                             A  A  G            C A

               2110                2130                2150
                .         .         .         .         .         .
adw2   : ACTCTAGCTACCTGGGTGGGTAATAATTTGGAAGATCCAGCATCTAGGGATCTTGTAGTA
adw    :                               C             C  A     A     C
adr(1) :  A T G  C           A G            C        C     AT A     C
adr(2) :  A   G  C           A G            C        C     AT A     C
ayr    :  A   G  C           A G            C        C     AT A     C
ayw(1) :                      GT                  G     A  C  A     C
ayw(2) :                      GG                AT   C     C  A     C
```

FIGURE 2 (continued).

```
              2170                2190                2210
                 .         .         .         .         .         .
adw2   : AATTATGTTAATACTAACGTGGGTTTAAAGATCAGGCAACTATTGTGGTTTCATATATCT
adw    :                   A
adr(1) :  GC     C   GT   TA    CC    A     A                    T  C
adr(2) :  GC     C   GT   TA    CC    A     A                 C  T  C
ayr    :  GC     C   GT   TA    CC    A  T  A                 C  T  C
ayw(1) :  G      C  C     TA    CC     T          C           C  T
ayw(2) :  G      C  C     TA    CC    AT                      C  T

              2230                2250                2270
                 .         .         .         .         .         .
adw2   : TGCCTTACTTTTGGAAGAGAGACTGTACTTGAATATTTGGTCTCTTTCGGAGTGTGGATT
adw    :
adr(1) :   T                 A     T     G        G     T
adr(2) :                     A     TT G  G        A     T
ayr    :                     A     C     G        G     T
ayw(1) :   T  C              A  A  TA A  G        G
ayw(2) :   T  C              A  A  TA A  G        G     T

              2290                2310                2330
                 .         .        P1         .         .         .
adw2   : CGCACTCCTCCAGCCTATAGACCACCAAATGCCCCTATCTTATCAACACTTCCGGAAACT
adw    :
adr(1) :            C  T  C
adr(2) :            C  T  C
ayr    :            C  T  C
ayw(1) :               T                        C                G
ayw(2) :               T                                         G

              2350                2370                2390
                 .         .         .         .         .         .
adw2   : ACTGTTGTTAGACGACGGGACCGAGGCAGGTCCCCTAGAAGAAGAACTCCCTCGCCTCGC
adw    :
adr(1) :                ......
adr(2) :                ......
ayr    :                ......
ayw(1) :                ......
ayw(2) :                ......

              2410                2430                2450
                 .         .         .         .         .       Ct.
adw2   : AGACGCAGATCTCCATCGCCGCGTCGCAGAAGATCTCAATCTCGGGAATCTCAATGTTAG
adw    :              A
adr(1) :      A  G    A
adr(2) :      A  G    A
ayr    :      A  G    A
ayw(1) :      A  G    A
ayw(2) :      A  G    A

              2470                2490                2510
                 .         .         .         .         .         .
adw2   : TATTCCTTGGACTCATAAGGTGGGAAACTTTACGGGGCTTTATTCCTCTACAGTACCTAT
adw    :                      C
adr(1) :    C                             T           T     T      G
adr(2) :    C                             T           T     T      G
ayr    :    C                             T           T     T      G
ayw(1) :                         G        T           T     T      G
ayw(2) :                                              T     T      G
```

FIGURE 2 (continued).

```
                        2530                2550                2570
                  .         .         .         .         .         .
adw2   : CTTTAATCCTGAATGGCAAACTCCTTCCTTTCCTAAGATTCATTTACAAGAGGACATTAT
adw    :
adr(1):   C          G CC        C           C           G
adr(2):           C  G           C         C C           G
ayr    :             T   A       C         C C           G
ayw(1):            C T   A    A  A  T        T  A        CC A
ayw(2):        C   C T   A    A  C  T        T  A        CC A

                        2590                2610                2630
                  .         .         .         .         .         .
adw2   : TAATAGGTGTCAACAATTTGTGGGCCCTCTCACTGTAAATGAAAAGAGAAGATTGAAATT
adw    :
adr(1):        A          A            T  A  T        A  G     A
adr(2):        A          A            T  G  T        A  G     A
ayr    :       A          A            G  A  T        A  G     A
ayw(1):  C  A AA   G    G     A     A     A  T     G  A         C
ayw(2):  C  A AA   G          A     A     A  C     G  A      C  C

                        2650                2670                2690
                  .         .         .         .         .         .
adw2   : AATTATGCCTGCTAGATTCTATCCTACCCACACTAAATATTTGCCCTTAGACAAAGGAAT
adw    :
adr(1):                 G          A  TT  C              G  T     C
adr(2):                 G          A  TT                          C
ayr    :                G          A  TT  C              G        C
ayw(1):  G           C  G  T     A AGGTT  C        A  A  G  T  G  T
ayw(2):  G        A     G  T     A ATGTT  C           A  G  T  G  T

                        2710                2730                2750
                  .         .         .         .         .         .
adw2   : TAAACCTTATTATCCAGATCAGGTAGTTAATCATTACTTCCAAACCAGACATTATTTACA
adw    :
adr(1):                 T  A  T C                A    T  G
adr(2):         G        T  A  T C                A    T  G
ayr    :       G        T  AT T C                A    T  G
ayw(1):                   A  TC                      T     C
ayw(2):                   AT TT                      T

                        2770                2790                2810
                  .         .         .         .         .         .
adw2   : TACTCTTTGGAAGGCTGGTATTCTATATAAGCGGGAAACCACACGTAGCGCATCATTTTG
adw    :                                A
adr(1):         G           C           AA A     T     C     T
adr(2):         G           C            A A     T     C     C
ayr    :       G           C            A A     T     C     C
ayw(1):  C     A        G     AT        A A     A    A      C
ayw(2):  C     A        G     AT    C   A A     T    A      C

                        2830                2850                2870
                  .         .         .    PS1^ia    .         .
adw2   : CGGGTCACCATATTCTTGGGAACAAGAGCTACAGCATGGGAGGTTGGTCATCAAAACCTC
adw    :                                      .....................
adr(1):  T                               G                T  C
adr(2):  T                                                T  C
ayr    : T                                                T  C
ayw(1):  T                          T     ...........................
ayw(2):  T                                ...........................
```

FIGURE 2 (continued).

```
                2890                2910                2930
                  PS1ib      .         .         .         .         .
adw2    : GCAAAGGCATGGGGACGAATCTTTCTGTTCCCAATCCTCTGGGATTCTTTCCCGATCATC
adw     :                                   C            C
adr(1)  :  AC                                                       C
adr(2)  :  AC                                                       C
ayr     :  AC                                                       C
ayw(1)  : ......        CA         CACCAG                        C  C
ayw(2)  : ......        CA         CACCAG                        C  C

                2950                2970                2990
                   .         .         .         .         .         .
adw2    : AGTTGGACCCTGCATTCGGAGCCAACTCAAACAATCCAGATTGGGACTTCAACCCCGTCA
adw     :                               CA                        A
adr(1)  :              G                                          AA
adr(2)  :              G                                          AA
ayr     :              G                                          AA
ayw(1)  :        T  A  C   A    A   A CGCA                    T   AA
ayw(2)  :        T  A  C   A    A   A CGCA                    T   AA

                3010                3030                3050
                   .         .         .         .         .         .
adw2    : AGGACGACTGGCCAGCAGCCAACCAAGTAGGAGTGGGAGCATTCGGGCCAAGGCTCACCC
adw     :      C                   G                        G
adr(1)  :     TC         AG  A T A G       C     AC         G  T
adr(2)  :     TC A       AG  A T   G       C                G  T
ayr     :     TC         AG  A T   G       C         T  T   G  T
ayw(1)  :      AC        AC      A G       CT            TGG TT
ayw(2)  :      AC        AC      A G       CT            TGG AT

                3070                3090                3110
                   .         .         .         .         .         .
adw2    : CTCCACACGGCGGTATTTTGGGGTGGAGCCCTCAGGCTCAGGGCATATTGACCACAGTGT
adw     :
adr(1)  :  A            C                                     A      C
adr(2)  :  A            C                                     A      C
ayr     :  A        A  CC                                     A   C  C
ayw(1)  :  A  G     A  CC                                C ACAA  TT  C
ayw(2)  :  A        A  CC                                C AGAA  GT  C

                3130                3150                3170
                   .         .         .         .         .         .
adw2    : CAACAATTCCTCCTCCTGCCTCCACCAATCGGCAGTCAGGAAGGCAGCCTACTCCCATCT
adw     :
adr(1)  :   G  GCG          TT                       A
adr(2)  :   G  GCA                                   A
ayr     :   G  GCA                                   A
ayw(1)  :   G   A   G                    C                    C  GC G
ayw(2)  :   G   A   G           T        C                    C  GC G

                3190                3210
                   .         .         .  PS2i   .
adw2    : CTCCACCTCTAAGAGACAGTCATCCTCAGGCCATGCAGTGG
adw     :
adr(1)  :
adr(2)  :
ayr     :
ayw(1)  :         T G   A   C
ayw(2)  :         T G   A   C
```

FIGURE 2 (continued).

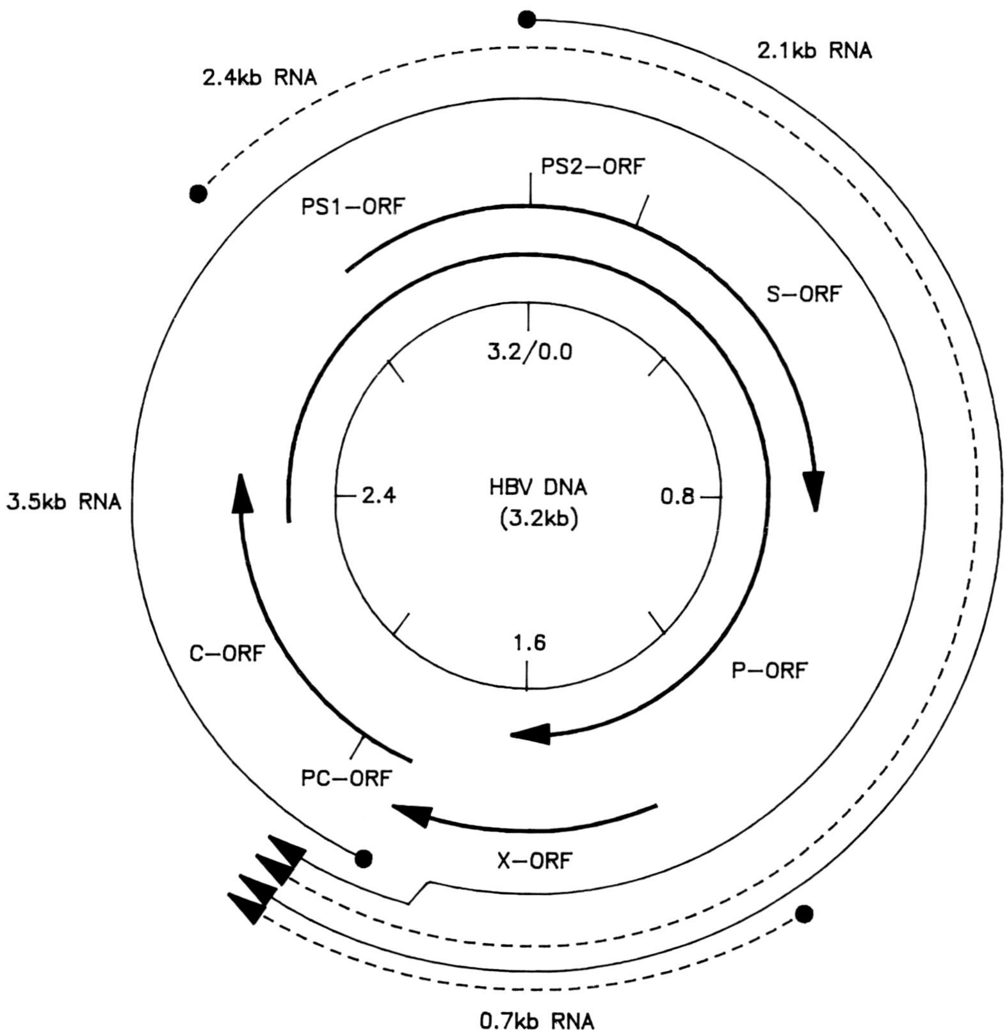

FIGURE 3. Organization of the HBV genome. The coordinates of the HBV DNA are the same as in Figure 2. PS, presurface; S, surface; P, polymerase; X, X gene; PC, precore; C, core; ORF, open reading frame.

III. ORGANIZATION OF HEPATITIS B VIRUS

A. VIRAL CODING CAPACITY

The sequences of the various HBV DNAs[47-61] revealed that there are four long open reading frames conserved between all the viral genomes (Figures 2 and 3). These encode the envelope or surface antigens (HBsAg),[72-75] the nucleocapsid antigens (HBcAg and HBeAg),[51,76,77] the DNA polymerase (P) gene product[78-80] and the X gene product.[66,81,82] The surface antigen (subtype *ayw*) open reading frame contains three in-frame translational initiation codons that permit the synthesis of HBsAg polypeptides of 25, 31, and 43 kDa.[50] These polypeptides are variably glycosylated, giving rise to the six polypeptides HBsAg/P25:GP28, HBsAg/GP33:GP36, and HBsAg/P43:GP46, respectively[75] (Figure 4). These surface antigen polypeptides are also known as the major, middle, and large HBsAg polypeptides, respectively.

The nucleocapsid antigen open reading frame contains two in-frame translation initiation codons that permit the synthesis of polypeptides of 18 and 21 kDa (Figure 5). Synthesis of the

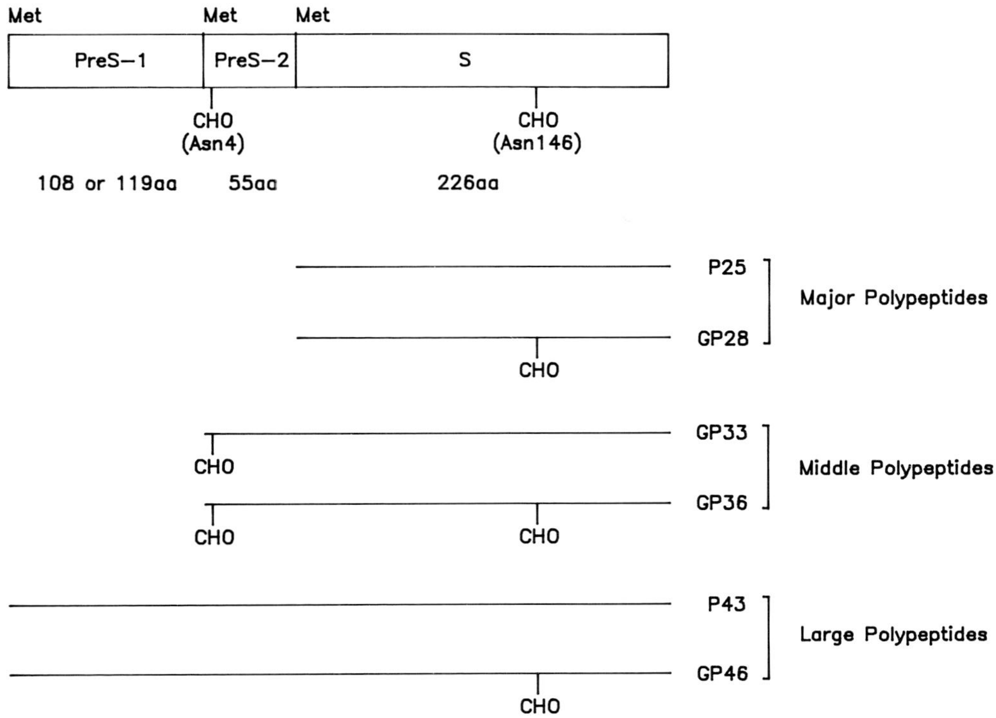

FIGURE 4. Organization of the HBsAg open reading frame. The methionine (Met) residues encoded by the initiation codons for the large, middle, and major surface antigen polypeptides are indicated. The asparagine (Asn) residues at which carbohydrate (CHO) can modify the HBsAg are shown.

HBeAg/P18 polypeptide results from the proteolytic processing of both the amino and carboxy termini of the primary translation product of the complete nucleocapsid open reading frame, which has the capacity to encode a polypeptide of 24 kDa.[76,77] The polypeptide product synthesized from the second in-frame translation initiation codon is the 21-kDa core antigen (HBcAg/P21) polypeptide.[51]

The DNA polymerase and X gene open reading frames have the capacity to code for polypeptides of 94 and 17 kDa, respectively. The molecular masses of the HBV polypeptides are derived from their predicted amino acid sequences determined from the nucleotide sequence of the HBV genome (subtype *ayw*).[50] A variation in the molecular mass of the large HBsAg polypeptide has been observed and ascribed to surface antigen subtype variation.[75] This difference in size results from an additional 11 amino-terminal amino acids present in the large envelope polypeptide, HBsAg/P44:GP47 (subtype adw_2)[52] (Figure 2).

B. STRUCTURE OF THE HBV PARTICLE

The envelope of the HBV Dane particle possesses the same antigenic determinants as are found on the 22-nm diameter subviral spheres and filaments present in the serum of infected individuals.[17,22] Characterization of the polypeptide composition of HBV and subviral particles demonstrated that the common antigenic determinants were located within the HBsAg polypeptides.[74,75] The HBsAg polypeptides compose a set of six coterminal polypeptides that differ by the extent of their glycosylation and amino-terminal sequence[75] (Figure 4). The major HBsAg (HBsAg/P25:GP28) comprises 226 amino acids and the HBsAg/P25 polypeptide differs from the HBsAg/GP28 polypeptide by the addition of Asn_{146}-linked complex glycan.[75,83,84] The glycan, representing approximately 75 μg carbohydrate per milligram HBsAg,

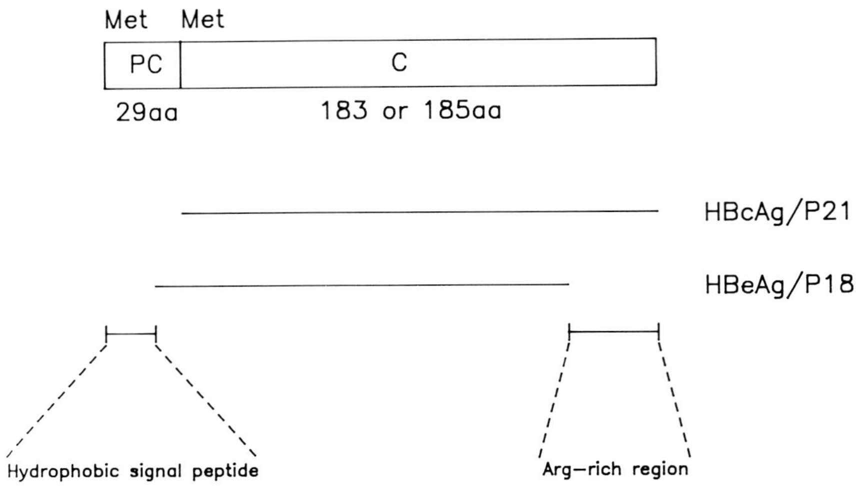

FIGURE 5. Organization of the nucleocapsid open reading frame. The methionine (Met) residues encoded by the precore (PC) and core (C) initiation codons are indicated. The 19-amino acid hydrophobic signal peptide and the 34-amino acid arginine-rich carboxy-terminal region cleaved from the HBeAg/P24 precursor to produce the secreted HBeAg/P18 are shown.

is composed of *N*-acetylglucosamine, mannose, galactose, and sialic acid residues plus fucose as a minor component.[83,85] The middle HBsAg (HBsAg/GP33:GP36) differs from the major HBsAg by an additional 55 amino-terminal amino acids encoded by the preS2 region.[74] In addition, there is evidence that Asn_4-linked (Asn_4 of the preS2 region) mannose-rich glycan is always found added to the middle HBsAg polypeptides, accounting for the absence of a nonglycosylated form of this polypeptide.[74,86] The large HBsAg (HBsAg/P43:GP46) differs from the middle HBsAg by an additional 108 or 119 amino-terminal amino acids, depending on the viral strain, encoded by the preS1 region. In addition, the preS2 glycosylation site is not substituted in the large HBsAg polypeptide, and the presence or absence of Asn_{146}-linked complex glycan in the major HBsAg domain of the large HBsAg polypeptide is considered to account for the two forms of this polypeptide.[75]

Characterization of the surface antigen composition of serum-derived HBV particles, filaments, and spheres demonstrated that they possessed different ratios of the various envelope polypeptides.[75] The composition of the filaments and Dane particles is approximately 10 to 20% of each of the middle and large polypeptides, with the remaining HBsAg contributed by the major surface antigen polypeptides. In contrast, the 22-nm spheres comprise only 1 to 2% large envelope polypeptide, 10 to 20% middle polypeptide, and the remaining envelope antigen is contributed by the major surface antigen polypeptide.[75] The level of glycosylation of Asn_{146} among the three envelope polypeptides was approximately the same, with about half of the molecules modified.[75] The contribution of carbohydrate to the total mass of the 22-nm spheres has been estimated to be between 3 and 8%,[85,87] and based on the relative abundance of the larger envelope polypeptides and the extent of their glycosylation, it is probable that the contribution of carbohydrate to the mass of the HBsAg filaments and envelope of the virion is similar. The HBV envelope contains approximately 400 subunits,[75] including 40 to 80 molecules of each of the middle and large polypeptides in the virion. In comparison, the 22-nm sphere consists of approximately 100 polypeptide subunits,[75,88] of which very few are the large envelope polypeptide. Eukaryotic expression studies of the three surface antigen open reading frames demonstrated that the middle and major envelope polypeptides can assemble into spherical 22-nm HBsAg particles,[89-99] whereas production of the large envelope polypeptide is necessary for the synthesis and assembly of HBsAg filaments.[94,100] These observations indicate that the large envelope polypeptide influences the nature of the assembled surface antigen particles and is an important component of the HBV envelope.[75] In addition to the surface antigen polypeptides, the 22-nm HBsAg particles contain cellular lipid.[101-103] The contribution of the lipid to the mass of the 22-nm spheres was determined to be approximately 25% and its composition is similar to other normal human serum lipoproteins.[101] The presence of glycolipid in HBsAg particles has been reported[102] but has not been observed in subsequent analyses.[85,101]

The predominant polypeptide of the nucleocapsid of HBV is the 21-kDa hepatitis B core antigen, HBcAg.[51,75,104-109] The nucleocapsid appears to be composed of approximately 180 of these polypeptide subunits, which have been proposed to be assembled with icosahedral symmetry.[110] The HBcAg is a phosphoprotein that possesses autophosphorylating serine protein kinase activity.[111-113] The significance of this activity is currently unclear. It has been shown that there is endogenous DNA polymerase activity within the nucleocapsid of HBV,[34,114] which is presumably encoded by the HBV DNA polymerase open reading frame.[78-80] As in the DHBV system, it is likely that the amino-terminal portion of the HBV DNA polymerase open reading frame encodes a polypeptide, the terminal protein, which is attached to the 5′ end of the HBV DNA minus strand[44] and probably serves as the primer for reverse transcription of the viral pregenomic RNA.[114-116] The HBV DNA polymerase open reading frame also encodes RNase H activity, which is likely to be responsible for the degradation of the pregenomic RNA as the minus strand of HBV DNA is being synthesized.[114]

The three envelope polypeptides, the P21 core polypeptide, and the HBV DNA polymerase polypeptide represent all of the protein components of the virion that have been identified as viral structural polypeptides. In addition to the protein, carbohydrate, and lipid, the virion also contains a 3.2-kb partially double-stranded circular DNA genome that is encapsidated within the nucleocapsid.[36,38-41]

IV. INTRACELLULAR LIFE CYCLE OF HBV IN THE HUMAN HEPATOCYTE

A. VIRUS ENTRY INTO HEPATOCYTES

The inability to infect permanent tissue culture cell lines with HBV is a major obstacle to identifying the mechanism of viral entry into hepatocytes.[117-119] This has prevented the identification of putative viral receptors and their ligands. However, evidence has been presented suggesting there may be selective interactions between surface antigen polypeptides and cell surface elements of several hepatic and nonhepatic tissue culture cell lines and also primary human hepatocyte plasma membrane components.[120-124] These interactions may be occurring directly through receptor–ligand interactions or indirectly through polyalbumin, the albumin receptors on the surface of the human hepatocytes,[125-127] and the receptor for polymerized human serum albumin present in the preS2 region of the envelope polypeptides.[92,93,128] The role of these interactions in the infection of hepatocytes and possibly other cell types is currently unclear.

B. GENERATION OF A TRANSCRIPTIONALLY ACTIVE HBV TEMPLATE

Once HBV has entered the hepatocyte, whether by a receptor-mediated process or by fusion with the plasma membrane, it must release the nucleocapsid from the surface antigen envelope (Figure 6). This process may be intimately linked with virus entry into the cell or may represent a separate step in the viral life cycle. The viral genome must then be translocated to the nucleus, presumably within the nucleocapsid and mediated by the nuclear translocation signal present in the HBcAg polypeptide.[94] Since the presumed template for the

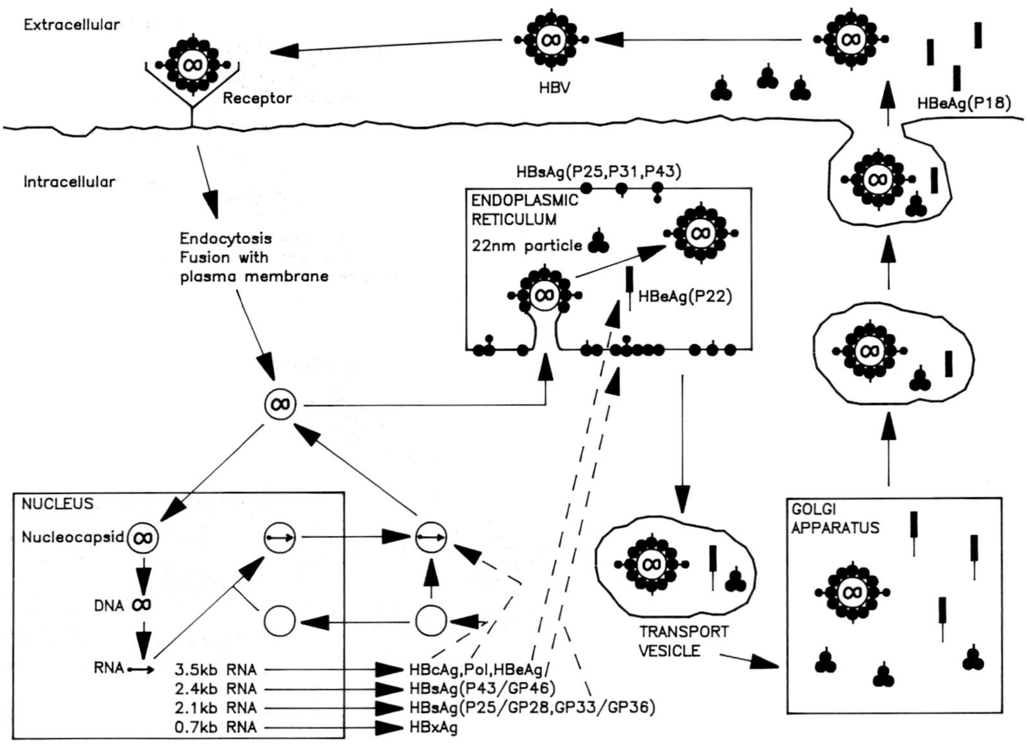

FIGURE 6. The intracellular pathway for the synthesis and secretion of HBV, HBsAg subviral particles, and HBeAg polypeptides.

transcription of HBV mRNAs is a covalently closed circular molecule of the HBV genome, several modifications of the infecting viral genome must be made before transcription can occur.[43] The single-stranded region of the HBV genome must be converted to double-stranded DNA. As it is known that the synthesis of this region of DNA can be completed *in vitro* by the endogenous HBV DNA polymerase present in the nucleocapsid of the virus,[35,37,38] it is likely that during infection the same enzyme may convert the partially double-stranded DNA into nicked circular double-stranded DNA. This event could occur within the nucleocapsid or might occur to the HBV genome after it has been released from the nucleocapsid into the nucleoplasm. In addition to the synthesis of DNA, removal of the terminal protein and oligoribonucleotide attached to the 5′ ends of the viral DNA strands, and ligation of the nicks in the double-stranded DNA must occur to generate a covalently closed, double-stranded HBV genome.[129]

C. TRANSCRIPTION OF THE HBV GENOME

HBV-infected liver in both acute and chronic disease expresses two predominant HBV-specific transcripts of 2.1 and 3.5 kb[130-134] (Figures 3 and 6). The 2.1-kb RNA has several transcription initiation sites that span the translation initiation codon for the middle envelope polypeptide,[67,135,136] and terminates at the single polyadenylation site in the HBV genome.[135-137] These transcripts encode the middle and major surface antigen polypeptides. The 3.5-kb RNA also has several transcription initiation sites that span the initiation codon for the precore signal sequence which precedes the HBcAg open reading frame.[67,135] The transcripts initiating before the precore initiation codon encode the secreted HBeAg/P18 polypeptide, whereas the transcripts initiating after the precore initiation codon encode the HBcAg/

P21 polypeptide, which is the major polypeptide of the nucleocapsid. The 3.5-kb transcripts are the only identified RNAs that can encode the HBV DNA polymerase. This is probably achieved by the internal entry of ribosomes onto the 3.5-kb transcript and initiation of translation from the first in-frame translation initiation codon of the HBV polymerase open reading frame[138] by a process similar to that described for the expression of the DHBV DNA polymerase.[139,140] The expression of the only other known HBV structural polypeptide, the large envelope polypeptide, occurs from a 2.4-kb transcript that initiates just upstream of the translation initiation site for this open reading frame [67,135,141] (Figures 3 and 6). Although the evidence for the existence of this transcript in infected liver tissue is limited,[43,131] it has clearly been identified in human hepatoma cell lines that are synthesizing HBV particles.[33,64,66,135] It is also likely that a 0.7-kb transcript initiating upstream of the X gene open reading frame is synthesized at some stage during HBV infection (Figures 3 and 6). Although there is no direct evidence for the synthesis of this transcript in human infection, the equivalent transcript has been identified in RNA isolated from woodchucks infected with WHV.[142] In addition, a 0.7-kb transcript derived from the X gene region of the HBV genome has been observed in transgenic mice replicating HBV[68] and cell lines transfected with HBV DNA.[137,143-146] In addition, this open reading frame is conserved between all of the mammalian hepadnaviruses, suggesting the functional importance of the X gene in the life cycle of the virus.[147] Antibodies to the X gene product have also been observed in the sera of infected individuals, indicating that the X gene product is synthesized during the course of at least some HBV infections.[82,146,148-153] The role of the X gene product in the HBV life cycle is currently unclear. However, it appears that the product of the X gene can trans-activate several heterologous promoters,[154-161] suggesting it may regulate the expression of various viral and cellular genes important in the production of viral particles. In addition to the unspliced transcripts, spliced HBV RNAs have been described.[137,162,163] Since it is not clear if these RNAs code for HBV polypeptides, their significance in the life cycle of HBV is currently unknown.

D. EXPRESSION AND SECRETION OF THE HBV SURFACE AND NUCLEOCAPSID POLYPEPTIDES

The various HBV transcripts encode the known viral polypeptide products. The surface antigen polypeptides contain signal sequences that direct their translocation into the endoplasmic reticulum membrane.[164-166] Since the major surface antigen promoter is considerably more active than the large surface antigen promoter, a large excess of the major envelope polypeptide is synthesized.[33,64,66,67,131,132,134,167,168] The middle surface antigen polypeptide is synthesized relatively inefficiently from the transcripts expressed from the major surface antigen promoter for two reasons. The translation initiation codon for the middle envelope polypeptide is located about 15 nucleotides from the 5′ end of the longest 2.1-kb surface antigen RNA and is absent from the shorter 2.1-kb RNAs.[67,135,136] In addition, the sequence around the translation initiation codon does not conform to the favored consensus sequence for efficient translational initiation.[74,94,169] This probably explains why the middle envelope polypeptide is present only at less than 20% of the level of the major surface antigen polypeptide in the sera of infected individuals.[74,75,86] The envelope polypeptides are probably expressed in excess over the other viral components. They have the capacity to assemble into 22-nm spheres and filaments that bud into the lumen of the endoplasmic reticulum and subsequently are secreted by passage through the endoplasmic reticulum and Golgi compartment.[164-166,170-172] This process accounts for the large excess of subviral particles over virions in the sera of viremic individuals.[168] The reason subviral particles are synthesized in excess over virions is unknown.

The HBeAg is produced from the translation product of the longest 3.5-kb viral RNAs. It is generated from a precursor polypeptide containing a signal peptide encoded by the precore sequence that targets this polypeptide to the endoplasmic reticulum.[77,94,173-176] After the cleavage of the amino-terminal signal sequence by the signal peptidase in the endoplasmic reticulum, the HBeAg/P22 polypeptide proceeds through the secretion pathway and the 34 carboxy-terminal amino acids are removed by proteolytic processing,[76] resulting in the secretion of the mature HBeAg/P18 polypeptide. The carboxy-terminal processing probably occurs in the late or post-Golgi compartment by cellular proteases possessing trypsin- and carboxypeptidase B-like specificities that could correctly modify the arginine-rich carboxy-terminal region of the unprocessed HBeAg/P22 polypeptide.[177] As with the secretion of the large excess of the subviral envelope antigen particles, the function of circulating HBeAg in the serum of viremic patients is not known. However, it has been speculated that HBeAg might have a role in the modulation of the host's immune response to the virus.[178] In a transgenic mouse model, a role for HBeAg in neonatal tolerance has been demonstrated, supporting the hypothesis that maternal HBeAg from a chronic carrier mother might tolerize the fetus *in utero*, predisposing it to chronic infection upon exposure to HBV at birth.[179]

E. REPLICATION OF HBV

The replication strategy of HBV requires that a 3.5-kb transcript serve an additional function as the replication template in the viral life cycle, besides encoding the nucleocapsid antigens and the HBV DNA polymerase[43] (Figure 7). It has not been demonstrated for HBV, but by analogy with GSHV, the shortest of the 3.5-kb HBV RNAs is probably the only transcript that serves as the template for viral replication.[180] This RNA is encapsidated by HBcAg/P21 polypeptides to form immature core particles that also contain the product of the HBV DNA polymerase open reading frame.[114,181] The reverse transcriptase and RNAse H activities of the HBV DNA polymerase synthesize the long (or minus) strand of HBV genomic DNA using the terminal protein as primer and the pregenomic RNA as template.[182] Since the terminal protein is probably encoded in the amino-terminal region of the HBV DNA polymerase open reading frame in a manner similar to that reported for the duck hepatitis B virus terminal protein,[115,116] this replication strategy ensures that both the primer and polymerase required for viral replication are packaged together into the nucleocapsid. Initiation of long-strand synthesis probably occurs at the 3′ direct repeat 1 (DR1) sequence present in the terminal redundancy (R) of the 3.5-kb transcript, as has been reported for replication of the WHV minus-strand DNA[183] (Figure 7). Extension of long-strand synthesis to the 5′ end of the 3.5-kb transcript generates a short terminal redundancy (r) of approximately 8 nucleotides in this strand.[43] The concomitant RNase H degradation of the transcript results in the generation of the 17-nucleotide capped oligoribonucleotide primer from the 5′ end of this RNA necessary for initiation of the short (or plus) strand synthesis.[43,45,114] This oligoribonucleotide is translocated by an unknown mechanism from the end of the newly synthesized long strand to the direct repeat 2 (DR2) sequence located 226 nucleotides from the 5′ end of the long strand. It serves as the primer for the short-strand synthesis that proceeds from the primer at DR2 to the end of the minus strand.[43] The eight-nucleotide redundancy in the long strand is presumed to permit the transfer of the 3′ end of the growing short strand from the protein-linked 5′ end to the 3′ end of the long strand. This would generate a circular genome structure that upon further extension of the short strand, would generate the partially double-stranded HBV genome found in the Dane particle[36,38-41,43] (Figures 1 and 7). The conversion of the pregenomic RNA in immature core particles to the partially double-stranded DNA found in the mature core particles probably occurs in the cytoplasm[129,184-188] in a manner similar to that described for DHBV[189] (Figure 6). The final step of the assembly of the virus particle involves the association of the mature core particle with an appropriately arranged assembly of envelope antigen molecules within the endoplasmic reticulum membrane. The surface antigen assembly subsequently buds into the lumen of the endoplasmic reticulum enveloping the mature core particle.[190-195] The complete virion is secreted from the cell through the endoplasmic reticulum

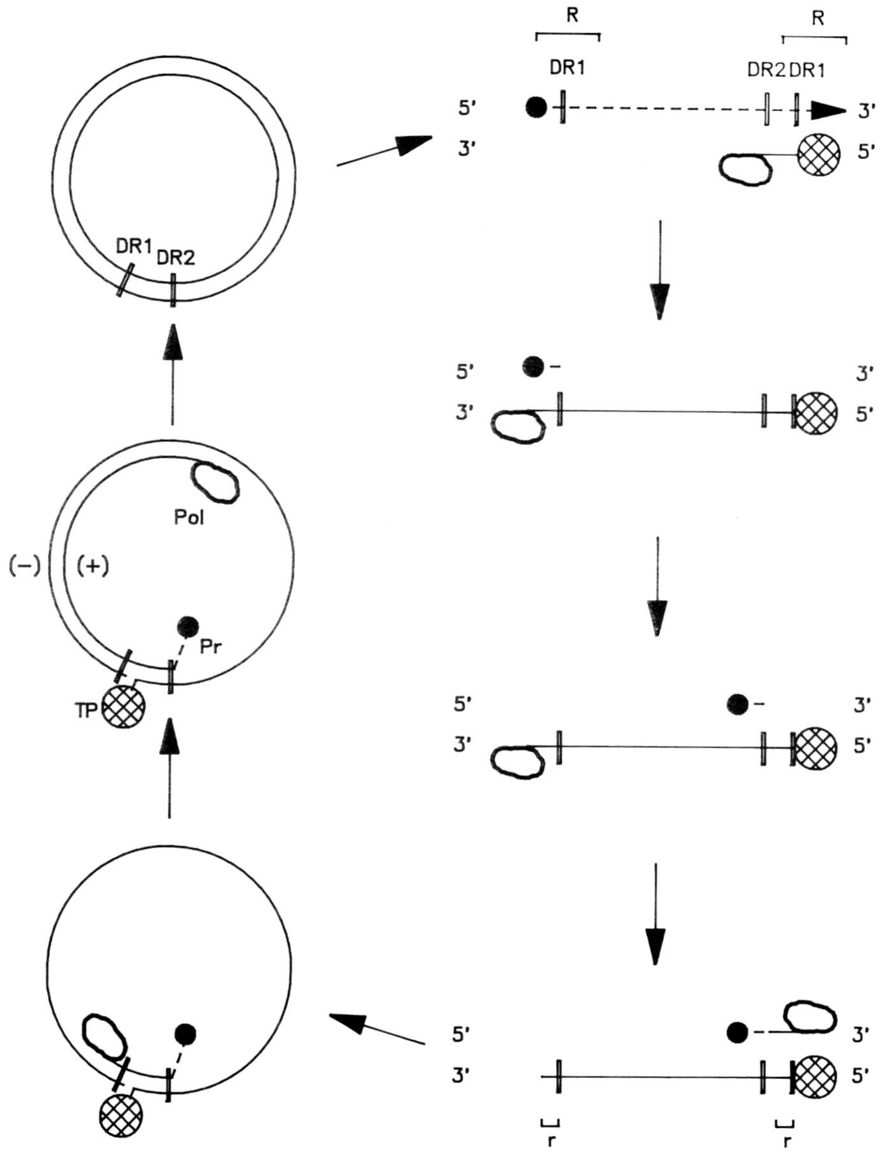

FIGURE 7. DNA and RNA intermediates in the replication cycle of HBV. DNA strands are indicated by solid lines and RNA is represented by a dashed line. R, terminal redundancy in the pregenomic RNA; r, terminal redundancy in the HBV minus strand; (+) and (–) indicate the plus (short) and minus (long) strands of the HBV genomic DNA; DR1 and DR2, direct repeats 1 and 2; Pol, HBV DNA polymerase; TP, terminal protein; Pr, oligoribonucleotide primer of plus-strand synthesis.

and Golgi apparatus pathway (Figure 6). The carbohydrate attached to the surface antigen of the virion and the subviral particles is added during their passage through this pathway.[170,171]

F. MAINTENANCE OF THE NUCLEAR COVALENTLY CLOSED CIRCULAR HBV DNA POOL

The replication cycle of HBV described in Section IV.E would suggest that the number of covalently closed circular HBV genomes present in the nucleus of an infected hepatocyte would be equal to the number of virus particles that infect that particular cell. Amplification of virus production would therefore depend on each nuclear covalently closed circular HBV DNA molecule producing multiple copies of the pregenomic RNA. This RNA is subsequently reverse transcribed to produce new HBV genomes that are released from the cell in virus particles. Therefore, one mechanism employed by HBV to amplify virus production is an intrinsic property of the viral replication cycle and depends on the cellular RNA polymerase II transcribing multiple copies of the pregenomic RNA before the HBV template is destroyed in the hepatocyte nuclei.

However, a second mechanism for the amplification of HBV production, similar to that described for DHBV,[196,197] may also be operative (Figure 6). Based on the known properties of the HBV polypeptides and the subcellular compartmentalization of the HBV viral and subviral particles, a speculative model for the process leading to the amplification of the nuclear covalently closed circular HBV genomic sequences can be proposed. The mature nucleocapsids that are enveloped by the surface antigen are secreted from the cell as virions. It is possible that this is not the only pathway that mature nucleocapsids follow in the infected cell. In the absence of the envelope antigen, the mature nucleocapsid is probably indistinguishable from the nucleocapsid released from the virion as it infects the hepatocyte. As the fate of this incoming nucleocapsid is probably to carry the HBV genome to the nucleus, it is likely that a newly synthesized mature nucleocapsid that fails to interact with the envelope antigen could return to the nucleus of the cell in which it was synthesized. This could lead to the amplification of the nuclear pool of HBV DNA within a particular cell. Therefore, the mature nucleocapsid may represent a branch point in the production of HBV particles (Figure 6). It may return to the nucleus of the cell to increase the covalently closed circular HBV DNA pool that is used to generate HBV transcripts, or it may interact with the envelope antigen in the endoplasmic reticulum membrane to produce virions that are secreted from the cell and enter the circulation, where they can infect additional hepatocytes. The regulation of the relative frequency with which mature core particles pass through these two pathways will determine the extent of the amplification of the nuclear HBV DNA pool and the level of virus secretion from a particular cell. It is interesting to speculate that initially after infection of a cell the recycling of HBV genomes to the nucleus predominates, whereas later in infection the large nuclear pool of HBV templates is used predominantly to direct the synthesis of virions that are secreted from the cell. The relative production of nucleocapsid and envelope antigens may be involved in determining which pathway predominates in the hepatocyte at different times during the life cycle of the virus.

V. HBV INFECTION

A. HOST RANGE AND TISSUE TROPISM

The host range of HBV is very restricted.[1] HBV has definitively been shown to infect man and chimpanzees.[58,198-201] Experimental transmission of HBV to chimpanzees has been demonstrated reproducibly.[70,71,202-204] In addition, there have been reports of experimental transmission of HBV to gibbons,[205] rhesus monkeys,[201,206] woolly monkeys,[201] and African Green monkeys.[206,207] Experimental transmission to additional primate and nonprimate species has been unsuccessful.[201]

The tissue tropism of HBV has been examined in man, chimpanzees, and transgenic mice. The primary site of virus replication is the hepatocyte, although evidence of viral infection and replication has been observed in other tissues. HBV nucleic acids and proteins have been observed in various nonhepatic tissues, including pancreas, skin, kidney, bile duct epithelium, and peripheral blood mononuclear cells (PBMC), with the most extensive studies performed on PBMC.[208-211]

In HBV-infected chimpanzees, viral DNA and RNA were detected in PBMC, including both T and B cells.[212] The DNA was present primarily in monomeric and multimeric episomal forms, although integrated forms of the DNA were observed. The pregenomic and greater than pregenomic length viral transcripts observed in the chimpanzee PBMC were expressed at levels approximately 30-fold lower than those seen in the hepatocytes.[212]

The forms of HBV DNA most commonly detected in human PBMC are monomers and high-molecular-weight oligomers, which are found in the episomal or integrated state, depending upon the patient.[208,213-228] Few groups have observed the lower molecular weight species indicative of viral replication in the PBMC.[216,217,222,225] The subpopulations of PBMC, including B and T cells (both T4 and T8 subsets), appear to contain HBV DNA.[219,220,224,225] The presence of HBV DNA in the PBMC does not seem to correlate with clinical categories of liver disease.[219,220] HBV DNA has also been detected in human spermatozoa,[229] pancreas, skin and kidney,[209] and Kaposi sarcoma.[230]

HBV RNA and viral antigens have been observed in human PBMC.[226,227,231-233] Viral antigens have also been detected in a B cell hybridoma made from the PBMC of a hepatitis patient.[228] Hepatitis B surface antigen has been detected in the tumor cells of human primary testicular cancer,[210] and HBsAg and HBcAg have been detected in human pancreas tissue[210] and bone marrow aspirates.[234-236] Cell cultures derived from bone marrow aspirates contain low levels of HBV DNA and Dane-like particles as well as HBsAg and HBcAg.[234,235] The significance of these observations is presently unclear.

The factors responsible for the highly restricted species and tissue tropism of HBV are currently unknown. However, replication of HBV may be restricted to hepatocytes for several reasons, including the absence of viral receptors on nonhepatic tissues and the requirement for liver-specific transcription factors for expression from some of the HBV promoters. Characterization of the cell-type specificity of the HBV enhancer and promoters has suggested that the enhancer and the core promoter may direct preferential gene expression in hepatoma cell lines,[237-241] whereas the major surface antigen gene appears to be expressed efficiently in a range of tissue culture cell types.[144,242,243] It has been demonstrated that production of virus particles can be achieved by transfection of hepatoma cells with HBV genomic DNA.[33,64-67] The requirement for hepatoma cells in the production of HBV and WHV by this approach can be circumvented by expressing the pregenomic RNA from an exogenous promoter rather than the core promoter.[244,245] These observations suggest that in transfection experiments a block to virus production in nonhepatic cells is at the level of transcription from the core promoter. In addition, the large surface antigen promoter has been shown to include a sequence element that is necessary for maximal transcriptional activity from this promoter.[167,246,247] This sequence element binds the highly liver-specific transcription factor, hepatocyte nuclear factor 1,[248-250] and probably explains the preferential expression from this promoter in hepatoma cell lines. In transgenic mice synthesizing HBV particles, it has been shown that the major HBV transcripts are predominantly synthesized in liver and kidney tissue. This indicates that there is a tissue-specific transcriptional restriction in the expression of the HBV genes in this system.[68,141] It also suggests that the block in the HBV life cycle in the murine system is at the level of virus entry into the hepatocyte, possibly indicating the absence of an appropriate receptor in this species. These data suggest that a major factor contributing to the hepatotropism of HBV may be its highly tissue-specific transcriptional regulation, and the species host range restriction may be at the level of the viral receptor.

B. ROUTES OF INFECTION

HBV infection is a worldwide health problem and is endemic in many regions of Asia and Africa.[251] Maintenance of HBV infection in the population occurs by both vertical and horizontal transmission of the virus.[1] Vertical transmission of infection occurs by the passage of HBV from an infected mother to her child at or prior to birth by mechanism(s) that currently are unclear.[252] However, the mixing of maternal blood with the fetal circulation at parturition is a likely source of viral infection of the neonate. Horizontal transmission of HBV infection can occur during sexual contact between infected and uninfected individuals and by percutaneous contact (e.g., injection) or other close contacts between the virus and man (e.g., mother to infant and between patient and healthcare worker).[1,253-258] Regardless of the route by which HBV gains access to the human circulation, the initial contact between the virus and the hepatocytes occurs in the liver sinusoids in the vicinity of the branches of the hepatic artery and the portal vein (i.e., the periportal area). This interaction between the virus and the hepatocyte represents the initial step in the viral life cycle.

C. CLINICAL CONSEQUENCES OF INFECTION

A consequence of infection with HBV is the induction of the host's immune response by the various viral antigens. These antigen–antibody responses include the induction of anti-HBs by the major surface antigen polypeptide,[259] anti-preS2 by the preS2 region of the middle and large surface antigen polypeptides,[260-262] anti-preS1 by the preS1 region of the large surface antigen polypeptide,[260,263] anti-HBc by the HBcAg,[259] anti-HBe by the HBeAg,[259] anti-HBx by the X gene product,[82,148-153] and anti-HBp by the DNA polymerase polypeptide.[264-267]

The first evidence of primary infection with HBV during a self-limited, HBsAg-positive infection is the observation of HBsAg in serum.[259] This is followed by the appearance of HBV particles with concomitant endogenous DNA polymerase activity and HBeAg antigenemia. As the levels of these viral particles and polypeptides in the serum start to decline, anti-HBc appears in the serum. At this time clinical hepatitis can be observed as elevated levels of liver transaminases in the serum. The decline of the clinical symptoms is followed by a rise in anti-HBs and anti-HBe titers in the serum, which subsequently decrease with time. The observation of these antiviral antibody responses indicates the resolution of the infection. Self-limiting HBV infection without detectable HBsAg can occur and is indicated by observable anti-HBs and anti-HBc titers in the serum in the absence of detectable HBV antigenemia.[259]

The course of a persistent or chronic HBV infection is initially similar to the HBsAg-positive, self-limited infection.[259] HBsAg, HBeAg, viral particles, and anti-HBc appear in the serum. The infection may be asymptomatic or characterized by liver damage. The infection fails to resolve, and circulating HBsAg and virus particles generally persist for years without seroconversion to an anti-HBs status. After an extended period of chronic infection, viral HBV DNA can often be found integrated into the host chromosomal DNA.[268,269] During this period of time seroconversion from HBeAg to anti-HBe may occur. The consequence of this event can be the generation of a chronic carrier state where HBsAg antigenemia exists in the absence of serum HBV particles.[270,271]

It is apparent that HBV infection can lead to hepatocellular necrosis in man.[272] The mechanism of liver damage and the factors determining the severity and outcome of infection are not fully understood. However, the observation that hepatocytes of asymptomatic chronic carriers can produce large quantities of HBV without apparent liver disease indicates that the virus is not directly cytopathic. In support of this conclusion, cell lines and transgenic mice producing HBV particles have demonstrated no cytopathic effect due to virus production.[33,64-69,141] In addition, the findings that (1) infected liver tissue reveals mononuclear inflammatory cells in close proximity to necrotic hepatocytes, (2) immunomodulation of infected patients affects the severity of liver damage, (3) the serology during and after infection correlates with the observed temporal elevation in serum transaminase levels, and (4) agammaglobulinemic patients can show liver damage during HBV infection implicate the host cellular immune response as a major factor contributing to the pathogenesis of liver disease.[273-275]

HBsAg chronic carrier status has additional clinical implications. Cirrhosis of the liver as a consequence of chronic HBV infection can lead to liver failure and death. In addition, the estimated relative risk of primary hepatocellular carcinoma (PHC) in chronic HBV carriers is approximately 100 times greater than in those who are serologically negative for HBsAg.[276,277] In addition, tumors and cell lines derived from chronic HBV carriers usually contain HBV DNA integrated in the chromosomal DNA,[278] suggesting a possible role for the virus in the transformation process occurring in the infected hepatocyte. In one tumor, the HBV sequences have been observed integrated into a homolog of the known cellular protooncogene, c-*erbA*, suggesting a direct role for viral DNA in the transformation process.[279] However, in the majority of tumors that have been examined, the integration sites for the HBV DNA are at different chromosomal locations, suggesting that the role of these sequences in the transformation process is likely to be more complex than in the simple insertional mutagenesis models.[278]

VI. HBV-RELATED INFECTIOUS AGENTS

HBV is a human pathogen and a member of a small family of related animal viruses.[1] The genomes of HBV from several different viral isolates have been cloned and sequenced.[47-61] The most distantly related genomes differ by 11% in their nucleotide sequence.[59] However, serological, clinical, and molecular evidence indicates the existence of additional HBV-related infectious agents.

The best characterized of these agents is the hepatitis delta virus (HDV), which has been analyzed in detail. HDV was first recognized as a new antigen/antibody system, hepatitis delta antigen (HDAg) and the antibody to HDAg (anti-HD), associated with a subgroup of chronic HBsAg carriers[280] who often display severe forms of clinical hepatitis.[281] Subsequent studies have revealed that HDV is a 36-nm particle comprising a 1679- or 1683-nucleotide single-stranded circular RNA genome encapsidated in HDAg and enveloped in HBsAg.[282-287] From the HDV sequence, it is apparent that the genome can form a rod structure by intramolecular base pairing and that a single antigenomic open reading frame codes for HDAg.[288,289] Structural elements of the HDV genome similar to those found in plant viroid genomes,[290] the ability of HDV RNA to catalyze self-cleavage and self-ligation reactions,[291-294] and the nature of the HDV transcripts[295,296] have led to the proposal that HDV uses a rolling circle replication strategy.[295,296] The only contribution HBV appears to have in the HDV life cycle is to supply HBsAg to envelope the HDAg and HDV genomic RNA.[297] However, this requirement for HBsAg limits HDV infection to coinfection with HBV or infection of chronic HBsAg carriers.

A putative variant form of HBV, HBV-2, has been described.[298] Infection with this virus is characterized by the presence of serum HBsAg but the absence of anti-HBc. The subsequent loss of HBsAg occurs without the appearance of HBeAg or the production of anti-HBs or anti-HBc.[298] Similar serological evidence has been reported from several regions of the world, suggesting this infectious agent may be relatively widespread.[299-304] HBV-2 has not been characterized at the molecular level, so its precise relationship to HBV is currently unknown.

VII. SUMMARY AND CONCLUSIONS

Hepatitis B virus is a worldwide health problem. The identification, isolation, and molecular characterization of the viral agent responsible for this health problem has led to a basic understanding of its cellular life cycle. In addition, the molecular mechanism(s) responsible for liver cell injury and hepatocellular carcinoma are now more amenable to detailed analyses.

ACKNOWLEDGMENTS

We are grateful to Dr. David R. Milich for critical reading of the manuscript. We thank Lynn Keyes for the preparation of the manuscript. The research studies of the authors are supported by Public Health Service Grant AI25183 from the National Institutes of Health.

REFERENCES

1. **Gust, I.D., Burrell, C.J., Coulepis, A.G., Robinson, W.S., and Zuckerman, A.J.,** Taxonomic classification of human hepatitis B virus, *Intervirology*, 25, 14, 1986.
2. **Summers, J., Smolec, J.M., and Snyder, R.,** A virus similar to human hepatitis B virus associated with hepatitis and hepatoma in woodchucks, *Proc. Natl. Acad. Sci. U.S.A.*, 75, 4533, 1978.
3. **Marion, P.L., Oshiro, L.S., Regnery, D.C., Scullard, G.H., and Robinson, W.S.,** A virus in Beechey ground squirrels that is related to hepatitis B virus of humans, *Proc. Natl. Acad. Sci. U.S.A.*, 77, 2941, 1980.
4. **Mason, W.S., Seal, G., and Summers, J.,** Virus of Pekin ducks with structural and biological relatedness to human hepatitis B virus, *J. Virol.*, 36, 829, 1980.
5. **Sprengel, R., Kaleta, E.F., and Will, H.,** Isolation and characterization of a hepatitis B virus endemic in herons, *J. Virol.*, 62, 3832, 1988.
6. **Feitelson, M.A., Millman, I., Halbherr, T., Simmons, H., and Blumberg, B.S.,** A newly identified hepatitis B type virus in tree squirrels, *Proc. Natl. Acad. Sci. U.S.A.*, 83, 2233, 1986.
7. **Blumberg, B.S., Alter, H.J., and Visnich, S.,** A new antigen in leukemia sera, *JAMA*, 191, 541, 1965.
8. **Prince, A.M.,** Relation of Australia and SH antigens, *Lancet*, 2, 462, 1968.
9. **Blumberg, B.S., Phil, D., Gerstley, B.J.S., Hungerford, D.A., London, W.T., and Sutnick, A.I.,** A serum antigen (Australia antigen) in Down's syndrome, leukemia, and hepatitis, *Ann. Intern. Med.*, 66, 924, 1967.
10. **Sutnick, A.I., London, W.T., and Blumberg, B.S.,** Australia antigen, Down's syndrome and hepatitis, *J. Clin. Invest.*, 46, 1122, 1967.
11. **Sutnick, A.I., London, W.T., Gerstley, B.J.S., Cronlund, M.M., and Blumberg, B.S.,** Anicteric hepatitis associated with Australia antigen, *JAMA*, 205, 670, 1968.
12. **Prince, A.M.,** An antigen detected in the blood during the incubation period of serum hepatitis, *PNAS*, 60, 814, 1968.
13. **Okochi, K. and Murakami, S.,** Observations on Australia antigen in Japanese, *Vox Sang.*, 15, 374, 1968.
14. **Giles, J.P., McCollum, R.W., Berndtson, L.W., and Krugman, S.,** Viral hepatitis: Relation of Australia/SH antigen to the Willowbrook MS-2 strain, *New Engl. J. Med.*, 281, 119, 1969.
15. **Blumberg, B.S., Sutnick, A.I., and London, W.T.,** Australia antigen and hepatitis, *JAMA*, 207, 1895, 1969.
16. **Bayer, M.E., Blumberg, B.S., and Werner, B.,** Particles associated with Australia antigen in the sera of patients with leukaemia, Down's syndrome and hepatitis, *Nature (London)*, 218, 1057, 1968.
17. **Dane, D.S., Cameron, C.H., and Briggs, M.,** Virus-like particles in serum of patients with Australia-antigen-associated hepatitis, *Lancet*, 1, 695, 1970.
18. **Almeida, J.D., and Waterson, A.P.,** Immune complexes in hepatitis, *Lancet*, 2, 983, 1969.
19. **Bond, H.E. and Hall, W.T.,** Separation and purification of hepatitis-associated antigen into morphologic types by zonal ultracentrifugation, *J. Infect. Dis.*, 125, 263, 1972.
20. **Robinson, W.S.,** The genome of hepatitis B virus, *Annu. Rev. Microbiol.*, 31, 357, 1977.
21. **Gust, I.D., Cross, G., Kaldor, J., and Ferris, A.A.,** Virus-like particles in Australia-antigen-associated hepatitis, *Lancet*, 1, 953, 1970.
22. **Jokelainen, P.T., Krohn, K., Prince, A.M., and Finlayson, N.D.C.,** Electron microscopic observations on virus-like particles associated with SH antigen, *J. Virol.*, 6, 685, 1970.
23. **Committee on Viral Hepatitis, and National Academy of Sciences,** Nomenclature of antigens associated with viral hepatitis type B, *Morbid. Mortal. Wkly. Rep.*, 23, 29, 1974.
24. **Gerin, J.L., Holland, P.V., and Purcell, R.H.,** Australia antigen: large-scale purification from human serum and biochemical studies of its proteins, *J. Virol.*, 7, 569, 1971.
25. **Almeida, J.D., Rubenstein, D., and Stott, E.J.,** New antigen-antibody system in Australia-antigen-positive hepatitis, *Lancet*, 2, 1225, 1971.

26. **Kim, C.Y. and Tilles, J.G.,** Purification and biophysical characterization of hepatitis B antigen, *J. Clin. Invest.*, 52, 1176, 1973.
27. **Barker, L.F., Smith, K.O., Gehle, W.D., and Shulman, N.R.,** Some antigenic and physical properties of virus-like particles in sera of hepatitis patients, *J. Immunol.*, 102, 1529, 1969.
28. **Howard, C.R. and Burrell, C.J.,** Structure and nature of hepatitis B antigen, *Prog. Med. Virol.*, 22, 36, 1976.
29. **Hirschman, S.Z., Gerber, M., and Garfinkel, E.,** Purification of naked intranuclear particles from human liver infected by hepatitis B virus, *Proc. Natl. Acad. Sci. U.S.A.*, 71, 3345, 1974.
30. **Barker, L.F., Almeida, J.D., Hoofnagle, J.H., Gerety, R.J., Jackson, D.R., and McGrath, P.P.,** Hepatitis B core antigen: immunology and electron microscopy, *J. Virol.*, 14, 1552, 1974.
31. **Kaplan, P.M., Ford, E.C., Purcell, R.H., and Gerin, J.L.,** Demonstration of subpopulations of Dane particles, *J. Virol.*, 17, 885, 1976.
32. **Moritsugu, Y., Gold, J.W.M., Wagner, J., Dodd, R.Y., and Purcell, R.H.,** Hepatitis B core antigen: detection of antibody by radioimmunoprecipitation, *J. Immunol.*, 114, 1792, 1975.
33. **Sureau, C., Romet-Lemonne, J.-L., Mullins, J.I., and Essex, M.,** Production of hepatitis B virus by a differentiated human hepatoma cell line after transfection with cloned circular HBV DNA, *Cell*, 47, 37, 1986.
34. **Kaplan, P.M., Greenman, R.L., Gerin, J.L., Purcell, R.H., and Robinson, W.S.,** DNA polymerase associated with human hepatitis B antigen, *J. Virol.*, 12, 995, 1973.
35. **Robinson, W.S. and Greenman, R.L.,** DNA polymerase in the core of the human hepatitis B virus candidate, *J. Virol.*, 13, 1231, 1974.
36. **Robinson, W.S., Clayton, D.A., and Greenman, R.L.,** DNA of a human hepatitis B virus candidate, *J. Virol.*, 14, 384, 1974.
37. **Lutwick, L.I. and Robinson, W.S.,** DNA synthesized in the hepatitis B Dane particle DNA polymerase reaction, *J. Virol.*, 21, 96, 1977.
38. **Hruska, J.F., Clayton, D.A., Rubenstein, J.L.R., and Robinson, W.S.,** Structure of hepatitis B Dane particle DNA before and after the Dane particle DNA polymerase reaction, *J. Virol.*, 21, 666, 1977.
39. **Summers, J., O'Connell, A., and Millman, I.,** Genome of hepatitis B virus: restriction enzyme cleavage and structure of DNA extracted from Dane particles, *Proc. Natl. Acad. Sci. U.S.A.*, 72, 4597, 1975.
40. **Landers, T.A., Greenberg, H.B., and Robinson, W.S.,** Structure of hepatitis B Dane particle DNA and nature of the endogenous DNA polymerase reaction, *J. Virol.*, 23, 368, 1977.
41. **Sattler, F. and Robinson, W.S.,** Hepatitis B viral DNA molecules have cohesive ends, *J. Virol.*, 32, 226, 1979.
42. **Siddiqui, A., Sattler, F., and Robinson, W.S.,** Restriction endonuclease cleavage map and location of unique features of the DNA of hepatitis B virus, subtype adw_2, *Proc. Natl. Acad. Sci. U.S.A.*, 76, 4664, 1979.
43. **Will, H., Reiser, W., Weimer, T., Pfaff, E., Buscher, M., Sprengle, R., Cattaneo, R., and Schaller, H.,** Replication strategy of human hepatitis B virus, *J. Virol.*, 61, 904, 1987.
44. **Gerlich, W.H. and Robinson, W.S.,** Hepatitis B virus contains protein attached to the 5′ terminus of its complete DNA strand, *Cell*, 21, 801, 1980.
45. **Lien, J.-M., Aldrich, C.E., and Mason, W.S.,** Evidence that a capped oligoribonucleotide is the primer for duck hepatitis B virus plus-strand DNA synthesis, *J. Virol.*, 57, 229, 1986.
46. **Seeger, C., Ganem, D., and Varmus, H.E.,** Biochemical and genetic evidence for the hepatitis B virus replication strategy, *Science*, 232, 447, 1986.
47. **Charnay, P., Pourcel, C., Louise, A., Fritsch, A., and Tiollais, P.,** Cloning in *Escherichia coli* and physical structure of hepatitis B virion DNA, *Proc. Natl. Acad. Sci. U.S.A.*, 76, 2222, 1979.
48. **Sninsky, J.J., Siddiqui, A., Robinson, W.S., and Cohen, S.N.,** Cloning and endonuclease mapping of the hepatitis B viral genome, *Nature (London)*, 279, 346, 1979.
49. **Valenzuela, P., Gray, P., Quiroga, M., Zaldivar, J., Goodman, H.M., and Rutter, W.J.,** Nucleotide sequence of the gene coding for the major protein of hepatitis B virus surface antigen, *Nature (London)*, 280, 815, 1979.
50. **Galibert, F., Mandart, E., Fitoussi, F., Tiollais, P., and Charnay, P.,** Nucleotide sequence of the hepatitis B virus genome (subtype ayw) cloned in *E. coli*, *Nature (London)*, 281, 646, 1979.
51. **Pasek, M., Goto, T., Gilbert, W., Zink, B., Schaller, H., MacKay, P., Leadbetter, G., and Murray, K.,** Hepatitis B virus genes and their expression in *E. coli*, *Nature (London)*, 282, 575, 1979.
52. **Valenzuela, P., Quiroga, M., Zaldivar, J., Gray, P., and Rutter, W.J.,** The nucleotide sequence of the hepatitis B viral genome and the identification of the major viral genes, in *Animal Virus Genetics*, Fields, B., Jaenisch, R., and Fox, C.F., Eds., Academic Press, New York, 1980, 57.
53. **Will, H., Kuhn, C., Cattaneo, R., and Schaller, H.,** Structure and function of the hepatitis B virus genome, in *Primary and Tertiary Structure of Nucleic Acids and Cancer Research*, Miwa, M., Nishimura, S., Rich, A., Soell, D.G., and Sugimura, T., Eds., Jap. Sci. Soc. Press, Tokyo, 1982, 237.

54. **Ono, Y., Onda, H., Sasada, R., Igarashi, K., Sugino, Y., and Nishioka, K.,** The complete nucleotide sequences of the cloned hepatitis B virus DNA: subtype adr and adw, *Nucleic Acids Res.*, 11, 1747, 1983.
55. **Fujiyama, A., Miyanohara, A., Nozaki, C., Yoneyama, T., Ohtomo, N., and Matsubara, K.,** Cloning and structural analyses of hepatitis B virus DNAs, subtype *adr*, *Nucleic Acids Res.*, 11, 4601, 1983.
56. **Bichko, V., Pushko, P., Dreilina, D., Pumpen, P., and Gren, E.,** Subtype ayw variant of hepatitis B virus DNA primary structure analysis, *FEBS Lett.*, 185, 208, 1985.
57. **Okamoto, H., Imai, M., Kametani, M., Nakamura, T., and Mayumi, M.,** Genomic heterogeneity of hepatitis B virus in a 54-year-old woman who contacted the infection through materno-fetal transmission, *Jpn. J. Exp. Med.*, 57, 231, 1987.
58. **Vaudin, M., Wolstenholme, A.J., Tsiquay, K.N., Zuckerman, A.J., and Harrison, T.J.,** The complete nucleotide sequence of the genome of a hepatitis B virus isolated from a naturally infected chimpanzee, *J. Gen. Virol.*, 69, 1383, 1988.
59. **Okamoto, H., Tsuda, F., Sakugawa, H., Sastrosoewignjo, R.I., Imai, M., Miyakawa, Y., and Mayumi, M.,** Typing hepatitis B virus by homology in nucleotide sequence: comparison of surface antigen subtypes, *J. Gen. Virol.*, 69, 2575, 1988.
60. **Okamoto, H., Imai, M., Shimozaki, M., Hoshi, Y., Izuka, H., Gotanda, T., Tsuda, F., Miyakawa, Y., and Mayumi, M.,** Nucleotide sequence of a cloned hepatitis B virus genome, subtype ayr: comparison with genomes of the other three subtypes, *J. Gen. Virol.*, 67, 2305, 1986.
61. **Kobayashi, M., and Koike, K.,** Complete nucleotide sequence of hepatitis B virus DNA of subtype adr and its conserved gene organization, *Gene*, 30, 227, 1984.
62. **Will, H., Cattaneo, R., Koch, H.-G., Darai, G., and Schaller, H.,** Cloned HBV DNA causes hepatitis in chimpanzees, *Nature (London)*, 299, 740, 1982.
63. **Will, H., Cattaneo, R., Darai, G., Deinhardt, F., Schellekens, H., and Schaller, H.,** Infectious hepatitis B virus from cloned DNA of known nucleotide sequence, *Proc. Natl. Acad. Sci. U.S.A.*, 82, 891, 1985.
64. **Tsurimoto, T., Fujiyama, A., and Matsubara, K.,** Stable expression and replication of hepatitis B virus genome in an integrated state in a human hepatoma cell line transfected with the cloned viral DNA, *Proc. Natl. Acad. Sci. U.S.A.*, 84, 444, 1987.
65. **Sells, M.A., Chen, M.-L., and Acs, G.,** Production of hepatitis B virus particles in HepG2 cells transfected with cloned hepatitis B virus DNA, *Proc. Natl. Acad. Sci. U.S.A.*, 84, 1005, 1987.
66. **Chang, C., Jeng, K.-s., Hu, C.-p., Lo, S.J., Su, T.-s., Ting, L.-P., Chou, C.-K., Han, S.-h., Pfaff, E., Salfeld, J., and Schaller, H.,** Production of hepatitis B virus *in vitro* by transient expression of cloned HBV DNA in a hepatoma cell line, *EMBO J.*, 6, 675, 1987.
67. **Yaginuma, K., Shirakata, Y., Kobayashi, M., and Koike, K.,** Hepatitis B virus (HBV) particles are produced in a cell culture system by transient expression of transfected HBV DNA, *Proc. Natl. Acad. Sci. U.S.A.*, 84, 2678, 1987.
68. **Araki, K., Miyazaki, J.-I., Hino, O., Tomita, N., Chisaka, O., Matsubara, K., and Yamamura, K.-I.,** Expression and replication of hepatitis B virus genome in transgenic mice, *Proc. Natl. Acad. Sci. U.S.A.*, 86, 207, 1989.
69. **Shih, C., Li, L.-S., Roychoudhury, S., and Ho, M.-H.,** *In vitro* propagation of human hepatitis B virus in a rat hepatoma cell line, *Proc. Natl. Acad. Sci. U.S.A.*, 86, 6323, 1989.
70. **Acs, G., Sells, M.A., Purcell, R.H., Price, P., Engle, R., Shapiro, M., and Popper, H.,** Hepatitis B virus produced by transfected HepG2 cells causes hepatitis in chimpanzees, *Proc. Natl. Acad. Sci. U.S.A.*, 84, 4641, 1987.
71. **Sureau, C., Eichberg, J.W., Hubbard, G.B., Romet-Lemonne, J.L., and Essex, M.,** A molecularly cloned hepatitis B virus produced *in vitro* is infectious in a chimpanzee, *J. Virol.*, 62, 3064, 1988.
72. **Peterson, D.L.,** Isolation and characterization of the major protein and glycoprotein of hepatitis B surface antigen, *J. Biol. Chem.*, 256, 6975, 1981.
73. **Peterson, D.L., Roberts, I.M., and Vyas, G.N.,** Partial amino acid sequence of two major component polypeptides of hepatitis B surface antigen, *Proc. Natl. Acad. Sci. U.S.A.*, 74, 1530, 1977.
74. **Stibbe, W. and Gerlich, W.H.,** Structural relationships between minor and major proteins of hepatitis B surface antigen, *J. Virol.*, 46, 626, 1983.
75. **Heermann, K.H., Goldmann, U., Schwartz, W., Seyffarth, T., Baumgarten, H., and Gerlich, W.H.,** Large surface proteins of hepatitis B virus containing the Pre-s sequence, *J. Virol.*, 52, 396, 1984.
76. **Takahashi, K., Machida, A., Funatsu, G., Nomura, M., Usuda, S., Aoyagi, S., Tachibana, K., Miyamoto, H., Imai, M., Nakamura, T., Miyakawa, Y., and Mayumi, M.,** Immunochemical structure of hepatitis B e antigen in the serum, *J. Immunol.*, 130, 2903, 1983.
77. **Standring, D.N., Ou, J.-H., Masiarz, F.R., and Rutter, W.J.,** A signal peptide encoded within the precore region of hepatitis B virus directs the secretion of heterogeneous populations of e antigens in *Xenopus* oocytes, *Proc. Natl. Acad. Sci. U.S.A.*, 85, 8405, 1988.

78. **Toh, H., Hayashida, H., and Miyata, T.,** Sequence homology between retroviral reverse transcriptase and putative polymerases of hepatitis B virus and cauliflower mosaic virus, *Nature (London)*, 305, 827, 1983.
79. **Bavand, M., Feitelson, M., and Laub, O.,** The hepatitis B virus-associated reverse transcriptase is encoded by the viral *pol* gene, *J. Virol.*, 63, 1019, 1989.
80. **Mack, D.H., Bloch, W., Nath, N., and Sninsky, J.J.,** Hepatitis B virus particles contain a polypeptide encoded by the largest open reading frame: a putative reverse transcriptase, *J. Virol.*, 62, 4786, 1988.
81. **Chisaka, O., Araki, K., Ochiya, T., Tsurimoto, T., Hiranyawasitte-Attatippaholkun, W., Yanaihara, N., and Matsubara, K.,** Purification of hepatitis B virus gene *X* product synthesized in *Escherichia coli* and its detection in a human hepatoblastoma cell line producing hepatitis B virus, *Gene*, 60, 183, 1987.
82. **Katayama, K., Hayashi, N., Sasaki, Y., Kasahara, A., Ueda, K., Fusamoto, H., Sato, N., Chisaka, O., Matsubara, K., and Kamada, T.,** Detection of hepatitis B virus X gene protein and antibody in type B chronic liver disease, *Gastroenterology*, 97, 990, 1989.
83. **Machida, A., Kishimoto, S., Ohnuma, H., Miyamoto, H., Baba, K., Oad, K., Nakamura, T., Funatsu, G., Miyakawa, Y., and Mayumi, M.,** A glycopeptide containing 15 amino acid residues derived from hepatitis B surface antigen particles: demonstration of immunogenicity to raise anti-HBs in mice, *Mol. Immunol.*, 19, 1087, 1982.
84. **Skelly, J., Howard, C.R., and Zuckerman, A.J.,** The labelling of galactose residues in hepatitis B surface antigen glycoprotein, *J. Gen. Virol.*, 41, 447, 1978.
85. **Shiraishi, H., Kohama, T., Shirachi, R., and Ishida, N.,** Carbohydrate composition of hepatitis B surface antigen, *J. Gen. Virol.*, 36, 207, 1977.
86. **Stibbe, W. and Gerlich, W.H.,** Variable protein composition of hepatitis B surface antigen from different donors, *Virology*, 123, 436, 1982.
87. **Chairez, R., Steiner, S., Melnick, J.L., and Dreesman, G.R.,** Glycoproteins associated with hepatitis B antigen, *Intervirology*, 1, 224, 1973.
88. **Aggerbeck, L.P. and Peterson, D.L.,** Electron microscopic and solution X-ray scattering observations on the structure of hepatitis B surface antigen, *Virology*, 141, 155, 1985.
89. **Moriarty, A.M, Hoyer, B.H., Shih, J.W.-K., Gerin, J.L., and Hamer, D.H.,** Expression of the hepatitis B virus surface antigen gene in cell culture by using a simian virus 40 vector, *Proc. Natl. Acad. Sci. U.S.A.*, 78, 2606, 1981.
90. **Stenlund, A., Lamy, D., Moreno-Lopez, J., Ahola, H., Petterson, U., and Tiollais, P.,** Secretion of the hepatitis B virus surface antigen from mouse cells using an extra-chromosomal eucaryotic vector, *EMBO J.*, 2, 669, 1983.
91. **Hsiung, N., Fitts, R., Wilson, S., Milne, A., and Hamer, D.H.,** Efficient production of hepatitis B surface antigen using a bovine papilloma virus-metallothionein vector, *J. Mol. Appl. Genet.*, 2, 497, 1984.
92. **Michel, M.-L., Pontisso, P., Sobczak, E., Malpiece, Y., Streeck, R.E., and Tiollais, P.,** Synthesis in animal cells of hepatitis B surface antigen particles carrying a receptor for polymerized human serum albumin, *Proc. Natl. Acad. Sci. U.S.A.*, 81, 7708, 1984.
93. **Persing, D.H., Varmus, H.E., and Ganem, D.,** A frameshift mutation in the *pres-S* region of the human hepatitis B virus genome allows production of surface antigen particles but eliminates binding to polymerized albumin, *Proc. Natl. Acad. Sci. U.S.A.*, 82, 3440, 1985.
94. **McLachlan, A., Milich, D.R., Raney, A.K., Riggs, M.G., Hughes, J.L., Sorge, J., and Chisari, F.V.,** Expression of hepatitis B virus surface and core antigens: influences of Pre-S and precore sequences, *J. Virol.*, 61, 683, 1987.
95. **Ou, J.-H. and Rutter, W.J.,** Regulation of secretion of the hepatitis B virus major surface antigen by the PreS-1 protein, *J. Virol.*, 61, 782, 1987.
96. **Cheng, K.-C. and Moss, B.,** Selective synthesis and secretion of particles composed of the hepatitis B virus middle surface protein directed by a recombinant vaccinia virus: induction of antibodies to Pre-S and S epitopes, *J. Virol.*, 61, 1286, 1987.
97. **Molnar-Kimber, K.L., Jarocki-Witek, V., Dheer, S.K., Vernon, S.K., Conley, A.J., Davis, A.R., and Hung, P.P.,** Distinctive properties of the hepatitis B virus envelope proteins, *J. Virol.*, 62, 407, 1988.
98. **Langley, K.E., Egan, K.M., Barendt, J.M., Parker, C.G., and Bitter, G.A.,** Characterization of purified hepatitis B surface antigen containing pre-S(2) epitopes expressed in *Saccharomyces cerevisiae*, *Gene*, 67, 229, 1988.
99. **Yoneyama, T., Akatsuka, T., and Miyamura, T.,** Stable expression of the hepatitis B virus surface antigen containing Pre-S2 protein in mouse cells using a bovine papillomavirus vector, *J. Gen. Virol.*, 69, 1931, 1988.
100. **Chisari, F.V., Filippi, P., Buras, J., McLachlan, A., Popper, H., Pinkert, C.A., Palmiter, R.D., and Brinster, R.L.,** Structural and pathological effects of synthesis of hepatitis B virus large envelope polypeptide in transgenic mice, *Proc. Natl. Acad. Sci. U.S.A.*, 84, 6909, 1987.

101. **Gavilanes, F., Gonzalez-Ros, J.M., and Peterson, D.L.,** Structure of hepatitis B surface antigen, *J. Biol. Chem.*, 257, 7770, 1982.
102. **Steiner, S., Huebner, M.T., and Dreesman, G.R.,** Major polar lipids of hepatitis B antigen preparations: evidence for the presence of a glycosphingolipid, *J. Virol.*, 14, 572, 1974.
103. **Kim, C.Y. and Bissell, D.M.,** Stability of the lipid and protein of hepatitis-associated (Australia) antigen, *J. Infect. Dis.*, 123, 470, 1971.
104. **Budkowska, A., Shih, J.W.-K., and Gerin, J.L.,** Immunochemistry and polypeptide composition of hepatitis B core antigen (HBcAg), *J. Immunol.*, 118, 1300, 1977.
105. **Hruska, J.F. and Robinson, W.S.,** The proteins of hepatitis B Dane particle cores, *J. Med. Virol.*, 1, 119, 1977.
106. **Petit, M.-A., and Pillot, J.,** HBc and HBe antigenicity and DNA-binding activity of major core protein P22 in hepatitis B virus core particles isolated from the cytoplasm of human liver cells, *J. Virol.*, 53, 543, 1985.
107. **Theilmann, L., Hofmann, W.J., and Pfaff, E.,** Isolation and characterization of cytoplasmic and nuclear particles of hepatitis B virus, *Virus Res.*, 14, 225, 1989.
108. **Theilmann, L., Pfaff, E., Kommerell, B., Gmelin, K., Schaller, H., and Salfeld, J.,** Detection of hepatitis B virus core gene products in sera and liver of HBV-infected individuals, *J. Hepatol.*, 8, 77, 1989.
109. **Yamaki, M., Ohori, H., Onodera, S., Ishida, N., and Maeda, H.,** Circular dichroism and biochemical properties of the hepatitis B virus core antigen, *Biochim. Biophys. Acta*, 706, 165, 1982.
110. **Onodera, S., Ohori, H., Yamaki, M., and Ishida, N.,** Electron microscopy of human hepatitis B virus cores by negative staining-carbon film technique, *J. Med. Virol.*, 10, 147, 1982.
111. **Gerlich, W.H., Goldmann, U., Muller, R., Stibbe, W., and Wolff, W.,** Specificity and localization of the hepatitis B virus associated protein kinase, *J. Virol.*, 42, 761, 1982.
112. **Albin, C. and Robinson, W.S.,** Protein kinase activity in hepatitis B virus, *J. Virol.*, 34, 297, 1980.
113. **Roossinck, M.J. and Siddiqui, A.,** *In vivo* phosphorylation and protein analysis of hepatitis B virus core antigen, *J. Virol.*, 61, 955, 1987.
114. **Radziwill, G., Tucker, W., and Schaller, H.,** Mutational analysis of the hepatitis B virus P gene product: domain structure and RNase activity, *J. Virol.*, 64, 613, 1990.
115. **Bosch, V., Bartenschlager, R., Radziwill, G., and Schaller, H.,** The duck hepatitis B virus P-gene codes for protein strongly associated with the 5′-end of the viral DNA minus strand, *Virology*, 166, 475, 1988.
116. **Bartenschlager, R. and Schaller, H.,** The amino-terminal domain of the hepadnaviral P-gene encodes the terminal protein (genome-linked protein) believed to prime reverse transcription, *EMBO J.*, 7, 4185, 1988.
117. **Hirschman, S.Z.,** Replication of hepatitis B virus in cell culture systems, in *Advances in Hepatitis Research*, Chisari, F.V., Ed., Masson, Paris, 1984, 54.
118. **Rijntjes, P.J.M., Moshage, H.J., and Yap, S.H.,** *In vitro* infection of primary cultures of cryopreserved adult human hepatocytes with hepatitis B virus, *Virus Res.*, 10, 95, 1988.
119. **Gripon, P., Diot, C., Theze, N., Fourel, I., Loreal, O., Brechot, C., and Guguen-Guillouzo, C.,** Hepatitis B virus infection of adult human hepatocytes cultured in the presence of dimethyl sulfoxide, *J. Virol.*, 62, 4136, 1988.
120. **Neurath, A.R., Kent, S.B.H., Strick, N., and Parker, K.,** Identification and chemical synthesis of a host cell receptor binding site on hepatitis B virus, *Cell*, 46, 429, 1986.
121. **Peeples, M.E., Komai, K., Radek, R., and Bankowski, M.J.,** A cultured cell receptor for the small S protein of hepatitis B virus, *Virology*, 160, 135, 1987.
122. **Komai, K., Kaplan, M., and Peeples, M.E.,** The vero cell receptor for the hepatitis B virus small S protein is a sialoglycoprotein, *Virology*, 163, 629, 1988.
123. **Pontisso, P., Petit, M.-A., Bankowski, M.J., and Peeples, M.E.,** Human liver plasma membranes contain receptors for the hepatitis B virus Pre-S1 region and, via polymerized human serum albumin, for the Pre-S2 region, *J. Virol.*, 63, 1981, 1989.
124. **Pontisso, P., Ruvoletto, M.G., Gerlich, W.H., Heermann, K.-H., Bardini, R., and Alberti, A.,** Identification of an attachment site for human liver plasma membranes on hepatitis B virus particles, *Virology*, 173, 522, 1989.
125. **Weisiger, R., Gollan, J.L., and Ockner, R.,** Receptor for albumin on the liver cell surface may mediate uptake of fatty acids and other albumin-bound substances, *Science*, 211, 1048, 1981.
126. **Ockner, R.K., Weisiger, R.A., and Gollan, J.L.,** Hepatic uptake of albumin-bound substances: albumin receptor concept, *Am. J. Physiol.*, 245, G13, 1983.
127. **Thung, S.N. and Gerber, M.,** Presence of receptors for polymerized albumin in HBsAg-containing hepatocytes and hepatoma cell line, *Hepatology*, 1, 132, 1981.
128. **Machida, A., Kishimoto, S., Ohnuma, H., Baba, K., Ito, Y., Miyamoto, H., Funatsu, G., Oda, K., Usuda, S., Togami, S., Nakamura, T., Miyakawa, Y., and Mayumi, M.,** A polypeptide containing 55 amino acid residues coded by the Pre-S region of hepatitis B virus deoxyribonucleic acid bears the receptor for polymerized human as well as chimpanzee albumins, *Gastroenterology*, 86, 910, 1984.

129. **Miller, R.H. and Robinson, W.S.,** Hepatitis B virus DNA forms in nuclear and cytoplasmic fractions of infected human liver, *Virology*, 137, 390, 1984.
130. **Cattaneo, R., Will, H., Hernandez, N., and Schaller, H.,** Signals regulating hepatitis B surface antigen transcription, *Nature (London)*, 305, 336, 1983.
131. **Cattaneo, R., Will, H., and Schaller, H.,** Hepatitis B virus transcription in the infected liver, *EMBO J.*, 3, 2191, 1984.
132. **Yokosuka, O., Omata, M., Imazeki, F., Ito, Y., and Okuda, K.,** Hepatitis B virus RNA transcripts and DNA in chronic liver disease, *N. Engl. J. Med.*, 315, 1187, 1986.
133. **Imazeki, F., Yaginuma, K., Omata, M., Okuda, K., Kobayashi, M., and Koike, K.,** RNA transcripts of hepatitis B virus in hepatocellular carcinoma, *Hepatology*, 7, 753, 1987.
134. **Su, T.-s., Lui, W.-Y., Lin, L.-H., Han, S.-h., and P'eng, F.-K.,** Analysis of hepatitis B virus transcripts in infected human livers, *Hepatology*, 9, 180, 1989.
135. **Sells, M.A., Zelent, A.Z., Shvartsman, M., and Acs, G.,** Replicative intermediates of hepatitis B virus in HepG2 cells that produce infectious virions, *J. Virol.*, 62, 2836, 1988.
136. **Standring, D.N., Rutter, W.J., Varmus, H.E., and Ganem, D.,** Transcription of the hepatitis B surface antigen gene in cultured murine cells initiates within the presurface region, *J. Virol.*, 50, 563, 1984.
137. **Simonsen, C.C., and Levinson, A.D.,** Analysis of processing and polyadenylation signals of the hepatitis B virus surface antigen gene by using simian virus 40-hepatitis B virus chimeric plasmids, *Mol. Cell. Biol.*, 3, 2250, 1983.
138. **Roychoudhury, S., and Shih, C.,** *Cis* rescue of a mutated reverse transcriptase gene of human hepatitis B virus by creation of an internal ATG, *J. Virol.*, 64, 1063, 1990.
139. **Schlicht, H.-J., Radziwill, G., and Schaller, H.,** Synthesis and encapsidation of duck hepatitis B virus reverse transcriptase do not require formation of core-polymerase fusion proteins, *Cell*, 56, 85, 1989.
140. **Chang, L.-J., Pryciak, P., Ganem, D., and Varmus, H.E.,** Biosynthesis of the reverse transcriptase of hepatitis B viruses involves *de novo* translational initiation, not ribosomal frameshifting, *Nature (London)*, 337, 364, 1989.
141. **Farza, H., Hadchouel, M., Scotto, J., Tiollais, P., Babinet, C., and Pourcel, C.,** Replication and gene expression of hepatitis B virus in a transgenic mouse that contains the complete viral genome, *J. Virol.*, 62, 4144, 1988.
142. **Kaneko, S. and Miller, R.H.,** X-region-specific transcript in mammalian hepatitis B virus-infected liver, *J. Virol.*, 62, 3979, 1988.
143. **Gough, N.M.,** Core and e antigen synthesis in rodent cells transformed with hepatitis B virus DNA is associated with greater than genome length viral messenger RNAs, *J. Mol. Biol.*, 165, 683, 1983.
144. **Siddiqui, A., Jameel, S., and Mapoles, J.,** Transcriptional control elements of hepatitis B surface antigen gene, *Proc. Natl. Acad. Sci. U.S.A.*, 83, 566, 1986.
145. **Saito, I., Oya, Y., and Shimojo, H.,** Novel RNA family structure of hepatitis B virus expressed in human cells, using a helper-free adenovirus vector, *J. Virol.*, 58, 554, 1986.
146. **Siddiqui, A., Jameel, S., and Mapoles, J.,** Expression of the hepatitis B virus X gene in mammalian cells, *Proc. Natl. Acad. Sci. U.S.A.*, 84, 2513, 1987.
147. **Tiollais, P., Pourcel, C., and Dejean, A.,** The hepatitis B virus, *Nature (London)*, 317, 489, 1985.
148. **Moriarty, A.M., Alexander, H., Lerner, R.A., and Thornton, G.B.,** Antibodies to peptides detect new hepatitis B antigen: serological correlation with hepatocellular carcinoma, *Science*, 227, 429, 1985.
149. **Kay, A., Mandart, E., Trepo, C., and Galibert, F.,** The HBV HBx gene expressed in *E. coli* is recognised by sera from hepatitis patients, *EMBO J.*, 4, 1287, 1985.
150. **Meyers, M.L., Trepo, L.V., Nath, N., and Sninsky, J.J.,** Hepatitis B virus polypeptide X: expression in *Escherichia coli* and identification of specific antibodies in sera from hepatitis B virus-infected humans, *J. Virol.*, 57, 101, 1986.
151. **Elfassi, E., Haseltine, W.A., and Dienstag, J.L.,** Detection of hepatitis B virus X product using an open reading frame *Escherichia coli* expression vector, *Proc. Natl. Acad. Sci. U.S.A.*, 83, 2219, 1986.
152. **Liang, X.-H., Stemler, M., Will, H., Braun, R., Tang, Z.-Y., and Schroder, C.H.,** Low incidence and high titers of antibodies to hepatitis B virus X-protein in sera of Chinese patients with hepatocellular carcinoma, *J. Med. Virol.*, 25, 329, 1988.
153. **Offensperger, W., Wahl, S., Neurath, A.R., Price, P., Strick, N., Kent, S.B.H., Christman, J.K., and Acs, G.,** Expression in *Escherichia coli* of a cloned DNA sequence encoding the pres-S2 region of hepatitis B virus, *Proc. Natl. Acad. Sci. U.S.A.*, 82, 7540, 1985.
154. **Spandau, D.F. and Lee, S.-D.,** Trans-activation of viral enhancers by the hepatitis B virus X protein, *J. Virol.*, 62, 427, 1988.
155. **Twu, J.-S. and Schloemer, R.H.,** Transcriptional trans-activating function of hepatitis B virus, *J. Virol.*, 61, 3448, 1987.

156. **Zahm, P., Hofschneider, P.H., and Koshy, R.,** The HBV X-ORF encodes a transactivator: a potential factor in viral hepatocarcinogenesis, *Oncogene*, 3, 169, 1988.
157. **Seto, E., Yen, T.S.B., Peterlin, B.M., and Ou, J.-H.,** Trans-activation of the human immunodeficiency virus long terminal repeat by the hepatitis B virus X protein, *Proc. Natl. Acad. Sci. U.S.A.*, 85, 8286, 1988.
158. **Twu, J.-S. and Robinson, W.S.,** Hepatitis B virus X gene can transactivate heterologous viral sequences, *Proc. Natl. Acad. Sci. U.S.A.*, 86, 2046, 1989.
159. **Siddiqui, A., Gaynor, R., Srinivasan, A., Mapoles, J., and Farr, R.W.,** Trans-activation of viral enhancers including long terminal repeat of the human immunodeficiency virus by the hepatitis B virus X protein, *Virology*, 169, 479, 1989.
160. **Colgrove, R., Simon, G., and Ganem, D.,** Transcriptional activation of homologous and heterologous genes by the hepatitis B virus X gene product in cells permissive for viral replication, *J. Virol.*, 63, 4019, 1989.
161. **Seto, E., Zhou, D.-X., Peterlin, B.M., and Yen, T.S.B.,** Trans-activation by the hepatitis B virus X protein shows cell-type specificity, *Virology*, 173, 764, 1989.
162. **Su, T.-s., Lai, C.-H., Huang, J.-L., Lin, L.-H., Yauk, Y.-K., Chang, C., Lo, S.J., and Han, S.-h.,** Hepatitis B virus transcript produced by RNA splicing, *J. Virol.*, 63, 4011, 1989.
163. **Chen, P.-J., Chen, C.-R., Sung, J.-L., and Chen, D.-S.,** Identification of a doubly spliced viral transcript joining the separated domains for putative protease and reverse transcriptase of hepatitis B virus, *J. Virol.*, 63, 4165, 1989.
164. **Eble, B.E., Lingappa, V.R., and Ganem, D.,** Hepatitis B surface antigen: an unusual secreted protein initially synthesized as a transmembrane polypeptide, *Mol. Cell. Biol.*, 6, 1454, 1986.
165. **Eble, B.E., MacRae, D.R., Lingappa, V.R., and Ganem, D.,** Multiple topogenic sequences determine the transmembrane orientation of hepatitis B surface antigen, *Mol. Cell. Biol.*, 7, 3591, 1987.
166. **Simon, K., Lingappa, V.R., and Ganem, D.,** Secreted hepatitis B surface antigen polypeptides are derived from a transmembrane precursor, *J. Cell Biol.*, 107, 2163, 1988.
167. **Raney, A.K., Milich, D.R., Easton, A.J., and McLachlan, A.,** Differentiation specific transcriptional regulation of the hepatitis B virus large surface antigen gene in human hepatoma cell lines, *J. Virol.*, 64, 2360, 1990.
168. **Robinson, W.S. and Lutwick, L.I.,** The virus of hepatitis, type B, *N. Engl. J. Med.*, 1168, 1976.
169. **Price, P.M., Mohamad, A., Zelent, A., Neurath, A.R., and Acs, G.,** Translational selection in the expression of the hepatitis B virus envelope proteins, *DNA*, 7, 417, 1988.
170. **Patzer, E.J., Nakamura, G.R., and Yaffe, A.,** Intracellular transport and secretion of hepatitis B virus surface antigen in mammalian cells, *J. Virol.*, 51, 346, 1984.
171. **Patzer, E.J., Nakamura, G.R., Simonsen, C.C., Levinson, A.D., and Brands, R.,** Intracellular assembly and packaging of hepatitis B surface antigen particles occur in the endoplasmic reticulum, *J. Virol.*, 58, 884, 1986.
172. **Eble, B.E., Lingappa, V.R., and Ganem, D.,** The N-terminal (pre-S2) domain of a hepatitis B virus surface glycoprotein is translocated across membranes by downstream signal sequences, *J. Virol.*, 64, 1414, 1990.
173. **Garcia, P.D., Ou, J.-H., Rutter, W.J., and Walter, P.,** Targeting of the hepatitis B virus precore protein to the endoplasmic reticulum membrane: after signal peptide cleavage translocation can be aborted and the product released into the cytoplasm, *J. Cell Biol.*, 106, 1093, 1988.
174. **Weimer, T., Salfeld, J., and Will, H.,** Expression of the hepatitis B virus core gene *in vitro* and *in vivo*, *J. Virol.*, 61, 3109, 1987.
175. **Bruss, V. and Gerlich, W.H.,** Formation of transmembranous hepatitis B e-antigen by cotranslational *in vitro* processing of the viral precore protein, *Virology*, 163, 268, 1988.
176. **Jean-Jean, O., Levrero, M., Will, H., Perricaudet, M., and Rossignol, J.-M.,** Expression mechanism of the hepatitis B virus (HBV) C gene and biosynthesis of HBe antigen, *Virology*, 170, 99, 1989.
177. **Docherty, K. and Steiner, D.F.,** Post-translational proteolysis in polypeptide hormone biosynthesis, *Annu. Rev. Physiol.*, 44, 625, 1982.
178. **Thomas, H.C., Jacyna, M., Waters, J., and Main, J.,** Virus-host interaction in chronic hepatitis B virus infection, Semin. Liver Dis. 8, 342, 1988.
179. **Milich, D.R., Jones, J.E., Hughes, J.L., Price, J., Raney, A.K., and McLachlan, A.,** Is a function of the secreted hepatitis B e antigen to induce immunologic tolerance: a transgenic model, *Proc. Natl. Acad. Sci. U.S.A.*, 87, 6599, 1990.
180. **Enders, G.H., Ganem, D., and Varmus, H.E.,** 5′-Terminal sequences influence the segregation of ground squirrel hepatitis virus RNAs into polyribosomes and viral core particles, *J. Virol.*, 61, 35, 1987.
181. **Hirsch, R.C., Lavine, J.E., Chang, L.-J., Varmus, H.E., and Ganem, D.,** Polymerase gene products of hepatitis B viruses are required for genomic RNA packaging as well as for reverse transcription, *Nature (London)*, 344, 552, 1990.

182. **Miller, R.H., Tran, C.-T., and Robinson, W.S.,** Hepatitis B virus particles of plasma and liver contain viral DNA-RNA hybrid molecules, *Virology*, 139, 53, 1984.
183. **Seeger, C. and Maragos, J.,** Identification and characterization of the woodchuck hepatitis virus origin of DNA replication, *J. Virol.*, 64, 16, 1990.
184. **Burrell, C.J., Gowans, E.J., Jilbert, A.R., Lake, J.R., and Marmion, B.P.,** Hepatitis B virus DNA detection by *in situ* cytohybridization: implications for viral replication strategy and pathogenesis of chronic hepatitis, *Hepatology*, 2, 85S, 1982.
185. **Gowans, E.J., Burrell, C.J., Jilbert, A.R., and Marmion, B.P.,** Patterns of single- and double-stranded hepatitis B virus DNA and viral antigen accumulation in infected liver cells, *J. Gen. Virol.*, 64, 1229, 1983.
186. **Blum, H.E., Hasse, A.T., and Vyas, G.N.,** Molecular pathogenesis of hepatitis B virus infection: simultaneous detection of viral DNA and antigens in paraffin-embedded liver sections, *Lancet*, 2, 771, 1984.
187. **Gowans, E.J., Burrell, C.J., Jilbert, A.R., and Marmion, B.P.,** Cytoplasmic (but not nuclear) hepatitis B virus (HBV) core antigen reflects HBV DNA synthesis at the level of the infected hepatocyte, *Intervirology*, 24, 220, 1985.
188. **Arakaki, Y., Toyoda, H., Watanabe, S., Seki, S., and Oda, T.,** Detection of hepatitis B virus-DNA in liver tissues in chronic type B liver diseases and its relevancy to viral antigens: a study by *in situ* hybridization using a biotinylated probe and streptavidin-alkaline phosphatase, *Acta Histochem. Cytochem.*, 21, 489, 1988.
189. **Summers, J. and Mason, W.S.,** Replication of the genome of a hepatitis B-like virus by reverse transcription of an RNA intermediate, *Cell*, 29, 403, 1982.
190. **Stein, O., Fainaru, M., and Stein, Y.,** Visualization of virus-like particles in endoplasmic reticulum of hepatocytes of Australia antigen carriers, *Lab. Invest.*, 26, 262, 1972.
191. **Huang, S.-N. and Groh, V.,** Immunoagglutination electron microscopic study on virus-like particles and Australia antigen in liver tissue, *Lab. Invest.*, 29, 353, 1973.
192. **Yamada, G. and Nakane, P.K.,** Hepatitis B core and surface antigens in liver tissue, *Lab. Invest.*, 36, 649, 1977.
193. **Yamada, G., Sakamoto, Y., Mizuno, M., Nishihara, T., Kobayashi, T., Takahashi, T., and Nagashima, H.,** Electron and immunoelectron microscopic study of Dane particle formation in chronic hepatitis B virus infection, *Gastroenterology*, 83, 348, 1982.
194. **Gerber, M.A., Sells, M.A., Chen, M.-L., Thung, S.N., Tabibzadeh, S.S., Hood, A., and Acs, G.,** Morphologic, immunohistochemical, and ultrastructural studies of the production of hepatitis B virus *in vitro*, *Lab. Invest.*, 59, 173, 1988.
195. **Kamimura, T., Yoshikawa, A., Ichida, F., and Sasaki, H.,** Electron microscopic studies of Dane particles in hepatocytes with special reference to intracellular development of Dane particles and their relation with HBeAg in serum, *Hepatology*, 1, 392, 1981.
196. **Tuttleman, J.S., Pourcel, C., and Summers, J.,** Formation of the pool of covalently closed circular viral DNA in hepadnavirus-infected cells, *Cell*, 47, 451, 1986.
197. **Summers, J., Smith, P.M., and Horwich, A.L.,** Hepadnavirus envelope proteins regulate covalently closed circular DNA amplification, *J. Virol.*, 64, 2819, 1990.
198. **Lichter, E.A.,** Chimpanzee antibodies to Australia antigen, *Nature (London)*, 224, 810, 1969.
199. **Maynard, J.E., Hartwell, W.V., and Berquist, K.R.,** Hepatitis-associated antigen in chimpanzees, *J. Infect. Dis.*, 123, 660, 1971.
200. **Hirschman, R.J., Shulman, R., Barker, L.F., and Smith, K.O.,** Virus-like particles in sera of patients with infectious and serum hepatitis, *JAMA*, 208, 1667, 1969.
201. **Barker, L.F., Maynard, J.E., Purcell, R.H., Hoofnagle, J.H., Berquist, K.R., and London, W.T.,** Viral hepatitis, type B, in experimental animals, *Am. J. Med. Sci.*, 270, 189, 1975.
202. **Barker, L.F., Chisari, F.V., McGrath, P.P., Dalgard, D.W., Kirschstein, R.L., Almeida, J.D., Edgington, T.S., Sharp, D.G., and Peterson, M.R.,** Transmission of type B viral hepatitis to chimpanzees, *J. Infect. Dis.*, 127, 648, 1973.
203. **Markenson, J.A., Gerety, R.J., Hoofnagle, J.H., and Barker, L.F.,** Effects of cyclophosphamide on hepatitis B virus infection and challenge in chimpanzees, *J. Infect. Dis.*, 131, 79, 1975.
204. **Maynard, J.E., Berquist, K.R., Krushak, D.H., and Purcell, R.H.,** Experimental infection of chimpanzees with the virus of hepatitis B, *Nature (London)*, 237, 514, 1972.
205. **Bancroft, W.H., Snitbhan, R., Scott, R.McN., Tingpalapong, M., Watson, W.T., Tanticharoenyos, P., Karwacki, J.J., and Scrimarut, S.,** Transmission of hepatitis B virus to gibbons by exposure to human saliva containing hepatitis B surface antigen, *J. Infect. Dis.*, 135, 79, 1977.
206. **London, W.T., Alter, H.J., Lander, J., and Purcell, R.H.,** Serial transmission in rhesus monkeys of an agent related to hepatitis-associated antigen, *J. Infect. Dis.*, 125, 382, 1972.
207. **London, W.T., Millman, I., Suttnick, A.I., and Blumberg, B.S.,** Transmission, replication, and passage of Australia antigen in African Green monkeys, *Clin. Res.*, 18, 536, 1970.

208. **Lamelin, J.-P. and Trepo, C.,** The hepatitis B virus and the peripheral blood mononuclear cells: a brief review, *J. Hepatol.*, 10, 120, 1990.
209. **Dejean, A., Lugassy, C., Zafrani, S., Tiollais, P., and Brechot, C.,** Detection of hepatitis B virus DNA in pancreas, kidney and skin of two human carriers of the virus, *J. Gen. Virol.*, 65, 651, 1984.
210. **Somlo, F.,** Localization of hepatitis B virus in a primary testicular cancer, *Arch. Pathol. Lab. Med.*, 113, 1184, 1989.
211. **Blum, H.E., Stowring, L., Figus, A., Montgomery, C.K., Haase, A.T., and Vyas, G.N.,** Detection of hepatitis B virus DNA in hepatocytes, bile duct epithelium, and vascular elements by *in situ* hybridization, *Proc. Natl. Acad. Sci. U.S.A.*, 80, 6685, 1983.
212. **Korba, B.E., Wells, F., Tennant, B.C., Yoakum, G.H., Purcell, R.H., and Gerin, J.L.,** Hepadnavirus infection of peripheral blood lymphocytes *in vivo*: woodchuck and chimpanzee models of viral hepatitis, *J. Virol.*, 58, 1, 1986.
213. **Laure, F., Zagury, D., Saimot, A.G., Gallo, R.C., Hahn, B.H., and Brechot, C.,** Hepatitis B virus DNA sequences in lymphoid cells from patients with AIDS and AIDS-related complex, *Science*, 229, 561, 1985.
214. **Lie-Injo, L.E., Balasegaram, M., Lopez, C.G., and Herrera, A.R.,** Hepatitis B virus DNA in liver and white blood cells of patients with hepatoma, *DNA*, 2, 301, 1983.
215. **Pontisso, P., Poon, M.C., Tiollais, P., and Brechot, C.,** Detection of hepatitis B virus DNA in mononuclear blood cells, *Br. Med. J.*, 288, 1563, 1984.
216. **Gu, J.-R., Chen, Y.C., Jiang, H.-Q., Zhang, Y.-L., Wu, S.-M., Jiang, W.-L., and Jian, J.,** State of hepatitis B virus DNA in leucocytes of hepatitis B patients, *J. Med. Virol.*, 17, 73, 1985.
217. **Noonan, C.A., Yoffe, B., Mansell, P.W.A., Melnick, J.L., and Hollinger, F.B.,** Extrachromosomal sequences of hepatitis B virus DNA in peripheral blood mononuclear cells of acquired immune deficiency syndrome patients, *Proc. Natl. Acad. Sci. U.S.A.*, 83, 5698, 1986.
218. **Lai, M.E., Farci, P., Figus, A., Mulargia, A.P., Orgiana, G., Cuboni, C., Angioni, G., and Balestrieri, A.,** Presenza di sequenze nucleotidiche del virus dell'epatite B nei linfociti periferici di pazienti con epatite acuta di tipo B, *Minerva Medica*, 80, 335, 1989.
219. **Pasquinelli, C., Laure, F., Chatenoud, L., Beaurin, G., Gazengel, C., Bismuth, H., Degos, F., Tiollais, P., Bach, J.F., and Brechot, C.,** Hepatitis B virus DNA in mononuclear blood cells. A frequent event in hepatitis B surface antigen-positive and -negative patients with acute and chronic liver disease, *J. Hepatol.*, 3, 95, 1986.
220. **Davison, F., Alexander, G.J.M., Anastassakos, Ch., Fagan, E.A., and Williams, R.,** Leucocyte hepatitis B virus DNA in acute and chronic hepatitis B virus infection, *J. Med. Virol.*, 22, 379, 1987.
221. **Hirschman, S.Z. and Zucker, M.L.,** Recombinant DNA related to hepatitis B and human immunodeficiency viruses in mononuclear cells of patients with AIDS, *J. Med. Virol.*, 26, 145, 1988.
222. **Morichika, S., Hada, H., Arima, T., Togawa, K., Watanabe, M., and Nagashima, H.,** Hepatitis B virus DNA replication in peripheral blood mononuclear cells, *Lancet*, 2, 1431, 1985.
223. **Shen, H.-D., Choo, K.-B., Lee, S.-D., Tsai, Y.-T., and Han, S.-h.,** Hepatitis B virus DNA in leukocytes of patients with hepatitis B virus-associated liver diseases, *J. Med. Virol.*, 18, 201, 1986.
224. **Laure, F., Chatenoud, L., Pasquinelli, C., Gazengel, C., Beaurain, G., Torchet, M.-F., Zagury, D., Bach, J.F., and Brechot, C.,** Frequent lymphocytes infection by hepatitis B virus in haemophiliacs, *J. Haematol.*, 65, 181, 1987.
225. **Yoffe, B., Noonan, C.A., Melnick, J.L., and Hollinger, F.B.,** Hepatitis B virus DNA in mononuclear cells and analysis of cell subsets for the presence of replicative intermediates of viral DNA, *J. Infect. Dis.*, 153, 471, 1986.
226. **Colucci, G., Jacobson, L., Lalatta, F., Vernance, S., Waksal, S.D., and Weksler, M.E.,** Detection of hepatitis B virus DNA, RNA and surface antigen in mononuclear cells from patients with type B hepatitis, *Clin. Res.*, 34, 648A, 1986.
227. **Zoulim, F., Vitvitski, L., Pichoud, C., Rougier, P., Bouffard, P., and Trepo, C.,** Expression of pre-S proteins by peripheral blood mononuclear cells (PBMC) in chronic hepatitis: a relevant noninvasive approach to HBV replication, *Hepatology*, 8, 1389, 1988.
228. **Colucci, G., Lyons, P., Beazer, Y., and Waksal, S.D.,** Production of hepatitis B virus-infected human B-cell hybridomas: transmission of the viral genome to normal lymphocytes in cocultures, *Virology*, 164, 238, 1988.
229. **Hadchouel, M., Scotto, J., Huret, J.L., Molinie, C., Villa, E., Degos, F., and Brechot, C.,** Presence of HBV DNA in spermatozoa: a possible vertical transmission of HBV via the germ line, *J. Med. Virol.*, 16, 61, 1985.
230. **Siddiqui, A.,** Hepatitis B virus DNA in Kaposi sarcoma, *Proc. Natl. Acad. Sci. U.S.A.*, 80, 4861, 1983.
231. **Hadchouel, M., Pasquinelli, C., Fournier, J.G., Hugon, R.N., Scotto, J., Bernard, O., and Brechot, C.,** Detection of mononuclear cells expressing hepatitis B virus in peripheral blood from HBsAg positive and negative patients by *in situ* hybridisation, *J. Med. Virol.*, 24, 27, 1988.
232. **Chong-Jin, O. and Jeak-Ling, D.,** Hepatitis B virus in lymphocytes of seronegative carriers, *Lancet*, 1, 395, 1984.

233. **Parvaz, P., Lamelin, J.-P., Vitvitski, L., Bouffard, P., Pichoud, C., Cova, L., and Trepo, C.,** Prevalence and significance of hepatitis B virus antigens: expression in peripheral blood mononuclear cells in chronic active hepatitis, *Clin. Immunol. Immunopathol.*, 43, 1, 1987.
234. **Elfassi, E., Romet-Lemonne, J.L., Essex, M., Frances-McLane, M., and Haseltine, W.A.,** Evidence of extrachromosomal forms of hepatitis B viral DNA in a bone marrow culture obtained from a patient recently infected with hepatitis B virus, *Proc. Natl. Acad. Sci. U.S.A.*, 81, 3526, 1984.
235. **Romet-Lemonne, J.-L., Frances-McLane, M., Elfassi, E., Haseltine, W.A., Azocar, J., and Essex, M.,** Hepatitis B virus infection in cultured human lymphoblastoid cells, *Science*, 221, 667, 1983.
236. **Romet-Lemonne, J.L., Elfassi, E., Haseltine, W.A., and Essex, M.,** Infection of bone marrow cells by hepatitis B virus, *Lancet*, 2, 732, 1983.
237. **Jameel, S. and Siddiqui, A.,** The human hepatitis B virus enhancer requires *trans*-acting cellular factor(s) for activity, *Mol. Cell. Biol.*, 6, 710, 1986.
238. **Karpen, S., Banerjee, R., Zelent, A., Price, P., and Acs, G.,** Identification of protein-binding sites in the hepatitis B virus enhancer and core promoter domains, *Mol. Cell. Biol.*, 8, 5159, 1988.
239. **Antonucci, T.K. and Rutter, W.J.,** Hepatitis B virus (HBV) promoters are regulated by the HBV enhancer in a tissue-specific manner, *J. Virol.*, 63, 579, 1989.
240. **Honigwachs, J., Faktor, O., Dikstein, R., Shaul, Y., and Laub, O.,** Liver-specific expression of hepatitis B virus is determined by the combined action of the core gene promoter and the enhancer, *J. Virol.*, 63, 919, 1989.
241. **Yee, J.-K.,** A liver-specific enhancer in the core promoter region of human hepatitis B virus, *Science*, 246, 658, 1988.
242. **Pourcel, C., Louis, A., Gervais, M., Chenciner, N., Dubois, M.-F., and Tiollais, P.,** Transcription of the hepatitis B surface antigen gene in mouse cells transformed with cloned viral DNA, *J. Virol.*, 42, 100, 1982.
243. **De-Medina, T., Faktor, O., and Shaul, Y.,** The S promoter of hepatitis B virus is regulated by positive and negative elements, *Mol. Cell. Biol.*, 8, 2449, 1988.
244. **Junker, M., Galle, P., and Schaller, H.,** Expression and replication of the hepatitis B virus genome under foreign promoter control, *Nucleic Acids Res.*, 15, 10117, 1987.
245. **Seeger, C., Baldwin, B., and Tennant, B.C.,** Expression of infectious woodchuck hepatitis virus in murine and avian fibroblasts, *J. Virol.*, 63, 4665, 1989.
246. **Chang, H.-K., Wang, B.-Y., Yuh, C.-H., Wei, C.-L., and Ting, L.-P.,** A liver-specific nuclear factor interacts with the promoter region of the large surface protein gene of human hepatitis B virus, *Mol. Cell. Biol.*, 9, 5189, 1989.
247. **Nakao, K., Miyao, Y., Ohe, Y., and Tamaoki, T.,** Involvement of an AFP1-binding site in cell-specific transcription of the pres-S1 region of the human hepatitis B virus surface antigen gene, *Nucleic Acids Res.*, 17, 9833, 1989.
248. **Courtois, G., Baumhueter, S., and Crabtree, G.R.,** Purified hepatocyte nuclear factor 1 interacts with a family of hepatocyte-specific promoters, *Proc. Natl. Acad. Sci. U.S.A.*, 85, 7937, 1988.
249. **Lichtsteiner, S. and Schibler, U.,** A glycosylated liver-specific transcription factor stimulates transcription of the albumin gene, *Cell*, 57, 1179, 1989.
250. **Frain, M., Swart, G., Monaci, P., Nicosia, A., Stampfil, S., Frank, R., and Cortese, R.,** The liver-specific transcription factor LF-B1 contains a highly diverged homeobox DNA binding domain, *Cell*, 59, 145, 1989.
251. **Nishioka, K., Levin, A.G., and Simons, M.J.,** Hepatitis B antigen, antigen subtypes, and hepatitis B antibody in normal subjects and patients with liver disease, *Bull. WHO*, 52, 293, 1975.
252. **Lee, A.K.Y., Ip, H.M.H., and Wong, V.C.W.,** Mechanisms of maternal-fetal transmission of hepatitis B virus, *J. Infect. Dis.*, 138, 668, 1978.
253. **Prince, A.M., Hargrove, R.L., Szmuness, W., Cherubin, C.E., Fontana, V.J., and Jeffries, G.H.,** Immunologic distinction between infectious and serum hepatitis, *N. Engl. J. Med.*, 282, 987, 1970.
254. **Lewis, T.L., Alter, H.J., Chalmers, T.C., Holland, P.V., Purcell, R.H., Alling, D.W., Young, D., Frenkel, L.D., Lee, S.L., and Lamson, M.E.,** A comparison of the frequency of hepatitis-B antigen and antibody in hospital and nonhospital personnel, *N. Engl. J. Med.*, 289, 647, 1973.
255. **Szmuness, W., Much, M.I., Prince, A.M., Hoofnagle, J.H., Cherubin, C.E., Harley, E.J., and Block, G.H.,** On the role of sexual behavior in the spread of hepatitis B infection, *Ann. Intern. Med.*, 83, 489, 1975.
256. **Dietzman, D.E., Harnisch, J.P., Ray, G., Alexander, E.R., and Holmes, K.K.,** Hepatitis B surface antigen (HBsAg) and antibody to HBsAg: prevalence in homosexual and heterosexual men, *JAMA*, 238, 2625, 1977.
257. **Lauer, J.L., VanDrunen, N.A., Washburn, J.W., and Balfour, H.H.,** Transmission of hepatitis B virus in clinical laboratory areas, *J. Infect. Dis.*, 140, 513, 1979.
258. **Tong, M.J., Thursby, M., Rakela, J., McPeak, C., Edwards, V.M., and Mosley, J.W.,** Studies on the maternal-infant transmission of the viruses which cause acute hepatitis, *Gastroenterology*, 80, 999, 1981.
259. **Ganem, D.,** Persistent infection of humans with hepatitis B virus: mechanisms and consequences, *Rev. Infect. Dis.*, 4, 1026, 1982.

260. **Neurath, A.R., Kent, S.B.H., Strick, N., Taylor, P., and Stevens, C.E.,** Hepatitis B virus contains pre-S gene-encoded domains, *Nature (London)*, 315, 154, 1985.
261. **Neurath, A.R., Kent, S.B.H., and Strick, N.,** Detection of antiviral antibodies with predetermined specificity using synthetic peptide-β lactamase conjugates: application to antibodies specific for the preS region of the hepatitis B virus envelope proteins, *J. Gen. Virol.*, 67, 453, 1986.
262. **Alberti, A., Pontisso, P., Tagariello, G., Cavalletto, D., Chemello, L., and Belussi, F.,** Antibody response to Pre-S2 and hepatitis B virus induced liver damage, *Lancet*, 1, 1421, 1988.
263. **Klinkert, M.-Q., Theilmann, L., Pfaff, E., and Schaller, H.,** Pre-s1 antigens and antibodies early in the course of acute hepatitis B virus infection, *J. Virol.*, 58, 522, 1986.
264. **Stemler, M., Hess, J., Braun, R., Will, H., and Schroder, C.H.,** Serological evidence for expression of the polymerase gene of human hepatitis B virus *in vivo*, *J. Gen. Virol.*, 69, 689, 1988.
265. **Feitelson, M.A., Millman, I., Duncan, G.D., and Blumberg, B.S.,** Presence of antibodies to the polymerase gene product(s) of hepatitis B and woodchuck hepatitis virus in natural and experimental infections, *J. Med. Virol.*, 24, 121, 1988.
266. **Weimer, T., Weimer, K., Tu, Z.-X., Jung, M.-C., Pape, G.R., and Will, H.,** Immunogenicity of human hepatitis B virus P-gene derived proteins, *J. Immunol.*, 143, 3750, 1989.
267. **Chang, L.-J., Dienstag, J., Ganem, D., and Varmus, H.,** Detection of antibodies against hepatitis B virus polymerase antigen in hepatitis B virus-infected patients, *Hepatology*, 10, 332, 1989.
268. **Shafritz, D.A., Shouval, D., Sherman, H.I., Hadziyannis, S.J., and Kew, M.C.,** Integration of hepatitis B virus DNA into the genome of liver cells in chronic liver disease and hepatocellular carcinoma, *N. Engl. J. Med.*, 305, 1067, 1981.
269. **Takada, S., Gotoh, Y., Hayashi, S., Yoshida, M., and Koike, K.,** Structural rearrangement of integrated hepatitis B virus DNA as well as cellular flanking DNA is present in chronically infected hepatic tissues, *J. Virol.*, 64, 822, 1990.
270. **Realdi, G., Alberti, A., Rugge, M., Bortolotti, F., Rigoli, A.M., Tremolada, F., and Ruol, A.,** Seroconversion from hepatitis B e antigen to anti-HBe in chronic hepatitis B virus infection, *Gastroenterology*, 79, 195, 1980.
271. **Hoofnagle, J.H., Dusheiko, G.M., Seeff, L.B., Jones, E.A., Waggoner, J.G., and Bales, Z.B.,** Seroconversion from hepatitis B e antigen to antibody in chronic type B hepatitis, *Ann. Intern. Med.*, 94, 744, 1981.
272. **Dienstag, J.L., Bhan, A.K., Klingenstein, R.J., and Savarese, A.M.,** Immunopathogenesis of liver disease associated with hepatitis B, in *Viral Hepatitis, 1981 International Symposium*, Vyas, G.N., Alter, H.J., and Maynard, J.E., Eds., Franklin Institute Press, Philadelphia, 1982, 221.
273. **Dudley, F.J., Fox, R.A., and Sherlock, S.,** Cellular immunity and hepatitis-associated, Australia antigen liver disease, *Lancet*, 1, 723, 1972.
274. **Alberti, A., Trevisian, A., Fattovich, G., and Realdi, G.,** The role of hepatitis B virus replication and hepatocyte membrane expression in the pathogenesis of HBV-related hepatic damage, in *Advances in Hepatitis Research*, Chisari, F.V., Ed., Masson, Paris, 1984, 134.
275. **Dienstag, J.L.,** Studies of cell-mediated immunity in chronic hepatitis B virus infection: the elusive goal of virus and host antigen specificity, in *Advances in Hepatitis Research*, Chisari, F.V., Ed., Masson, Paris, 1984, 163.
276. **Beasley, R.P., Hwang, L.-Y., Lin, C.-C., and Chien, C.-S.,** Hepatocellular carcinoma and hepatitis B virus — A prospective study of 22707 men in Taiwan, *Lancet*, 2, 1129, 1981.
277. **Beasley, R.P.,** Hepatitis B virus: the major etiology of hepatocellular carcinoma, *Cancer*, 61, 1942, 1988.
278. **Ganem, D. and Varmus, H.E.,** The molecular biology of the hepatitis B viruses, *Annu. Rev. Biochem.*, 56, 651, 1987.
279. **Dejean, A., Bougueleret, L., Grzeschik, K.-H., and Tiollais, P.,** Hepatitis B virus DNA integration in a sequence homologous to v-erb-A and steroid receptor genes in a hepatocellular carcinoma, *Nature (London)*, 322, 70, 1986.
280. **Rizzetto, M., Canese, M.G., Arico, S., Crivelli, O., Trepo, C., Bonino, F., and Verme, G.,** Immunofluorescence detection of new antigen-antibody system associated to hepatitis B virus in liver and in serum of HBsAg carrier, *Gut*, 18, 997, 1977.
281. **Smedile, A., Verme, G., Cargnel, A., Dentico, P., Opolon, P., Vergani, D., Farci, P., Caredda, F., Caporaso, N., Trepo, C., Gimson, A., Williams, R., and Rizzetto, M.,** Influence of delta infection on severity of hepatitis B, *Lancet*, 2, 945, 1982.
282. **Rizzetto, M., Hoyer, B., Canese, M.G., Shih, J.W.-K., Purcell, R.H., and Gerin, J.L.,** δ agent: Association of δ antigen with hepatitis B surface antigen and RNA in serum of δ-infected chimpanzees, *Proc. Natl. Acad. Sci. U.S.A.*, 77, 6124, 1980.
283. **Bonino, F., Hoyer, B., Shih, J.W.-K., Rizzetto, M., Purcell, R.H., and Gerin, J.L.,** Delta hepatitis agent: structural and antigenic properties of the delta-associated particle, *Infect. Immunol.*, 43, 1000, 1984.

284. **Wang, K.-S., Choo, Q.-L., Weiner, A.J., Ou, J.-H., Najarian, R.C., Thayer, R.M., Mullenbach, G.T., Denniston, K.J., Gerin, J.L., and Houghton, M.,** Structure, sequence and expression of the hepatitis delta (δ) viral genome, *Nature (London)*, 323, 508, 1986.
285. **Wang, K.-S., Choo, Q.-L., Weiner, A.J., Ou, J.-H., Najarian, R.C., Thayer, R.M., Mullenbach, G.T., Denniston, K.J., Gerin, J.L., and Houghton, M.,** Corrigendum: structure, sequence and expression of the hepatitis delta (δ) viral genome, *Nature (London)*, 328, 456, 1987.
286. **Kos, A., Dijkema, R., Arnberg, A.C., van der Meide, P.H., and Schellekens, H.,** The hepatitis delta (δ) virus possesses a circular RNA, *Nature (London)*, 323, 558, 1986.
287. **Makino, S., Chang, M.-F., Shieh, C.-K., Kamahora, T., Vannier, D.M., Govindarajan, S., and Lai, M.M.C.,** Molecular cloning and sequencing of a human hepatitis delta (δ) virus RNA, *Nature (London)*, 329, 343, 1987.
288. **Weiner, A.J., Choo, Q.-L., Wang, K.-S., Govindarajan, S., Redeker, A.G., Gerin, J.L., and Houghton, M.,** A single antigenomic open reading frame of the hepatitis delta virus encodes the epitope(s) of both hepatitis delta antigen polypeptides $p24^{\delta}$ and $p27^{\delta}$, *J. Virol.*, 62, 594, 1988.
289. **Kuo, M.Y., Goldberg, J., Coates, L., Mason, W., Gerin, J., and Taylor, J.,** Molecular cloning of hepatitis delta virus RNA from an infected woodchuck liver: sequence, structure, and application, *J. Virol.*, 62, 1855, 1988.
290. **Branch, A.D., Benenfeld, B.J., Baroudy, B.M., Wells, F.V., Gerin, J.L., and Robertson, H.D.,** An ultraviolet-sensitive RNA structural element in a viroid-like domain of the hepatitis delta virus, *Science*, 243, 649, 1989.
291. **Kuo, M.Y.-P., Sharmeen, L., Dinter-Gottlieb, G., and Taylor, J.,** Characterization of self-cleaving RNA sequences on the genome and antigenome of human hepatitis delta virus, *J. Virol.*, 62, 4439, 1988.
292. **Wu, H.-N. and Lai, M.M.C.,** Reversible cleavage and ligation of hepatitis delta virus RNA, *Science*, 243, 652, 1989.
293. **Sharmeen, L., Kuo, M.Y.-P., and Taylor, J.,** Self-ligating RNA sequences on the antigenome of human hepatitis delta virus, *J. Virol.*, 63, 1428, 1989.
294. **Wu, H.-N., Lin, Y.-J., Lin, F.-P., Makino, S., Chang, M.-F., and Lai, M.M.C.,** Human hepatitis δ virus RNA subfragments contain an autocleavage activity, *Proc. Natl. Acad. Sci. U.S.A.*, 86, 1831, 1989.
295. **Negro, F., Korba, B.E., Forzani, B., Baroudy, B.M., Brown, T.L., Gerin, J.L., and Ponzetto, A.,** Hepatitis delta virus (HDV) and woodchuck hepatitis virus (WHV) nucleic acids in tissues of HDV-infected chronic WHV carrier woodchucks, *J. Virol.*, 63, 1612, 1989.
296. **Chen, P.-J., Kalpana, G., Goldberg, J., Mason, W., Werner, B., Gerin, J., and Taylor, J.,** Structure and replication of the genome of the hepatitis δ virus, *Proc. Natl. Acad. Sci. U.S.A.*, 83, 8774, 1986.
297. **Kuo, M.Y.-P., Chao, M., and Taylor, J.,** Initiation of replication of the human hepatitis delta virus genome from cloned DNA: role of delta antigen, *J. Virol.*, 63, 1945, 1989.
298. **Coursaget, P., Bourdil, C., Adamowicz, P., Chotard, J., Mar, I.D., Yvonnet, B., Mevelec, M.N., Barres, J.L., N'Doye, R., and Chiron, J.P.,** HBsAG positive reactivity in man not due to hepatitis B virus, *Lancet*, 2, 1354, 1987.
299. **Budkowska, A., Dubreuil, P., Ouatarra, A., and Pillot, J.,** Anti-Pre-S2 as only serum HBV marker: possible relation to HBV-2 infection, *Lancet*, 1, 656, 1988.
300. **Redeker, A.G. and Govindarajan, S.,** Possible case of HBV-2 infection, *Lancet*, 1, 656, 1988.
301. **Wu, J.-S., Ko, Y.-C., and Liu, W.T.,** Hepatitis B virus type 2, *Lancet*, 1, 990, 1988.
302. **Tagariello, G., Traldi, A., Cavalletto, D., Pontisso, P., and Alberti, A.,** Hepatitis B virus type 2 in Italian hemophiliacs, *Transfusion*, 29, 558, 1989.
303. **Echevarria, J.M., Leon, P., Domingo, C.J., and Lopez, J.A.,** Atypical hepatitis B virus in Spain?, *Lancet*, 2, 1315, 1988.
304. **Wands, J.R., Fujita, Y.K., Isselbacher, K.J., Degott, C., Schellekens, H., Dazza, M.-Ch., Thiers, V., Tiollais, P., and Brechot, C.,** Identification and transmission of hepatitis B virus-related variants, *Proc. Natl. Acad. Sci. U.S.A.*, 83, 6608, 1986.

Chapter 2

GROUND SQUIRREL HEPATITIS VIRUS

Patricia L. Marion

TABLE OF CONTENTS

I. INTRODUCTION

Ground squirrel hepatitis virus (GSHV) was the second hepatitis B-related virus found in nonprimate animal species. The discovery of GSHV in Beechey ground squirrels in 1979[1] followed that of woodchuck hepatitis virus (WHV)[2] in eastern woodchucks by a year and was the result of a search for a hepatitis B-like virus in California relatives of the woodchuck. GSHV remains one of only five genomically defined members of the hepadnavirus family,[3,4] despite a considerable search for virus in the sera of many animal species.

The relationship of the ground squirrel virus to hepatitis B was established by the following observations: (1) sera from some ground squirrels contained particles with the unique physical and morphological properties of hepatitis B virus (HBV). (2) These particles, like HBV, harbored a DNA polymerase that repaired a single-stranded region in a viral DNA similar in size to the 3200-base pair (bp) HBV DNA. (3) Most of the repaired GSHV DNA in virions was held in circular configuration by cohesive ends resembling those of HBV DNA. (4) Finally, the surface antigen of the ground squirrel virus particles was found to cross react with that of hepatitis B virus.

Since its discovery, study of the ground squirrel hepadnavirus model has helped to define the hepadnavirus family and identify properties unique to HBV. The GSHV model has been used to elucidate many steps in hepadnaviral replication, providing fresh infected tissues that were difficult to obtain in HBV infections. The course of infection of GSHV has been studied and found to be similar to that of HBV. The pathology associated with GSHV infection has been clarified following several years of observation of carrier animals. We now know that the association of persistent hepadnaviral infection with both hepatitis and primary hepatocellular carcinoma is common to all the mammalian hepadnaviruses.

II. NATURAL OCCURRENCE AND HOST RANGE

GSHV infection in the wild has been detected only in the original host, Beechey ground squirrels (*Spermophilus beecheyi*).[1,5-7] These squirrels are found on the Pacific Coast of North America ranging from Baja California in Mexico north into the state of Washington in the U.S. The known geographical range of the virus is extremely limited, being confined to squirrels living on the grounds of Stanford University, Stanford, CA and adjacent areas. Ground squirrels trapped 3 mi distant from the virus-infected population have neither virus nor antibody to viral antigens in their sera. We have been unable to detect virus in hundreds of sera of Beechey ground squirrels from many other parts of the state of California, the major habitat of this species. Virus has not been found in other ground squirrel species in California nor in seven animal species trapped in the area endemic for GSHV. The reasons for the strict geographical isolation of GSHV are not known. While the travels of both man and the domestic duck explain the worldwide spread of HBV and duck hepatitis B virus (DHBV), the natural occurrence of the other hepadnavirus of wild mammals, WHV, is restricted to a geographical range of a few states on the East Coast of the United States.[8]

The host range, defined by development of viremia following injection of virus into adult animals, is as limited with GSHV as with HBV. Experimental transmission of GSHV by injection of whole virus has resulted in a GSHV viremia only in the original host and in two other members of the sciuromorph family, the chipmunk (a *Eutamias* species)[7] and the woodchuck.[8a] No viremia resulted from injection of GSHV into several more closely related *Spermophilus* species, nor from injection into several species unrelated to the squirrel family: Sprague-Dawley rats, Syrian hamsters, guinea pigs, New Zealand rabbits, and two strains of mice.[5,6]

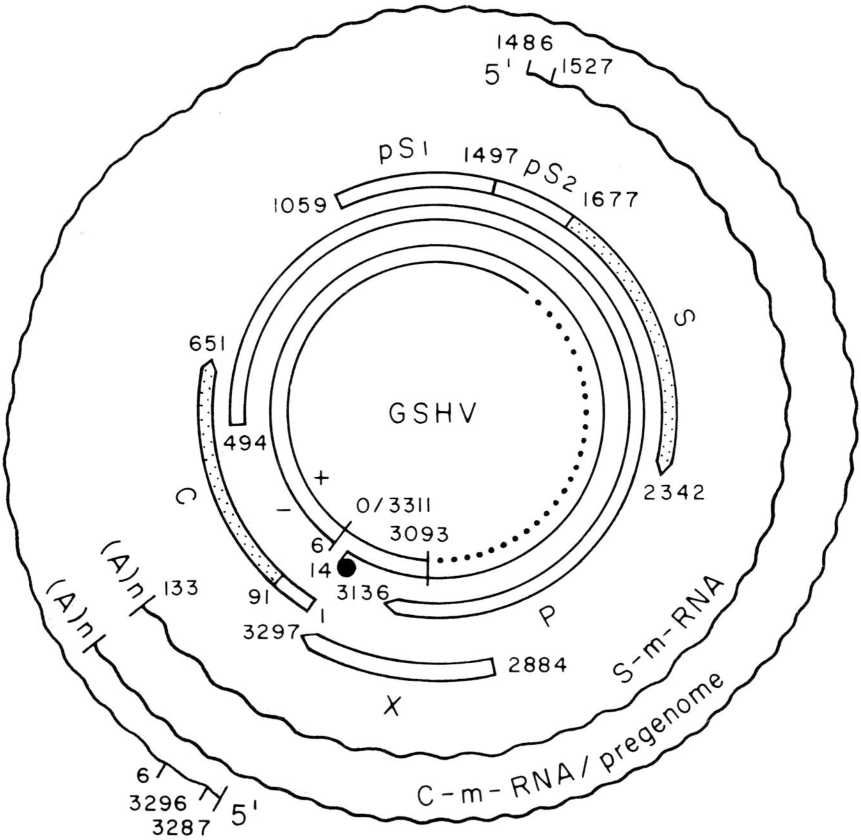

FIGURE 1. Structure and genetic organization of ground squirrel hepatitis virus (GSHV). The (–) and short (+) strands of virion DNA are shown in the inner circles, with the dotted line representing the heterogeneous 3′ ends of the short strand. The protein covalently bound to the 5′ end of the minus strand is indicated by the filled circle. The positions of the open reading frames are indicated by the wide arrows, and those of the major transcripts by the wavy lines. The positions of the open reading frames and virion DNA termini are from References 12 and 13, while the termini of the RNAs are from References 14 and 15.

III. VIRUS STRUCTURE

Electron microscopy shows the GSHV virion to be about 47 nm in diameter, slightly larger than the 42-nm HBV virion.[1] Like HBV, the GSHV virion consists of a nucleocapsid containing a DNA genome. The nucleocapsid is surrounded by an envelope containing lipid, protein, and carbohydrate.

A. VIRAL GENOME

The GSHV genome is similar in size, physical structure, and sequence to the genomes of the other mammalian hepadnaviruses. The nucleic acid of the ground squirrel hepatitis virion was observed in the initial publication to be DNA of about 3200 bp in length, partially single stranded, and held circular by cohesive ends (Figure 1). Like HBV DNA, the long, complete strand has a protein bound to the 5′ end.[9] The virion DNA was subsequently cloned into plasmids[10,11] and the complete nucleotide sequence published in 1984 by Seeger et al.[12] The

TABLE 1
Comparison of Mammalian Hepadnavirus Open Reading Frames

Virus	Genome size (bp)	Number of coded amino acids/ORF					
		PreS	S	PreC	C	P	X
GSHV	3311	206	222	30	188	881	138
WHV	3308–3323	204–209	222	30	187	879–884	141
HBV	3182–3221	163–174	226	29	183–185	838–845	145

Note: GSHV, Ground squirrel hepatitis virus; WHV, woodchuck hepatitis virus; HBV, hepatitis B virus; S, C, P, and X, genes encoding viral proteins.

DNA of the sequenced clone was 3311 bp long and contained four major, partially overlapping open reading frames (ORFs) in an arrangement previously described for HBV and WHV (see Figure 1).[13-15] The S and C ORFs encode the surface (envelope) and core (nucleocapsid) proteins, respectively. The long ORF (P) gene product has not yet been fully elucidated, but is believed to code for reverse transcriptase and RNase H activities as well as the protein bound to the 5′ end of the virion DNA minus strand, if analogous to the P gene structure of DHBV.[16] A gene product of the remaining short ORF, the X ORF, has recently been found to function as a transcriptional trans-activator.[17,18] As with the four other hepadnaviruses, additional ORFs precede and are contiguous with the S and C ORFs of GSHV DNA and are known as the preS and preC coding regions. Interestingly, GSHV has the only hepadnavirus genome with an intergenic region. There are 11 nucleotides between the termination codon of the X gene and the AUG of the preC gene.

As predicted by earlier DNA hybridization and serology experiments,[19,20] GSHV is closely related to WHV, the hepadnavirus infecting another member of the sciuromorph family, with a nucleotide homology of 82% and an amino acid homology of 78%. GSHV is related to HBV less closely, but still significantly, with nucleotide and amino acid homologies of 55 and 46%, respectively.

A comparison of the sizes of the ORF products of the mammalian hepadnaviruses is shown in Table 1. The gene products of GSHV are very similar in size to those of WHV. WHV and GSHV have their greatest difference in amino acid homology in the preS and X regions (71% each), and their greatest similarity in the C and S genes, with 92 and 90% homology, respectively. As the preS and X genes are thought to interact with host proteins (see Sections III.B and III.D), this divergence may reflect the species specificity of the viruses. Compared to HBV, GSHV shows the greatest difference in gene size in the overlapping preS and P ORFs. GSHV also has the least amino acid homology with HBV in the preS region (17%) and the most homology in the structural C (68%) and S (61%) genes.

There are two 11-nucleotide direct repeats (DRs) in the GSHV genome, beginning at nucleotide positions 11 (DR1) and 3100 (DR2). These sequence features, similarly present in the other hepadnaviruses, play an important role in viral replication. Seeger et al.,[13] using primer extension and S1 nuclease protection studies, located the three fixed termini of the long and short GSHV DNA strands, determining in a mammalian hepadnavirus those loci that had been determined earlier for DHBV virion DNA.[21-23] The 5′ end of the long strand (after removal of the bound protein and linearization) was mapped to position 14 within the first direct repeat sequences (DR1), with the 3′ end located only 8 nucleotides upstream. The long strand therefore has a terminal redundancy of nine nucleotides and is longer than the cloned GSHV genome by the same amount. The 5′ end of the short strand was determined to reside at position 3093, 232 nucleotides upstream from the 5′ end of the long strand and 7 nucleotides upstream of the second direct repeat (DR2). The 5′ end of the short strand is not DNA, but rather consists of 17 or 18 ribonucleotides covalently linked to DNA. This short 5′ sequence of RNA will be further discussed in Section IV. The 3′ ends of the short strand are heterogeneous, with the short strand spanning only 40 to 60% of the long strand.

B. VIRAL ENVELOPE

All studies of the GSHV envelope to date indicate that it is similar in physical, chemical, and antigenic properties to that of HBV. In the HBV genome, the S gene encodes a 25-kilodalton (kDa) polypeptide that is found in the virion envelope.[24] This surface antigen (HBsAg) elicits protective antibody appearing at the end of acute HBV infections. The most abundant forms of virus particles in sera during HBV infections are not complete virions, but are defective spherical and tubular forms 22 nm in diameter that contain only viral envelope and no nucleocapsid. These defective forms, known as sAg particles, also exist in GSHV infections, with the tubular forms particularly abundant and elongated into a filamentous form.[1] These numerous, unusually long filaments are not seen in the other hepadnavirus infections.

Purified GSHsAg particles have a buoyant density equivalent to that of HBsAg particles, and contain a pattern of polypeptides similar to those of HBsAg particles on sodium dodecyl sulfate-polyacrylamide gel electrophoresis (SDS-PAGE).[25] In HBV this pattern consists of two major polypeptides, the 25-kDa species encoded by the viral genome and a 29-kDa species, plus several slower migrating minor polypeptide species ranging to 65 kDa. All the polypeptides can be precipitated individually by antibody to HBsAg. The 29-kDa polypeptide is a glycosylated form of the 25-kDa polypeptide. In similar analysis of GSHsAg particles, the major polypeptides migrate as 23- and 27-kDa species. The sAg particles of GSHV, HBV, and WHV are all serologically crossreactive, with the strongest reactivity between those of GSHV and WHV.[19] The sAg particles may not play a major role in hepadnavirus species specificity. Ground squirrels, which do not become infected when injected with WHV, can be infected by GSHV in which the native sAg gene has been replaced with that of WHV.[26]

The 31- to 45-kDa polypeptides in HBV particles showing HBsAg reactivity are either glycosylated or unglycosylated products of the regions encoded by portions of the preS ORF contiguous with the entire S gene.[24,27,28] Polypeptides encoded by the complete preS have both the preS1 and the preS2 regions as well as the S region, while other polypeptides have only the preS2 region adjacent to the S gene. In HBV, the preS2 region encodes a polyalbumin-binding site, possibly involved in viral infection.[29-31] Polypeptide species containing preS1 appear to be more numerous on complete virions than on defective particles.[24,27] There is evidence that the preS polypeptides are more immunogenic than the S polypeptides alone[27,32,33] and that anti-preS is a neutralizing antibody.[34] The presence of 45- and 47-kDa preS-encoded polypeptides in GSHV particles has been demonstrated by Schaeffer et al,.[35] using Western blot analysis with an antibody to a bacterially synthesized fusion molecule containing 89 WHV preS-encoded amino acids. A similar analysis by Persing et al.,[36] using bacterially synthesized GSHV preS1, identified a 43-kDa polypeptide that carried preS1 determinants. Persing et al.[36] observed three species of polypeptides ranging from 40 to 49 kDa in purified GSHsAg particles, using silver staining and SDS-PAGE. Some of these polypeptides may have been reactive to the anti-preS sera used by Schaeffer et al.[35]

C. NUCLEOCAPSID

By electron microscopy, the nucleocapsid of HBV appears to be a sphere of about 27 nm in diameter.[37] These nucleocapsids or cores can be isolated from virions by removal of the virus envelope with a nonionic detergent.[38] An antibody to the surface of the cores (anti-HBc) is produced early in infection. HBV cores contain a major polypeptide of 19 kDa[39,40] that is encoded by the C ORF.[41,42] Similarly sized particles with the same antigenic reactivity can be isolated from the liver of infected individuals.

GSHV virions contain a nucleocapsid of about 27 nm in diameter, equivalent in size to that of HBV.[1] Antibody to HBV cores cross reacts with GSHV cores isolated from virus-infected ground squirrel liver, using the procedures used for HBV cores.[43] These GSHV cores also contain one major polypeptide component, which migrates on SDS gels slightly faster than the HBV core polypeptide. As with HBV, the GSHV core particles contain the viral DNA and a DNA polymerase activity. In addition, purified core particles of both GSHV and HBV have an associated protein kinase activity that phosphorylates the core polypeptide predominantly at serine, but also at threonine residues.[44] While the precise size and sequence of GSHV and HBV cores have been shown to differ, no morphological or functional differences in the nucleocapsids have been detected.

D. X GENE PRODUCT

Expression of the HBV X ORF can increase expression of reporter genes under the control of a variety of promoters.[17] Colgrove et al.[18] have shown that the GSHV X gene expresses a trans-activating activity similar to that of HBV, using three heterologous promoters. As with HBV, this activity appeared to be present in similar levels in both nonliver cells and permissive cells derived from liver.

IV. VIRAL REPLICATION

Following the seminal work by Summers, Mason, and co-workers on hepadnavirus replication in ducklings infected with DHBV,[45,46] experiments with GSHV-infected ground squirrels confirmed that, in the mammalian hepadnaviruses, replication also occurs through reverse transcription of an RNA intermediate. In addition, the GSHV model was used by Seeger, Enders, Ganem, Varmus, and co-workers to elucidate many additional steps of hepadnaviral replication.

GSHV-infected ground squirrel liver, like all hepadnavirus-infected liver, contains viral DNA species not seen in virions.[9,47] A protein-free supercoiled viral DNA in the nucleus of infected cells is made from incoming virion relaxed circular DNA and serves as the template for synthesis of viral RNA. Relaxed circular viral DNA with variable amounts of repair of the short strand is found in the cytoplasm along with single-stranded species ranging in size from the lowest limit of detection on a Southern blot of an agarose gel to the full-length 3300-nucleotide species. All of these cytoplasmic DNAs have covalently bound protein and all are found in core particles. Phage-derived single-stranded probes of hepadnaviral DNA were used to demonstrate that the short strand of virion DNA shares the same polarity as viral mRNA. The long strand is therefore minus-stranded DNA, as is the single-stranded viral DNA found in infected liver.

Plus-stranded viral RNAs found in the liver are three different 3.5-kb species with terminal redundancies of 130 to 160 nucleotides, and 2.3-kb RNAs (Figure 1).[14] All these RNAs are polyadenylated at the same location and appear to be unspliced. The 2.3-kb RNAs cover the S gene and possibly the preS2 region, providing the mRNA for expression of some of the sAg particles. The other ORFs are complete only in the 3.5-kb RNAs. Expression of the remaining genes is thought to be from some form of these RNAs, with the possible exception of the X gene, which may have its own message.[48] Enders et al.[15] determined that the shortest of three

3.5-kb RNAs is found predominantly in the core particles. This species is assumed to be the template for reverse transcription leading to the synthesis of the minus or long strand of viral DNA. The primer for this reverse transcription reaction appears to be the protein bound to the 5′ end of the minus-strand DNA. Degradation of the RNA template, by an RNase H activity, leaves an approximately 18-nucleotide RNA oligomer that serves as a primer for plus DNA strand synthesis and that remains in the virion after export from the cell. The direct repeats (DRs) play a role in viral DNA synthesis. The RNA oligomer remaining at DR1 moves to DR2 before extension of the strand by the polymerase, thus forming the cohesive ends of the virion DNA. By mutating both GSHV DNA DRs by one base and infecting animals with viral DNA with one or the other mutation, Seeger et al.[13] showed that the plus strand begins with an RNA oligomer having the sequence of DR1 but continues with DNA synthesized with the sequence of DR2. This experiment also indicated that successful viral replication of GSHV is not dependent on the two DRs being of strictly identical sequence.

V. TISSUE TROPISM

GSHV is predominantly a hepatotropic virus. Although little liver pathology was observed in early studies of GSHV infection, viral cores containing DNA polymerase activity were found to be as abundant in GSHV-infected ground squirrel liver as in HBV-infected liver.[43] A study by Ganem et al.[5] of various tissues of GSHV-infected squirrels demonstrated that viral DNA was found in highest amounts in the liver (600 to 6000 genome equivalents per cell). GSHV DNA equivalent to less than 1% of that in the liver was found in other tissues. Trace amounts of viral DNA found in the kidney and spleen may have been from contamination with the virions in the blood or possibly from low-level replication, as has been demonstrated in DHBV-injected ducklings.[49] With an assay sensitivity of two viral genome equivalents per cell, no viral DNA was detected in the pancreas, contrary to studies of DHBV infection.[50] Replication in lymphocytes, reported to occur in the other mammalian hepadnaviruses,[51,52] has not been examined.

VI. COURSE OF THE VIRUS INFECTION

GSHV infection in the ground squirrel has been monitored by measuring viral DNA polymerase, GSHsAg, anti-GSHs, and anti-GSHc levels in sera or by detection of viral DNA in the serum or liver.[1,5,6] In our study of experimental transmission of GSHV,[6] virions in sera were measured by DNA polymerase assay, anti-GSHc by enzyme immunoassay, and both GSHsAg and anti-GSHs by radioimmunoassay. All antibody assays were done using reagents purified from GSHV particles or ground squirrel sera. The sAg assay was a commercial assay for HBsAg. GSHV-infected serum was injected into Beechey ground squirrels having no detectable markers of GSHV, and serum samples were taken at 2-week intervals. As with experimental transmission of HBV, experimental infection of ground squirrels with GSHV led to three types of response: (1) self-limited or transient GSHsAg-positive infection, (2) GSHsAg-positive infection that became persistent, and (3) primary anti-GSHs and anti-GSHc responses without detectable GSHsAg or virion DNA polymerase. Anti-GSHc was a marker that appeared in all ground squirrels with primary GSHV infection and never appeared when other species were injected with GSHV. With increasing dilution of the inoculum there was a longer delay before appearance of anti-GSHc. Onset of infection ranged from 4 to 16 weeks after inoculation. There were proportionally fewer self-limited and more primary antibody responses without GSHsAg detection (45 and 45%) than in one study of experimental transmission of HBV (70 and 23%).[53] These differences may be due simply to the low number of squirrels studied or the less frequent blood sampling of the squirrels, and may not reflect any true difference in response of humans and squirrels to hepadnavirus injection.

In naturally infected chronic carriers of GSHV, the levels of virion-associated DNA polymerase activity are higher by 10- to 20-fold than those of nonexperimental HBV carriers.[53a] These high levels appear to be due to large amounts of virions in the blood rather than to increased DNA polymerase activity per virion, as the amount of viral DNA detectable by blotting techniques is also high. Why GSHV infections are of such a high titer is not known. Clearance of viremia in naturally occurring persistent infection is a rare event. Only 1 of 39 carrier ground squirrels studied over a 2-year period cleared viremia and became positive for anti-GSHs. This animal had been trapped only a few months prior to seroconversion, however, and the infection may have been acute and not chronic. In one study, replicative forms of GSHV DNA were found in the liver of all of seven squirrels trapped as carriers and observed in captivity for 4 to 6 years until death. Thus, viremia probably continued in all these animals that survived until middle or old age. We have never observed a carrier squirrel to have GSHsAg in the absence of viral DNA or DNA polymerase activity in the blood in late infection, as is seen in HBV infection of man. The longer life span of man compared to that of a squirrel may explain these differences, though a difference in other host or virus genetics cannot be excluded. Prenatal transmission of GSHV infection from mother to offspring was examined in a study of six newborn squirrels born to an animal with a high titer of GSHV.[54] Due to the small size of the newborns, livers and not sera were tested for markers of GSHV. No viral DNA was detected in any of the newborns, suggesting that transmission of virus from mother to offspring does not routinely occur *in utero*.

In a study by Persing et al.,[36] peptides encoded by both the preS and the X regions of the GSHV genome were produced in bacteria. Sera from GSHV-infected and postresolution squirrels were used to immunoprecipitate radiolabeled, bacterially synthesized preS- and S-encoded peptides. Only the postresolution sera precipitated the preS peptide, with 6 of 12 anti-S positive and 2 of 7 anti-S negative animals showing an anti-preS response. It is not known whether antibody to these preS determinants or to the synthesized S determinants contribute to either resolution of infection or immunity to reinfection. Four of twelve postresolution ground squirrel sera immunoprecipitated bacterially synthesized peptide from the X region, indicating that this coding area is expressed during GSHV infection.

VII. PATHOLOGY OF GSHV INFECTION

Early studies of GSHV infection indicated that this hepadnavirus was less pathogenic than either HBV or WHV.[5,6] Our accumulated data from captive squirrels observed over a period of 9 years support these early observations. However, significant to moderately severe hepatitis is associated with GSHV infection. A coded study showed that at necropsy 59% of 32 GSHV carriers had hepatitis rated significant or greater, while only 17% of 47 uninfected ground squirrels exhibited any hepatitis above the background level.[55] Unlike observations of WHV and HBV carriers,[54,56] severe hepatitis extending beyond the portal area and into the parenchyma of the liver is rarely observed in GSHV infection. The cirrhosis associated with HBV infection of man has not been seen in GSHV and only occasionally in WHV infection.

The primary hepatocellular carcinoma (PHC) associated with HBV infection in man usually occurs after 10 to 30 years of persistent infection or more and in only some of those infected.[54] WHV-infected woodchucks are more likely to develop PHC, with about one of three captive adult carriers developing liver tumors per year.[55,56] To examine the association of GSHV with PHC, we have observed virus-infected and uninfected ground squirrels held in separate quarters since 1980. Some animals have been added into the colony since its inception and some squirrels were experimentally infected and studied. Necropsies were done on all animals at death and histology performed under code. For 4 years no PHC was observed. There has been a dramatic increase in the number of tumors observed in our colony in the past 5 years, however, with PHC emerging as the predominant type of tumor.[55,58]

Tumors have appeared only in squirrels estimated to be 4.4 years of age or older. In this age group neoplasms of various types developed in 16 of 24 carrier squirrels, 11 of 20 postresolution squirrels, and 12 of 26 GSHV marker-free squirrels. There have been a total of 18 cases of PHC (45% of all neoplasms): 11 of these were in carrier animals, 5 in convalescent, and 2 in GSHV-free animals. The association of PHC with the GSHV carrier state is significant ($p = 0.0016$). Various tumors other than PHC have been observed, with no tumor type predominating. The average age of the squirrels when PHC is detected is 6.45 years, older than the average age of WHV carriers when PHC develops, but considerably younger than that of HBV carriers. The average age of PHC detection in postresolution squirrels was similar (6.8 years), while that of the marker-free squirrels was considerably higher at 8.5 years, near the end of the 9.5-year maximum life span. Of the carrier animals developing PHC, one was an experimentally infected animal, while the others became GSHV carriers before capture. As with PHC in woodchucks, viral replication is active in the liver of ground squirrels that have developed PHC. High-molecular-weight viral DNA was detected in the tumor DNA of three of seven carrier animals analyzed by Southern blots. This high-molecular-weight viral DNA is assumed to be integrated into host DNA, as has been observed in tumors associated with the other hepadnaviruses.[59]

PHCs appearing in the younger squirrels (ages 4.5 to 5.4 years) were all single, small tumors, and those in the older squirrels (ages 6.0 years and older) were all multinodular. The diameter of the major tumor was generally larger in older squirrels. The histological type of all but one PHC was trabecular, with the arrangement of the tumor cells maintaining some aspects of liver architecture. The remaining PHC was a more anaplastic medullary hepatocellular carcinoma, and this tumor arose in one of the squirrels with anti-GSHs and not GSHsAg. Trabecular PHC is the type of liver tumor most often associated with both HBV and WHV.[54,57] Thus, development of PHC, especially of the trabecular type, appears to be a common element of long-term mammalian hepadnavirus infection.

An exciting experiment is currently being conducted by Seeger and colleagues[59a] to test whether the host or the virus is responsible for the more rapid development of PHC in WHV-infected woodchucks compared to GSHV-infected ground squirrels. These animal species have similar life spans. Taking advantage of the susceptibility of newborn woodchucks to both WHV and GSHV, Seeger and colleagues injected 16 newborn woodchucks with WHV and 17 with GSHV, and all became chronically infected. After 40 months of observation, eight WHV-infected animals (50%) have died with PHC, compared to only one (6%) of the GSHV-infected animals. Of the surviving animals, only 4 of the WHV-injected animals remain without indications of PHC development (γ-glutamyl transpeptidase levels of 10 U/ml or less), while 13 of the GSHV-injected animals remain without symptoms. Animals infected with GSHV have been similar to those infected with WHV in levels of sAg and viral DNA in serum and in amount of viral DNA in the liver. No difference between the two groups was noted in liver histopathology in biopsy specimens at 2 years after infection. These results indicate that the greater oncogenic potential of WHV lies in the virus itself. Similar experiments using recombinant genomes of WHV and GSHV should eventually reveal the portion of the WHV genome that enhances oncogenicity.

VIII. USE OF GSHV AS A MODEL OF HBV INFECTION IN MAN

The ways in which study of GSHV has advanced knowledge of HBV and the hepadnavirus family have been described in the preceding sections. The concordance of the morphology and structure of GSHV with those of the other hepadnaviruses has helped to define the virus family. Development of PHC in long-term carriers of GSHV has affirmed the association of hepadnaviruses with primary liver cancer. Use of fresh tissues from GSHV-infected ground squirrels has been invaluable in analyzing the various steps of hepadnaviral infection and also in determining whether phenomena discovered in the avian hepadnavirus, the most divergent of the four well-defined hepadnaviruses, are also present in the mammalian viruses. It is with the GSHV-ground squirrel model that mutational analysis of the genetic organization of hepadnaviruses was first tested *in vivo*.[26] We have also demonstrated the usefulness of GSHV carrier ground squirrels for evaluating the effects of antiviral agents.[60]

The limitations of the GSHV model of HBV have been revealed by these early studies. Beechey ground squirrels are available only in certain locations of the West Coast of the United States. Naturally infected animals are no longer easy to obtain because the original locale of high incidence is now more inhabited by man. Production of chronic viremia occurs in only 10% of injected animals. Acute viremia does not always result from injection of GSHV into susceptible squirrels, as many animals respond with only a primary antibody response to both surface and core antigens. The lapse of time between injection of high titer virus into ground squirrels and development of viremia is several weeks, with a great deal of variation seen among individual squirrels.[5,6] Injection of similar amounts of DHBV into young ducklings usually results in a persistent viremia in only a few days, with little variation in response among the ducklings.[61] It is difficult to obtain newborn ground squirrels that might more readily develop persistent virus infection, as there is currently no breeding facility for these animals, and newborn squirrels in the wild remain underground until they are 2 months old. Development of GSHV-related PHC in squirrels is a slow process compared to the woodchuck model, and is, therefore, more difficult to study.

Nevertheless, there are ways in which the GSHV model of HBV can be useful in the future. Experimental observations of DHBV must be confirmed in one of the mammalian systems before application to hepadnaviruses in general and HBV in particular. *In vitro* culture of GSHV appears possible, using either woodchuck or ground squirrel primary hepatocyte cultures.[62] Culture by transfection of viral DNA into susceptible cell lines should be feasible. GSHV readily infects newborn woodchucks available in breeding facilities. For those with access to infected ground squirrels, GSHV is as convenient a mammalian model as WHV for studies of the effects of new antivirals and immune modulators on mammalian hepadnavirus infection. Perhaps the most interesting and important use of GSHV in coming years will be in the determination of which areas of hepadnaviral genomes influence the rate of development of primary hepatocellular carcinoma.

ACKNOWLEDGMENT

This work was supported by Public Health Service grant AI–20551.

REFERENCES

1. **Marion, P. L., Oshiro, L. S., Regnery, D. C., Scullard, G. H., and Robinson, W. S.**, A virus in Beechey ground squirrels that is related to hepatitis B virus of humans, *Proc. Natl. Acad. Sci. U.S.A.*, 77, 2941, 1980.
2. **Summers, J., Smolec, J. M., and Snyder, R.**, A virus similar to human hepatitis B virus associated with hepatitis and hepatoma in woodchucks, *Proc. Natl. Acad. Sci. U.S.A.*, 75, 4533, 1978.
3. **Mason, W. S., Seal, G., and Summers, J.**, Virus of Pekin ducks with structural and biological relatedness to human hepatitis B virus, *J. Virol.*, 36, 829, 1980.
4. **Sprengel, R., Kaleta, E. F., and Will, H.**, Isolation and characterization of a hepatitis B virus endemic in herons, *J. Virol.*, 62, 3832, 1988.
5. **Ganem, D., Weiser, B., Barchuk, A., Brown, R. J., and Varmus, H. E.**, Biological characterization of acute infection with ground squirrel hepatitis virus, *J. Virol.*, 44, 366, 1982.
6. **Marion, P. L., Knight, S. S., Salazar, F. H., Popper, H., and Robinson, W. S.**, Ground squirrel hepatitis virus infection, *Hepatology*, 3, 519, 1983.
7. **Trueba, D., Phelan, M., Nelson, J., Beck, F., Pecha, B.S., Brown, R. J., Varmus, H. E., and Ganem, D.**, Transmission of ground squirrel hepatitis virus to homologous and heterologous hosts, *Hepatology*, 5, 435, 1985.
8. **Tyler, G. V., Summers, J. W., and Synder, R. L.**, Woodchuck hepatitis virus in natural woodchuck populations, *J. Wildl. Dis.*, 17, 297, 1981.

8a. **Seeger, C., et al.**, personal communication, 1990.

9. **Weiser, B., Ganem, D., Seeger, C., and Varmus, H. E.**, Closed circular viral DNA and asymmetrical heterogeneous forms in livers from animals infected with ground squirrel hepatitis virus, *J. Virol.*, 48, 1, 1983.
10. **Siddiqui, A., Marion, P. L., and Robinson, W. S.**, Ground squirrel hepatitis virus DNA: molecular cloning and comparison with hepatitis B virus DNA, *J. Virol.*, 38, 393, 1981.
11. **Ganem, D., Greenbaum, L., and Varmus, H. E.**, Virion DNA of ground squirrel hepatitis virus: structural analysis and molecular cloning, *J. Virol.*, 44, 374, 1982.
12. **Seeger, C., Ganem, D., and Varmus, H. E.**, Nucleotide sequence of an infectious molecularly cloned genome of ground squirrel hepatitis virus, *J. Virol.*, 51, 367, 1984.
13. **Seeger, C., Ganem, D., and Varmus, H. E.**, Biochemical and genetic evidence for the hepatitis B virus replication strategy, *Science*, 232, 477, 1986.
14. **Enders, G. H., Ganem, D., and Varmus, H.**, Mapping the major transcripts of ground squirrel hepatitis virus: the presumptive template for reverse transcriptase is terminally redundant, *Cell*, 42, 297, 1985.
15. **Enders, G. H., Ganem, D., and Varmus, H. E.**, 5′-Terminal sequences influence the segregation of ground squirrel hepatitis virus RNAs into polyribosomes and viral core particles, *J. Virol.*, 61, 35, 1987.
16. **Bosch, V., Bartenschlager, R., Radziwill, G., and Schaller, H.**, The duck hepatitis B virus P-gene codes for protein strongly associated with the 5′-end of the viral minus DNA strand, *Virology*, 166, 475, 1988.
17. **Spandau, D. F. and Lee, C. H.**, Trans-activation of viral enhancers by the hepatitis B virus X protein, *J. Virol.*, 62, 427, 1988.
18. **Colgrove, R., Simon, G., and Ganem, D.**, Transcriptional activation of homologous and heterologous genes by the hepatitis B virus X gene product in cells permissive for viral replication, *J. Virol.*, 63, 4019, 1989.
19. **Feitelson, M. A., Marion, P. L., and Robinson, W. S.**, Antigenic and structural relationships of the surface antigens of hepatitis B virus, ground squirrel hepatitis virus, and woodchuck hepatitis virus, *J. Virol.*, 39, 447, 1981.
20. **Marion, P. L. and Rogler, C. E.**, unpublished observations, 1991.
21. **Molnar-Kimber, K. L., Summers, J., Taylor, J. M., and Mason, W. S.**, Protein covalently bound to minus-strand DNA intermediates of duck hepatitis B virus, *J. Virol.*, 45, 165, 1983.
22. **Molnar-Kimber, K. L., Summers, J. W., and Mason, W. S.**, Mapping of the cohesive overlap of duck hepatitis B virus DNA and of the site of initiation of reverse transcription, *J. Virol.*, 51, 181, 1984.
23. **Lien, J. M., Aldrich, C. E., and Mason, W. S.**, Evidence that a capped oligoribonucleotide is the primer for duck hepatitis B virus plus-strand DNA synthesis, *J. Virol.*, 57, 29, 1986.
24. **Stibbe, W. and Gerlich, W. H.**, Structural relationships between minor and major proteins of hepatitis B surface antigen, *J. Virol.*, 46, 626, 1983.
25. **Gerlich, W. H., Feitelson, M. A., Marion, P. L., and Robinson, W. S.**, Structural relationships between the surface antigens of ground squirrel hepatitis virus and human hepatitis B virus, *J. Virol.*, 36, 787, 1980.
26. **Seeger, C., Marion, P. L., Ganem, D., and Varmus, H. E.**, *In vitro* recombinants of ground squirrel and woodchuck hepatitis viral DNAs produce infectious virus in squirrels, *J. Virol.*, 61, 3241, 1987.

27. **Heermann, K. H., Goldmann, U., Schwartz, W., Seyffarth, T., Baumgarten, H., and Gerlich, W. H.**, Large surface proteins of hepatitis B virus containing the pre-s sequence, *J. Virol.*, 52, 396, 1984.
28. **Wong, D. T., Nath, N., and Sninsky, J. J.**, Identification of hepatitis B virus polypeptides encoded by the entire pre-s open reading frame, *J. Virol.*, 55, 223, 1985.
29. **Machida, A., Kishimoto, S., Ohumura, H., Miyamoto, H., Baba, K., Oda, K., Nakamura, T., and Miyakawa, Y.**, A hepatitis B surface antigen polypeptide (P31) with the receptor for polymerized human as well as chimpanzee albumins, *Gastroenterology*, 85, 268, 1983.
30. **Machida, A., Kishimoto, S., Ohnuma, H., Baba, K., Ito, Y., Miyamoto, H., Funatsu, G., and Oda, K.**, A polypeptide containing 55 amino acid residues coded by the pre-S region of hepatitis B virus bears the receptor for polymerized human as well as chimpanzee albumins, *Gastroenterology*, 86, 910, 1984.
31. **Persing, D. H., Varmus, H. E., and Ganem, D.**, A frameshift mutation in the pre-S region of the human hepatitis B virus genome allows production of surface antigen particles but eliminates binding to polymerized albumin, *Proc. Natl. Acad. Sci. U.S.A.*, 82, 3440, 1985.
32. **Milich, D. R., McNamara, M. K., McLachlan, A., Thornton, G. B., and Chisari, F. V.**, Distinct H-2-linked regulation of T-cell responses to the pre-S and S regions of the same hepatitis B surface antigen polypeptide allows circumvention of non-responsiveness to the S region, *Proc. Natl. Acad. Sci. U.S.A.*, 82, 8168, 1985.
33. **Milich, D. R., Thornton, G. B., Neurath, A. R., Kent, S.B., Tiollais, P., and Chisari, F. V.**, Enhanced immunogenicity of the pre-S region of hepatitis B surface antigen, *Science*, 228, 1195, 1985.
34. **Neurath, A. R., Kent, S., Parker, K., Prince, A., Strick, N., Brotman, B., and Sproul, P.**, Antibodies to a synthetic peptide from the pre-S 120-145 region of the HBV envelope are virus neutralizing, *Vaccine*, 4, 35, 1986.
35. **Schaeffer, E., Snyder, R. L., and Sninsky, J. J.**, Identification and localization of pre-S encoded polypeptides from woodchuck and ground squirrel hepatitis viruses, *J. Virol.*, 57, 173, 1986.
36. **Persing, D. H., Varmus, H. E., and Ganem, D.**, Antibodies to pre-S and X determinants arise during natural infection with ground squirrel hepatitis virus, *J. Virol.*, 60, 177, 1986.
37. **Almeida, J. D., Rubenstein, D., and Stott, E. J.**, New antigen-antibody system in Australia antigen-positive hepatitis, *Lancet*, 2, 1225, 1971.
38. **Robinson, W. S. and Greenman, R. L.**, DNA polymerase in the core of the human hepatitis B virus candidate, *J. Virol.*, 13, 1231, 1974.
39. **Budkowska, A., Shih, J.W.-K., and Gerin, J. L.**, Immunochemistry and polypeptide composition of hepatitis B core antigen (HBcAg), *J. Immunol.*, 118, 1300, 1977.
40. **Hruska, J. F. and Robinson, W. S.**, The proteins of hepatitis B Dane particle cores, *J. Med. Virol.*, 1, 119, 1977.
41. **Galibert, F., Mandart, E., Fitoussi, F., Tiollais, P., and Charnay, P.**, Nucleotide sequence of the hepatitis B virus genome (subtype ayw) cloned in E. coli, *Nature (London)*, 281, 646, 1979.
42. **Pasek, M., Goto, T., Gilbert, W., Zink, B., Schaller, H., MacKay, P., Leadbetter, G., and Murray, K.**, Hepatitis B virus genes and their expression in E. coli, *Nature (London)*, 282, 575, 1979.
43. **Feitelson, M. A., Marion, P. L., and Robinson, W. S.**, Core particles of hepatitis B virus and ground squirrel hepatitis virus. I. Relationship between hepatitis B core antigen- and ground squirrel hepatitis core antigen-associated polypeptides by sodium dodecyl sulfate-polyacrylamide gel electrophoresis and tryptic peptide mapping, *J. Virol.*, 43, 687, 1982.
44. **Feitelson, M. A., Marion, P. L., and Robinson, W. S.**, Core particles of hepatitis B virus and ground squirrel hepatitis virus. II. Characterization of the protein kinase reaction associated with ground squirrel hepatitis virus and hepatitis B virus, *J. Virol.*, 43, 741, 1982.
45. **Summers, J. and Mason, W. S.**, Replication of the genome of a hepatitis B-like virus by reverse transcription of an RNA intermediate, *Cell*, 29, 403, 1982.
46. **Mason, W. S., Aldrich, C., Summers, J., and Taylor, J. M.**, Asymmetric replication of duck hepatitis B virus DNA in liver cells: free minus-strand DNA, *Proc. Natl. Acad. Sci. U.S.A.*, 79, 3997, 1982.
47. **Miller, R. H., Marion, P. L., and Robinson, W. S.**, Hepatitis B viral DNA-RNA hybrid molecules in particles from infected liver are converted to viral DNA molecules during an endogenous DNA polymerase reaction, *Virology*, 139, 64, 1984.
48. **Kaneko, S. and Miller, R. H.**, X-region-specific transcript in mammalian hepatitis B virus-infected liver, *J. Virol.*, 62, 3979, 1988.
49. **Tagawa, M., Omata, M., Yokosuka, O., Uchiumi, K., Imazeki, F., and Okuda, K.**, Early events in duck hepatitis B virus infection. Sequential appearance of viral deoxyribonucleic acid in the liver, pancreas, kidney, and spleen, *Gastroenterology*, 89, 1224, 1985.

50. **Mason, W.S., Taylor, J.M., Seal, G., and Summers, J.**, An HBV-like virus of domestic ducks, in *Viral Hepatitis, 1981 Int. Symp.*, Szmuness, W., Alter, H. J., and Maynard, J. E., Eds., Franklin Institute Press, Philadelphia, 1981, 107.

51. **Romet-Lemonne, J. L., McLane, M. F., Elfassi, E., Haseltine, W., Azocar, J., and Essex, M.**, Hepatitis B virus infection in cultured human lymphoblastoid cells, *Science*, 221, 667, 1983.

52. **Korba, B. E., Wells, F., Tennant, B. C., Cote, P. J., and Gerin, J. L.**, Lymphoid cells in the spleens of woodchuck hepatitis virus-infected woodchucks are a site of active viral replication, *J. Virol.*, 61, 1318, 1987.

53. **Hoofnagle, J.H., Seeff, L.B., Bales, Z.B., et al,** Serologic responses in hepatitis B, in *Viral Hepatitis*, Vyas, G. N., Cohen, S. N., and Schmid, R., Eds., Franklin Institute Press, Philadelphia, 1978, 219.

53a. **Marion, P.L., et al.**, unpublished observations, 1990.

54. **Popper, H., Gerber, M. A., and Thung, S. N.**, The relation of hepatocellular carcinoma to infection with hepatitis B and related viruses in man and animals, *Hepatology*, 2, 1s, 1982.

55. **Marion, P.L., Popper, H., Azcarraga, R. R., Steevens, C., Van Davelaar, M. J., Garcia, G., and Robinson, W. S.**, Ground squirrel hepatitis virus and hepatocellular carcinoma, in *Hepadna Viruses*, Robinson, W., Koike, K., and Will, H., Eds., Alan R. Liss, New York, 1987, 337.

56. **Popper, H., Shih, J. W., Gerin, J. L., Wong, D. C., Hoyer, B. H., London, W. T., Sly, D. L., and Purcell, R. H.**, Woodchuck hepatitis and hepatocellular carcinoma: correlation of histologic with virologic observations, *Hepatology*, 1, 91, 1981.

57. **Snyder, R.L. and Summers, J.**, Woodchuck hepatitis virus and hepatocellular carcinoma, in *Viruses in Naturally Occurring Cancers*, Cold Spring Harbor Laboratory, Cold Spring Harbor, New York, 1980, 447.

58. **Marion, P. L., Van Davelaar, M. J., Knight, S. S., Salazar, F. H., Garcia, G., Popper, H., and Robinson, W. S.**, Hepatocellular carcinoma in ground squirrels persistently infected with ground squirrel hepatitis virus, *Proc. Natl. Acad. Sci. U.S.A.*, 83, 4543, 1986.

59. **Robinson, W. S., Miller, R. H., Klote, L., Marion, P. L., and Lee, S. C.**, Hepatitis B virus and hepatocellular carcinoma, in *Viral Hepatitis and Liver Diseases*, Vyas, G. N., Dienstag, J. L., and Hoofnagle, J. H., Eds., Grune & Stratton, New York, 1984, 245.

59a. **Seeger, C., et al.**, personal communication, 1991.

60. **Smee, D. F., Knight, S. S., Duke, A. E., Robinson, W. S., Matthews, T. R., and Marion, P. L.**, Activities of arabinosyladenine monophosphate and 9-(1,3-dihydroxy-2-propoxymethyl)guanine against ground squirrel hepatitis virus *in vivo* as determined by reduction in serum virion-associated DNA polymerase, *Antimicrob. Agents Chemother.*, 27, 277, 1985.

61. **Marion, P. L., Cullen, J. M., Azcarraga, R. R., Van Davelaar, M. J., and Robinson, W. S.**, Experimental transmission of duck hepatitis B virus to Pekin ducks and to domestic geese, *Hepatology*, 7, 724, 1987.

62. **Aldrich, C. E., Coates, L., Wu, T. T., Newbold, J., Tennant, B. C., Summers, J., Seeger, C., and Mason, W. S.**, *In vitro* infection of woodchuck hepatocytes with woodchuck hepatitis virus and ground squirrel hepatitis virus, *Virology*, 172, 247, 1989.

Chapter 3

THE BIOLOGY OF AVIAN HEPATITIS B VIRUSES

Florian Schödel, Thomas Weimer, Doris Fernholz, Ralf Schneider, Rolf Sprengel, Gerhild Wildner, and Hans Will

TABLE OF CONTENTS

I. INTRODUCTION

The avian hepadnaviruses so far contribute two members to the *Hepadnaviridae*: duck hepatitis B virus (DHBV) and heron hepatitis B virus (HHBV). Both are small (≈3 kb), partially double-stranded enveloped DNA viruses similar in genome organization and mode of replication to the other hepadnaviruses found in man, woodchucks, and ground squirrels[1,2] (Chapters 1 and 2, this volume; Figure 1). On the basis of nucleotide or deduced amino acid sequence divergence, they can be regarded as a separate group only distantly related to the mammalian hepadnaviruses. By molecular phylogenetic comparison, the divergence of the avian hepadnaviruses from the mammalian hepadnaviruses seems to be a relatively recent event (≈30,000 years ago) and thus would not coincide with the host species divergence.[3]

DHBV was the first hepadnavirus for which *in vitro* infection of cultured primary hepatocytes was established.[4] Later it was found that DHBV can replicate in human hepatoma cell lines after transfection of cloned viral DNA.[5-7] This and the possibility of infecting ducklings or transfecting them intrahepatically with cloned viral DNA[8] explains why current interest in the avian hepadnaviruses focuses mainly on their use as an experimental system for the molecular dissection of the hepadnaviral life cycle and host–virus interactions. In contrast to the infection of mammals with hepadnaviruses, DHBV infection of ducks seems not to cause major pathogenic sequelae and is therefore not an ideal system to study hepadnavirus-induced disease. It is, however, an appropriate model to study the influence of pharmacological agents on the hepadnavirus life cycle *in vivo* and *in vitro*, and some of these studies will be briefly mentioned.

The aim of this chapter will be to describe the incidence and course of DHBV infection and its use as an experimental system, to present and compare in some detail hitherto unpublished DHBV DNA sequences, and to point to some peculiarities of the DHBV/HHBV system.

II. INCIDENCE AND HOST RANGE

DHBV infection seems to be endemic in most parts of the world. DHBV was first discovered in sera from Pekin ducks in mainland China[9] and then in commercial flocks of Pekin ducks in the U.S.[10-12] In the following years, DHBV was described and molecularly cloned from Pekin ducks and geese *(Anser domesticus)* in Germany.[13] A hepatitis B virus isolated from geese[14] (DHBV1, Figure 2[14a]) proved to be so closely related to the previously cloned DHBV isolates that it is reasonable to regard it as a DHBV variant and not as a new avian hepadnavirus. It may have been horizontally transmitted from ducks to geese or vice versa. In France, 1 to 6% of Pekin ducklings of different breeds were found to be DHBV infected.[15] DHBV was also identified in Pekin ducks in Australia[16] and Taiwan, and in other duck breeds (Indian Runner and Khaki Campbell) in the U.S.[10-12] Between 3 and 10% of wild migrating mallards *(Anas platyrhynchos)*, which are the common ancestors of domestic ducks *(Anas domesticus)*, were also found infected with DHBV in France.[17] Recently a DHBV-related virus has been found in maned ducks *(Chenonetta jubata)*, a wild duck species confined to Australia and phylogenetically distinct from domestic ducks and their progenitors.[18]

Several strains of ducks and geese have been experimentally infected with DHBV.[1,19,20] The only duck breed identified to date that seems not to be susceptible to experimental DHBV infection is the Muscovy *(Cairina)* duck, which is not descended from mallards but belongs to the same subfamily *(Anatinae)* and is more closely related to domestic ducks than to geese.[19]

A virus related to DHBV was detected in 11 of 54 sera of German grey herons *(Ardea cinerea)*.[21] The virus (HHBV) DNA was molecularly cloned and sequenced. The nucleotide

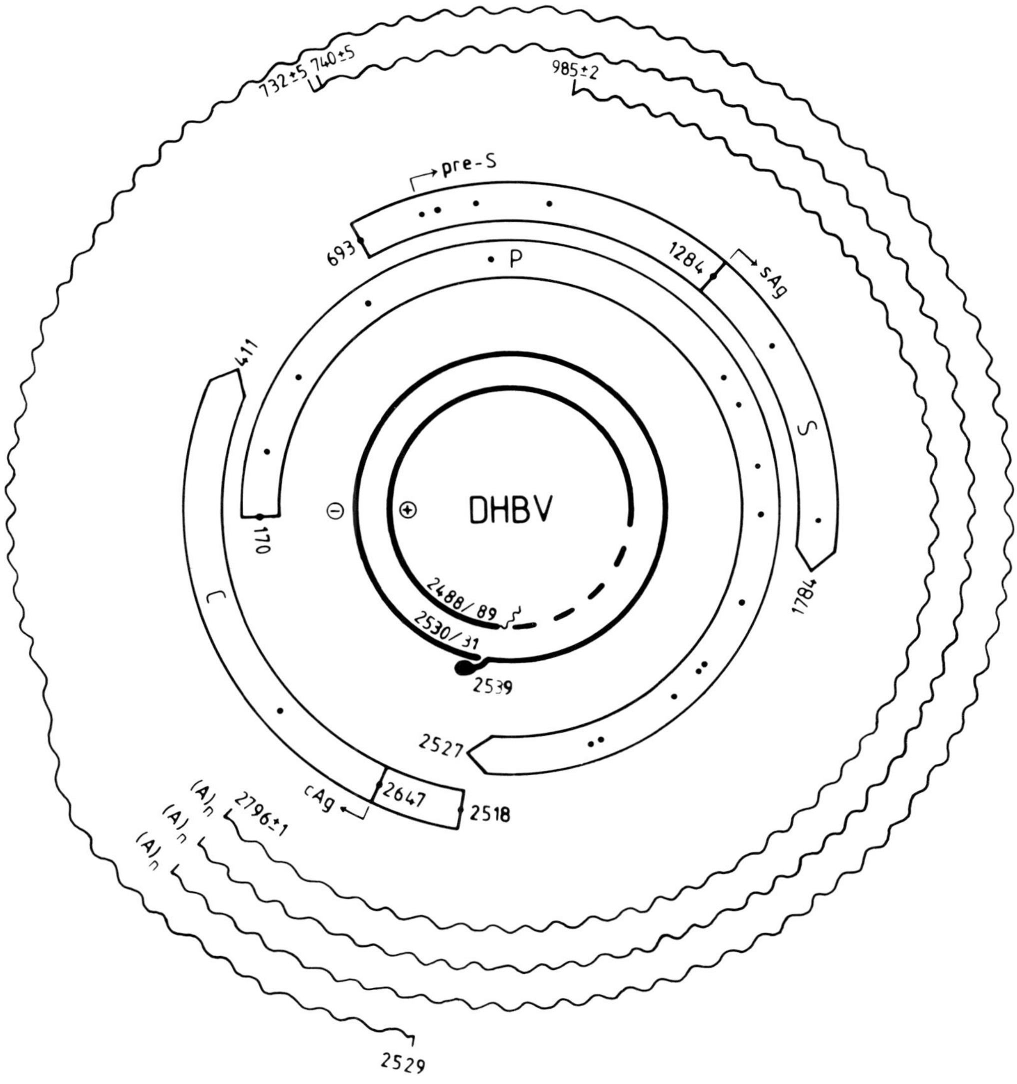

FIGURE 1. Gene organization, genome structure, and transcription map of duck hepatitis B virus (DHBV). The inner circles represent the two DNA strands of the virion-encapsidated genome. A short wavy line indicates the oligoribonucleotide priming DNA plus-strand synthesis; the P gene-derived protein covalently linked to the 5′ end of the DNA minus strand is indicated by a filled circle. Open arrows symbolize the long open reading frames (ORFs) encoded on the DNA minus strand, and in-frame ATGs are indicated by dots. The three major transcripts are represented by the outer wavy lines. Positions of 5′ ends and 3′ ends of ORFs, transcripts, and DNA plus and minus strands are for DHBV3 (for references see text).

sequence of HHBV differs from DHBV by 21.6%. HHBV is similar to DHBV in morphology, genome organization, and size of the major viral envelope and core proteins. Whereas the core protein and the major viral envelope protein are highly conserved between DHBV and HHBV, a marked divergence exists between the preS and P protein sequences where the two genes overlap[21] (Figures 2 through 5). The translated sequence of preS and P genes is also most variable in the overlap region of several sequenced HHBV isolates.[21,22] Virus-positive sera from herons induced no infection in Pekin ducklings, and the cloned HHBV could not be transmitted to ducklings by intrahepatic DNA transfection.[21] Further DHBV isolates were recently molecularly cloned and partially sequenced from German and Chinese Pekin ducks.[23-26]

III. TRANSMISSION AND EXPERIMENTAL INFECTION

The major route of natural transmission of DHBV seems to be vertical, from the chronic carrier dam to the developing embryo.[27] Chronic carrier dams transmit the virus to close to 100% of their offspring.[28] Whether or not horizontal transmission plays a role in DHBV maintenance is not established, although a situation in which it possibly occurred has been described.[29]

Maternally transmitted virus particles are found in large quantities in the yolk of eggs prior to incubation.[30] Active DHBV transcription and replication in the endodermal layer of the yolk sac tissue are identified by d 4 of incubation.[31] Virions are then supposed to be transported to the nascent embryonic liver with yolk sac tissue. During embryogenesis, the amount of DHBV RNA found in yolk sac tissue exceeds that found in embryonic liver.[32] By d 6 of incubation, single-strand DHBV DNA is synthesized in the embryonic liver, coinciding with liver formation. By d 8, accumulation of covalently closed circular DNA (cccDNA) and viral antigen is detected in hepatocytes.[31,33] The virus replicates also in cells of the bile duct epithelium. Only from d 8 to 10 of incubation are immature virus particles released into the serum of congenitally infected duck embryos. At later stages of postembryonic development (10 to 20 d posthatch), viral antigens are discovered in pancreatic and kidney cells.[30,34-39] Viral antigen and DNA are also discovered in splenic cells.[10,40,41] Viral RNA and single-stranded viral DNA (ssDNA), indicative of viral replication, are detected in the pancreas, but not or in very small amounts in the kidney. Recently viral RNA and DNA have also been described in the lung, heart, and intestine.[42] Most of the congenitally infected ducklings become chronic carriers,[28,30] the level of serum DNA peaking at about the fourth to fifth week posthatch and reaching levels of approximately 10^{11} to 10^{13} DNA genome equivalents per milliliter of blood.

The kinetics of the appearance of viral transcripts, replicative intermediates, and viral antigens were more closely studied using experimental infection of embryonic or 1-d posthatch ducklings than during the natural vertical transmission. When ducks are infected with viremic serum at d 1 posthatch, cccDNA is first discovered in liver cells by 6 h, together with the first RNA transcripts; at 12 h all three major RNA transcripts are identified, followed by ssDNA. The amount of RNA transcripts increases prior to the increase of DHBV DNA by d 3 to 6.[33] In most of the animals we have experimentally infected, the highest titer of DHBV DNA in the serum was found at d 7 postinfection. The viral titer then often declined during the next few weeks to a lower level, which then remained constant in most cases for at least several months.[43] PreS protein could be detected in the serum of infected ducklings by enzyme-linked immunosorbent assay (ELISA) or Western blotting with antisera against recombinant proteins simultaneously with the appearance of viral DNA detected using a dot-blot hybridization technique.[43,44] In contrast, the level of DHBeAg did not correlate with viremia. DHBeAg was detected at higher titers when the DHBV viremia had already declined and sometimes when no viral DNA could be detected.[43] Thus preS but not DHBeAg in the serum correlates with viremia. When the inoculated virus disappears from the circulation of a parenterally infected duckling, hepatocytes are the first cells where viral antigen and DNA is discovered. Infection spreads first within the liver and subsequently to extrahepatic tissues.[45,46,47] As mentioned above, DHBV nucleic acid and protein are also detected in extrahepatic tissues. In one study DHBV DNA was discovered in pancreatic cells by d 4, in serum by d 6, in splenic red and white pulp by d 7, and in renal glomerula by d 14.[40] In naturally

```
H4         A C  C   ACGT T      A   AT  G   T AC   G  TT    C A A   C C                          T  CC C  C
D22          A  CC                A     T   T         A      C                            C      T
D26          A  CC                A     T             A                                          T
D16                               A A   T             A
D1               C                  A   T             A
D3    1 CATGCTCATT TGAAAGCTTA TGCTAAGATC AACGAGGAAT CCCTGGATAG GGCTAGGAGA TTGCTTTGGT GGCATTACAA CTGTTTACTG TGGGGAGAAG  100

H4       C C   T   A  T             CC C          C  T  T     C   A A    C     C AG
D22         C  T      T             GC T          C        A                   CC
D26         C  T      T             GC T   C      C        A            C      CC
D16        C A T                    T                G              A                       G
D1          C  T                    G
D3  101 CTAATGTAAC TAACTATATT TCTCGATTGC GTACTTGGTT ATCAACTCCT GAGAGATATA GAGGTAGAGA TGCCCCAACC ATTGAAGCAA TCACTAGACC  200
                                                                                     ▶ P

H4            G          A       A    TCA     AA   A                                     C    G
D22                              A         T
D26                              A         T
D16           G                  A            A
D1
D3  201 AATCCAAGTG GCTCAGGGAG GCCGAAAAAC AACTTCGGGT ACTAGAAAAC CTCGTGGACT CGAACCTAGA AGAAGAAAAG TTAAAACCAC AGTTGTCTAT  300

H4                      T       C A GC       T  T  C   T                                 A CG G  A AC  GACA C    C
D22                                  T             T                                                C   AGA
D26                                  T             T                               G                C   AGA
D16                             C                                                                   C     T
D1                                           T              G                                       C       C
D3  301 GGGAGAAGAC GTTCAAAGTC CAGGGAAAGG AGAGCCCCTA CACCCCAACG TGCGGGCTCC CCTCTCCCAC GTAGTTCGAG CAGTCACCAC AGATCTCCCT  400

H4       A      G     GGGAC       AC      A A  G         AA T   GT G                   TGAG          C C      GA
D22                                               T             AT G               A   T  T         T  T            T
D26                   G                        T  T             AT A               A   T            T               T
D16                                          T    T             A  A                   T  T
D1                                                              A
```

FIGURE 2. Computer-based alignments of DHBV and HHBV DNA sequences. D1, DHBV1; D3, DHBV3; D16, DHBV16; D22, DHBV22; D26, DHBV26; H4, HHBV4 (for references see text). The alignment was computed, as were the alignments of protein sequences translated from these DNA sequences (Figures 3 to 5), using software of the University of Wisconsin Genetics Computer Group (UWGCG, version 6.1)[14a] on a Vax. DHBV3 DNA is given in the bottom line as the matrix to which the other DNAs are aligned, nucleotide positions are for DHBV3, and only nucleotides differing from the DHBV3, sequence are shown. Gaps introduced to maximize the number of matches are indicated by black dots. For easier orientation, ATGs and stop codons probably important in the synthesis of preS/S, preC/C, and P proteins are indicated by black arrows, with the respective symbols underneath the sequence. The direct repeats (DR1 and DR2) are underlined.

```
D3  401 CGCCTAGGAA ATAAATTACC TGCTAGGCAT CACCTAGGAA AATTGTCAGG TCTCTATCAA ATGAAGGGCT GCACATTTAA CCCAGAATGG AAAGTACCAG  500
                   C ◄

H4       C       C    AA C  C TC CAG  GA     T  C                                C             G   T
D22                              A GC  A                                         C
D26                              A GC  A   G                                     C
D16                           A       T    G T
D1                                          T                                                      T
D3  501 ATATTTCGGA TACTCATTTT GATTTAGAAG TAATAAATGA GTGCCCTTCC CGAAATTGGA AATATTTGAC TCCAGCCAAA TTCTGGCCCA AGAGCATTTC  600

H4         TC    A      TT             C  G  C  A   ATTCCA C AAA T  T   G   C G    AA G T   C  C AT A      T
D22                 A                G T  G  C               G  T                                          C
D26                 A                G T  G  C                  T
D16                      T                G                         T
D1                                   A       C                      T      T
D3  601 CTACTTTCCT GTCCAGGCAG GGGTTAAACC AAAATATCCT GACAATGTGA TGCAACACGA ATCAATAGTA GGTAAATATT TAACCAGGCT CTATGAAGCA  700

H4              C         AG  T                     C  G  C C  TA  TC C          AA  G          CCG     A  .........
D22                                                                              TT   T            G          G
D26                                                                              TT   T            G          G
D16                                                                                                          C   TA
D1                                                       A  C  C  G          A   GT                          T  AA A
D3  701 GGAATCCTTT ATAAGCGGAT ATCTAAACAT TTGGTCACAT TTAAAGGTCA GCCTTATAAT TGGGAACAGC AACACCTTGT CAATCAACAT CAAATTCCTG  800

H4              TA C  A                C ACA  A       G                 T AT  C GC A  T   A  A  A        T  T CCCGAATTTT
D22          T     A  T                        GC         C            GA    G C    C   T  A  A   C C     C
D26          T     A  T                        GC         C                  G C        T  A  A   C C     C
D16                                               AC G                 A A  G      C             A  G     C
D1                                                                            T                                 T
D3  801 ATGGGGCAAC ATCCAGCAAA ATCAATGGAC GTCAGGAGAA TAGAAGGAGG AGAACTCCTA TTAAATCAAC TTGCCGGCAG AATGATACC. ........AA  891
        ► pre-S

H4      CG   C CA  A   AC G A   A  A     T  C  TC  A     GA G   T  CA CG  CT  G     A  AC TCGG      A  AG C GG   GCA
D22             G  C                                                                  A          T  T  AA     A
D26            AG  C               A                   A                              A          T  T  AA  C  A  G
D16             T           A               C             G                                             A  T
D1          A  A             A
D3  892 AAGGGACTCT GACATGGTCG GGCAAGTTTC CAACAATAGA TCACGTATTA GACCATGTGC AAACAATGGA GGAGATAAAC ACCCTCCAGC AACAGGGAGC  991
```

FIGURE 2 (continued).

```
H4      C     G AG    A   C C  C   T G   CC ...GG    AAC G GG  CCACCCC    A CCATCAC      AGAA         CA         AAG
D22     A               G       A   CA       A C      T                         C                         G  A  T  C
D26                     G       A   CA   A   A C      T                         C                         G  A  T  C
D16                     G       A    A   A             A                        C                G    C      A  A  C
D1                      G                                                                                    A  T  C
D3  992 TTGGCCTGCT GGGGCAGGAA GGCGAGTCGG GTTATCAAAT CCGGCTCCTC AAGAGATTCC TCAGCCTCAG TGGACTCCCG AAGAAGATCA AAAGGCCCGA 1091

H4         TTC  CA AG AA  T    GAGAACCGT CCGAAACCCG CC   AC GC       TCCC  TC  GAGT   G ACG TG   GAA  TC            A
D22       A         T     T                                                 C         A            A     G          A
D26       A         T     T                                          A  G   C      A        C            G          A
D16       A         C     T                                A                C                T  T  C  TC
D1        A               T   G                                      A                                 C
D3 1092 GAGGCTTTTC GACGTTACCA A......... .......... ..GAAGAGAG ACCACCGGAA ACGACCACCA TTCCTCCGAC ATCTCCAACT CAGTGGAAGC 1170

H4       TTC     A     CC TT     AAA CA   ......... ...  C CT  ATC  CG CA  GG A  C   A  GGT C G A C TCAA G  CCCA AG T
D22      G     C   G                  A    CA G  T  AC C         C   C T       AAT   C  G     C      T         A
D26      G     C   G                  A    CA G  T  GC C         C  GC T       AAT   C  G     C    T T  A
D16            C   G                  A       G  T   C C            GC            T  A        G                A
D1       G                                CC  C                     GC T          C                     A   C
D3 1171 TACAACCAGG AGACGATCCA CTCCTGGGGA ATCAATCACT CTTGGAGACT CATCCATTAT ACCAG...AC GGAACCAGCT GTGCCAGTGA TAAAGACTCC 1267

H4      CA  AAC             G GA C                                    G          G A               C
D22        A  G  A
D26        A  G GA
D16     C   T G
D1      C  A  G
D3 1268 TCCCCTAAAG AAGAAAATGT CTGGTACCTT CGGGGGAATA CTAGCTGGCC TAATCGGATT ACTGGTAAGC TTTTTCTTGT TGATAAAAAT TCTAGAAATA 1367
                         ► S

H4        GA   A                                          A         T              A        T        C   A
D22       T            T                                                          C
D26       T            T                                                                             C
D16       GA           T
D1
D3 1368 CTACGGAGGC TAGACTGGTG GTGGATTTCT CTCAGTTCTC CAAAGGGAAA AATGCAATGC GCTTTCCAAG ATACTGGAGC CCAAATCTCT CCACATTACG 1467

H4      T                                    C      T                                GC                 A           T
D22     TC     T                             C      T                                                   A
D26     TC     T                             C      T                                                   A
D16     T      T                                                                                        A
D1      T                                                                                               A
```

FIGURE 2 (continued).

```
D3 1468 CAGGATCCTG CCCGTGGGGA TGCCCAGGAT TTCTTTGGAC CTATCTCAGG CTTTTTATCA TCTTCCTCTT AATCCTGCTA GTAGCAGCCG GCTTGCTGTA 1567

H4               A   CAA             C A                A                   C            A             A      G
D22                  C                           G      A         T         C
D26                  C                           G      A         T         C
D16                  C
D1                   C
D3 1568 TCTGACGGAC AATGGGTCTA CTATTTTAGG AAAGCTCCAA TGGGCGTCGG TCTCAGCCCT TTTCTCCTCC ATCTCTTCAC TACTGCCCTC GGATCCGAAA 1667

H4        G  A                        TT            C               T        GT                 A
D22       G              T            T
D26       G              T            T
D16                                                                                             G
D1                                                                                              G       T    T
D3 1668 TCTCTCGTCG CTTTAACGTT TGGACTTTCA CTTATATGGA TGACTTCCTC CTCTGCCACC CAAACGCTCG TCACCTTAAC TCAATTAGCC ACGCTGTCTG 1767

H4       A     C T      A  C   G     G   C     T  C     TG  T   T  C  G   G  AACG         T      C  CG G   CAGCA    A
D22                      C         C C   C        C      T  G             A AT  G  A                C          C T GA
D26                      C                     T  C      T  T      A        A      T                           C   GA
D16                                            T  C      C  G   T     G   G AT     A                C  G          A T
D1                        C         C        T                                      T                                A
D3 1768 CTCTTTTTTA CAAGAGTTAG GAATAAGAAT AAACTTCGAT AAAACAACAC CATCTCCAGT CACCGAAATC AGATTCCTCG GTTATCAAAT TGATGAACAC 1867
                 Pre-S/S ◄

H4      CAT         C                  T    C    A         C         C          C                        C         CA
D22                 C                                                                     A                                  G
D26                                                   C               A  G       A                 C                G
D16                                                            C                 A               C
D1                                          C                             A                 C                 C
D3 1868 TTCATGAAGA TTGAAGAAAG CAGATGGAAA GAATTAAGGA CTGTAATTAA GAAAATAAAA GTTGGAGAAT GGTATGATTG GAAATGTATT CAAAGATTTG 1967

H4       A  C   C  C  T  T  T         T   G  A  A   C          C       T      C A    CTG    C T AG   T  C    GC   T
D22      C         A     T            T   A  A  G   C  A            G                     A   C T A       C
D26      C         A     T      G  A  T   C  A      C  A                                  A   C T A    C
D16      G            T  T  T   G     T      A                                                             C
D1       C            T  T
D3 1968 TTGGGCATTT GAACTTCGTG TTACCTTTCA CTAAGGGTAA TATTGAAATG TTAAAACCAA TGTATGCTGC TATTACTAAC CAAGTAAATT TTAGCTTCTC 2067
```

FIGURE 2 (continued).

```
H4          AGG     AA TC      G      T     T     A   A     G   G C C GG         GT     T                     C  G
D22                         C  G  C   T     T         T          TCG  C   A     A       A G                   C
D26           C             C  G  C   T     T         T          TC   C   A                                      G
D16        AT C                                                       A   G
D1                                                       C                A
D3 2068 TTCTGCTTAT AGGACTTTGT TATATAAACT AACAATGGGT GTGTGTAAAT TAAGAATTAA TCCAAAGTCC TCTGTACCTT TGCCACGTGT AGCTACAGAT 2167

H4           ATT                                                   A C                   C      C          T GC     G
D22          T                                                                    T                        T GC     G
D26                                                                                                        T GC     G
D16                                                                                              G T                T
D1                                                                                                           T  A
D3 2168 GCTACCCCAA CACATGGCGC AATATCCCAT ATCACCGGCG GGAGCGCAGT GTTTGCTTTT TCAAAGGTCA GAGATATACA TATACAGGAA CTATTGATGA 2267

H4      TA     A   C A T              CA     TA AC  GA       T     A          T GGAA A TCTCT A  C    G            AA
D22     TA     A     A T           G   A    A A AC   A C  T  T   T                                       A        G
D26     TA     A     AAT           G   A      A AC   A C  T  T             T                             A        A
D16            A   C A                A   T  T            T      T            T  G                                 A
D1             A                          G  T            T      T         T                                       A
D3 2268 CTTGTTTGGC TAGGATAATG ATTAAACCTC GATGCCTCTT ATCTGACTCA ACCTTTGTTT GCCACAAACG TTATCAGACG TTACCATGGC ATTTTGCTGT 2367

H4        AT         A  AT A  CCAGGC GAC  T      A    A  C  T                   T       A          A   T
D22        T         AC    AT CTT      C  T           A            C                               A
D26        T         A   T AT CTT      C  T           A                                            A
D16                  A            G       T                                                        A   T
D1                   A                    T                                                        A
D3 2368 GTTGGCCAAA CAGTTGCTCA AACCTATACA ACTGTACTTT GTCCCGAGCA AATATAATCC TGCTGACGGC CCATCCAGGC ACAGACCGCC TGATTGGACG 2467

H4        CG  A                     GCAT         A                   TCTC      T ACA C   TCTC A             G
D22        G  A                                                                                             C
D26        G  A                                                                                             C
D16
D1             T                                                                 C
D3 2468 GCTTTTCCAT ACACCCCTCT CTCGAAAGCA ATATATATTC CACATAGGCT ATGTGGAACT TAAGAATTAC ACCCCTCTCC TTCGGAGCTG CTTGCCAAGG 2567
                                                          ► preC P ◄
```

FIGURE 2 (continued).

```
H4                   G  C C         TCCT                                                             G           A
D22                                   T
D26                                   T
D16
D1
D3 2568 TATCTTTACG TCTACATTGC TGTTGTCGTG TGTGACTGTA CCTTTGGTAT GTACCATTGT TTATGATTCT TGCTTATATA TGGATATCAA TGCTTCTAGA 2667
                                                                                               →   C

H4        T     A      A         G                   TC G  T                      T   G                        A G C
D22             T      A       T G                         T            A   G     T      C
D26                    T       T G               C         T            A   G     G   G
D16                                                                               T      C         G           A
D1                                                     C                A
D3 2668 GCCTTAGCCA ATGTGTATGA TCTACCAGAT GATTTCTTTC CAAAAATAGA TGATCTTGTT AGAGATGCAA AAGATGCTTT AGAACCTTAT TGGAGATCAG 2767

H4       AA                           T                 T G      G               C   T          AGCC    T   A  C  CC
D22                                                                                  T                  C   T     T
D26                                                                       A      T                      T     C
D16                             G                         C                                C                    T
D2                                        G                                                                    G   A
D3 2768 ATTCAATAAA GAAACATGTT TTAATTGCAA CTCACTTTGT GGATCTTATT GAAGACTTCT GGCAGACTAC ACAGGGTATG CATGAAATAG CCGAAGCATT 2867

H4      CC       A   T            CGTA   A     GG G   T T        ACT  TAG T               C  C    AAC   C  C   TC A
D22           T      T            A  A   A  T   GC          G             A     T  G          C               C
D26                               A  A   A  T   GC          G             A     T  G          C
D16           T         T  C                T  A CG   T     T             G  A  T         A                         C
D1                             G               T  G             C               T            C   A
D3 2868 AAGAGCAGTT ATACCACCTA CTACTACTCC TGTACCCCAA GGATATCTCA TTCAGCACGA CGAGGCAGAA GAGATTCCTT TGGGAGATTT ATTTAAACAT 2967

H4        A  G  G         C A              C     C      A
D22                 A                  A         T      T  AC
D26                                    G         T      T  AC
D16       A               A            C   C     G
D1
D3 2968 CAGGAAGAAA GGATAGTTAG TTTCCAACCT GATTATCCAA TTACGGCTAG AATT 3021
```

FIGURE 2 (continued).

```
H4    MIISISSL     S  EF       SRARISL   KS   R . ..N  HTQ    TTD  V        QH        PEFS PI  TA      Q   M  IDSV
D22          T                           F    S         Q                I            ...   V
D26                                      F    S         Q                             ...   V             I
D16                                                 T M                    I            ...             L
D1                                NPTS   NS       NLI                                 ...
D3           MKQ ESFISGYLNI WSHLKVSLII GNSNTLSINI KFLMGQHPAK SMDVRRIEGG ELLLNQLAGR MIP...KGTL TWSGKFPTID HVLDHVQTME
                                                      ► Pre-S

H4     LR   AG H   E TA  L   DQ R .TP  PIT  E   K   K F KQ    N  KPAE A   PITELHAA E PQWKISPE DP  KAKA I PV K  E  I
D22         K            A   T                                                                K        .   N
D26         N            A   T                                                                K        .   N     S
D16         N                  T                                         S  P                          .   .S
D1                                                                          P              T H         .   .P
D3    EINTLQQQGA WPAGAGRRVG LSNPAPQEIP QPQWTPEEDQ KAREAFRRYQ EERPPETTTI PPTSPTQWKL QPGDDPLLGN QSLLETHPL. YQ.TEPAVPV

H4    L V K TN     GA               G     T        K             E  L     N    T     V                         L
D22         ..                                                           H           V
D26         ..R                                                                 T     V
D16         ..                                                                       V
D1          ..                                                                       V
D3    IKTPPL..KK KMSGTFGGIL AGLIGLLVSF FLLIKILEIL RALDWWWISL SSPKGKMQCA FQDTGAQISP HYAGSCPWGC PGFLWTYLRL FIIFLLILLV
                  ► S

H4         F  E  K   FE     E     S    Y     E            F    T    V   V               F NSG*
D22                      R  E  L  S                   M   L
D26                      R  E  L  S                   M   L
D16
D1                                                                             V
D3    AAGLLYLTDN GSTILGKLQW ASVSALFSSI SSLLPSDPKS LVALTFGLSL IWMTSSSATQ TLVTLTQLAT LSALFYKS*
```

FIGURE 3. Alignment of preS/S protein sequences. Abbreviations and alignments are as in Figure 2.

```
H4        S  LH SP              S  FL                    V                    Q                  KAET
D22                                L
D26                                L
D16                                                                                          K
D1                                                                    T
D3      MWNLRITPLS FGAACQGIFT STLLLSCVTV PLVCTIVYDS CLYMDINASR ALANVYDLPD DFFPKIDDLV RDAKDALEPY WRSDSIKKHV LIATHFVDLI
        ▶ preC                                         ▶ C

H4                 SQ  D           V   E  F  T S        N   SN       N             T   RV T L   QA  K
D22                                    A       E
D26                                    A       E
D16                     S              P       E
D1                       I
D3      EDFWQTTQGM HEIAEALRAV IPPTTTPVPQ GYLIQHDEAE EIPLGDLFKH QEERIVSFQP DYPITARIHA HLKAYAKINE ESLDRARRLL WWHYNCLLWG

H4        T                 K    K                     NQ K                                G  S SPS          NRGNQT
D22                                                      S                                 D     S                R
D26                                                      S                                 D     S                R
D16       Q                 K                             T
D1                                                                                            S
D3      EANVTNYISR LRTWLSTPER YRGRDAPTIE AITRPIQVAQ GGRKTTSGTR KPRGLEPRRR KVKTTVVYGR RRSKSRERRA PTPQRAGSPL PRSSSSHHRS

H4          E
D22
D26
D16
D3
D1      PSPRK*
```

FIGURE 4. Alignment of preC/C proteins. Abbreviations and alignments are as in Figure 2.

```
H4    ....MNKL T K A        IASS   PLL QI   AS    FL PKN  AR              R  K   I K  E           D R            L
D22    Q LM  H           T  I       TL  I    GF     H     A                        H
D26    Q L   H              I       TL  I    GFA    H   T A                        H
D16    QKL   H                      KL      V     C    N       R          R
D1      KL   H                      TL      G
D3    MLRSTANPWI GLGDCFGGIT TVYCGEKLM* LTIFLDCVLG YQLLRDIEVE MPQPLKQSLD QSKWLREAEK QLRVLENLVD SNLEEEKLKP QLSMGEDVQS

H4     EA D              IG ETRH P H   RD     K     T         E   H  I     A N SQ  I                         L  HS
D22     I                    A E                                  D            MQ  I
D26     I              R     A E        R                                      MQ  IV
D16                          A I                                               N D  V                             V
D1        D    S             A
D3    PGKGEPLHPN VRAPLSHVVR AVTTDLPRLG NKLPARHHLG KLSGLYQMKG CTFNPEWKVP DISDTHFDLE VINECPSRNW KYLTPAKFWP KSISYFPVQA

H4           EFQ QN   L ND   NK F         V          PYFT    K   P ...Q H  Y     D    S    II TATSS K  S S IFGAHNN
D22      A                                                 L  Y  K   V    S TC      A      D  A PIS PH P       CH
D26      A                                                 L  Y  K   V    S TC      A         A  IS PH P       H
D16                                                                H Y              TD    NT V P   KD P      F   R
D1       T                                               K    Y      FN N                   L          S     NT
D3    GVKPKYPDNV MQHESIVGKY LTRLYEAGIL YKRISKHLVT FKGQPYNWEQ QHLVNQHQIP DGATSSKING RQENARRRTP IKSTCRQNDT KR..DSDMVG

H4    RKISYH TRD GSHRLS RTS D TSRGALAG  DSTPIGPG  TAAHP THH   R   Q  QG VLQAIS EP  ETR NG  SH HRVARCRTS    DF  PF
D22                     N   SSK R M     G E   NQ  C       P        ES        S S   S S        H  T   T A   TA  G
D26    I     H          N   SSKPRG      G E  INQ  C       P        ES        S S   S S     RD H N  H   A   TA  G
D16       T   V               ES        G E  I    D       P     G P S        T S   P S  K     H    VFPS        G
D1    K                                 G                           S        S S     SG    R          T     A
D3    QVSNNRSRIR PCANNGGDKH PPATGSLACW GRKASRVIKS GSSRDSSASV DSRRRSKG.. .....PRGFS TLPRRETTGN DHHSSDISNS VEATTRRRST

H4    ....QSK AY PRQGT RGTD PQGPKAHQQ    GS            V  RI            E                       K  C   T
D22     KQVSTR   PTL EYRA R  C  N  TEK                                  S
D26     KQVSAR   PAL EYRA R VCN    TEK                                  S
D16     K VSPR    A  V R  G     N PLE                                   E
D1      DPL       AL A        NQ  PTE
D3    PGESITLGDS SIIPD.GTSC ASDKDSSPKE ENVWYLRGNT SWPNRITGKL FLVDKNSRNT TEARLVVDFS QFSKGKNAMR FPRYWSPNLS TLRRILPVGM

H4                       A             K Q                          I A   AS       S            S     T       T   F
D22                                    R                                A
D26                                    R                                A
D16                                    R                                                           A
D1                                     R                                                           A    S
```

FIGURE 5. Alignment of P protein sequences. Abbreviations and alignments are as in Figure 2.

```
D3    PRISLDLSQA FYHLPLNPAS SSRLAVSDGQ WVYYFRKAPM GVGLSPFLLH LFTTALGSEI SRRFNVWTFT YMDDFLLCHP NARHLNSISH AVCSFLQELG

H4         M       T       E SKQH          N            Q           I                        D C HR    A   R KI
D22               N           HR                      I                                          HK
D26               N           QR                      I                                          HK
D16               N            N                                                                           S
D1                                Q                     I
D3    IRINFDKTTP SPVTEIRFLG YQIDEHFMKI EESRWKELRT VIKKIKVGEW YDWKCIQRFV GHLNFVLPFT KGNIEMLKPM YAAITNQVNF SFSSAYRTLL

H4               TLD  V L           L                 T               V    KL       I T   Y    RKFSK       MY     T
D22              S K    M                                             V    KL      YI T                N    F     S
D26              S K                                                  V    KL       I T                N    F     S
D16                K                                            V     S    K                               M
D1                 K                                                                                       M
D3    YKLTMGVCKL RINPKSSVPL PRVATDATPT HGAISHITGG SAVFAFSKVR DIHIQELLMT CLARIMIKPR CLLSDSTFVC HKRYQTLPWH FAVLAKQLLK

H4     RLT  Y            T    K     VT       H          L
D22    S P                    K     VT
D26    S P                    K     VT
D16                           K
D1                            K      L
D3     PIQLYFVPSK YNPADGPSRH RPPDWTAFPY TPLSKAIYIP HRLCGT*
```

FIGURE 5 (continued).

infected animals, DHBV and antigen are first confined to endocrine cells in islets, then found in exocrine acinar cells.[39] In experimentally infected ducklings posthatch, the sequence of events seems to be reversed.[39] In the pancreas and in splenic cells, ssDNA and dsDNA are found, suggesting ongoing replication.[31] In one study, only ssDNA was found in splenic cells, suggesting incomplete replication restricted to minus-strand synthesis.[46] The time window during which ducklings can be infected with DHBV, resulting in chronic carrier status, is short.[20,29] It was first noted by Mason et al. that 2- to 3-month-old ducklings do not support a measurable viremia when intravenously infected with 200 times the dose of viremic serum that successfully induced chronic carrier status or transient viremia in 15-d-old embryos and 1-d-old ducklings.[29] Within the first 3 to 5 d posthatching, a chronic carrier status can be attained in virtually all ducklings infected with viremic serum. In one report, a transient (1 month) viremia was the outcome of inoculation at d 7 posthatch.[48] The congenital or early infection with DHBV may induce a status of partial tolerance to viral antigens that could explain the susceptibility to chronic infection.[49] In contrast to HBV infection, where naked viral cores are found predominantly in the nucleus of infected hepatocytes, DHBV core particles are primarily located in the cytoplasm of infected cells, sometimes in association with the cisternae of the rough endoplasmic reticulum.[40,47,50] Naked cores seem to assemble with the viral envelope when the virions are budding from endoplasmic reticulum membranes.

IV. PATHOLOGY AND IMMUNOLOGY

The information about the immunological consequences of natural or experimental DHBV infection in ducks is scarce and the information about the respective liver pathology is controversial. Originally a transient antibody response to core antigen was detected in a majority of ducklings experimentally infected at d 8 posthatch and only in a minority of ducklings infected at d 1, but not in congenitally infected animals.[49] In this experiment, no antibodies to surface antigens were demonstrated, all of the infected animals became viremic. In another report, adult Pekin ducks inoculated repeatedly with viremic serum developed antibodies to envelope proteins (here collectively called anti-S), but no detectable viremia.[51] It was not determined whether the antibodies were directed against DHBsAg or preS proteins. Anti-S was also discovered in single, noninfected adult ducks and ducklings in a naturally infected flock.[51] The anti-S in hatchlings was virus neutralizing. This shows that ducks can mount an anti-S response after contact with naturally occurring virus and probably transmit the protective antibodies to their offspring. When newly hatched ducklings were infected with a low-titer DHBV-positive serum so that a proportion of the animals could clear the viremia, those that became viremic and later cleared the virus developed antibodies to preS and DHBsAg.[52] Antigen-antibody complexes were detected in the kidneys of congenitally infected ducklings,[53] which raises the question whether congenital infection induces a split tolerance against viral antigens. Virus-neutralizing antibodies recognizing the preS region of the DHBV surface antigens have been raised by immunization with recombinant proteins or DHBV virions in rabbits and mice.[54,55] These will be further discussed in Section IX.A. The cellular immune response to DHBV antigens has not been studied, whereas the importance of cellular immune defense mechanisms in the control of human hepatitis B virus infection is increasingly recognized.[56]

The description of liver pathology associated with natural or experimental DHBV infection ranges from nonexistent to the observation of hepatocellular carcinoma in Chinese Pekin ducks. Most authors find a mild to moderately severe hepatitis associated with the peak of viral replication at 3 to 6 weeks after infection in some of the animals.[16,19,40,41,45,57,58] It consists mainly of cellular infiltrates in the periportal region. There is evidence that the incidence and degree of hepatitis encountered after experimental infection with DHBV are linked to the time of infection; the sooner posthatch the animals are infected, the less liver pathology is observed.[59] Direct cytopathic effects on the hepatocytes that support viral replication are usually not described. Both the decreasing susceptibility to DHBV infection and the increased pathology induced by DHBV infection with age may be linked to the maturation of the immune system. Cases of hepatocellular carcinoma (HCC) with chromosomally integrated DHBV DNA were described in Pekin ducks from a certain area in mainland China.[60-62] The chromosomally integrated DHBV DNAs characterized so far have a structure similar to those observed in woodchuck or human hepatocellular carcinoma.[62,63] Environmental and genetic factors may contribute to the divergent pathogenicity. One of the incriminated factors is the high aflatoxin B_1 content in food, another is additional liver parasites.[64,65] Aflatoxin B_1 alone can cause development of HCCs in ducks.[64,65] A direct link between DHBV infection and carcinogenesis has so far not been established. In congenitally infected ducklings, the arginine-stimulated insulin response appears to be impaired.[66]

V. TRANSFECTION AND TISSUE CULTURE

The three major advances in the study of the hepadnaviral life cycle by molecular genetic means have been (1) the demonstration that a productive DHBV infection can be initiated by intrahepatic transfection with cloned viral DNA,[8] (2) the development of tissue culture systems that allow DHBV infection *in vitro*,[4] and (3) the recent finding that cultured human hepatoma cells support the replication of DHBV on transfection with cloned viral DNA.[5-7]

When cloned viral DNA dissolved in a suitable medium supplemented with DEAE dextran is injected into the livers of ducklings approximately 1 to 3 d posthatch, it can initiate a productive DHBV infection indistinguishable from natural or experimental infection with virus.[8] This finding has for the first time allowed us to genetically manipulate the viral genome and test its viability and biochemistry *in vivo*, and has since become an invaluable tool for the dissection of the viral life cycle. The time window in which intrahepatic transfection of viral DNA induces a productive infection is similar to that found for experimental infection.

Tuttleman et al. found that primary hepatocytes cultured from perfused duck livers can be infected with DHBV *in vitro* and imitate all stages of viral infection.[4,67] The percentage of hepatocytes recruited to viral replication *in vitro* is relatively low and the cultures can only be infected for a period of some days after establishment of the culture. Recently a more efficient *in vitro* infection has been described in primary hepatocyte cultures grown in the presence of dimethyl sulfoxide (DMSO).[68] In addition, the DMSO treatment expands the time in which the cultured hepatocytes are susceptible to infection to at least 2 weeks. The length of time that *in vitro*-infected hepatocytes can be kept in culture can be prolonged by cocultivation with rat hepatocytes.[69] Yolk sac cells also appear to be susceptible to *in vitro* infection with DHBV.[70]

Transfection of primary hepatocyte cultures is inefficient. For this reason, it has been an experimental breakthrough to show that cell lines of human hepatoma origin efficiently support the replication of DHBV and produce infectious DHBV virions after transfection of cloned viral DNA.[5-7] A chicken hepatoma cell line, LMH,[71] was found to produce 10 times more virus than the human hepatoma cell lines upon transfection with DHBV DNA.[72] Apart from being a valuable tool for molecular genetic studies, this shows that the relatively narrow host range of hepadnaviruses is not caused by intracellular factors affecting replication, transcription, assembly, packaging, or export of virions. Some of these functions, however, require tissue-specific rather than species-specific factors. Primary hepatocytes can be infected *in vitro* for a certain time after cultivation regardless of the age of the duck they come from,[73] so that the differentiation of hepatic tissue *in vivo* may be only one of the factors determining susceptibility. One of the causes for host and possibly tissue restriction is probably in viral uptake and/or unpackaging functions. For HBV, the preS region of the envelope proteins has been implied in the mediation of virus binding to hepatocytes.[74,75] In DHBV, the role of the preS region should be amenable to genetic analysis, as *in vitro* and *in vivo* infection systems are available.

VI. MORPHOLOGY

Mature virions in the liver or serum of infected animals are spherical and have a diameter of ≈40 nm by electron microscopy and band in sucrose or CsCl density gradient centrifugation at 1.16 g/cm^3.[10,11] When viral particles are purified from serum, a second fraction can be identified banding at 1.15 g/cm^3, consisting of empty viral particles that are highly pleiomorphic and have diameters ranging from 35 to 60 nm.[11] No filamentous forms of the empty viral particles are found. This points to a difference in the structure of the preS proteins of HBV and DHBV, since it has been shown that the filamentous form of empty viral particles in HBV is due to the presence of the large (preS1) envelope protein.[76] Naked core particles found in the cytoplasm and the endoplasmic reticulum of infected hepatocytes have a diameter of ≈35 to 37.5 nm.[11,46]

VII. GENOME STRUCTURE, REPLICATION, AND VIRAL TRANSCRIPTS

DHBV and HHBV have the characteristic genome structure of all hepadnaviruses. They share a small (3021 to 3027 bases), circular, partially single-stranded DNA genome of fixed polarity.[77] The relaxed circular conformation of the genome is maintained by a short (50 bp) complementary cohesive overlap region comprising the 5′ ends of both DNA strands.[78] A protein derived from the amino terminus of the P gene product is covalently linked to a G at the 5′ end of the DNA minus strand, which is at position 2539 (position 1 is the C of the unique *Eco*RI site).[79-82] This genome-linked protein is believed to be the primer for reverse transcription of the RNA pregenome. The DNA minus strand is terminally redundant by 6 to 8 bp. All known viral proteins are encoded by the DNA minus strand. The short or plus strand has a fixed 5′ end (position 2488), but heterogeneous 3′ ends. Unlike mammalian hepadnavirions, which usually have an incomplete DNA plus-strand, most of the DHBV plus strand DNAs in mature virions are almost full length. Only a minority of virions has a DNA plus strand of half the genome length or less. A virion-encapsidated polymerase can fill in the gap, and this reaction is exploited to radioactively label the genome *in vitro* and to characterize the genome by restriction enzyme fragmentation.[11,13] A short capped oligonucleotide (≈17 ribonucleotides) is linked to the 5′ end of the DNA plus strand and presumably serves as a primer for its synthesis.[83] The replication of hepadnaviruses is described in detail elsewhere[2] (Chapter 11, this volume). Briefly, the 3.5-kb C mRNA or pregenome serves as the template for synthesis of the DNA minus strand.[84] DNA minus strand synthesis, probably primed by the genome-linked protein, starts at the first direct repeat (DR1) in the terminal redundancy located at the 3′ end of the pregenomic RNA. Reverse transcription is mediated by the product of the P open reading frame (ORF). The region probably encoding the reverse transcriptase has been identified by amino acid sequence similarity of the translated sequence to retroviral enzymes[85,87] and by genetic analysis,[86] but is not yet characterized biochemically. An RNase H activity probably removes the transcribed RNA from the DNA/RNA hybrid.[87] Synthesis of DNA plus strand is primed by a capped 17-base oligoribonucleotide originating from the 5′ end of the pregenomic RNA.[83]

This oligoribonucleotide is transposed to DR2 on the minus-strand DNA, where DNA plus-strand synthesis is initiated. DNA plus-strand synthesis is not completed to cccDNA in virions. A small percentage of virions harbor DNA plus-strands of only half-genome length, resulting in the typical partially single-stranded DNA molecule. For the generation of cccDNA in cells that support viral replication, further enzymatic activities must be postulated: the genome-linked protein and the terminal redundancy in the minus strand must be removed and the open circular DNA must be ligated, after completion of the plus strand. cccDNA amplification in the productively infected hepatocyte in cell culture proceeds by a completely intracellular pathway[88] not requiring reinfection. The regulation of viral cccDNA synthesis is dependent on the presence of viral envelope proteins.[89] Mutant viruses defective in envelope synthesis but competent for DNA synthesis lead to accumulation of high levels of cccDNA in infected cells.

Three major species of unspliced, polyadenylated coterminal RNA transcripts have been identified in DHBV-infected livers: the aforementioned 3.5-kb RNA, initiating at position 2529, a 2.35-kb RNA, initiating at position 732 ± 5 and 740 ± 5, and a 2.13-kb RNA at position 985 ± 5.[90] In addition, three minor RNA species (1 to 5%) of more twice genome length were found.[90] Other minor RNAs seen in Northern blots[90] might correspond to spliced transcripts as recently described for HBV.[91-93] The function of the different RNA species for translation will be discussed in the paragraphs on the respective viral proteins. Since the life cycle of the HHBV is incompletely studied and only deduced from the sequence comparison with DHBV, it will not be separately discussed.

VIII. DNA SEQUENCE COMPARISON

The DNA sequences of DHBV isolated from German geese (DHBV1),[14] Pekin ducks in Germany (DHBV3)[8] and the U.S. (DHBV16),[94] and from Chinese brown spotted ducks (DHBV22 and DHBV26)[14,95] are aligned with the heron hepatitis B virus (HHBV4)[21] (Figure 2). The recently cloned and sequenced viral isolates from German (DHBVF1-6) and Chinese (DHBVS18-B, DHBVQ49-1) Pekin ducks[24,25] have not yet been tested for viability and are therefore not included in the comparison. DHBVF1-6 is similar in sequence to DHBV16 and DHBVS18-B is similar to DHBV22/26. DHBV1 and DHBV22 DNAs are noninfectious by intrahepatic transfection of Pekin ducklings and thus seem to represent defective viruses. The finding that two of five DHBV viruses shown in Figure 2 are defective viruses is not obvious from the gene organization. The defect may be caused by mutations in regulatory elements or by mutations affecting structurally or enzymatically important protein domains. The percentage of defective viral genomes among other cloned hepadnaviruses is not easy to test and therefore unknown but might well be in a similar range. An example of a high percentage of HBV virions defective in gene organization in a chronic carrier has recently been published.[96] The DHBV genomes cloned from Chinese isolates have in common a genome three nucleotides longer than the German isolates. The alignment (Figure 2) shows positional and base conservation of the direct repeats DR1 and DR2, which are important for replication of the DHBVs. The HHBV4 DRs are also conserved, but the distance between DR1 and DR2 is 45 instead of the 46 nucleotides found in DHBV. If HHBV is a viable genome, this difference affects the precore amino acid sequence (five amino acid exchanges at the amino terminus and predictably the structure of the replicative intermediates).[21] The same difference in distance between the direct repeats has been verified in five other HHBV clones and appears therefore characteristic for HHBV. Most sequence diversity between DHBVs and HHBV is found in the preS region and the overlapping P gene region. TATA boxes upstream of the pregenome/C mRNA and the preS mRNA initiation sites are conserved. No TATA box or sequence similar to the simian virus 40 late promoter, as suggested for HBV, is found in DHBVs or HHBV upstream of the (for HHBV putative) start of the 2.13-kb RNA thought to be responsible for synthesis of the major surface antigen. In contrast to mammalian hepadnaviruses, which appear to produce little preS1 mRNA in infected tissue (the corresponding mRNAs have not been identified for GSHV, and the data for HBV and WHV are ambiguous), the corresponding DHBV preS transcript is produced in large amounts in infected liver (ratio of pregenome to preS to S RNAs is ≈1:1:1).[90] A downstream sequence element that seems to be responsible for downregulation of the preS1 mRNA transcription in HBV[97] is therefore not functional or is absent in DHBV. An enhancer consensus motif (AGTGTTTGCT) is conserved between most DHBV isolates compared (position 2218 in DHBV3, Figure 2), but shows one base exchange in DHBVF1-6 and in HHBV. It enhances transcription from the core promoter as shown by chloramphenicol acetyltransferase (CAT) assays.[43] The AATAAA sequence involved in polyadenylation and processing of the coterminal RNAs is conserved among the DHBVs as well as in HHBV. Further conserved base sequences probably important for structural and functional integrity of the deduced viral proteins will be discussed in Section IX.

IX. PROTEINS

A. SURFACE PROTEINS

As in the other hepadnaviruses, DHBV and probably also HHBV encode all known envelope proteins in one open reading frame (preS/S ORF). PreS and S proteins are amino-terminally distinct but share a common carboxyl-terminal sequence (Figures 2 and 3). In contrast to the mammalian hepadnaviruses, only two major DHBV envelope proteins (of 36 and 17 kDa) have been characterized in detail.[98-101] There is, however, accumulating evidence for the existence of further envelope proteins. The preS proteins in DHBV are probably synthesized from the 2.35-kb RNA and the S, or major surface protein, from the 2.13-kb RNA.[90]

Two major species of proteins of 36 and 28 kDa from the preS region are consistently found in DHBV-infected liver. In DHBV-positive serum, the 28-kDa protein is not always seen and is sometimes seen only in small quantities.[44,102] In addition, minor bands of 37 to 38, 33, and 30 kDa are identified by immunoblot analysis of infected liver and serum. The weak 37- to 38-kDa band could represent myristilated forms. The start of translation for the 36-kDa preS protein is at the conserved AUG at position 801, as determined by mutation analysis and expression studies in cell culture.[44] The first AUG in the ORF at position 693 is probably not used for preS protein synthesis based on the following observations: (1) it is not conserved in DHBV22 (Figures 2 and 3); (2) a downstream stop codon is found in DHBVF1-6;[24] (3) mutation of the U at position 694 to a C does not affect the viability of the virus nor the size or amount of the preS proteins in transfected duckling liver or cell culture;[44] and (4) a preS mRNA has not been identified that initiates upstream of the AUG at position 693.[90] The second AUG speculated as a possible site for initiation of preS protein synthesis at position 798 is present only in DHBV16 and DHBVF1-6. Mutants with a stop codon UAG at position 801 or an ACG instead of the AUG are not viable in the duckling, and we therefore assume that the AUG at position 801 is the start site of preS translation.[44] Several other mutants at downstream AUGs located at positions 825, 882, and 957 did not influence virus viability and size of the major p36 and p28 preS proteins in the livers of transfected ducklings.[44] These data also infer that the 28-kDa protein is probably not a protein derived from a distinct start of translation, but a proteolytic product of the 36-kDa protein, or a product of an as yet unidentified spliced mRNA.

The preS proteins are the most heterogeneous proteins among the avian hepadnaviruses (e.g., 50.3% amino acid differences between DHBV3 and HHBV in the preS region). They possibly mediate virus–cell attachment. Given the high variation in primary sequence found in natural isolates of DHBV and HHBV (gaps must be introduced in the preS region to align the preS regions; Figures 2 and 3), it is not too astonishing to find that deletions and insertions can be made at some points of the preS ORF without affecting virus viability.[103] In this report,[103] ducklings were transfected with DHBV DNA in which deletion of a restriction site in the overlapping preS/P gene region interrupted both reading frames. The only viable mutants in this experiment had either wild-type sequence or an insertion that restored the continuity of both reading frames.[103] This further supports the notion that continuity of the preS and P gene frames appear essential to virus viability. Homologous recombination between transfected DHBV DNA molecules has been described,[104] but the mechanism for both the generation of a frame-restoring insertion and the regeneration of the wild-type genotype in this experiment[103] is unclear.

Some preS protein sequences are exposed at the virus-surface and can be recognized by virus-neutralizing antibodies.[54,55] We have identified a conserved preS epitope (among DHBV isolates, not HHBV) that is recognized by a virus-neutralizing monoclonal antibody raised in mice.[54] Several other virus-neutralizing monoclonal antibodies recognizing different epitopes are also directed against preS proteins, but the mapping of the preS amino acid sequences corresponding to these epitopes is not complete.[55]

The major surface antigen is synthesized from the AUG at position 1284 and has an apparent molecular mass of 17 kDa.[99] SAg from DHBV and HHBV viremic sera comigrate in sodium dodecyl sulfate polyacrylamide gel electrophoresis (SDS-PAGE). With the nucleocapsid protein, it is the best conserved polypeptide sequence among DHBV/HHBV isolates (16.8% amino acid exchanges between the S amino acid sequences of DHBV3 and HHBV4). In contrast to the mammalian hepadnaviruses, pre-S and S proteins of DHBV are not glycosylated. S protein also constitutes most of the mass of the empty viral envelopes found in abundance in the sera of DHBV-infected animals.

B. e ANTIGEN AND CORE PROTEINS

The nucleocapsid or core proteins of DHBV (DHBcAg) and HHBV are highly conserved (Figure 4) and have an apparent molecular mass of 32 kDa. In the precore/core ORF, as in the preS/S ORF, the viruses have the same compact organization, utilizing more than one in-frame AUG for the synthesis of distinct proteins from one ORF. Synthesis of DHBcAg initiates at the second AUG of the C gene, the third AUG on the C mRNA.[105] A translational start at the precore AUG at position 2518 leads to the synthesis of a precore protein that is glycosylated, processed at the amino and carboxyl termini, and secreted into blood or tissue culture supernatant as DHBeAg.[106] A minor messenger RNA initiating upstream of the precore AUG at position 2518 has been identified *in vivo* from DHBV-positive livers that present in amounts of less than 5% of the pregenomic RNA.[43] However, when the precore region alone is placed upstream of a CAT gene and transfected into HepG2 cells, a major RNA initiating at position 2450/2451 can be detected by primer extension,[43] indicating a strong precore promoter. This RNA species was not detected when the core promoter was simultaneously present on the same construct. A similar negative transcriptional interference phenomenon has been noted for the HBV preS promoter.[97] Since DHBeAg is expressed *in vivo* and in cell culture, we assume that it is efficiently translated from this minor RNA. In addition, DHBeAg could be synthesized by internal initiation of translation on one of the twice genome-length RNAs. Synthesis of DHBeAg seems conserved among the virus isolates; the precore and core ORF AUGs of several molecular clones, including the cloned HHBV genome, are conserved. Yet when synthesis of DHBeAg is artificially abolished by introduction of a stop codon or a frameshift mutation in the precore region or by mutation of the precore AUG, virus viability and titers seem unaffected, both *in vivo* and in cell culture,[43,106,107] and both the physical mutations and the DHBeAg-negative phenotype are stably transmitted (over six animal passages in our hands).[43] An artificial DHBeAg-negative virus mutant[107] is transmitted vertically without apparent differences in titer as compared to wild-type virus.[108] DHBcAg, a partially phosphorylated[105,109] protein with a heterogeneous migration behavior in SDS-PAGE between 32 and 34 kDa, is translated from the pregenomic/C mRNA. The heterogeneity is removed by phosphatase treatment.[109] The primary sequence of core antigen is highly conserved among all avian hepadnaviruses (16.5% divergence between DHBV3 and HHBV). Short parts of the amino terminus, the middle, and the carboxyl terminus are conserved among avian and mammalian hepadnaviruses. DHBcAg is considerably larger than the mammalian core antigens (262 versus 183 amino acids). Copies of the same molecule (180) assemble to form the viral nucleocapsid. DHBcAg has a spikey appearance, possibly due to a 39-amino acid insert located in an external loop displaying the core particle antigenicity in HBV.[110,111] As this sequence diversity in the external loop, predicted by homology with picornavirus nucleocapsid proteins, seems not to interfere with the nucleocapsid assembly function, we have used this region in HBV core particles to express heterologous epitopes on the surface of core particles.[112] DHBcAg is phosphorylated.[105,109] When the 12 carboxyl-terminal amino acids are removed by the introduction of a stop codon, the resulting mutants are still phosphorylated, are able to package viral RNA, assemble into particles, and replicate in cell culture, but are nonviable *in vivo*.[105] Removal of the last 36 C-terminal amino acids abolishes phosphorylation and protein heterogeneity but not particle formation and packaging of viral RNA measured after transient expression in HepG2 cells.[105] DHBcAg produced in *Escherichia coli* is not phosphorylated, and mutations in the P ORF do not interfere with phosphorylation. Phosphorylation is therefore likely to be catalyzed by cellular enzymes. It has not been shown whether phosphorylation of DHBcAg is important in the viral life cycle.

The signals responsible for encapsidation of viral nucleic acid and P protein(s) are incompletely understood. Mutant viral DNAs unable to direct the synthesis either of core gene products or of P gene products can complement these functions in trans upon cotransfection into hepatoma cells,[86,113] leading to the formation of core particles with partially double-stranded DNA that can be filled in by the endogenous polymerase. This implies that the signals for packaging and translation of the P gene must be independent. Pregenomic RNA seems preferentially packaged compared to preC mRNA in GSHV[114] and recent data indicate that a short sequence in the precore region and the absence of the precore AUG are crucial for encapsidation of the HBV pregenomic RNA.[115] As mentioned above, artificial DHBeAg-negative DHBV mutants with a mutated precore AUG are viable *in vivo* and *in vitro* without apparent differences in replication from wild-type virus.[43] We have recently identified an HBV mutant in a chronic HBeAg-negative, anti-HBe-positive HBV carrier with a mutation of the precore AUG.[116] A system has been presented in which DHBcAg is translated from a mRNA with a heterologous 5′ end that is itself not or inefficiently packaged and thus also allows analysis of packaging functions on other transcripts simultaneously provided in *trans*.[117] Transducible, but replication-defective, DHB viruses were generated by transfecting HuH7 with cloned DHBV DNA in which the direct repeat 1 was deleted.[118] These replication-defective viruses were capable of rescuing other mutants with mutations in viral genes. A mutant DNA encoding truncated core and envelope genes was only poorly rescued and interfered with the synthesis of wild-type DNA from an infectious virus.[118]

C. PRODUCTS OF THE P OPEN READING FRAME

The longest open reading frame in the hepadnavirus genomes is assumed to code for the genome-linked protein, the viral reverse transcriptase, and the RNase H (Figure 4). Apart from the genome-linked protein,[77,78] no protein product of the DHBV or HHBV gene has so far been identified from DHBV infected cells. Thus, it is not clear whether the P translation product must be proteolytically processed to release the diverse enzymatic activities it is endowed with, or whether it is operative as a single long polypeptide. Failure to purify an enzymatically active polymerase protein from virions or infected cells may be due to the special structural requirements of the hepadnaviral transcription complex in core particles and to its low abundance.[119] Recently it has been found that antisera raised to the amino terminus of the P gene product expressed in heterologous hosts precipitate the viral DNA minus strand.[80,81] The genome-linked protein thus seems to be derived from the P ORF. The P ORF reading frame is assumed to code for the viral reverse transcriptase based on amino acid sequence similarities between retroviral reverse transcriptases and hepadnaviral P gene products.[85] Later a homology was also found between retroviral and *E. coli* RNase H and a sequence at the carboxyl terminus of the putative P products, suggesting a similar organization of the P genes of some retroviruses, reverse transcribing elements, and hepadnaviruses (5′-[integrase]-reverse transcriptase-RNase H-3′).[87] These homology data were confirmed by genetic analysis for HBV, since frameshift mutations in the 3′ end of the P ORF seem to abolish second-strand (DNA plus strand) synthesis.[115] The region between the reverse transcriptase homology and the RNase H homology is not conserved between the avian hepadnaviruses and is believed to serve a tether function without encoding enzymatic activity.[13,87] It was originally speculated that hepadnaviruses synthesized their P ORF products by translational frameshifting from the core gene into the polymerase ORF similar to some retroviruses. Mutational analysis of the DHBV core-P overlap has established that neither spliced messages nor translational frameshifting are necessary for the production of the P gene products and that both core and polymerase polypeptides are vital for the viral life cycle.[86,113,120] A P gene product containing all functional domains is required as a structural compound for encapsidation of the pregenomic RNA.[121-123] Both core and P functions can be complemented in trans when mutants are cotransfected that are defective in either core or P production (Chapter 8, this volume). Neither a P-specific mRNA nor a spliced mRNA fusing the core gene to the P gene has been identified. Initiation of P synthesis appears to be cap dependent[124] and is initiated internally at the AUG at position 170 or further downstream on the C mRNA or preC mRNA. An AUG codon upstream of position 170, conserved among avian hepadnaviruses, was already improbable as a start of translation as there is a stop codon after 29 amino acids in the viable DHBV3 clone (Figure 2).

D. X OPEN READING FRAME

An X ORF found in the mammalian hepadnaviruses is absent from the genome of avian hepadnaviruses. A complete DHBV genome, in contrast to complete HBV, WHV, or GSHV genomes, was not able to activate in *trans* the transcription of an indicator gene placed under Rous sarcoma virus long terminal repeat (RSV LTR) control and cotransfected into a cell line (HuH7) that supports replication of DHBV.[125] Thus, the notion that a truncated functional X gene is to be found elsewhere in the DHBV genome[126] is not supported by experimental evidence.

X. ANTIVIRAL AGENTS

The influence of pharmacological agents on the life cycle of DHBV can easily be tested *in vitro* and *in vivo,* which makes the DHBV infection system attractive for such studies. An in-depth description of the various agents tested would be beyond the scope of this chapter. Suramin, foscarnet (PFA), and various nucleoside analogs have been found to interfere with viral replication *in vitro* and/or *in vivo*.[127-138] Suramin blocks infection of hepatocytes with

DHBV.[131] This property has been exploited to provide evidence that the viral cccDNA is amplified after infection by intracellular reverse transcription without requiring intra–extracellular traffic.[88]

ACKNOWLEDGMENTS

We thank H. Blum, L. Cova, R. Dixon, D. Ganem, P. Marion, B. Mason, J. Newbold, P. O'Connell, and H. Schaller for sharing unpublished information. Work in the authors' laboratory was supported by grants of the Deutsche Forschungsgemeinschaft to F.S. and H.W. and by grants of the Walter Schulz and the Wilhelm Sander Stiftung to F.S.

REFERENCES

1. **Sprengel, R. and Will, H.,** Duck hepatitis B virus, in *Virus Diseases in Laboratory and Captive Animals*, Darai, G., Ed., Martinius Nijhoff, The Hague, 1988, 363.
2. **Schödel, F., Sprengel, R., Weimer, T., Fernholz, D., Schneider, R., and Will, H.,** Animal hepatitis B viruses, *Adv. Viral Oncol.*, 8, 73, 1989.
3. **Orito, E., Mizokami, M., Ina, Y., Moriyama, E. N., Kameshima, N., Yamamoto, M., and Gojobori, T.,** Host-independent evolution and a genetic classification of the hepadnavirus family based on nucleotide sequences, *Proc. Natl. Acad. Sci. U.S.A.*, 86, 7059, 1989.
4. **Tuttleman, J. S., Pugh, J. C., and Summers, J. W.,** *In vitro* experimental infection of primary duck hepatocyte cultures with duck hepatitis B virus, *J. Virol.*, 58, 17, 1986.
5. **Pugh, J. C., Yaginuma, K., Koike, K., and Summers, J.,** Duck hepatitis B virus (DHBV) particles produced by transient expression of DHBV DNA in a human hepatoma cell line are infectious *in vitro*, *J. Virol.*, 62, 3513, 1988.
6. **Galle, P. R., Schlicht, H. J., Fischer, M., and Schaller, H.,** Production of infectious duck hepatitis B virus in a human hepatoma cell line, *J. Virol.*, 62, 1736, 1988.
7. **Hirsch, R., Colgrove, R., and Ganem, D.,** Replication of duck hepatitis B virus in two differentiated human hepatoma cell lines after transfection with cloned viral DNA, *Virology*, 167, 136, 1988.
8. **Sprengel, R., Kuhn, C., Manso, C., and Will, H.,** Cloned duck hepatitis B virus DNA is infectious in Pekin ducks, *J. Virol.*, 52, 932, 1984.
9. **Zhou, Y. Z.,** A virus possibly associated with hepatitis and hepatoma in ducks, *Shanghai Med. J.*, 3, 641, 1980.
10. **Mason, W. S., Seal, G., and Summers, J.,** Virus of Pekin ducks with structural and biological relatedness to human hepatitis B virus, *J. Virol.*, 36, 829, 1980.
11. **Mason, W. S., Taylor, J. M., Seal, G., and Summers, J.,** An HBV-like virus of domestic ducks, in *Viral Hepatitis, 1981 International Symposium*, Szmuness, W., Alter, H. J., and Maynard, J. E., Eds., Franklin Institute Press, Philadelphia, 1981, 107.
12. **Summers, J.,** Three recently described animal virus models for human hepatitis B virus, *Hepatology*, 1, 179, 1981.
13. **Sprengel, R., Kuhn, C., Will, H., and Schaller, H.,** Comparative sequence analysis of duck and human hepatitis B virus genomes, *J. Med. Virol.*, 15, 323, 1985.
14. **Sprengel, R., Schneider, R., Marion, P.L., Fernholz, D., Wildner, G., and Will, H.,** Comparative sequence analysis of defective and infectious avian hepadnavirus genomes, *Nucleic Acid Res.*, submitted, 1991.

14a. **Devereux, J., Haeberli, P., and Smithies, O.,** A comprehensive set of sequence analysis programs for the VAX, *Nucleic Acids Res.*, 12, 387, 1984.

15. **Cova, L., Hantz, O., Arliaud-Gassin, M., Chevalier, A., Berthillon, P., Boulay, J., Jacquet, C., Chomel, B., Vitvitski, L., and Trépo, C.,** Comparative study of DHBV DNA levels and endogenous DNA polymerase activity in naturally infected ducklings in France, *J. Virol. Methods*, 10, 251, 1985.
16. **Freiman, J. S. and Cossart, Y. E.,** Natural duck hepatitis B virus infection in Australia, *Aust. J. Exp. Biol. Med. Sci.*, 64, 477, 1986.

17. **Cova, L., Lambert, V., Chevallier, A., Hantz, O., Fourel, I., Jacquet, C., Pichoud, C., Boulay, J., Chomel, B., Vitvitski, L., and Trépo, C.,** Evidence for the presence of duck hepatitis B virus in wild migrating ducks, *J. Gen. Virol.*, 67, 537, 1986.
18. **Dixon, R. J., Jones, N. F., Campbell, M., Brooker, A., and Bowden, D. S.,** The Australian maned duck hepatitis B virus: a potential novel hepatitis B virus of ducks, in *Hepatitis B Viruses, Abstracts,* Marion, P., and Schaller, H., Eds., Cold Spring Harbor Laboratory, Cold Spring Harbor, New York, 1989, and personal communication, 1990.
19. **Marion, P. L., Cullen, J. M., Azcarraga, R. R., Van Davelaar, M. J., and Robinson, W. S.,** Experimental transmission of duck hepatitis B virus to Pekin ducks and to domestic geese, *Hepatology*, 7, 724, 1987.
20. **Omata, M., Yokosuka, O., Imazeki, F., Matsuyama, Y., Uchiumi, K., Ito, Y., Mori, J., and Okuda, K.,** Transmission of duck hepatitis B virus from Chinese carrier ducks to Japanese ducklings: a study of viral DNA in serum and tissue, *Hepatology*, 4, 603, 1984.
21. **Sprengel, R., Kaleta, E. F., and Will, H.,** Isolation and characterization of a hepatitis B virus endemic in herons, *J. Virol.*, 62, 3832, 1988.
22. **Sprengel, R., Wildner, G., and Will, H.,** Unpublished observations, 1990.
23. **Oexle, K., Blum, H. E., Walter, E., Offensperger, W. B., Offensperger, S., Grundmann, H., Teubner, K., and Gerok, W.,** Duck hepatitis B virus: cloning and subcloning of the viral genome, *Zb. Bakteriol. Hyg. A*, 270, 424, 1989.
24. **Mattes, F., Tong, S., Teubner, K., and Blum, H. E.** Complete nucleotide sequence of a German duck hepatitis B virus, *Nucl. Acid Res.*, 18, 6140, 1990.
25. **Tong, S., Mattes, F., Teubner, K., and Blum, H. E.** Complete nucleotide sequence of a Chinese duck hepatitis B virus, *Nucl. Acid Res.*, 18, 6139, 1990.
26. **Uchida, M., Esumi, M., and Shikata, T.,** Molecular cloning and sequence analysis of duck hepatitis B virus genomes of a new variant isolated from Shanghai ducks, *Virology*, 173, 600, 1989.
27. **O'Connell, A. P., Urban, M. K., and London, W. T.,** Naturally occurring infection of Pekin duck embryos by duck hepatitis B virus, *Proc. Natl. Acad. Sci. U.S.A.*, 80, 1703, 1983.
28. **Tsiquaye, K. N., McCaul, T. F., and Zuckerman, A. J.,** Maternal transmission of duck hepatitis B virus in pedigree Pekin ducks, *Hepatology*, 5, 622, 1985.
29. **Mason, W. S., Halpern, M. S., England, J. M., Seal, G., Egan, J., Coates, L., Aldrich, C., and Summers, J.,** Experimental transmission of duck hepatitis B virus, *Virology*, 131, 375, 1983.
30. **Urban, M. K., O'Connell, A. P., and London, W. T.,** Sequence of events in natural infection of Pekin duck embryos with duck hepatitis B virus, *J. Virol.*, 55, 16, 1985.
31. **Tagawa, M., Omata, M., Yokosuka, O., Uchiumi, K., Imazeki, F., and Okuda, K.,** Early events in duck hepatitis B virus infection, *Gastroenterology*, 89, 1224, 1985.
32. **Tagawa, M., Robinson, W. S., and Marion, P. L.,** Duck hepatitis B virus replicates in the yolk sac of developing embryos, *J. Virol.*, 61, 2273, 1987.
33. **Tagawa, M., Omata, M., and Okuda, K.,** Appearance of viral RNA transcripts in the early stage of duck hepatitis B virus infection, *Virology*, 152, 477, 1986.
34. **Halpern, M. S., England, J. M., Deery, D. T., Petcu, D. J., Mason, W. S., and Molnar-Kimber, K. L.,** Viral nucleic acid synthesis and antigen accumulation in pancreas and kidney of Pekin ducks infected with duck hepatitis B virus, *Proc. Natl. Acad. Sci. U.S.A.*, 80, 4865, 1983.
35. **Halpern, M. S., Egan, J., Mason, W. S., and England, J. M.,** Viral antigen in endocrine cells of the pancreatic islets and adrenal cortex of Pekin ducks infected with duck hepatitis B virus, *Virus Res.*, 1, 213, 1984.
36. **Halpern, M. S., England, J. M., Flores, L., Egan, J., Newbold, J., and Mason, W. S.,** Individual cells in tissues of DHBV-infected ducks express antigens cross-reactive with those on virus surface antigen particles and immature viral cores, *Virology*, 137, 408, 1984.
37. **Halpern, M. S., Egan, J., McMahon, S. B., and Ewert, D. L.,** Duck hepatitis B virus is tropic for exocrine cells of the pancreas, *Virology*, 146, 157, 1985.
38. **Tsiquaye, K. N., Rapicetta, M., McCaul, T. F., and Zuckerman, A. J.,** Experimental *in ovo* transmission of duck hepatitis B virus, *J. Virol. Methods*, 11, 49, 1985.
39. **Halpern, M. S., McMahon, S. B., Mason, W. S., and O'Connell, A. P.,** Viral antigen expression in the pancreas of DHBV-infected embryos and young ducks, *Virology*, 150, 276, 1986.
40. **Freiman, J. S., Jilbert, A. R., Dixon, R. J., Holmes, M., Gowans, E. J., Burrell, C. J., Wills, E. J., and Cossart, Y. E.,** Experimental duck hepatitis B virus infection: pathology and evolution of hepatic and extrahepatic infection, *Hepatology*, 8, 507, 1988.
41. **Cullen, J. M., Marion, P. L., and Newbold, J. E.,** A sequential histologic and immunohistochemical study of duck hepatitis B virus infection in Pekin ducks, *Vet. Pathol.*, 26, 164, 1989.

42. **Hosoda, K., Omata, M., Uchiumi, K., Imazeki, F., Yokosuka, O., Ito, Y., Okuda, K., and Ohto, M.,** Extrahepatic replication of duck hepatitis B virus: more than expected, *Hepatology*, 11, 44, 1990.
43. **Schneider, R., Fernholz, D., Wildner, G., and Will, H.,** Mechanism, kinetics and role of duck hepatitis B virus e-antigen expression *in vivo*, *Virology*, 182, 503, 1991; and unpublished observations, 1990.
44. **Fernholz, D. and Will, H.,** Unpublished observations, 1990.
45. **Jilbert, A. R., Freiman, J. S., Burrell, C. J., Holmes, M., Gowans, E. J., Rowland, R., Hall, P., and Cossart, Y. E.,** Early virus cell interactions in duck hepatitis B virus infection, in *Viral Hepatitis and Liver Disease*, Zuckermann, A., Ed., Alan R. Liss, New York, 1988, 497.
46. **Jilbert, A. R., Freiman, J. S., Gowans, E. J., Holmes, M., Cossart, Y. E., and Burrell, C. J.,** Duck hepatitis B virus DNA in liver, spleen, and pancreas: analysis by *in situ* and Southern blot hybridization, *Virology*, 158, 330, 1987.
47. **Uchida, T., Suzuki, K., Esumi, M., Arii, M., Oomura, M., and Shikata, T.,** Occurrence and ultrastructural localization of duck hepatitis B virus in the liver of ducks after experimental infection, *Hepatology*, 7, 29, 1987.
48. **Fukuda, R., Okinaga, S., Akagi, S., Hidaka, M., Ono, N., Fukumoto, S., and Shimada, Y.,** Alteration of infection pattern of duck hepatitis B virus by immunomodulatory drugs, *J. Med. Virol.*, 26, 387, 1988.
49. **Halpern, M. S., Mason, W. S., Coates, L., O'Connell, A. P., and England, J. M.,** Humoral immune responsiveness in duck hepatitis B virus-infected ducks, *J. Virol.*, 61, 916, 1987.
50. **McCaul, T. F., Tsiquaye, K. N., and Zuckerman, A. J.,** Studies by electron microscopy on the assembly of duck hepatitis B virus in the liver, *J. Med. Virol.*, 16, 77, 1985.
51. **Vickery, K., Freiman, J. S., Dixon, R. J., Kearney, R., Murray, S., and Cossart, Y. E.,** Immunity in Pekin ducks experimentally and naturally infected with duck hepatitis B virus, *J. Med. Virol.*, 28, 231, 1989.
52. **Chenng, R. C., Trujillo, D. E., Robinson, W. S., Greenberg, H. B., and Marion, P. L.,** Epitope-specific antibody response to the surface antigen of duck hepatitis B virus in infected ducks, *Virology*, 176, 546, 1990.
53. **Halpern, M. S., Mason, W. S., and England, J. M.,** Viral antigen accumulation in kidney glomeruli of duck hepatitis B virus infected ducks, in *UCLA Symp. Molecular and Cellular Biology, New Series Hepadna Viruses*, Robinson, W., Koike, K., and Will, H., Eds., Alan R. Liss, New York, 1987, 411.
54. **Lambert, V., Fernholz, D., Sprengel, R., Fourel, I., Deléage, G., Wildner, G., Peyret, C., Trépo, C., Cova, L., and Will, H.,** Virus neutralizing monoclonal antibody to a conserved epitope on the duck hepatitis B virus pre-S protein, *J. Virol.*, 64, 1290, 1990.
55. **Cheung, R. C., Robinson, W. S., Marion, P. L., Greenberg, H. B.,** Epitope mapping of neutralizing monoclonal antibodies against duck hepatitis B virus. *J. Virol.*, 63, 2445, 1989.
56. **Milich, D. R.,** T- and B-cell recognition of hepatitis B viral antigens, *Immunol. Today*, 9, 380, 1989.
57. **Jilbert, A. R., Freiman, J. S., Burrell, C. J., Holmes, M., Gowans, E. J., Rowland, R., Hall, P., and Cossart, Y. E.,** Virus-liver cell interactions in duck hepatitis B virus infection, *Gastroenterology*, 95, 1375, 1988.
58. **Marion, P. L., Knight, S. S., Ho, B.-K., Guo, Y.-Y., Robinson, W. S., and Popper, H.,** Liver disease associated with duck hepatitis B virus infection of domestic ducks, *Proc. Natl. Acad. Sci. U.S.A.*, 81, 898, 1984.
59. **Fukuda, R., Fukumoto, S., and Shimada, Y.,** A sequential study of viral DNA in serum in experimental transmission of duck hepatitis B virus, *J. Med. Virol.*, 21, 311, 1987.
60. **Omata, M., Uchiumi, K., Ito, Y., Yokosuka, O., Mori, J., Terao, K., Wei-Fa, Y., O'Connell, A. P., London, W. T., and Okuda, K.,** Duck hepatitis B virus and liver diseases, *Gastroenterology*, 85, 260, 1983.
61. **Uchida, T., Suzuki, K., Arii, M., Shikata, T., Fukuda, R., and Tao, Y.,** Geographical pathology of duck livers infected with duck hepatitis B virus from Chiba and Shimane in Japan and Shanghai in China, *Cancer Res.*, 48, 1319, 1988.
62. **Imazeki, F., Yaginuma, K., Omata, M., Okuda, K., Kobayashi, M., and Koike, K.,** Integrated structures of duck hepatitis B virus DNA in hepatocellular carcinoma, *J. Virol.*, 62, 861, 1988.
63. **Yokosuka, O., Omata, M., Zhou, Y.-Z., Imazeki, F., and Okuda, K.,** Duck hepatitis B virus DNA in liver and serum of Chinese ducks: integration of viral DNA in a hepatocellular carcinoma, *Proc. Natl. Acad. Sci. U.S.A.*, 82, 5180, 1985.
64. **Uchida, T., Suzuki, K., Esumi, M., Arii, M., and Shikata, T.,** Influence of aflatoxin B_1 intoxication on duck livers with duck hepatitis B virus infection, *Cancer Res.*, 48, 1559, 1988.
65. **Cova, L., Wild, C. P., Mehrotra, R., Turusov, V., Shirai, T., Lambert, V., Jacquet, C., Tomatis, L. Trépo, C., and Montesano, R.,** Contribution of aflatoxin B_1 and hepatitis B virus infection in the induction of liver tumors in ducks, *Cancer Res.*, 50, 2156, 1990.
66. **England, J. M. and Halpern, M. S.,** The arginine-stimulated insulin response is impaired in congenital duck hepatitis B virus infection. *Virology*, 165, 299, 1988.

67. **Tuttleman, J. S., Pourcel, C., and Summers, J.,** Formation of the pool of covalently closed circular viral DNA in hepadnavirus-infected cells, *Cell*, 47, 451, 1986.
68. **Pugh, J. C. and Summers, J. W.,** Infection and uptake of duck hepatitis B virus by duck hepatocytes maintained in the presence of dimethyl sulfoxide, *Virology*, 172, 564, 1989.
69. **Fourel, I., Gripon, P., Hantz, O., Cova, L., Lambert, V., Jacquet, C., Watanabe, K., Fox, J., Guillouzo, C., and Trépo, C.,** Prolonged duck hepatitis B virus replication in duck hepatocytes cocultivated with rat epithelial cells: a useful system for antiviral testing, *Hepatology*, 10, 186, 1989.
70. **O'Connell, A. P. and London, W. T.,** Duck hepatitis B virus in yolk sac endoderm and in cultured yolk sac cells, in *Viral Hepatitis and Liver Disease,* Alan R. Liss, New York, 1988, 486.
71. **Kawaguchi, T., Nomura, K., Hirayama, Y., and Kitagawa, T.,** Establishment and characterization of a chicken hepatocellular carcinoma cell line, LMH, *Cancer Res.*, 47, 4460, 1987.
72. **Condreay, L. D., Aldrich, C., Coates, L., Mason, W. S., and Wu, T.-T.,** Efficient duck hepatitis B virus production by an avian liver tumor cell line, *J. Virol.*, 64, 3249, 1990.
73. **Uchida, T., Suzuki, K., Okuda, Y., and Shikata, T.,** *In vitro* transmission of duck hepatitis B virus to primary duck hepatocyte cultures, *Hepatology*, 8, 760, 1988.
74. **Neurath, A. R., Kent, S. B. H., Strick, N., and Parker, K.,** Identification and chemical synthesis of a host cell receptor binding site on hepatitis B virus, *Cell*, 46, 429, 1986.
75. **Petit, M.-A., Dubanchet, S., and Capel, F.,** A monoclonal antibody specific for the hepatocyte receptor binding site on hepatitis B virus, *Mol. Immunol.*, 26, 531, 1989.
76. **Heermann, K. H., Goldmann, U., Schwartz, W., Seyffarth, T., Baumgarten, H., and Gerlich, W. H.,** Large surface proteins of hepatitis B virus containing the pre-S sequence, *J. Virol.*, 52, 396, 1984.
77. **Mason, W. S., Aldrich, C., Summers, J., and Taylor, J. M.,** Asymmetric replication of duck hepatitis B virus DNA in liver cells: free minus-strand DNA, *Proc. Natl. Acad. Sci. U.S.A.*, 79, 3997, 1982.
78. **Molnar-Kimber, K. L., Summers, J. W., and Mason, W. S.,** Mapping of the cohesive overlap of duck hepatitis B virus DNA and of the site of initiation of reverse transcription, *J. Virol.*, 51, 181, 1984.
79. **Molnar-Kimber, K. L., Summers, J., Taylor, J. M., and Mason, W. S.,** Protein covalently bound to minus-strand DNA intermediates of duck hepatitis B virus, *J. Virol.*, 45, 165, 1983.
80. **Bartenschlager, R. and Schaller, H.,** The amino-terminal domain of the hepadnaviral P-gene encodes the terminal protein (genome-linked protein) believed to prime reverse transcription, *EMBO J.*, 7, 4185, 1988.
81. **Bosch, V., Bartenschlager, R., Radziwill, G., and Schaller, H.,** The duck hepatitis B virus P-gene codes for protein strongly associated with the 5′-end of the viral DNA minus strand, *Virology*, 166, 475, 1988.
82. **Lien, J. M., Petcu, D. J., Aldrich, C. E., and Mason, W. S.,** Initiation and termination of duck hepatitis B virus DNA synthesis during virus maturation, *J. Virol.*, 61, 3832, 1987.
83. **Lien, J.-M., Aldrich, C. E., and Mason, W. S.,** Evidence that a capped oligoribonucleotide is the primer for duck hepatitis B virus plus-strand DNA synthesis, *J. Virol.*, 57, 229, 1986.
84. **Summers, J. and Mason, W. S.,** Replication of the genome of a hepatitis B-like virus by reverse transcription of an RNA intermediate, *Cell*, 29, 403, 1982.
85. **Toh, H., Kikuno, R., Hayashida, H., Miyata, T., Kugimiya, W., Inouye, S., Yuki, S., and Saigo, K.,** Close structural resemblance between putative polymerase of a *Drosophila* transposable genetic element 17.6 and pol gene product of Moloney murine leukaemia virus, *EMBO J.*, 4, 1267, 1985.
86. **Schlicht, H. J., Radziwill, G., and Schaller, H.,** Synthesis and encapsidation of duck hepatitis B virus reverse transcriptase do not require formation of core-polymerase fusion proteins, *Cell*, 56, 85, 1989.
87. **Schödel, F., Weimer, T., Will, H., and Sprengel, R.,** Amino acid sequence similarity between retroviral and *E. coli* RNaseH and hepadnaviral gene products, *AIDS Res. Human Retrovir.*, 4, 9, 1988.
88. **Wu, T.-T., Coates, L., Aldrich, C. E., Summers, J., and Mason, W. S.,** In hepatocytes infected with duck hepatitis B virus, the template for viral RNA synthesis is amplified by an intracellular pathway, *Virology*, 175, 255, 1990.
89. **Summers, J., Smith, P. M., and Horwich, A. L.,** Hepadnavirus envelope proteins regulate covalently closed circular DNA amplification, *J. Virol.*, 64, 2819, 1990.
90. **Büscher, M., Reiser, W., Will, H., and Schaller, H.,** Transcripts and putative RNA pregenome of duck hepatitis B virus: implications for reverse transcription, *Cell*, 40, 717, 1985.
91. **Chen, P.-J, Chen, C.-R, Sung, J.-L., and Chen, D.-S.,** Identification of a doubly spliced viral transcript joining the separated domains for putative protease and reverse transcriptase of hepatitis B virus, *J. Virol.*, 63, 4165, 1989.
92. **Su, T.-S., Lai, C. J., Huang, J.-L., Lin, L.-H., Yauk, Y.-K., Chang, C., Lo, S. J., and Han, S. H.,** Hepatitis B virus transcript produced by RNA splicing, *J. Virol.*, 63, 4011, 1989.
93. **Suzuki, T., Masui, N., Kajino, K., Saito, I., and Miyamura, T.,** Detection and mapping of spliced RNA from transfected hepatitis B virus genome in a human hepatoma cell line, *Proc. Natl. Acad. Sci. U.S.A.*, 86, 8422, 1989.

94. **Mandart, E., Kay, A., and Galibert, F.,** Nucleotide sequence of a cloned duck hepatitis B virus genome: comparison with woodchuck and human hepatitis B virus sequences, *J. Virol.*, 49, 782, 1984.
95. **Marion, P.,** Unpublished observations, 1990.
96. **Okamoto, H., Tsuda, F., and Mayumi, M.,** Defective mutants of hepatitis B virus in the circulation of symptom-free carriers, *Jpn. J. Exp. Med.*, 57, 217, 1987.
97. **Bulla, G. A. and Siddiqui, A.,** The hepatitis B virus enhancer modulates transcription of the hepatitis B virus surface antigen gene from an internal location, *J. Virol.*, 62, 1437, 1988.
98. **Pugh, J. C., Sninsky, J. J., Summers, J. W., and Schaeffer, E.,** Characterization of a pre-S polypeptide on the surfaces of infectious avian hepadnavirus particles, *J. Virol.*, 61, 1384, 1987.
99. **Marion, P. L., Knight, S. S., Feitelson, M. A., Oshiro, L. S., and Robinson, W. S.,** Major polypeptide of duck hepatitis B surface antigen particles, *J. Virol.*, 48, 534, 1983.
100. **Feitelson, M. A., Marion, P. L., and Robinson, W. S.,** The nature of polypeptides larger in size than the major surface antigen components of hepatitis B virus and like viruses in ground squirrels, woodchucks and ducks, *Virology*, 130, 76, 1983.
101. **Schlicht, H.-J., Kuhn, C., Guhr, B., Mattaliano, R. J., and Schaller, H.,** Biochemical and immunological characterization of the duck hepatitis B virus envelope proteins, *J. Virol.*, 61, 2280, 1987.
102. **Yokosuka, O., Omata, M., and Ito, Y.,** Expression of pre-S1, pre-S2, and C proteins in duck hepatitis B virus infection, *Virology*, 167, 82, 1988.
103. **Li, J.-S., Cova, L., Buckland, R., Lambert, V., Deléage, G., and Trépo, C.,** Duck hepatitis B virus can tolerate insertion, deletion, and partial frameshift mutation in the distal pre-S region, *J. Virol.*, 63, 4965, 1989.
104. **Sprengel, R., Varmus, H. E., and Ganem, D.,** Homologous recombination between hepadnaviral genomes following *in vivo* DNA transfection: implications for studies of viral infectivity, *Virology*, 159, 454, 1987.
105. **Schlicht, H. J., Bartenschlager, R., and Schaller, H.,** The duck hepatitis B virus core protein contains a highly phosphorylated C terminus that is essential for replication but not for RNA packaging, *J. Virol.*, 63, 2995, 1989.
106. **Schlicht, H. J., Salfeld, J., and Schaller, H.,** The duck hepatitis B virus pre-C region encodes a signal sequence which is essential for synthesis and secretion of processed core proteins but not for virus formation, *J. Virol.*, 61, 3701, 1987.
107. **Chang, C., Enders, G., Sprengel, R., Peters, N., Varmus, H. E., and Ganem, D.,** Expression of the precore region of an avian hepatitis B virus is not required for viral replication, *J. Virol.*, 61, 3322, 1987.
108. **Newbold, J.,** Personal communication, 1990.
109. **Pugh, J. C., Zweidler, A., and Summers, J.,** Characterization of the major duck hepatitis B virus core particle protein, *J. Virol.*, 63, 1371, 1989.
110. **Salfeld, J., Pfaff, E., Noah, M., and Schaller, H.,** Antigenic determinants and functional domains in core antigen and e antigen from hepatitis B virus, *J. Virol.*, 63, 798, 1989.
111. **Argos, P. and Fuller, S. D.,** A model for the hepatitis B virus core protein: prediction of antigenic sites and relationship to RNA virus capsid proteins, *EMBO J.*, 7, 819, 1988.
112. **Schödel, F., Weimer, T., Will, H., and Milich, D.,** Recombinant hepatitis B virus (HBV) core particles carrying immunodominant B-cell epitopes of the HBV pre-S(2) region, in *Vaccines 90*, Brown, F., Chanock, R., Ginsberg, R., and Lerner, R., Eds., Cold Spring Harbor Laboratory, Cold Spring Harbor, New York, in press.
113. **Chang, L.-J., Pryciak, P., Ganem, D., and Varmus, H. E.,** Biosynthesis of the reverse transcriptase of hepatitis B viruses involves *de novo* translational initiation not ribosomal frameshifting, *Nature (London)*, 337, 364, 1989.
114. **Enders, G. H., Ganem, D., and Varmus, H. E.,** 5′-terminal sequences influence the segregation of ground squirrel hepatitis virus RNAs into polyribosomes and viral core particles, *J. Virol.*, 61, 35, 1987.
115. **Junker-Niepmann, M., Bartenschlager, R., and Schaller, H.,** A short *cis*-acting sequence is required for hepatitis pregenome encapsidation and sufficient for packaging of foreign RNA, *EMBO J.*, 9, 3389, 1990.
116. **Raimondo, G., Schneider, R., Stemler, M., Smedile, B., Podino, H., and Will, H.,** A new hepatitis B virus variant in a chronic carrier with multiple episode of viral reactivation and acute hepatitis, *Virology*, 178, 64, 1990.
117. **Lavine, J., Hirsch, R., and Ganem, D.,** A system for studying the selective encapsidation of hepadnavirus RNA, *J. Virol.*, 63, 4257, 1989.
118. **Horwich, A. L., Furtak, K., Pugh, J., and Summers, J.,** Synthesis of hepadnavirus particles that contain replication defective duck hepatitis B virus genomes in cultured Huh7 cells, *J. Virol.*, 64, 642, 1990.
119. **Radziwill, G., Zentgraf, H., Schaller, H., and Bosch, V.,** The duck hepatitis B virus DNA polymerase is tightly associated with the viral core structure and unable to switch to an exogenous template, *Virology*, 163, 123, 1988.

120. **Radziwill, G., Tucker, W., and Schaller, H.,** Mutational analysis of the hepatitis B virus P gene product: domain structure and RNase H activity, *J. Virol.*, 64, 613, 1990.
121. **Bartenschlager, R., Junker-Niepmann, M., and Schaller, H.,** The P gene product of hepatitis B virus is required as a structural component for genomic RNA encapsidation, *J. Virol.*, 64, 5324, 1990.
122. **Chang, L.-J., Hirsch, R. C., Ganem, D., and Varmus, H. E.,** Effects of insertional and point mutations on the functions of the duck hepatitis B virus polymerase, *J. Virol.*, 64, 5553, 1990.
123. **Hirsch, R. C., Lavine, J. E., Chang, L.-J., Varmus, H. E., and Ganem, D.,** Polymerase gene products of hepatitis B viruses are required for genomic RNA packaging as well as for reverse transcription, *Nature*, (London) 344, 552, 1990.
124. **Chang, L.-J., Ganem, D., and Varmus, H.,** Studies on the translational mechanism of the DHBV polymerase gene, in *Hepatitis B Viruses, Abstracts,* Marion, P. and Schaller, H., Eds., Cold Spring Harbor Laboratory, Cold Spring Harbor, New York, 1989, 14.
125. **Colgrove, R., Simon, G., and Ganem, D.,** Transcriptional activation of homologous and heterologous genes by the hepatitis B virus X gene product in cells permissive for viral replication, *J. Virol.*, 63, 4019, 1989.
126. **Feitelson, M. A. and Miller, R. H,** X gene-related sequences in the core gene of duck and heron hepatitis B viruses, *Proc. Natl. Acad. Sci. U.S.A.*, 85, 6162, 1988.
127. **Fourel, I., Hantz, O., Cova, L., Allaudeen, and Trépo, C.,** Main properties of duck hepatitis B virus DNA polymerase: comparison with the human and woodchuck hepatitis B virus DNA polymerases, *Antiviral Res.*, 8, 189, 1987.
128. **Hirota, K., Sherker, A. H., Omata, M., Yokosuka, O., and Okuda, K.,** Effects of adenine arabinoside on serum and intrahepatic replicative forms of duck hepatitis B virus in chronic infection, *Hepatology*, 7, 24, 1987.
129. **Lee, B., Luo, W., Suzuki, S., Robins, M. J., and Tyrrell, D. L.,** *In vitro* and *in vivo* comparison of the abilities of purine and pyrimidine 2′,3′-dideoxynucleosides to inhibit duck hepadnavirus, *Antimicrob. Agents Chemother.*, 33, 336, 1989.
130. **Mason, W. S., Lien, J.-M., Petcu, D. J., Coates, L., London, W. T., O'Connell, A., Aldrich, C., and Custer, R. P.,** *In vivo* and *in vitro* studies on duck hepatitis B virus replication, in *Hepadna Viruses,* Alan R. Liss, New York, 1987, 3.
131. **Petcu, D. J., Aldrich, C. E., Coates, L., Taylor, J. M., and Mason, W. S.,** Suramin inhibits *in vitro* infection by duck hepatitis B virus, Rous sarcoma virus, and hepatitis Delta virus, *Virology*, 167, 385, 1988.
132. **Sherker, A. H., Hirota, K., Omata, M., and Okuda, K.,** Foscarnet decreases serum and liver duck hepatitis B virus DNA in chronically infected ducks, *Gastroenterology*, 91, 818, 1986.
133. **Suzuki, S., Lee, B., Luo, W., Tovell, D., Robins, M. J., and Tyrrell, D. L. J.,** Inhibition of duck hepatitis B virus replication by purine 2′, 3′-dideoxynucleotides, *Biochem. Biophys. Res. Commun.*, 157, 1144, 1988.
134. **Tao, P.-Z., Löfgren, B., Lake-Bakaar, D., Johansson, N. G., Datema, R., and Öberg, B.,** Inhibition of human hepatitis B virus DNA polymerase and duck hepatitis B virus DNA polymerase by triphosphates of thymidine analogs and pharmacokinetic properties of the corresponding nucleosides, *J. Med. Virol.*, 26, 353, 1988.
135. **Tsiquaye, K. and Zuckerman, A.,** Suramin inhibits duck hepatitis B virus DNA polymerase activity, *J. Hepatol.*, 1, 663, 1985.
136. **Zuckerman, A. J.,** Screening of antiviral drugs for hepadna virus infection in Pekin ducks: a review, *J. Virol. Methods*, 17, 119, 1987.
137. **Kassianides, C., Hoofnagle, J. H., Miller, R. H., Doo, E., Ford, H., Broder, S., and Mitsuya, H.,** Inhibition of duck hepatitis B virus replication by 2′,3′-dideoxycytidine, *Gastroenterology*, 97, 1275, 1989.
138. **Offensperger, W.-B., Walter, E., Offensperger, S., Zeschnigk, C., Blum, H. E., and Gerok, W.,** Duck hepatitis B virus: DNA polymerase and reverse transcriptase activities of replicative complexes isolated from liver and their inhibition *in vitro*, *Virology*, 164, 48, 1988.

Chapter 4

HEPATITIS B VIRUS TRANSGENIC MOUSE MODEL

Christine Pourcel

TABLE OF CONTENTS

I. INTRODUCTION

Hepatitis B virus (HBV) belongs to a family of DNA viruses, the hepadnaviruses, that preferentially infect the liver and are responsible for disease states ranging from acute hepatitis to chronic hepatitis, cirrhosis, and hepatocellular carcinoma. Other members are the woodchuck hepatitis virus (WHV), the ground squirrel hepatitis virus (GSHV), and the duck hepatitis virus (DHBV).[1] They have a very restricted host specificity; HBV infects only man and chimpanzee. Since the discovery of the animal hepadnaviruses, the study of many aspects of the biology of the virus has been possible. The viral replication mechanism was described in DHBV-infected ducks[2] and then confirmed with the other viruses. A correlation between WHV infection and the development of hepatocellular carcinoma was demonstrated. The ability to infect animals with cloned DNA has also allowed some mutational analysis. For example, hybrids between WHV and GSHV genomes have been used to analyze some aspects of the restricted host range.[3] However, several questions have remained unanswered, in particular the basis for the host and tissue specificity and the pathogenesis linked to the infection.

Numerous attempts to transfect cells *in vitro* with HBV DNA have shown that only differentiated hepatocytes can support the complete viral life cycle.[4,5] The cellular receptors needed for the virus adsorption are probably absent in hepatoma cell lines, even the most differentiated lines, and constitute an essential element of the host and tissue specificity. An approach toward understanding the tissue specificity of viral synthesis is the search for viral cis-acting sequences that would be activated by factors present in specific cells. Enhancer sequences have been located in HBV but it is not clear from cell culture experiments whether they require trans-activation by factors present in the human liver only or if factors present in other tissues or in other species can also efficiently trans-activate transcription through these sequence elements.[6,7] The detection of viral DNA and proteins in other organs in man (lymphocytes, kidney, pancreas), and even of replicative intermediates in spleen of woodchucks and kidney and pancreas of ducks, shows that the virus can penetrate and multiply in nonliver cells.

In order to develop a better animal model for the study of the pathogenesis linked to HBV infection, attempts have been made to introduce the cloned HBV DNA into mice. Feitelson et al.[8] injected HBV into nude mice and observed injuries resembling hepatitis but there was no evidence of replication. However, when DNA is injected into the liver, only a limited number of cells become transfected and the DNA is not efficiently maintained in the absence of replication. Newly synthesized virions are not able to reinfect murine hepatocytes and therefore do not produce a detectable level of virus multiplication. The development of the transgenic mouse technology at the beginning of the 1980s introduced a new system of analysis in this animal.[9] In these mice, the DNA injected into the mouse egg at the one-cell stage integrates into the cellular genome and can be transmitted to the following generations, usually without rearrangement. The mouse has in each of its cells the new genetic material, and it is possible to analyze the expression of the foreign DNA in different tissues and in response to host factors (sex, age) or exogenous substances given to the animal.

II. PRODUCTION OF THE HBV-TRANSGENIC MICE

A. THE TRANSGENIC MOUSE SYSTEM

A mouse is transgenic when it is able to transmit to its progeny new genetic material, the transgene, in a manner identical to the endogenous genes in its chromosomes. Thus, the DNA must be introduced into the mouse egg and become integrated before differentiation of the

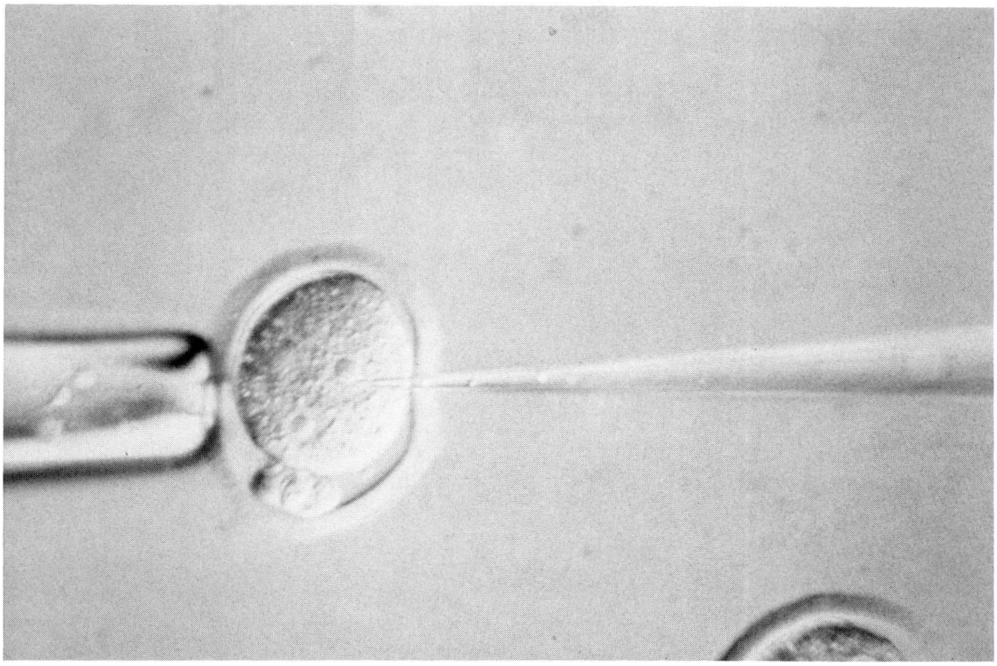

FIGURE 1. Microinjection of DNA into a pronucleus of a one-cell mouse egg. (From Charles Babinet, Institut Pasteur, Paris, France). The egg is maintained on the left by a holding pipet. The injection pipet penetrates into the larger pronucleus (the male pronucleus) and the DNA solution is expelled.

germ cells. The most successful method is the microinjection of DNA (about 500 molecules) into the pronuclei of fertilized eggs (Figure 1). In 10 to 40% of the surviving embryos the foreign DNA is stably integrated, generally at one locus. The number of copies that are retained can range from one to several hundred and are arranged primarily in a head-to-tail array, although some integration events may be more complicated. The sites of integration within the host genome appear to be diverse.[10] Microinjection of DNA requires skill and rather expensive equipment but is a fast procedure compared to the use of retroviral vectors. Large fragments of DNA (50 kb or more) can be introduced into the mouse egg by microinjection, thus allowing the study of large transcriptional units or hybrid molecules. One must keep in mind when analyzing the expression of a transgene that it may be influenced by the adjacent genomic sequences (position effect). Unexpected patterns of tissue specificity are sometimes observed and the analysis of independent lineages of mice is necessary. In the case of natural HBV infection, the integration of the viral DNA into the cellular genome is observed only during the chronic carrier stage. It is not a necessary step for virus multiplication since during the acute phase of infection, episomal molecules, the covalently closed circular DNA (cccDNA), are the templates for transcription. When the DNA is integrated, viral gene expression may be subjected to another type of regulation, such as the inhibition of structural gene expression in simian virus 40 (SV40) or adenoviruses.

B. STRUCTURE OF THE INJECTED SEQUENCES

In deciding which sequences to inject into the mouse egg, different groups have asked three questions: (1) How is viral gene expression regulated in the animal and is it possible to obtain a complete viral multiplication cycle in the mouse? (2) What are the effects of the viral gene

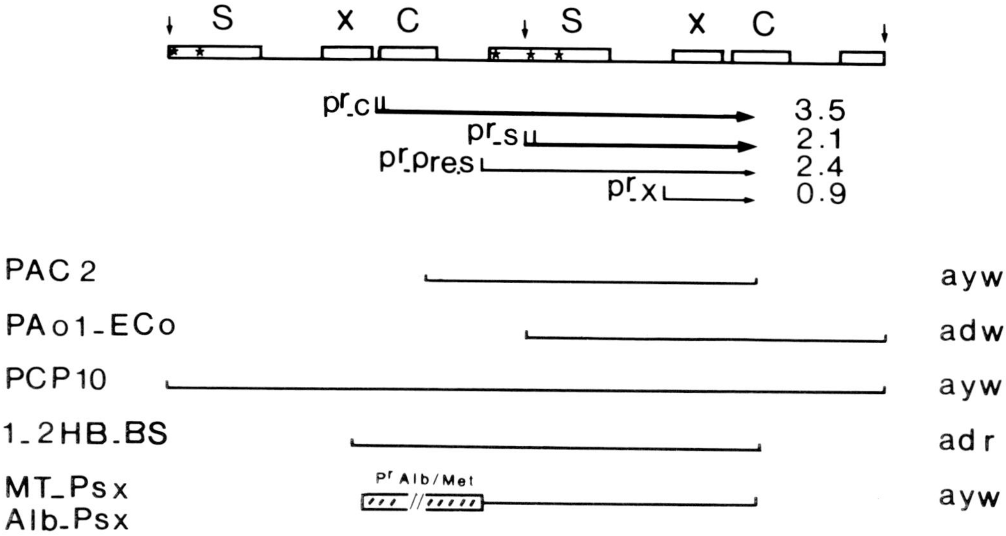

FIGURE 2. Structure of the injected HBV sequences. Two copies of HBV in a tandem head-to-tail arrangement are represented at the top of the figure. Below are the viral RNA transcripts (the 3.5-kb core/pregenome RNA, the 2.1-kb S RNA, the 2.4-kb preS RNA, and the 0.9-kb X RNA) and the respective promoters (pr-C, pr-S, pr-PreS, and pr-X). The different HBV sequences injected into mouse eggs are represented in the lower half of the figure. pr-Alb/Met, promoters for albumin and metallothionein genes.

products on the animal? (3) Are HBV sequences, cloned from a human hepatocellular carcinoma, able to induce rearrangements of the mouse genome and will this lead to tumor formation? To answer the first question, it is important not to include in the injected sequences any exogenous regulatory regions, whereas when one is interested only in the function of a viral gene product, a strong exogenous promoter can be used. Figure 2 shows the different types of HBV molecules that have been used to produce transgenic mice. In the following sections we will analyze the data obtained with independent transgenic mice concerning the different aspects of the viral cycle and the pathology induced in the animals.

III. REGULATION OF VIRAL GENE EXPRESSION

A. TISSUE SPECIFICITY

To address the problem of the tissue specifity of gene expression, either the complete genome or viral genes under the control of their own promoters were injected by three different groups. Mice were obtained that contained the intact viral DNA integrated into their genomes (Table 1), but many of them did not express the viral genes. This is probably due to the influence of the flanking genomic sequences on viral gene regulation (position effect) and may be related in some cases to the presence of plasmid sequences. Indeed, in other transgenic mice containing different constructs, plasmid sequences have been shown to repress gene expression.[9]

The first mice produced by our group contained only part of the viral genome cloned into a plasmid vector (PAC2). We had shown in a previous study that this plasmid was able to direct S gene expression in cultured mouse fibroblasts.[11] Deletion of the C gene from this plasmid eliminated the risk of virus production but did not remove any of the known regulatory sequences. Among six lineages containing PAC2 sequences, two were shown to produce HBsAg in their serum (E36 and E11).[12] In both lineages, S gene transcription was

TABLE 1
Production and Characterization of Transgenic Mice

Recombinant plasmid	Plasmid sequences[a]	Number of HBsAg+/HBV[b]	Mouse line	Number of copies of viral DNA per genome	Ref.
		HBV alone			
				4	14
			E36		
				<1	14
PAC2	+	2/7			
			E11	>1	14
PAC2	+	1/10	7-2	5–10	34
			G7	<1	15
PA OI-Eco	–	2/2			
			G26	>1	15
pCP10	+	1/9	pC21	Multiple	16
1-2 HB-BS	–	2/8	1/2 HB-BS10	ND[c]	17
Exogenous promoter					
			TM 18-11	Multiple	34
Met-HBV	–	2/6			
			TM 23-3	Multiple	34
			Alb 45-2	Multiple	35
Alb-HBV	–	8/8	Alb 45-3	Multiple	35
			Alb 50-4	Multiple	35
		Rearranged sequences			
pBC4-5.4B	–	0/3	Hep 2-2P	Multiple	39

[a] The vector sequences were removed before injection in some experiments.
[b] Number of mice expressing the hepatitis B surface antigen vs. HBV-containing mice.

restricted to the liver (Table 2) and started at 15 d of development, approximately at the same time as the albumin gene. Burk et al., using complete linear HBV genomes, obtained two mice, both of which expressed HBsAg and gave rise to two lineages (G7 and G26).[13] In the G7 lineage, the S coding region is expressed from a cellular promoter. In the two lineages, the S gene was preferentially transcribed in the liver starting at d 14 of development. A high level of expression was also found in the kidney and pancreas of mice from the G26 lineage.

In a second set of experiments we used a recombinant plasmid containing two HBV genomes (PCP10) that was shown to be infectious in the chimpanzee.[14] Nine mice were shown to have integrated tandem copies of HBV, but only one expressed both HBsAg and HBeAg in the serum.[15] Transcription of the major viral RNAs (2.1-kb S RNA and 3.5-kb C/pregenome RNA) was observed in the liver, the kidney, and the heart. The expression in the kidney was restricted to cells of the proximal tubules. Araki et al. have used a partial dimer containing a few hundred-base pair redundancy in the core promoter region. This is enough to preserve all of the viral transcriptional units. They obtained two mice producing serum HBsAg and HBeAg. In the one case they analyzed further, expression of the viral RNA was observed mainly in the liver and the kidney.[16]

The transgenic mice described by the different groups have HBV sequences from three different subtypes inserted at unique genomic locations. Thus, it appears that HBV contains

TABLE 2
Regulation of HBV Gene Expression and Viral Replication

	Tissue specificity					
	2.1-kb S RNA and 3.5-kb C/preC RNA					
Mice expressing	**Liver**	**Kidney**	**Other**	**0.9-kb X RNA**	**Replication**	**Regulation**
S RNA						
E36	+++	–	–	Testis Liver	–	Sexual hormones and glucocorticoids
E11	+++	–	–	Testis Liver	–	
G26	+++	++	Lung Stomach Intestine Pancreas	–	–	
G7	+++	+	Stomach	–	–	
S and C RNA						
PC21	++	++	Heart	Testis	+	ND[a]
1-2 HB-BS	+++	++	–	Testis	+	ND
Hybrid RNAs						
MT-PSX	+++	++	Testis Heart Lung Spleen Thymus	Testis	–	Zinc and glucocorticoids
Alb-PSX	+++	++	ND	ND	–	Constitutive

[a] ND, Not determined.

cis-acting elements that are responsible for preferential transcription of the viral genes in the liver and the kidney. Halpern et al.[17] have shown that DHBV can multiply in the hepatocytes, the proximal tubules of the kidney, and in acinar cells of the pancreas. Expression in the pancreas of mice was reported in the G26 lineage, but it might have been underestimated in others since it is technically difficult to purify intact RNA from this organ. No preferential expression was found in the spleen, although in the woodchuck this organ can support WHV multiplication.[18] It is not known at present if the transcription factors present in the different organs recognize the same regulatory sequences on HBV or independent ones. The fragment deleted in PAC2 might contain a kidney-specific element, which could explain the lack of S gene expression in the kidney of transgenic mice carrying this plasmid. Studies done in cell culture had suggested that S gene expression was not strictly dependent on the presence of liver-specific factors, unlike C gene/pregenome expression. In transgenic mice, the 2.1-kb S RNA and the 3.5-kb C/preC RNA species are always found in the same organs, although at different ratios. The fact that the viral sequences are present throughout mouse development may explain why strict patterns of expression are imposed in each lineage and it is always reproduced in the progeny. Possible mechanisms for the specificity of expression of these transcripts may include differential methylation and chromatin arrangement (see Section III.C). Results from different groups show that the 2.4-kb preS RNA and the 0.9-kb X RNA are expressed (when the transcription units are intact) together with the other viral RNAs, although at a much lower level. X RNA is also found at high levels in the testis of most of the HBV mice. The basis for this preferential expression is not understood since the testis is not considered a target for the virus multiplication. The presence of viral DNA in spermatozoa was reported but there was no evidence of viral gene expression in these cells.[19]

B. REGULATION BY STEROIDS

In four independent lineages, serum HBsAg levels were 5 to 10 times higher in adult males than in females. Castration resulted in a 50 to 80% decrease in S RNA and HBsAg levels, showing that the sex steroids naturally stimulate S gene expression.[20,21] Injection of pharmacological amounts of testosterone or estradiol restored 60% of the S gene expression in males and increased the levels in females by more than 100%. Dexamethasone also activated S gene transcription in normal females. On the contrary, X gene expression was not affected in castrated mice or following hormonal treatment.

The presence of a glucocorticoid-responsive element in HBV was described by Tur-Kaspa et al.[22] using cell cultures transfected with cloned HBV DNA. In addition, it is known that in man glucocorticoid treatment stimulates viral replication. The effect of sex steroids, either in cell culture systems or during natural infection, has not been described. The fact that men become chronic carriers with a higher frequency than women is the only argument for the involvement of sex hormones in the control of the viral cycle. The results obtained with transgenic mice should encourage the search for cis-acting elements responding to these hormones and for the promoters they regulate.

C. METHYLATION

It is now well established that differential methylation of DNA is involved in the control of cellular and viral gene expression.[23] Genes are generally hypermethylated in tissues in which they are not expressed. Modification of the methylation status of a gene occurs in germ cells and during development. Demethylation can follow activation by transcriptional factors or, alternatively, repressed genes become *de novo* methylated. *De novo* methylation is also found in tumors, probably as a consequence of reprogramming gene expression in the transformed cells.

Methylation of the HBV core gene region has been reported in the hepatocellular carcinoma cell line PLC/PRF/5, which produces only HBsAg, indicating that core gene transcription might have been repressed during tumor progression.[24] Synthesis of HBcAg was obtained in these cells following treatment with 5-azacytidine.[25] We have investigated the methylation status of HBV in different organs of the HBsAg-expressing lineages E36 and E11 at different times during development. We have found that the S gene and its promoter were hypermethylated in all the organs except the liver. This region is methylated in sperm and at early embryonic stages. It becomes progressively demethylated during liver maturation, probably in response to activation of the S gene by liver-specific factors.[26] Akamal et al.[27] have also observed differential methylation of the transgene in the G7 and G26 lineages in relation to S gene expression. Complete methylation of the transgene was found in the E36 lineage after transmission by females.[28] The *de novo* methylation imposed on the transgene during female gametogenesis or early development resulted in the irreversible loss of S and X gene expression. This is probably a consequence of integration at a particular locus in the mouse genome. Other examples of total HBV methylation were reported by the author's laboratory and Yamamura in transgenic mice that never expressed the viral genes. Araki et al.[29] showed that treatment of newborn HBsAg-negative transgenic mice with 5-azacytidine resulted in the expression of HBsAg and partial demethylation of the viral sequences.[29] Such results confirm that methylation maintains a repressive state of transcription, imposed during development, which can be reversed under certain conditions.

Finally, we found evidence that during cellular transformation S gene expression is lost and the gene becomes hypermethylated. In one set of experiments we immortalized primary hepatocytes of HBsAg-producing mice using SV40.[30] The transformed foci retained liver functions such as albumin synthesis, but the S gene was no longer transcribed. In these cells the transgene was not rearranged but *de novo* methylation in the S region was detectable. Similarly, S gene expression was lost in tumors induced by carcinogens, and the S region became methylated.[31] These results suggest the existence of negative regulation of S gene expression in tumors, possibly similar to that described by De-Medina et al.[32] in HeLa cells.

IV. REPLICATION

During a normal multiplication cycle, the viral DNA entering the cell is converted in the nucleus into a covalently closed circular molecule (cccDNA), which serves as the template for transcription. The core/pregenomic RNA is used for the synthesis of the core polypeptide and as the template for minus-strand DNA synthesis. The replicative intermediates include complexes containing the pregenomic RNA, the polymerase, and core polypeptides. Nucleocapsids involved in replication are found in the cytoplasm of infected cells. They are enveloped and secreted before the DNA plus strand is completed.[2] Replicative intermediates were found in two transgenic mouse lineages containing complete HBV genomes, and in one case the secretion of virions into the serum was demonstrated.[15,16] Replication correlated with transcription of the pregenomic RNA (demonstrated by S1 mapping analysis) regardless of the organ in which transcription occurred (liver, kidney, and heart). This shows that the synthesis of the pregenomic RNA correlates with the replicative cycle. However, the level of replication was much lower than during a normal infection and only small amounts of virions were produced. This may be due to the fact that integrated sequences are the template for transcription since no cccDNA could be found in the nuclei. Similarly, in hepatocellular carcinoma cell lines supporting HBV replication no cccDNA is detectable.[4,5] The problem of cccDNA formation is still not completely understood. The penetration of the virus into the cell and the migration of the viral DNA to the nucleus may be necessary steps for the formation of the first cccDNA molecule. Thereafter, a pool of about 50 cccDNA molecules is maintained in an infected cell. Experiments with infected duck hepatocytes in culture have shown that cccDNA is produced from the relaxed circular intermediates of replication.[33] This is acheived when elevated amounts of replication intermediates are present in the cell. In transgenic mice and in transfected cells the low levels of replication may explain why no recycling of the viral DNA to the nucleus takes place.

V. PROTEIN SYNTHESIS

A. THE ENVELOPE POLYPEPTIDE

In the mice described above, serum HBsAg levels ranged between about 10 ng/ml and 10 mg/ml and was independent of the number of integrated copies of HBV DNA. The highest levels, obtained in mice containing only one copy per genome, were similar to that of human chronic carriers in the absence of viral replication. Significant amounts of preS2 determinants were found on the HBsAg particles in mice of the E36 lineage. In these mice, although serum HBsAg levels were high, no signal could be detected inside the hepatocytes using immunoperoxidase assays, suggesting that secretion was very efficient. Chisari et al. produced mice containing the S region (including the preS1, preS2, and S coding regions) under the control of either the mouse metallothionein (Met)[34] or the mouse albumin (Alb)[35] promoter. When maximal expression of the foreign promoter was achieved in these mice (constitutively with Alb and after induction by zinc with Met), low levels of HBsAg were monitored in the serum, whereas immunofluoresence was strongly positive in the cytoplasm of the hepatocytes. This was due to the overproduction of the large envelope polypeptide, which resulted in abnormal HBsAg particles being trapped in the endoplasmic reticulum. In Section VI, the consequences of this accumulation on the cell will be discussed.

B. THE CORE POLYPEPTIDE

Core polypeptide synthesis was investigated in mouse PC21, in which all the viral transcripts and replicative intermediates had been detected. Immunoperoxidase assays using anti-core antibodies produced strong signals in the nuclei of about 10% of the hepatocytes and in many tubular cells of the kidney. In addition, cytoplasmic staining (together with nuclear staining) was found in clustered hepatocytes. Electron microscopy studies revealed the presence of capsids in the nuclei together with an amorphous unaggregated material. The detection of large amounts of core polypeptides is in contrast with the low level of replicative complexes found in the liver and the kidney. Transcription of RNAs starting in the preC region was demonstrated by S1 mapping analysis. These RNA species are believed to encode the capsid polypeptides and to be packaged by these same polypeptides to form replication complexes. The nuclear localization of core polypeptides, observed by staining with anti-core antibodies, suggests that packaging of the pregenome might occur in the nucleus.

VI. PATHOGENICITY

The mice produce the viral antigens before birth[20,21] and consequently are immunologically tolerant of these antigens. This is demonstrated by the absence in their serum of antibodies against HBsAg or of immune complexes. The injection of purified HBsAg particles of the subtype produced by the mouse does not induce an immune response. In contrast, the mice produce antibodies against particles of a different subtype.[34] Because of the immunotolerance, no injury can be attributed to the immune response and the mice are models for a category of chronic carriers immunologically nonresponsive to HBsAg. Any potential pathology seen in these mice would thus be a direct effect of viral gene expression. No cytotoxicity was detected in the different lineages containing only HBV sequences, even when large amounts of core polypeptides were stored in the nuclei. In contrast, the storage of HBsAg particles in the liver of the mice containing the Alb-PSX construct resulted in severe liver injury.[36] Due to an enlargement of the endoplasmic reticulum, the cells become enlarged, resembling typical ground-glass hepatocytes observed in long-term HBsAg-producing chronic carriers. This leads to cell death and emergence of nodules of regenerative hyperplasia. Eventually hepatocellular carcinoma can develop in these mice. Although such a storage problem does not account for most of the hepatocellular carcinomas in human patients, it confirms that cell death and regeneration are circumstances favoring the development of tumors by stimulating cell division.

In the absence of significant cell necrosis, the long-term presence of HBV multiplication in the liver may be responsible for modifications of hepatic metabolism leading to an enhanced susceptibility to carcinogens. We tested this hypothesis by analyzing the incidence of tumor development in response to carcinogen treatment in transgenic mice expressing HBsAg vs. nontransgenic mice. We found that tumor progression was more rapid in HBsAg-positive mice, suggesting that HBV gene expression could have a promoting effect on the chemical carcinogenesis.

VII. INSTABILITY OF REARRANGED SEQUENCES

In hepatocellular carcinomas, DNA rearrangements are often associated with HBV integration and this may be one process leading to cellular transformation.[37] An HBV fragment cloned from a hepatocellular carcinoma, which had an HBV-induced translocation, was used by Hino et al. to produce transgenic mice. The viral DNA consisted of two identical fragments in an inverted orientation, capable of forming a stem-loop structure. In one lineage the instability of the transgene was reported on transmission to the offspring.[38] This was reproducibly observed in several generations but did not result in tumor development. However, such mice may be useful to analyze the chromosome aberrations induced by viral DNA integration.

VIII. CONCLUSIONS

HBV-containing transgenic mice have helped to clarify several aspects of the viral life cycle and of the pathogenic effects linked to HBV infection. The regulation by sex steroid hormones and the potential direct cytopathic effect of the viral infection might have escaped investigation in other animal models.

Many possibilities are open to future experimentation. In particular, the mechanisms of immunologically mediated cell death can be studied by transferring viral antigen-stimulated lymphoid cells into transgenic mice (Figure 3). It should also be possible to use an exogenous promoter, active only after birth, to direct the expression of the viral genes. In this way a normal immune response should be induced in the transgenic mouse.

Antiviral therapy protocols can be applied to mice in which the virus replicates, with the primary goal of inactivating transcription from the integrated sequences. This may help in testing therapies aimed at modulating the disease state in chronic carriers who have integrated forms of HBV. In human virology, experiments such as the ones decribed here have not been conducted previously. It may not be possible to generalize this approach to every virus but it shows that transgenic mouse models can be used when other classical methods have failed to answer specific questions.

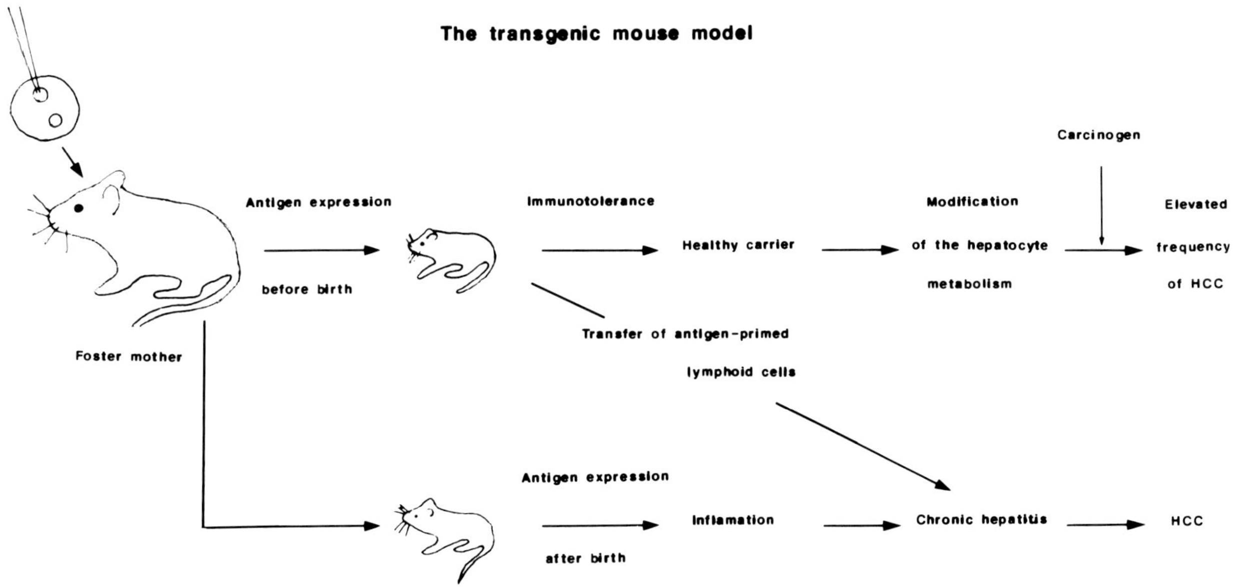

FIGURE 3. Schematic representation of the transgenic mouse model. The HBV genes injected into the mouse egg may be expressed in the liver either before or after the birth of the founder animals and their progeny. Induction of an immune response, naturally or following transfer of antigen-primed lymphoid cells, results in chronic inflammation leading to liver injury. In the absence of an immune response the mice are models for the healthy carrier state. HCC, hepatocellular carcinoma.

REFERENCES

1. **Ganem, D. and Varmus, H. E.**, The molecular biology of the hepatitis B viruses, *Annu. Rev. Biochem.*, 56, 651, 1987.
2. **Summers, J. and Mason, W. S.**, Replication of the genome of a hepatitis B-like virus by reverse transcription of an RNA intermediate, *Cell*, 29, 403, 1982.
3. **Seeger, C., Marion, P. L., Ganem, D., and Varmus H. E.**, *In vitro* recombinants of ground squirrel and woodchuck hepatitis viral DNAs produce infectious virus in squirrels, *J. Virol.*, 61, 3241, 1987.
4. **Sureau, C., Romet-Lemonne, J.-L., Mullins, J. I., and Essex, M.**, Production of hepatitis B virus by a differentiated human hepatoma cell line after transfection with cloned circular HBV DNA, *Cell*, 29, 37, 1986.
5. **Tsurimoto, T., Fujiyama, A., and Matsubara, K.**, Stable expression and replication of hepatitis B virus genome in an integrated state in a human hepatoma cell line transfected with the cloned viral DNA, *Proc. Natl. Acad. Sci. U.S.A.*, 84, 444, 1987.
6. **Shaul, Y., Rutter, W. R., and Laub, O.**, A human hepatitis B viral enhancer element, *EMBO J.*, 4, 427, 1985.
7. **Jameel, S. and Siddiqui, A.**, The human hepatitis B virus enhancer requires trans-acting cellular factor(s) for activity, *Mol. Cell. Biol.*, 6, 710, 1986.
8. **Feitelson, M. A., DeTolla, L. J., and Zhou, X.**, A chronic carrierlike state is established in nude mice injected with cloned hepatitis B virus DNA. *J. Virol.*, 62, 1408, 1988.
9. **Palmiter, R. D. and Brinster, R. L.**, Germline transformation of mice, *Annu. Rev. Genet.*, 20, 465, 1986.
10. **Hogan, B., Costantini, F., and Lacy, E.**, *Manipulating the Mouse Embryo*, Cold Spring Harbor Laboratory, Cold Spring Harbor, New York, 1986, xxx.
11. **Pourcel, C., Louise, A., Gervais, M., Chenciner, N., Dubois, M.-F., and Tiollais, P.**, Transcription of the hepatitis B surface antigen in mouse cells transformed with cloned viral DNA, *J. Virol.*, 42, 100, 1982.
12. **Babinet, C., Farza, H., Morello, D., Hadchouel, M., and Pourcel, C.**, Specific expression of hepatitis B surface antigen (HBsAg) in transgenic mice, *Science*, 230, 1160, 1985.
13. **Burk, R. D., DeLoia, J. A., El Awady, M. K., and Gearhart, J. D.**, Tissue preferential expression of the hepatitis B virus (HBV) surface antigen gene in two lines of HBV transgenic mice, *J. Virol.*, 62, 649, 1988.
14. **Will, H., Cattaneo, H., Koch, H., Darai, G., Schaller, H., Schellekens, P., van Eerd, C., and Deinhardt, F.**, Cloned HBV DNA causes hepatitis in chimpanzees, *Nature* (London), 299, 740, 1982.
15. **Farza, H., Hadchouel, M., Scotto, J., Tiollais, P., Babinet, C., and Pourcel, C.**, Replication and gene expression of hepatitis B virus in a transgenic mouse that contains the complete viral genome, *J. Virol.*, 62, 4144, 1988.
16. **Araki, K., Miyazaki, J., Hino, O., Tomita, N., Chisaka, O., Matsubara, K., and Yamamura, K.**, Expression and replication of hepatitis B virus genome in transgenic mice. *Proc. Natl. Acad. Sci. U.S.A.*, 86, 207, 1989.
17. **Halpern, M. S., England, J. M., Deery, D. T., Petcu, D. J., Mason, W. S., and Molnar-kimber, K. L.**, Viral nucleic acid synthesis and antigen accumulation in pancreas and kidney of Pekin ducks infected with duck hepatitis B virus, *Proc. Natl. Acad. Sci. U.S.A.*, 80, 4865, 1983.
18. **Korba, B. E., Wells, F., Tennant, B., Cote, P. J., and Gerin, J. L.**, Lymphoid cells in the spleens of woodchuck hepatitis virus-infected woodchucks are a site of active viral replication, *J. Virol.*, 61, 1318, 1987.
19. **Hadchouel, M., Scotto, J., Huret, J. L., Molinie, C., Villa, E., Degos, F., and Brechot, C.**, Presence of HBV DNA in spermatozoa: a possible vertical transmission of HBV via the germ line, *J. Med. Virol.*, 16, 61, 1985.
20. **Farza, H., Salmon, A.-M., Hadchouel, M., Moreau, J.-L., Babinet, C., Tiollais, P., and Pourcel, C.**, Hepatitis B surface antigen gene expression is regulated by sex-steroids and glucocorticoids in transgenic mice, *Proc. Natl. Acad. Sci. U.S.A.*, 84, 1187, 1987.
21. **DeLoia, J. A., Burk, R. D., and Gearhart, J. D.**, Developmental regulation of HBsAg expression in two lines of HBV transgenic mice, *J. Virol.*, 63, 4069, 1989.
22. **Tur-Kaspa, R., Shaul, Y., Moore, D. D., Burk, R. D., Okret, S., Poellinger, and L., Shafritz, D. A.**, The glucocorticoid receptor recognizes a specific nucleotide sequence in hepatitis B virus DNA causing increased activity of the HBV enhancer, *Virology*, 167, 630, 1988.
23. **Cedar, H.**, DNA methylation and gene activity, *Cell*, 53, 3, 1988.
24. **Miller, R. H. and Robinson, W. S.**, Integrated hepatitis B virus DNA sequences specifying the major viral core polypeptide are methylated in PLC/PRF/5 cells. *Proc. Natl. Acad. Sci. U.S.A.*, 80, 2534, 1983.
25. **Korba, B. E., Wilson, V. L., and Yoakum, G. H.**, Induction of hepatitis B virus core gene in human cells by cytosine demethylation in the promoter, *Science*, 228, 1103, 1985.
26. **Pourcel, C. and Farza, H.**, Transcription of the S gene in transgenic mice is associated with hypomethylation at specific sites and DNAse I sensitivity; submitted.

27. **Akmal, M., El-Ghor, A., and Burk, R. D.**, DNase I hypersensitive site maps to the HBV enhancer; submitted.
28. **Hadchouel, M., Farza, H., Simon, D., Tiollais, P., and Pourcel, C.**, Maternal inhibition of hepatitis B surface antigen gene expression in transgenic mice correlates with *de novo* methylation, *Nature (London)*, 329, 454, 1987.
29. **Araki, K., Miyazaki, J.-I., Tsurimoto, T., Inomoto, T., Iwanaga, T., Matsubara, K., Yamamura, K.-I.**, Demethylation by 5-azacytidine results in the expression of hepatitis B virus surface antigen in transgenic mice, *Jpn. J. Cancer Res.*, in press.
30. **Paul, D., Farza, H., Höhne, M., and Pourcel, C.**, Production of hepatitis B virus surface antigen in normal and immortalized transgenic mouse hepatocytes, in *Viral Hepatitis and Liver Disease*, Zuckerman, A. J., Ed., Alan R. Liss, New York, 1988, 379.
31. **Farza, H., Dragani, T., Manenti, G. T., Tiollais, P., Della-Porta, G., and Pourcel, C.**, Hepatocarcinogenesis in transgenic mice producing HBsAg, paper presented at the Symp. Molecular Biology of Hepatitis B Viruses, San Diego, 1988.
32. **De-Medina, T., Faktor, O., and Shaul, Y.**, The S promoter of hepatitis B virus is regulated by positive and negative elements, *Mol. Cell. Biol.*, 8, 2449, 1988.
33. **Tuttleman, J. S., Pourcel, C., and Summers, J.**, Formation of the pool of covalently closed circular viral DNA in hepadnavirus-infected cells, *Cell.*, 47, 451, 1986.
34. **Chisari, F. V., Pinkert, C. A., Milich, D. R., Fillipi, P., McLachlan, A., Palmiter, R., and Brinster, R.**, A transgenic mouse model of the chronic hepatitis B surface antigen carrier state, *Science*, 230, 1157, 1985.
35. **Chisari, F. V., Filippi, P., McLachlan, A., Milich, D. R., Riggs, M., Lee, S., Palmiter, R. D., Pinkert, C. A., and Brinster, R. L.**, Expression of hepatitis virus large envelope polypeptide inhibits hepatitis B surface antigen secretion in transgenic mice, *J. Virol.*, 60, 880, 1986.
36. **Chisari, F. V., Filippi, P., Buras, J., McLachlan, A., Popper, H., Pinkert, C. A., Palmiter, R. D., and Brinster, R. L.**, Structural and pathological effects of synthesis of hepatitis B virus large envelope polypeptide in transgenic mice, *Proc. Natl. Acad. Sci. U.S.A.*, 84, 6909, 1987.
37. **Tokino, T., Fukushige, S., Nakamura, T., Nagaya, T., Murotsu, T., Shiga, K., Aoki, N., and Matsubara, K.**, Chromosomal translocation and inverted duplication associated with integrated hepatitis B virus in hepatocellular carcinoma, *J. Virol.*, 61, 3848, 1987.
38. **Hino, O., Nomura, K., Ohtake, K., Kitagawa, T., Sugano, H., Kimura, S., Yokoyama, M., and Katsuki, M.**, Rearrangement of integrated HBV DNA in descendants of transgenic mice, *Proc. Jpn. Acad., Ser. B*, 62, xxx, 1986.

Chapter 5

TRANSCRIPTION OF HEPADNAVIRUSES

Aleem Siddiqui

TABLE OF CONTENTS

I. INTRODUCTION

The human hepatitis B virus (HBV) represents the prototype of a group of liver-tropic viruses called hepadnaviruses.[1-3] Lack of an efficient system of *in vitro* propagation for HBV has precluded biological studies and investigation of its mechanism of pathogenesis. Despite such a limitation, considerable progress has been made in elucidation of the transcriptional schemes utilized by this virus to regulate gene expression in a tissue-specific manner. Its liver specificity is due largely to the enhancer[4-6] element, which resides in a strategic location within the genome and which may also play a central role in the overall gene expression of HBV. Molecular cloning[7-9] and determination of the complete nucleotide sequence[9-11] of the viral genome marked the beginning of modern studies of HBV. Most of the information about transcription and its regulation has been derived from transfection of cultured cells using recombinant vectors containing defined sequences.[2,3]

Information on the replication of the virus[12] has been inferred from animal viruses that share common features with HBV.[1-3] The ability of the whole genome, through transient or stable transformation/expression systems,[13-16] to complete the viral replication cycle is an important and welcome addition to these studies. It enables investigators to view transcription of individual genes in the context of the entire genome.

The HBV genome consists of a small, approximately 3200-bp, partially double-stranded DNA that is maintained in a circular conformation, enclosed in a nucleocapsid core structure of 28 nm, which is contained in a hepatocyte membrane-derived envelope[1-3] (Figure 1). Determination of the nucleotide sequence[9-11] of the genome led to the assignment of four open reading frames (ORFs).[2,3] These include the S (surface) region, the C (core) region, and the X and P (or *pol*, polymerase) genes PreS1, preS2, and S polypeptides (HBsAg) are all encoded by the S ORF. Core (HBcAg) and e (HBeAg) antigen polypeptides are derived from the C ORF. The X gene product is a 16-kDa protein encoded by the X ORF. A putative reverse transcriptase protein most likely encoded by the P (*pol*) ORF has not been identified. The HBV genome utilizes three translational reading frames, and some of the open reading frames overlap with each other. The complete long strand serves as the coding strand.

The substrate for viral transcription *in vivo* is a covalently closed circular DNA (cccDNA) found in the infected hepatocytes.[17] In light of the replication pathway, which requires amplification of superhelical DNA by reverse transcription,[12,17] viral cccDNA is the principal molecule that should efficiently serve as a template for synthesis of mRNA and pregenomic RNA. Pregenomic RNA is reverse transcribed to generate the HBV genome. Efficient synthesis of viral mRNAs and their subsequent translation are crucial for the continuance of the viral replication cycle. This scheme was first demonstrated for duck hepatitis virus (DHBV). HBV and other animal viruses are believed to follow a similar pathway of replication.[2,12,18,19]

Viral transcripts identified so far fall into two major classes: 3.5-kb and 2.1/2.4-kb RNAs.[19-29] The 3.5-kb RNA species includes a mixture of RNAs with heterogeneous 5′ termini.[14,18,19,22,23,25,28] These include transcripts for the core/e polypeptides and perhaps the viral polymerase (*pol*) gene product.[18,29,31,32] One of the 3.5-kb RNA species functions as pregenomic RNA, a template for reverse transcription that is distinguished by distinct 5′ end structure(s).[18,27] The 2.1-kb mRNAs encode the preS2 and S polypeptides. The 2.6-kb transcripts include the coding information for the preS1 polypeptide or the large HBsAg.[2,20-26,29] Other subgenomic RNA species that may originate from a promoter located upstream of the X ORF have been identified in infected livers and transfected cells.[30-33] All of the viral transcripts described above are unspliced.[2,3] However, the surprising discovery of spliced transcripts in HBV-infected livers[33,34] adds a new twist to the transcriptional strategies used by the virus. If confirmed, these results would invoke a series of new schemes of transcription

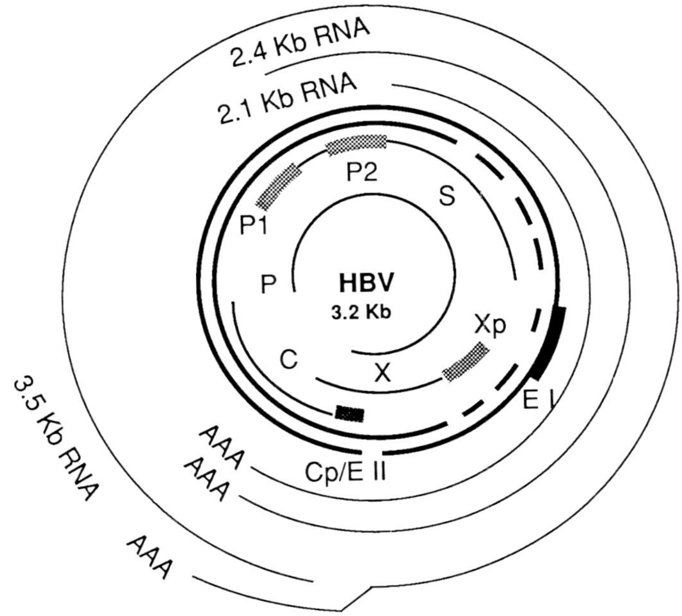

FIGURE 1. DNA genome of hepadnaviruses. Structural features include partially double-stranded molecule with a nick in the long strand. Four open reading frames are represented. S, surface antigen; C, core/e antigens; X, X gene; P, polymerase. The promoters are as follow: P1, preS1; P2, preS2; Cp, core/pregenomic; Xp, X; EI and EII, the enhancer elements I and II. Major viral transcripts are indicated as outer circles with a poly(A) (AAA) tail at the 3′ end of each mRNA.

and various alternative possibilities of regulation the virus might employ to establish infection in liver cells. Since these transcripts have not been fully characterized, speculation of their coding potential is premature at the present time.

A variant polyadenylation signal (TATAAA) located at nucleotide 1919 relative to the *Eco*RI site (arbitrarily set at 1) controls the formation of the 3′ end of all the viral RNAs. During synthesis of the 3.5-kb RNA the polyadenylation signal sequence is ignored on the first passage and only at the second readthrough is it used as a termination signal.[2,18] This strategy, while permitting transcription of the core/e gene, also generates a terminally redundant template for reverse transcription. Recent studies by Rusnak and Ganem[35] identified multiple cis-acting elements that are required for the efficient utilization of the polyadenylation signal in the HBV genome. The mechanism by which this selective 3′ end processing occurs is not known.

The cis-acting regulatory elements of HBV that control the synthesis of viral transcripts are currently being characterized in molecular detail and are being studied at the level of DNA–protein interactions. Some of the preliminary data that have been recently reported will be mentioned but by no means do they represent the complete picture of interactions between nuclear factors and viral control elements. The following sections include information obtained from all of the hepadnaviruses in order to provide an overview of the transcriptional program utilized by these viruses. Map positions of the nucleotide sequences derived from the HBV DNA (subtype adw_2) are used here to maintain uniformity (sequence obtained from GenBank database).

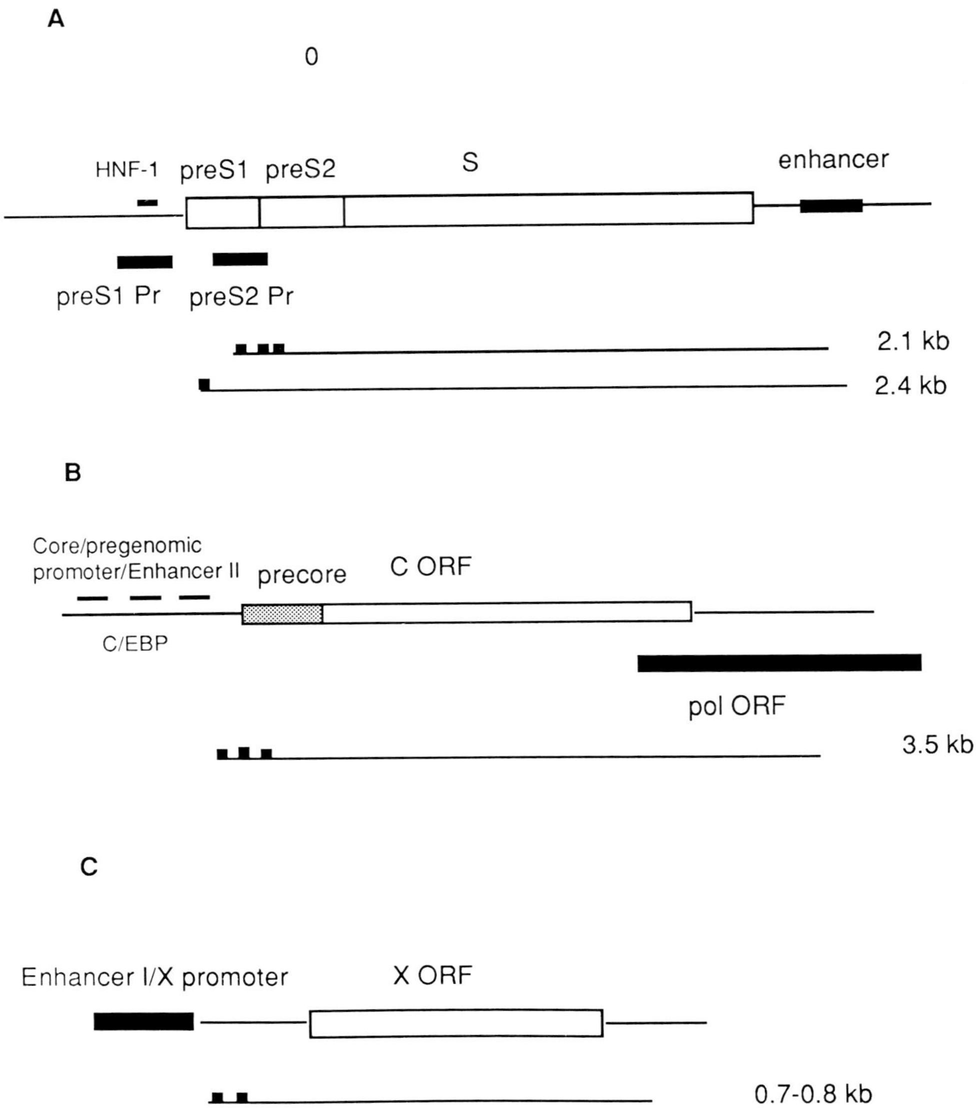

FIGURE 2. Fine structures of the 5′ ends of viral transcripts. Solid bars indicate the initiation sites. (A) mRNA (2.1 and 2.4 kb) of the surface antigen gene with multiple initiation sites from respective promoters. Hepatocyte nuclear factor 1 (HNF-1)-binding site in the preS1 promoter is indicated. (B) Core/pregenomic promoter containing three binding sites for C/EBP is shown. The 5′ end heterogeneous RNAs initiate from this promoter. Overlapping *pol* ORF is shown. (C) Putative X promoter and adjacent enhancer element are located upstream of the X ORF. At least two initiation sites have been demonstrated.

II. S OPEN READING FRAME

The hepatitis B surface antigen (HBsAg) is a term assigned to a group of viral structural proteins that constitute the principal components of subviral 22-nm particles and filaments, and of the envelope of the virus, the Dane particle.[1-3,36] The S ORF contains three conserved in-frame translation initiation codons that serve as initiation codons for a series of HBsAg polypeptides designated preS1 (large), preS2 (middle), and S (major). These proteins share common carboxyl termini but have variable amino termini (Figure 2A). The most abundant form is the major S protein, which exists in both glycosylated (GP27) and nonglycosylated (P24) forms.[2,3,36] The preS2 proteins (GP33/GP36) and preS1 (P39/GP41) proteins are less abundant and constitute about 15 and 2% of the S proteins produced from this ORF, respectively. These differential levels of S proteins correlate with the relative abundance of the corresponding transcripts.[20,22,24,25,33] Synthesis of the 2.1- and 2.4-kb viral transcripts, which contain the coding sequences from this region, is directed by two distinct promoter regions, each with a regulatory scheme dictating variable levels of transcription. The heterogeneity of the 5′ ends of the 2.1-kb transcripts produces separate mRNAs with the preS2 ATG and the S ATG as their first initiation codons. These transcripts encode the corresponding preS2 and S polypeptides, respectively.[2,20-22,29,36]

The two promoters responsible for transcription of the S ORF exhibit markedly different promoter acitivities.[20-22,37,38] The preS1 promoter, which controls the transcription of the 2.4-kb mRNA encoding the preS1 polypeptide, contains a canonical TATA sequence[2,3,37] at –66 relative to the PS1 ATG and a sequence motif for binding hepatocyte nuclear factor 1 (HNF-1) (5′-GTTAATCATTACT-3′). HNF-1 is a liver-specific protein, purified from rat livers, that binds to upstream sequences of the fibrinogen family of genes and several other liver-specific genes.[39] Courtois et al.[39] first demonstrated binding of purified HNF-1 to this promoter region between –62 and –87, upstream of the cap site for the preS1 mRNA. This has been subsequently confirmed by Chang et al.[40] Appropriate mutational analysis will be required to establish the functional significance of this interaction. The minimal sequences defined between –90 and +35 relative to the cap site (+1) of the preS1 transcript that retain the promoter I (P1) activity contain the HNF-1-binding site, implicating a functional role for this interaction in promoter function.[41] The preS1 promoter, which is relatively weak compared to the preS2 promoter,[20,22,40] appears to exhibit preferential activity in liver cells.[40] Further, the studies with transgenic mice revealed the restricted synthesis of the preS1 transcripts in the liver, whereas preS2 (2.1 kb) mRNA synthesis was also seen in other tissues.[42,43] Since HNF-1 is a liver-specific protein, its interaction with the preS1 promoter may play an important role in conferring liver-specific characteristics to this promoter.

The preS2 promoter is located within the preS1 coding sequence and controls the synthesis of the 2.1-kb transcripts encoding preS2 and S proteins. The preS2 promoter exhibits partial homology to the simian virus 40 (SV40) late promoter region.[21] The demonstrated 5′ end microheterogeneity of viral transcripts originating from the preS2 promoter is consistent with the GC-rich sequence and lack of a TATA element in the promoter. Sequential deletion mutations generated upstream of the preS2 promoter defined the 5′ flanking 103 nucleotides as the basal promoter region and the adjacent upstream 85 nucleotides as the modulating domain of that promoter.[44] Within the modulating domain lies an NF-1-binding sequence motif and binding of purified NF-1 protein[45] to this sequence has been demonstrated.[46] Disruption of the NF-1-binding sequence causes a four- to fivefold reduction in promoter activity.[29,44,46] Raney et al. further identified several nucleotide sequence domains that function as binding sites for nuclear proteins extracted from HepG2 cells.[44] The functional significance of these interactions, including NF-1 binding, needs to be determined.

Initiation sites for 2.1-kb transcripts have been mapped by a number of investigators using different HBV subtypes.[14,20-22,29] In general, there are three transcription initiation sites in the immediate vicinity of the preS2 ATG at nucleotide position (nt) 3213. Some of these initiation sites lie within the preS2 coding region and translation of this RNA may utilize the S ATG (nt 157) as initiation codon (Figure 2A). The major initiation site for the 2.4-kb mRNA has been mapped at –35 with respect to the preS1 ATG (nt 2856).[22,29,40]

During natural infection by the virus or transient expression of HBV genes in cell culture, the majority of transcripts (about 98%) initiate from the preS2 promoter region. The levels of transcripts derived from the preS1 promoter are barely detectable. Although the low level of preS1 transcripts could be attributed to weak promoter activity, fusion of the preS1 promoter to a reporter gene or the S protein coding sequences resulted in expression comparable to that observed from the preS2 promoter.[29,47] This suggests that transcription from the preS1 promoter may be repressed in the context of the native S gene. A detailed mutational analysis by Bulla and Siddiqui[29,47] identified a stretch of 60 bp, located within the preS region, the deletion of which elevated preS1 transcription, thus relieving the negative constraints on the preS1 promoter. The presence of this sequence does not appear to affect preS2 promoter activity. Whether this down regulation of preS1 synthesis is mediated by cellular or viral factors is not known. Furthermore, the mechanism by which this internal cis-acting element functions in the overall regulation of synthesis of the S proteins needs to be elucidated.

Utilizing two promoters to produce different levels of the various S proteins represents a unique form of eukaryotic gene regulation. Cellular genes whose synthesis is controlled by more than one promoter have been identified. The myosin light chain gene utilizes two promoters to generate two distinct polypeptides.[48] In the majority of genes that have two promoters, however, the differential use of promoters appears to be regulated either developmentally or in a tissue-specific manner.[48-50] The mouse α-amylase gene contains a relatively weak promoter that is active in liver, the parotid gland, and the pancreas.[50] Following weaning, however, a second promoter located 2.8 kbp upstream is efficiently used specifically in parotid tissue.[50] Several lines of evidence suggest that the relative levels of the S ORF polypeptides are important during HBV infection. First, the presence of preS1 proteins in the formation of HBV virions plays an important part in infectivity.[2,3] Second, overproduction of preS1 proteins apparently interferes with the secretion of HBsAg.[51,52] In addition, Neurath et al.[53] identified an amino acid sequence located within the preS1 amino-terminal region which may bind to a hepatocyte-specific receptor and thus may play a role in attachment and/or penetration of HBV into hepatocytes.

Attempts to express HBsAg in cultured cells using eukaryotic vectors have been extremely successful.[20,21,22,29,37,38,54] These studies utilized either strong heterologous promoters/enhancers or replication-efficient COS cells.[21,38] Liver cells proved to be extremely efficient in the synthesis and secretion of HBsAg. In a variety of cell lines tested following transient transfection with S gene-encoding vectors or the whole HBV genome, it is only in the liver cell lines HepG2 and Huh-7 that efficient HBsAg synthesis can be observed.[29,54] The S promoters assayed independently using the reporter gene chloramphenicol acetyltransferase (CAT) display comparable levels of activities in both liver and nonliver cells.[22,29] However, in the context of S gene expression, the promoters (preS1 and preS2) exhibit a preference for the liver cell environment and display variable efficiencies of preS1, preS2, and S expression.[14,29,55] With the removal of the enhancer the S gene transcription is reduced about five- to sevenfold fold without affecting the mRNA stability, as examined by nuclear run on analysis.[54,55] Similar conclusions were reached by other investigators.[46,56,57] These observations imply a role for the enhancer in the S gene expression. The HBV enhancer is strategically located within the 3′ untranslated region of S gene. It is the only enhancer known whose sequence is a component of the body of mature transcripts, being included in both the S gene transcripts and the 3.5-kb RNAs. The liver-specific function of the enhancer has been thoroughly demonstrated.[4,6,58,59] It is apparent that the enhancer imposes a modest level of influence in the liver-specific modulation of S gene expression.[29,44,54,57] The enhancer alone, however, does not appear to account for the large difference in S gene expression between liver and nonliver cells. S gene expression, in the context of the genome, may be controlled by interaction of several cis- and trans-acting elements facilitating efficient expression in the liver cell environment. For instance, binding of two liver-specific factors, hepatocyte nuclear factor 1 (HNF-1) and C/EBP, has been demonstrated[39,57-63] and might be functionally relevant. The overall liver specificity must result from the influence of the enhancer and the binding of liver cell-specific factors to the transcriptional control elements of the S promoters. In summary, transcription of the S ORF represents an interesting and tightly regulated process.[33,55] This is particularly important because altered levels of preS1 polypeptide can interfere in the normal processes of the viral infectious cycle.

III. C OPEN READING FRAME

The C ORF codes for the core (HBcAg) and e (HBeAg) antigen polypeptides (Figure 1). About 500 bp downstream from the start of the C ORF is a translational open reading frame overlapping the carobxyl terminus of C ORF. This ORF, the *pol* ORF, has been strongly implicated as encoding the viral polymerase. Rall et al.,[64] using an *in vitro* transcription system from HeLa extracts, identified sites of RNA polymerase II initiations upstream of the C ORF. Subsequently, it was shown by Roossinck et al.[28] that a promoter functional *in vivo* controlled the synthesis of core and e polypeptides (Figure 2B). The C ORF contains two ATG codons at nucleotide positions 1816 and 1903, immediately following the promoter region. The sequence between the two ATG codons constitutes the precore region, which contains a signal peptide sequence. Precore sequences included in the HBeAg result in its secretion during infection. In HBV-infected livers and cells transfected with recircularized or dimerized HBV genomes, major transcripts originating upstream of the C region are represented by the terminally redundant 3.5-kb RNA species. Mapping of initiation sites revealed 5′ end heterogeneity consistent with the lack of a TATA element.[28,42,60,65] In GSHV, the three initiation sites are mapped at –26, –16, and +6 relative to the first ATG in the C ORF.[27] Using transient expression of the whole HBV genome (*adr* strain), Yaginuma and Koike[65] identified cap sites at –30, –23, and +4 relative to the precore ATG. Several investigators have identified similar initiation sites in this region using other strains of HBV and hepadnaviruses (Figure 2B).[19,23,25-28,42,65] The 3.5-kb RNA represents a mixture of transcripts that is composed of mRNAs for the core and e polypeptides, the putative viral polymerase, and an RNA species that serves as the pregenome. The shortest 3.5-kb RNA, identified as the pregenome,[27] is packaged in the core particles where it is converted to HBV genomic DNA. Thus, the 5′ end structures and the cap sites are the features that distinguish functions of the core-region RNAs in HBV infection.

Biosynthesis of the *pol* protein has been addressed by two independent studies[66,67] in which a model of *de novo* initiation at the *pol* AUG was proposed. Internal initiation requires recognition by the ribosomes of the *pol* AUG at nucleotide 2333, which is located about 533 nucleotides downstream from the cap sites of the 3.5-kb RNAs. Several AUG codons exist between the cap sites and the *pol* AUG. Several examples of internal initiation have been reported. Poliovirus, with an approximately 750-nucleotide 5′ untranslated region and seven preceding AUGs, initiates at an internal AUG.[68] The *antennapedia* gene of *Drosophila*[69] uses a similar strategy. Internal translational initiation is believed to be facilitated by features in the 5′ end structures of RNA that permit direct entry of ribosomes and cellular factors. The logical choice for the transcript which encodes *pol* is a 3.5-kb RNA. Since the *pol* ORF is preceded by the core/e gene in a different reading frame, the translational strategy for *pol* was addressed by genetic analysis. This study firmly establishes the use of internal initiation as a mechanism for the biosynthesis of the polymerase as opposed to ribosomal frameshifting employed by other retroviruses.[66]

A recent model[34] proposed that a doubly spliced RNA of 2.1 kb may encode the Pol protein.

In this model, the core/pregenomic promoter would be used to direct the synthesis of this mRNA and two splicing events would be required to generate the 2.1-kb RNA. The first splicing event would remove 282 bp of the C ORF. The second splice would eliminate the termination codon in the C ORF, the previously assigned *pol* ATG, and the hypervariable region of preS1 and preS2, and finally join the exonic sequences to the S region while creating a new initiation codon in the truncated *pol* translational frame. Since a protein encoded by the P ORF has not been identified, it is difficult to identify conclusively the correct model for *pol* expression, despite some convincing genetic data in which mutation at the first ATG of the duck hepatitis virus *pol* ORF was negative for viral growth.[66] Additional data are required to confirm these observations. Future experiments should include expression of P ORF or the cDNA of the proposed spliced transcripts in hosts suitable (yeast or *Escherichia coli*) for characterization of reverse transcriptase and other enzymatic activities. In light of these controversies, a definite scheme of Pol biosynthesis needs to be established.

The promoter region involved in the genesis of the heterogeneous mixture of 3.5-kb RNA molecules has been the subject of investigations in several laboratories. It is designated here as the core/pregenomic promoter. Sequence analysis of its 5′ boundaries did not reveal any canonical sequence motifs. Binding sites for trans-acting cellular proteins have been identified using mutational analyses, DNase I protection, and methylation interference methods.[58,65,70,71] These data demonstrate the nucleotide sequences that may play functional roles in the transcriptional regulation of this region. The core/pregenomic promoter is a weak promoter both in the context of the core/e antigen gene it controls in the expression of the viral genome and heterologous reporter genes (CAT or luciferase).[28,60,71] Even at this low level of expression it does display preferential activity in liver cells. Its role in regulating the transcription of the core, e, and polymerase coding sequences in addition to the pregenomic RNA (replicative intermediate) makes this promoter one of the most important control elements of the virus. The activity of the core promoter in the context of a reporter gene such as luciferase is stimulated 15 to 20-fold in the presence of the enhancer. In contrast, in the context of the HBV genome, removal of the enhancer causes only a fourfold reduction in the synthesis of core/e protein.[72] It was further observed that the enhancer is absolutely required for the synthesis of the shortest RNA species, which has been identified as the pregenomic RNA, the template for reverse trancription. Since the promoter exhibits dependence on the enhancer, this suggests a direct role for the enhancer in controlling replication function. The influence of the enhancer on other functions associated with the transcripts originating from this promoter, such as Pol synthesis, have not been analyzed. Thus, aside from the cell surface receptor for the virus, the hepatotropism of the virus may be directly controlled by the liver-specific activity of the core promoter/enhancer complex. The enhancer in conjunction with core promoter may regulate transcription of various mRNAs at different levels.

In order to understand fully how the core promoter directs liver-specifc expression, it will be important to view the activity of this promoter in light of DNA-protein interactions. The interaction of trans-acting factors with the core promoter has been addressed recently. Karpen et al.[70] demonstrated that binding of a protein protects a palindromic sequence of 12 nucleotides (5′-AAGAGGACTCTTT-3′) at nucleotide 1657 from DNase I digestion. Yaginuma and Koike[65] demonstrated two additional sites of binding within the core promoter region, one by a liver-specific factor and the other by a transcription factor from a HeLa nuclear extract. Another study from this laboratory[71] demonstrated seven distinct areas of DNA–protein interactions in the presence of liver nuclear extracts, which includes the footprints described above. Lopez et al.[71] demonstrated C/EBP binding at several sites. C/EBP, a liver-specific protein purified from rat livers, binds to several liver gene promoters/enhancers[61,62] as a dimer maintained through a leucine zipper interaction. Cotransfection studies with C/EBP-encoding DNA and core promoter constructs indicate that C/EBP interactions result in trans-activation of these promoters, thus establishing the functional significance of C/EBP binding.[71] In the HBV enhancer there are three major sites of binding for C/EBP.[61,63] The fact that almost all of the footprints in the core promoter are due to liver-specific proteins strengthens the earlier observations demonstrating the liver-specific characteristics of this promoter.

A second enhancer-like sequence within the core promoter was first identified by Yee.[73] Subsequently it was confirmed by others.[74-76] Although the enhancer II element is located within the core/pregenomic promoter region, it appears to influence the S ORF synthesis from its promoters.[74,76] Lopez et al. described the DNA-protein interactions in this element and have localized several binding sites for C/EBP.[76] Cotransfection with C/EBP leads to trans-activation of transcription mediated through the enhancer II sequence elements. In summary, the core promoter plays a pivotal role in the viral life cycle. It controls liver-specific activity and its dependence on the enhancer element provides insight into the overall regulation of HBV gene expression.

IV. X OPEN READING FRAME

The fourth HBV ORF, X, encodes a protein of 154 amino acids (16.5 kDa) that has been identified in infected livers. Several groups have demonstrated the role of X protein as a transcriptional trans-activator of several viral and cellular genes.[77-82] Transcriptional activity controlled by a variety of viral and cellular promoters/enhancers has been shown to be increased *in trans* by the X protein. Among the celluar genes, MHC class I and II gene promoters are the targets of transactivation by the X protein.[96,98] Other approaches to demonstrate transactivation included linking the entire X gene to the DNA binding domain of proteins such as C/EBP or AP-2.[95,97] In addition to RNA polymerase II (Pol II) promoters, X also seems to activate transcription from Pol III promoters.[83] Its role in the viral life cycle, among other possible functions, is to regulate gene expression. The mechanism of this activity is not known at present. In a recent study from this laboratory, the X gene substituted for the transcriptional activation of the EIA in adenovirus infection, suggesting that pX may function by a mechanism similar to EIA.[99]

Studies on trans-activation by the X protein revealed two X-responsive targets, an NF~KB site[79,81] and a CRE-like variant sequence motif.[94] Although, the mechanism of how activation through NF~KB occurs remains to be understood, Maguire et al.[94] recently demonstrated that pX engages CRE-like sequences in the HBV enhancer by direct protein-protein interactions, resulting in an altered DNA binding specificity. Direct protein-protein interaction between pX and CREB or ATF-2 was demonstrated. pX does not bind DNA directly. The ability of pX to form complexes with X cellular factors identitfies the potential of activation of celllular genes during viral infection. pX has also been recently shown to possess a novel serine/threonine kinase activity. So far the only substrate for the pX kinase is histone 1. Interaction of pX with CREB or ATF *in vitro* occurs independent of phosphorylation. The identification of protein kinase activity associated with pX[93] and its ability to complex with cyclic AMP–inducible proteins[94] implicates its role in signal transduction mechanisms of regulation of cellular functions.

The biosynthesis of mRNA encoding X protein may be regulated by an upstream element whose functional domain seems to overlap with that of the enhancer element.[30,31] S1 nuclease mapping of subgenomic transcripts originating from the putative X promoter places initiation sites near a DNase I footprint (FPI),[58] the sequence motif of which resembles an NF-1 binding site at nucleotide 1229 (Figure 2C).[84,85] Several regions of DNA–protein interaction have been demonstrated in the enhancer element. If each of these is considered as a functional module,[86] then a few of these in some combinatorial fashion may be required to initiate transcription of the X gene, while others may be involved in conferring liver specificity.[62,86,87] Based on this rationale, Trujillo et al.[63] identified the FPII and FPI regions[58] (or the "E" element)[59] of the enhancer element I as the basal promoter for the X ORF. Multiple factors are known to bind in this domain, including C/EBP, CREB/ATF, AP-1 (Jun/Fos), eH-TF, and NF-1. According to this scheme of X gene transcription, a subgenomic mRNA species of about 0.7 to 0.8 kb would be sufficient to encode the X protein.[34,37] This message would utilize the unique polyadenylation signal located immediately downstream of the termination codon in the X ORF. Although an independent promoter for X has been demonstrated using reporter gene assays[30,31] and by examining the synthesis of X protein,[30] and subgenomic X-related transcripts have been identified in transfected cultured cells,[30,32,33] other possibilities for the biosynthesis of the X may also exist. Translation could initiate internally from either a 2.1- or 3.5-kb mRNA or a spliced transcript initiated from the S or core promoter. The reason for searching for other models of transcription for the X ORF is because of the failure to detect subgenomic transcripts of 0.7 to 0.8 kb in infected livers. It is quite likely that X is produced temporally and its synthesis is regulated tightly by factors of cellular origin, or that X is autoregulated.

V. CONCLUSIONS

A great wealth of information is available on the mechanism of eukaryotic/viral transcription.[62,86,87] Recently, success in the purification of certain RNA polymerase II transcription factors has come through the use of sequence-specific DNA affinity chromatography.[88] This technique has allowed the identification and characterization of several trans-acting cellular factors.[62,86,89] Additionally, the concept of protein-protein interaction[62,90,91] has been invoked to clarify the details of affinities of DNA-protein interactions. DNA-protein interaction studies shed new light on the intricate molecular aspects of viral transcription. These studies identify specific nucleotide sequences that bind proteins and bring about regulatory responses of viral transcription. Transcription factors have been purified and their cognate genes cloned and sequenced.[62,91] Transient introduction of expression vectors for the transcription factors and a promoter/enhancer linked to a conveniently assayable reporter gene into cells have allowed the positive regulatory activity to be reproduced, thus establishing the functional relationship of the interaction at the DNA-protein level.[92] Despite the lack of an efficient *in vitro* culture system or a suitable animal model, a clear understanding of the transcriptional program of HBV genes is beginning to emerge. This is largely due to recent methodological advances in the area of transcription of eukaryotic genes. The sensitive techniques described above have all had a major impact on the elucidation of eukaryotic gene expression. These methods have been adapted for utilization with HBV and it is hoped that with the use of these approaches we will learn a great deal about how the various promoters and their interactions with the enhancer element bring about the efficient regulation of HBV gene expression.

55. **Antonucci, T. K. and Rutter, W. J.**, Hepatitis B virus (HBV) promoters are regulated by the HBV enhancer in a tissue specific manner, *J. Virol.*, 63, 579, 1989.
56. **de-Medina, T., Faktor, O., and Shaul, Y.**, The S promoter of hepatitis B virus is regulated by positive and negative elements, *Mol. Cell. Biol.*, 8, 2449, 1988.
57. **Chang, H. K., Chou, C. K., Chang, C., Su, T. S., Hu, C. P., Yoshida, M., and Ting, L.P.**, The enhancer sequence of human hepatitis B virus can enhance the activity of its surface gene promoter, *Nucleic Acids Res.*, 15, 2261, 1987.
58. **Patel, N. U., Jameel, S., Isom, H., and Siddiqui, A.**, Interactions between nuclear factors and the hepatitis B virus enhancer, *J. Virol.*, 63, 5293, 1989.
59. **Shaul, Y. and Ben-Levy, R.**, Multiple nuclear proteins in liver cells are bound to hepatitis B virus element and its upstream sequences, *EMBO J.*, 6, 1913, 1987.
60. **Honigwachs, J., Faktor, O., Dikstein, R., Shaul, Y.**, and **Laub, O.**, Liver specific expression of hepatitis B virus is determined by the combined action of the core gene promoter and the enhancer, *J. Virol.*, 63, 919, 1989.
61. **Landschulz, W., Johnson, P., Adashi, E., Graves, B., and McKnight, S. L.**, Isolation of a recombinant copy of the gene encoding C/EBP, *Genes Dev.*, 2, 786, 1988.
62. **Johnson, P. F. and McKnight, S. L.**, Eukaryotic transcriptional regulatory proteins, *Annu. Rev. Biochem.*, 58, 799, 1989.
63. **Trujillo, M., Letovsky, J., Maguire, H., Lopez-Cabrera, M., and Siddiqui, A.,** Functional analysis of a liver-specific enhancer of the hepatitis B virus, Proc. *Natl. Acad. Sci. U.S.A.,* 88, 3797, 1991.
64. **Rall, L. B., Standring, D. N., Laub, C.**, and **Rutter, W. J.**, Transcription of hepatitis B virus by RNA polymerase II, *Mol. Cell. Biol.*, 3, 1766, 1983.
65. **Yaginuma, K. and Koike, K.**, Identification of a promoter region for 3.6 kilobase mRNA of hepatitis B virus and specific cellular binding protein, *J. Virol.*, 63, 2914, 1989.
66. **Chang, L. J., Pryciak, P., Ganem, D., and Varmus, H. E.**, Biosynthesis of the reverse transcriptase of hepatitis B viruses involves *de novo* translational initiation not ribosomal frameshifting, *Nature (London)*, 337, 364, 1989.
67. **Schlicht, H., Radziwill, G., and Schaller, H.**, Synthesis and encapsidation of duck hepatitis B virus reverse transcriptase do not require formation of core-polymerase fusion proteins, *Cell*, 56, 85, 1989.
68. **Pelletier, J. and Sonenberg, N.**, Internal initiation of translation of eukaryotic mRNA directed by a sequence derived from poliovirus RNA, *Nature (London)*, 334, 320, 1989.
69. **Stroehrer, V. L., Jorgensen, E. M., and Garger, R. L.**, Multiple transcripts from the antennapedia gene of *Drosophila melanogaster*, *Mol. Cell. Biol.*, 6, 4667, 1986.
70. **Karpen, S., Banerjee, R., Zelent, A., Price, P., and Acs, G.**, Identification of protein-binding sites in the hepatitis B virus enhancer and core promoter domains, *Mol. Cell. Biol.*, 8; 5159, 1988.
71. **Lopez-Cabrera, M., Letovsky, J., Hu, K.-Q., and Siddiqui, A.**, Multiple liver-specific factors bind to the hepatitis B virus core/pregenomic promter: trans-activation and repression by CCAAT/enhancer binding protein, *Proc. Natl. Acad. Sci. U.S.A.,* 87, 5069, 1990.
72. **Hu, K.-Q. and Siddiqui, A.**, Regulation of hepatitis B virus gene expression by enhancer element I, *Virology*, 181, 721, 1991.
73. **Yee, J. K.**, A liver-specific enhancer in the core promoter region of human hepatitis B virus, *Science*, 246, 658, 1989.
74. **Wang, Y., Chen, P., Wu, X., Sun, A., Wang, H., Zhu, Y., and Li, Z.**, A new enhancer element ENII identified in the X gene of hepatitis B virus, *J. Virol.*, 64, 3977, 1990.
75. **Yuh, C. and Ting, L.**, The genome of hepatitis B virus contains a second enhancer: cooperation of two elements within this enhancer is required for its function, *J. Virol.*, 64, 4281, 1990.
76. **Lopez-Cabrera, M., Letovsky, J., Hu, K.-Q., and Siddiqui, A.,** Transcriptional factor C/EBP binds to and transactivates the enhancer element II of the hepatitis B virus, *Virology*, 183, 825, 1991.
77. **Twu, J. S. and Schloemer, R. H.**, Transcriptional trans-activating function of hepatitis B virus, *J. Virol.*, 64, 3448, 1987.
78. **Spandau, D. and Lee., C. H.**, Trans-activation of viral enhancers by the hepatitis B virus X protein, *J. Virol.*, 62, 427, 1988.
79. **Twu, J. S. and Robinson, W. S.**, Hepatitis B virus X gene can transactivate heterologous viral sequences, *Proc. Natl. Acad. Sci. U.S.A.*, 86, 2046, 1989.
80. **Seto, E., Benedict Yen, T. S., Peterlin, B. M., and Ou, J. H.**, Trans-activation of the human immunodeficiency virus long terminal repeat by the hepatitis B virus X protein, *Proc. Natl. Acad. Sci. U.S.A.*, 85, 8286, 1988.
81. **Siddiqui, A., Gaynor, R., Srinivasan, A., Mapoles, J., and Farr, W.**, Trans-activation of viral enhancers including long terminal repeat of the human immunodeficiency virus by the hepatitis B virus X protein, *Virology*, 169, 2179, 1989.

82. **Colgrove, R., Simon, G., and Ganem, D.**, Transcriptional activation of homologous and heterologous genes by the hepatitis B virus X gene product in cells permissive for viral replication, *J. Virol.*, 63, 4019, 1989.
83. **Aufiero, B. and Schneider, R.**, The hepatitis B virus Z-gene product trans-activates both RNA polymerase II and III promoters, *EMBO J.*, 9, 497, 1990.
84. **Ben-Levy, R., Faktor, O., Berger, I., and Shaul, Y.**, Cellular factors that interact with the hepatitis B virus enhancer, *Mol. Cell. Biol.*, 9, 1804, 1989.
85. **Jameel, S. and Siddiqui, A.**, The HBV enhancer and trans-acting factors, UCLA Symp. Hepadna Viruses, Vol. 70, 65, 1987.
86. **Dynan, W. S.**, Modularity of promoters and enhancers, *Cell*, 58, 1, 1989.
87. **Jones, N., Rigby, P., and Ziff, E.**, Trans-acting protein factors and the regulation of eukaryotic transcription: lessons from studies on DNA tumor viruses, *Genes Dev.*, 2, 267, 1988.
88. **Kadonaga, J. T. and Tjian, R.**, Affinity purification of sequence-specific DNA binding proteins, *Proc. Natl. Acad. Sci. U.S.A.*, 83, 5889, 1986.
89. **Bohmann, D., Bos, T. J., Admon, A., Nishimura, T., Vogt, P. K.**, and **Tjian, R.,** Human proto-oncogene c-jun encodes a DNA binding protein with structural and functional properties of transcription factor Ap-1, *Science*, 238, 1386, 1987.
90. **Stern, S., Tanaka, M., and Herr, W.**, The oct-1 homeodomain directs formation of a multiprotein-DNA complex with the HSV transactivator VP16, *Nature (London)*, 341, 624, 1989.
91. **Struhl, K.**, Molecular mechanisms of transcriptional regulation in yeast, *Annu. Rev. Biochem.*, 58, 1051, 1989.
92. **Friedman, A. D., Landschulz, W. H., and McKnight, S. L.,** CCAAT/enhancer binding protein activates the serum albumin promoter in cultured hepatoma cells, *Genes Dev.*, 3, 1314, 1989.
93. **Wu, J. Y., Zhou, Z. Y., Judd, A., Cartwright, C. A., and Robinson, W. S.,** The hepatitis B virus HBV-encoded transcriptional transactivator hbx appears to be a novel protein serine/threonine Kinase, *Cell*, 63, 687, 1990.
94. **Maguire, H., Hoeffler, J., and Siddiqui, A.,** HBV X protein alters the DNA binding specificity of CREB and ATF-2 by protein–protein interaction, *Science*, 252, 842, 1991.
95. **Unger, T. and Shaul, Y.,** The X protein of the hepatitis B virus acts as a transcription factor when targeted to its responsive element, *EMBO J.*, 9, 1889, 1991.
96. **Hu, K.-Q., Vierling, J., and Siddiqui, A.,** Trans-activation of HLA-DR gene by the hepatitis B virus X gene product, *Proc. Natl. Acad. Sci. U.S.A.,* 87, 7140, 1990.
97. **Seto, E., Mitchell, P., and Yen, T. S. B.,** Transactivation by the hepatitis B virus X protein depends on AP–2, *Nature (London)*, 344, 72, 1990.
98. **Zhou, D., Tarabolulos, A., Ou, J.-H., and Yen, T. S. B.,** Activation of class I major histocompatibility complex gene expression by Hepatitis B Virus, *J. Virol*, 64, 4025, 1990.
99. **Schaack, J., Maguire, H., and Siddiqui, A.,** Hepatitis B virus x gene can substitute EIA function during adenovirus infection, submitted.

Chapter 6

SURFACE PROTEINS OF HEPATITIS B VIRUSES*

K.-H. Heermann and W. H. Gerlich

TABLE OF CONTENTS

* Dedicated to the 60th birthday of R. Thomssen.

I. HISTORICAL PERSPECTIVE AND ORIGIN OF NOMENCLATURE

The surface antigen of human hepatitis B virus (HBV) was discovered even before the virus itself was identified. The history of its discovery reflects some of the biological features that are shared by the surface antigens of all known hepadnaviruses. All hepadnaviruses can maintain persistent infection in their host without inducing apparent disease. The persistently infected hepatocyte produces and secretes the viral surface antigen proteins into the blood of the virus carrier. For poorly understood reasons, many individuals are not able to mount a sufficient immune response against the surface antigen and against the surface antigen-expressing cells to clear the viral infection. Thus, such virus carriers often have surprisingly high levels of the viral surface antigens in their blood. However, other subjects who produce a good immune response to HBV develop high-titer antibodies to the viral surface proteins.

In the early 1960s little was known about the human hepatitis viruses. All efforts to grow them in cell culture or to transmit the disease to experimental animals had failed. At this time, the population geneticist B. S. Blumberg started to search for polymorphic serum proteins that he wanted to use as genetic markers. As probes for such polymorphic serum proteins he used sera from persons who had received multiple blood transfusions. The rationale was that the polytransfused persons would develop antibodies against those polymorphic serum proteins that were different from their own set of corresponding proteins. This approach indeed allowed identification of the polymorphism in β-lipoproteins. The second "polymorphic protein" was detected first in the serum of an Australian aborigine and correspondingly it was named "Australia antigen" or "AuAg."[1] The frequent presence of AuAg in tropical populations but not in Caucasians, the familial clustering, and even the detection in patients with Down's syndrome seemed to be consistent with the assumption that AuAg was a polymorphic protein encoded by the human genome.

However, in further studies appearance of AuAg was noted in some persons who were previously AuAg negative, and finally it was noted that some of these persons developed hepatitis. At that time viral hepatitis was divided into two "types" on epidemiological grounds. Type A was transmitted by the fecal-oral route, whereas type B was often transmitted by percutaneous contact with human serum. Thus, hepatitis type B was also named serum hepatitis. AuAg was usually found to occur during and even before acute hepatitis B and to disappear soon after the onset of the disease.[2] A name that related better to the disease seemed now more appropriate for AuAg: the terms hepatitis-associated antigen (HAAg), serum hepatitis antigen (SHAg), and Au/SHAg were used, before an international group suggested the term hepatitis B antigen (HBAg).[3]

Initially, it was not clear if HBAg was a product of the virus genome or a product that was produced by the host in response to the virus. The first conclusive evidence that HBAg was indeed a viral antigen came from the discovery of the mutually exclusive HBAg subtype alleles *d* or *y*[4] and *w* or *r*.[5] The subtype remained constant within a given chain of infections. This suggested that the virus and not the host encoded HBAg.

The first efforts to purify HBAg from carrier blood led to the isolation of pleiomorphic, round particles of approximately 20-nm diameter that were devoid of nucleic acid.[6] However, D. S. Dane detected larger, double-shelled particles ("Dane particles") in the serum of virus carriers, which were coprecipitated together with the 20-nm HBAg particles after the addition of antibodies against HBAg.[7] J. D. Almeida et al. showed that the outer shell of the Dane particle could be stripped off by nonionic detergent, leaving a new antigen specificity that was termed hepatitis B core antigen (HBcAg).[8] Consequently, the outer shell was termed hepatitis B surface antigen (HBsAg).[3] The hypothesis that the Dane particle was the hepatitis B virion was strongly supported by the detection of an endogenous DNA polymerase activity[9] and DNA within the Dane particle.[10] A more complete account of the discovery of HBsAg and HBV is given by A. J. Zuckerman[3] and by W. S. Robinson.[11]

The discovery of hepatitis viruses did not follow the classical pathways of virology and, as a consequence, the nomenclature of hepatitis viruses is not completely consistent with the usual nomenclature of other viruses. HBV is an enveloped virus and, normally, HBs proteins would be considered as envelope proteins (Env) and the "core" protein as capsid protein or nucleoprotein.

The discovery of HBsAg laid the ground for the identification of further hepatitis viruses. Currently, five human hepatitis viruses are known. They are distinguished by the letters A to E. Again, in contrast to the usual nomenclature in virology, these letters do not refer to the taxonomical relatedness of these viruses, but only to the fact that these very different viruses may all cause clinical hepatitis.[12] However, viruses similar to human HBV have been found in several animal species and thus the virus family *Hepadnaviridae* was defined.[13,14] The envelope proteins of animal hepadnaviruses are currently also named HBs proteins. If distinction between the different animal viruses is necessary, the following abbreviation may be used: DHBs for HBs proteins from duck hepatitis B virus, GHBs for HBs protein from ground squirrel hepatitis (B) virus, WHBs for HBs protein from woodchuck hepatitis (B) virus.

The HBs proteins were detected by their antigenicity, but being an antigen is not their sole property. Thus, we suggest that the term HBsAg be used preferably in a context where reactions with the immune system are considered. Otherwise, the term HBs protein(s), surface protein(s) or envelope protein(s) should be used.

II. MORPHOLOGY AND OCCURRENCE OF NATURAL HBsAg PARTICLES

As mentioned before, HBs proteins are not only found on the complete virus particles (virions) but also on pleiomorphic particles of approximately 20-nm diameter, which will be referred to as HBs spheres. Moreover, filamentous particles of approximately 20-nm diameter and variable length (HBs filaments) are also present in the blood of virus carriers.[15] Electron micrographs of the three purified particles are shown in Figure 1. All three particle species are coprecipitated by antibody to HBs proteins (anti-HBs) and were thus shown to be HBsAg. There is no evidence that soluble, nonparticulate HBs protein is circulating in the blood.[15a] Although HBV may reside in cells other than hepatocytes, it appears that most or all HBs protein in the blood is produced in and secreted from the infected liver. The composition of the lipid in HBsAg particles is reported to be similar to that of the liver-derived, high-density lipoproteins.[16] Secretion of HBs protein into the bile has not yet been reported. Apparently, HBs protein contains a signal for its secretion into the blood.

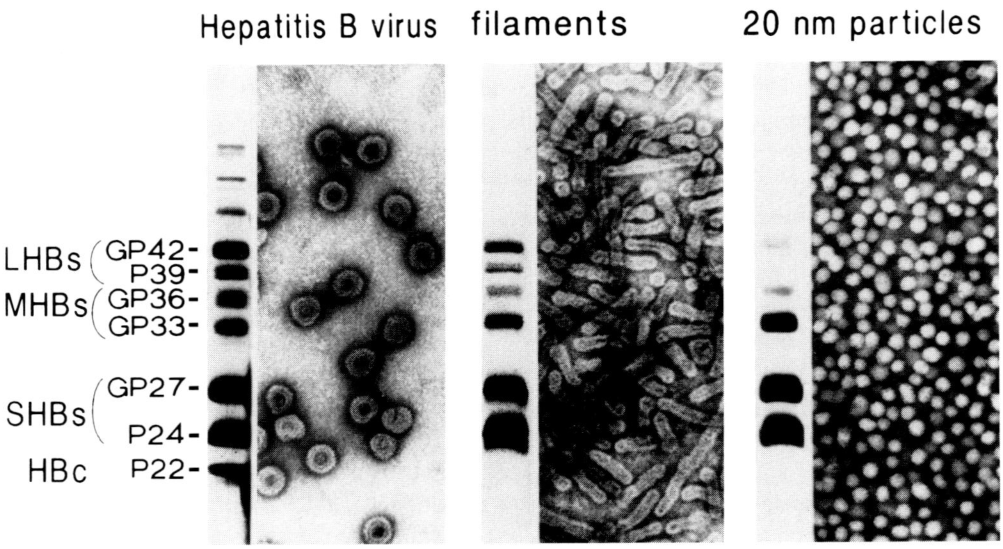

FIGURE 1. Morphology of (A) HBV particles, (B) HBs filaments, and (C) HBs spheres (20-nm particles). Particles were isolated from the blood of a persistently infected highly viremic asymptomatic carrier. Compared to the original volume of blood, virions and filaments were enriched approximately 100-fold, while the concentration of HBs spheres was approximately the same. The protein composition was analyzed by sodium dodecyl sulfate (SDS) gel electrophoresis and silver staining of the gels. Protein amount per electrophoresis lane was adjusted to approximately 2 μg. GP, glycoprotein; P, protein; numbers refer to the apparent size in kilodaltons. Virion diameter is 42 to 45 nm. LHBs, large hepatitis B surface protein; MHBs, medium hepatitis B surface protein; SHBs, small hepatitis B surface protein; HBc, hepatitis B core protein.

The amount of HBs particles circulating in the blood of virus carriers is extremely variable. It depends on the number of infected hepatocytes, the degree and nature of viral gene expression, and the intensity and specificity of the immune response. If massive infection of hepatocytes occurs during the late incubation period of acute hepatitis B or during chronic asymptomatic infection, typical levels are 100 to 200 μg HBs protein per milliliter serum.[17,18] In rare cases up to 1000 μg/ml is reached.[18a] Typically, 98 to 99% (wt/wt) of the HBs protein is present in 20-nm particles, 1 to 2% in filaments, and less than 0.2% in virions. For the production of the electron micrographs in Figure 1, 250 ml of plasma from viremic carriers was fractionated into approximately 30 μg purified virions, 400 μg HBs filaments, and 30 mg purified HBs spheres. The typical particle numbers in carrier plasma would be 10^{13}/ml for the spheres, 10^{10}/ml for the filaments, and 10^{9}/ml for the HBV particles, assuming a 10-fold lower molecular weight of the spheres.

The morphology of the HBs particles is very similar in hepadnavirus-infected woodchucks[19] or ground squirrels.[20] The total HBs concentration in serum is even higher in the chronically infected animals.[21] In hepadnavirus-infected ducks the morphology of HBs particles is considerably different because no filaments were found.[22] The morphology of DHBV is similar to that of mammalian HBVs, but the excess DHBs protein forms pleiomorphic membranous particles with an approximately 60-nm diameter (Figure 2). It is likely that empty DHBs particles form regularly shaped spheres in solution, and acquire their irregular shape due to the collapse of the empty shells on the electron microscopy grid. The typical concentration of DHBs protein in the serum of young carrier ducks is again approximately 100 μg/ml and the number of virions is approximately 10^{9}/ml.[22a] Later in the life of the carrier ducks, the concentration of DHBs and the number of virions decrease gradually to 10 μg/ml and 10^{7}/ml, respectively.[22a]

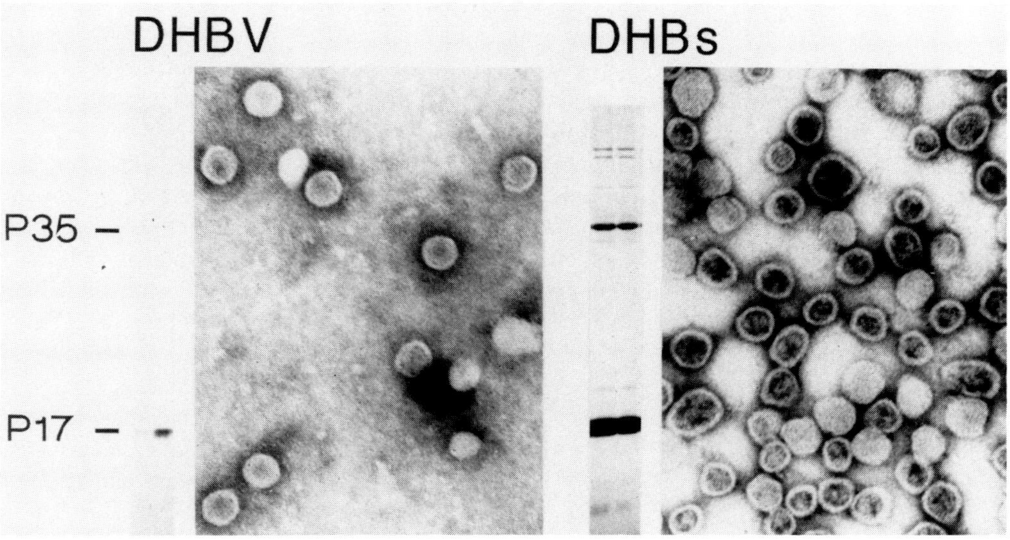

FIGURE 2. Morphology and protein composition of (A) DHBV and (B) DHBs particles. Compared to the original volume of blood, virions and DHBs particles were enriched approximately 10-fold. Concentration of virions was lower than that of DHBs particles. Thus, DHBc protein P32c is not visible. P17 corresponds to SHBs, P35 to LHBs. Virion diameter is approximately 42 nm. (We thank S. Lottmann and W. Schaldt for this figure.)

In human carriers with mild or no clinical symptoms, a steady state of HBs antigenemia and viremia may persist for years and decades. Many asymptomatic carriers develop, however, a second state of chronic infection where titers of virions in the blood are much lower. Typically, HBV DNA is no longer detectable in such carriers by direct dot-blot hybridization (i.e., $<10^5$ virions per milliliter)[23] but may be detectable after amplification by polymerase chain reaction (PCR). The HBs concentrations in such viremic carriers are often between 5 and 25 μg/ml, but lower values frequently occur. Patients with chronic active hepatitis B often also have lower serum HBs concentrations but have easily detectable virion titers. In low-titer viremic carriers, HBs protein usually forms spheres rather than filaments. Thus, it appears that filament production is somehow coupled to conditions that also allow for virion production, whereas production of spheres may occur in conditions where virion production is repressed or impossible. Due to this fact, HBsAg in serum is a highly sensitive serological marker for expression of HBV genomes in an infected host, but detection of HBs protein does not imply that infectious HBV is produced by the HBV carrier.

Extraction of HBs particles from infected liver is inefficient even if some HBs protein is stored in hepatocytes (see Section XII). HBsAg has been found in vascular immune complex deposits. Low amounts of HBsAg have also been detected in many kinds of body secretions[24] or excretions including saliva,[25] tears, breast milk,[26] semen,[27] and urine.[28] It is not completely clear if this is due to leakage of blood serum or active secretion into these fluids.

III. PREPARATION OF HBs PARTICLES

Due to their abundance, size, and typical density, HBs spheres are relatively easy to isolate and purify from the blood of highly viremic individuals. Gel chromatography through 4 or 6% agarose bead columns (preferably high-resolution type) removes more than 98% of the serum proteins. Only immunoglobulin M (IgM) and β-lipoprotein coelute with HBs spheres under these conditions.[18] HBs filaments and virions elute before the HBs spheres.[29] The opposite is true for DHBs; here, the DHBs particles elute slightly before the DHBV virions.[29a] The nonspecific absorption of HBs protein by column material must be saturated by pretreatment with serum or serum albumin. Even after such a pretreatment, many HBs filaments or virions may be irreversibly absorbed during the first pass of HBsAg-containing serum through a new column.

The diluted column eluate may be concentrated by ultrafiltration or pelleting in an ultracentrifuge. If the latter technique is to be applied, it improves the result to place a 20% sucrose cushion between the sample and the bottom of the tube. The concentrated HBs sample (spheres, or virions and filaments) is then placed on top of a continuous gradient with low steepness around the density of the HBs particles or virions. After overnight centrifugation, fractions can be collected that contain essentially pure HBs particles or virions. The HBs spheres and filaments resist the high ionic strength of cesium chloride gradients. Moreover, noncovalently bound serum proteins like part of the bound albumin are dissociated from the HBs particles and removed. However, virions are destabilized by CsCl and most of the endogenous DNA polymerase in the virions is inactivated. Thus, sucrose gradients are preferable for the preparation of virions.[30] A perfect separation of HBs filaments and virions is, however, not possible by this procedure.

The concentrations of HBs particles in supernatants from transfected cell cultures are usually much lower (10- to 1000-fold) than in sera of virus-infected individuals. Here, precipitation with polyethylene glycol (PEG) at neutral pH is a practical method of enrichment.[13] Virions and filaments are precipitated by 6% (wt/wt) PEG, while HBs spheres precipitate in 12% PEG. For large-scale preparation, zonal rotors are advisable.[32] For most laboratory purposes swinging bucket rotors are sufficient. Of course, many other precipitating reagents can also be used, and all types of gradient materials, affinity adsorbents, and electrophoresis devices have been applied to the purification of HBs particles, but the yield, antigen activity, and purity of the resulting preparations have usually not been characterized completely. Recently, five typical purification procedures for HBs spheres from human plasma have been compared.[33] A less advisable procedure of prepurification is digestion with proteases.[34] Using this method, the morphology and a major part of the antigenicity is retained, but the preS sequences are cleaved off and lost during further purification. Furthermore, HBs filaments are degraded to HBs spheres by preS-cleaving proteases.[34a] Many purification procedures seem to activate proteases transiently in the HBs-containing sample, which results in a partially degraded HBs protein. This can be noted by a blurred appearance of the HBs protein bands in gel electrophoresis and by the absence of the larger HBs proteins.

If purified HBs particles are handled in low concentration, siliconized and bovine serum albumin-coated vials and pipets should be used, because HBs is extremely sticky. Addition of nonionic detergents in low concentrations (e.g., 0.1% Tween 20) is possible, but more active detergents should be avoided, or their effect on the property to be studied must be carefully evaluated. Repeated freezing and thawing should also be avoided, because it can cause spontaneous aggregation. Disulfide-cleaving agents should be avoided under all circumstances, if intact HBs particles or enveloped virions are to be isolated. In contrast, HBsAg (except virions) is relatively stable against heat, e.g., heating to 60°C for 10 h is well tolerated. These and many other early results on the properties of HBsAg have been reviewed by Howard and Burell.[35]

IV. LABORATORY HAZARDS

Handling of human or primate sera requires precautions against potential infectivity. The laboratory where HBsAg is isolated from natural sources or from virion-producing cells should be qualified and licensed for work with infectious pathogenic materials. The staff involved should have completed successful vaccination with a licensed hepatitis B vaccine or be naturally immune. Human or primate material should be tested for occurrence of other pathogens (human immunodeficiency viruses, hepatitis C virus) and excluded from use, if possible. Materials from nonprimates (woodchucks, ground squirrels, ducks) are not dangerous as far as transmission of their hepadnaviruses to man is concerned. The CsCl-purified HBs spheres are usually not dangerous during normal laboratory handling, even if they are derived from virion-containing sources. Suspected residual infectivity can be completely removed by treatment with 1:1000 formalin for 3 d at 37°C,[34] if the specific purpose of the study does not prevent this.

V. PROTEIN COMPOSITION OF NATURAL HBs PARTICLES — EARLY RESULTS

The major components of HBs spheres were described during the early 1970s as proteins of 19 to 26 kilodaltons (kDa) and of 22 to 30 kDa in standard sodium dodecyl sulfate (SDS) gel electrophoresis.[35] The larger protein was glycosylated, while the smaller one was not.[36] D. L. Peterson showed that both proteins had the same amino- and carboxyl-terminal sequence and an identical overall amino acid composition.[37-39] Thus, it became clear that the larger protein was the glycosylated form of the smaller protein.

Some investigators did not find additional proteins in HBs spheres at that time. This was the case with protease-treated HBs preparations,[37] but also with certain untreated preparations from healthy HBsAg carriers.[40] However, most investigators found additional larger proteins in HBs preparations.[41,42] A 68-kDa protein was shown to be human serum albumin. The size and nature of the other protein bands were not well defined, but their serological cross reactivity suggested that they were somehow related to the major surface protein.[43]

These early studies were plagued by several technical problems. HBs proteins are not well stained by Coomassie Blue and, thus, high protein loads had to be applied to SDS gels to visualize the HBs proteins. Contaminating proteins were often better stained than the minor HBs proteins. Due to their high cysteine content, HBs proteins tended to form mixed dimers or oligomers after dissociation with mercaptoethanol, which lead to complicated patterns of "large" HBs proteins. Since there were no simple techniques available to determine the immunological specificity of the nonspecifically stained bands, there was no way to distinguish true HBs proteins from gel artifacts, degradation products, and contaminants.

Subsequently, cloning of the HBV genome rapidly allowed the identification of the gene for the major HBs protein. The size and the amino acid composition of the natural HBs protein agreed very well with the amino acid sequence predicted by a DNA segment within one of the four open reading frames (ORFs) of the HBV genome.[44,45] Consequently, this DNA segment of 226 codons was named the surface or S gene. Expression of the S gene under the control of foreign promoters showed that its product was sufficient to generate HBs spheres[46,47] and, moreover, such "recombinant" HBs spheres were found suitable as vaccines against HBV, like the natural HBs spheres from virus-carrier blood.

The DNA sequence of the HBV genome suggested immediately that at least one larger HBs protein should exist in addition to the well-characterized major HBs protein. The start codon of the major HBs protein is 274 or 263 codons (depending on the HBV subtype) downstream of the first start codon in the ORF. The DNA segment of the ORF that was upstream of the S gene was named preS.[48] Although this nomenclature seemed to be logical at the level of the DNA sequence analysis, it was unfortunate, because in biochemistry and cell biology the prefix "pre" is usually reserved for precursors of a final product. At no time was there an indication that the product of the whole ORF would be a precursor of the major HBs proteins.

VI. IDENTIFICATION OF PreS2-CONTAINING HBs PROTEINS

Although several previous articles had shown a slightly larger protein pair of 30 and 36 kDa in HBs spheres,[17,49-51] these data were initially neglected by the molecular biologists who had cloned the HBV genome, possibly because their sizes were not compatible with the 389 or 400 codons of the whole ORF. Moreover, the specificity of these protein bands was not proven, and some investigators could not find them at all.

Improved purification and gel analysis finally allowed identification of two glycoproteins, GP33 and GP36, as components of HBs spheres and as products of the S ORF. One important improvement was the omission of proteases and prevention of early denaturation during the purification procedure. A second point became apparent when HBs spheres were prepared separately from various donors. Readily detectable amounts of GP33/GP36 were found only in highly viremic HBV carriers. Sensitivity of HBs protein detection in SDS gels was significantly improved by silver staining, which allowed unequivocal identification of GP33/GP36. Radioimmunoprecipitation of ^{125}I-labeled HBs spheres precipitated not only the major HBs proteins but also GP33/GP36.[52]

The overall amino acid composition of GP33 and GP36 was similar to that of the major HBs protein. However, the amino-terminal sequence of GP33/GP36 seems to be blocked for Edman degradation and amino-terminal labeling,[52a] while the major HBs protein has a free methionine at the amino terminus. Partial proteolysis of GP33/GP36 produced several fragments that were also generated by proteolysis of the major proteins. These common fragments came from the carboxyl-terminal part of the major protein. The amino-terminal fragments were different. Short-time digestion of GP33 or GP36 with trypsin generated fragments which were 5 and 8 kDa smaller, respectively. The glutamine-specific V8 protease generated a 10-kDa fragment from GP33 and GP36 while the major protein was not cleaved. Treatment of GP33 or GP36 with endoglycosidase H reduced their apparent size by 3 kDa, which meant that it had removed an N-linked mannose-rich glycan, while the major HBs protein was unaffected. The predicted amino acid sequence upstream of the S gene shows a conserved start codon 55 amino acids upstream from the major S protein initiation site (the preS2 initiation codon), and 4 codons downstream of the preS2 initiation codon is a signal sequence N-X-S or T for addition of N-linked glycan, which is conserved in all mammalian hepadnaviruses (Figure 3). Arginines at positions 15, 18, and 47 of the predicted protein could readily explain the cleavage pattern of GP33/GP36 by trypsin or clostripain,[52a] and a glutamic acid immediately after the start of the major HBs protein explained the generation of the 10-kDa fragment.[53]

Independent from this work, Machida et al.[54] found a protein pair of 31 and 34 kDa in protein blots of purified HBsAg that was able to bind polymerized human serum albumin (pHSA) while the major HBs protein did not bind pHSA. Since the three carboxyl-terminal amino acids of their P31 and P34 were identical to those of the major protein it appeared likely that P31 and P34 contained the sequence of the major HBs proteins, and that an additional amino-terminal sequence had the ability to bind pHSA. A cyanogen bromide fragment from HBs spheres had indeed the composition that was predicted by the preS2 sequence, it was able to bind pHSA, and it carried glycoside.[55]

Additional evidence on the presence of preS sequences in HBs spheres came from expression experiments in transfected animal cells. In transfected mouse fibroblasts the intact S ORF expressed HBs spheres that were able to react with pHSA, but when a frameshift was introduced in the preS region the secreted HBs spheres no longer bound pHSA.[56] pHSA binding and the presence of P31 and P34 together with the major HBs protein were also observed in HBs spheres synthesized by transfected Chinese hamster ovary (CHO) cells.[57] Neurath et al. synthesized peptides of preS sequence 120–153,[58] raised antibodies against

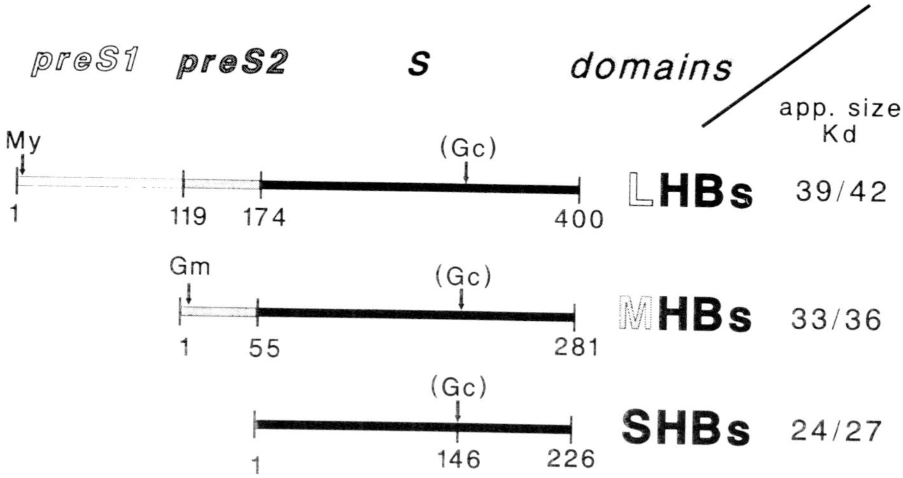

FIGURE 3. Genetic map of the three HBs proteins shown for subtype *adw*$_2$. Numbers refer to amino acids of the primary translation products. Methionine-1 is missing in LHBs and is replaced by myristic acid (My). Gm, mannose-rich mixed-type glycan linked to asparagine-4; (Gc), facultative glycan of complex type linked to asparagine-146 of SHBs or S domain.

these peptides, and showed that these anti-preS peptide antibodies reacted with P31 and P34, but not with the major HBs proteins.

VII. IDENTIFICATION OF PreS1-CONTAINING HBs PROTEINS

Traces of larger proteins than GP33/GP36 were seen in early studies on the protein composition of HBs particles. These studies were, however, confronted with the problem that dimeric major HBs protein would generate bands of approximately 44 to 48 kDa. GP33 and GP36 are even more prone to generating dimers with an apparent size of 68 kDa. Using improved methods a faint protein band (P41) was visible in certain preparations of HBs spheres but not in others.[59,32] In preparations of purified HBs filaments and virions, the additional P41 and a further protein, P38, were more readily detectable. Copurification of this protein pair in sucrose gradients with HBs filaments and virions suggested that it was possibly the product of the whole S ORF.[60] Further studies confirmed that HBs filaments and virions always contained the large HBs protein pair and that its proportion in these particles was significantly higher than in HBs spheres from the same donors (Figure 1).[30]

A reliable identification of the large HBs protein pair was possible by comparison of its immunological and biological properties with the sequence of the S ORF. Immunization of mice with purified virions led to a monoclonal antibody (MA18/7) that bound approximately 20 times better to virions than to HBs spheres from the same donor. HBs filaments bound approximately 10 times better. In immune blots this antibody bound selectively to the large protein pair, which was now considered to migrate in SDS electrophoresis at an apparent size of 39 and 42 kDa. The antibody bound to all HBs and virion preparations isolated from various HBV carriers irrespective of the HBs subtype. The binding to both intact virions and to denatured large HBs proteins suggested that the epitope recognized by MA18/7 was sequential and located at the surface of the virions, HBs filaments, and HBs spheres.[30]

Binding of the lectin concanavalin A to P42 but not to P39 in protein blots showed that P42 was glycosylated (i.e., GP42). Treatment of virions or HBs filaments with endoglycosidase F removed GP42 and increased the amount of P39. Thus, GP42 was the glycosylated version of the nonglycosylated P39. The cleavage by endoglycosidase F showed furthermore that the glycoside was linked to an asparagine residue. Treatment of virions with the glutamic acid-specific V8 protease generated one fragment of 18 kDa that bound MA18/7. This cleavage pattern was consistent with the absence of glutamic acid in the preS sequence and a cleavage at glutamic acid residue 2 of the S gene. The 18-kDa fragment generated by V8 protease treatment did not bind concanavalin A, and so it was inferred that the glycan of GP42 was bound to the S polypeptide region.[30] Further evidence that GP42 and P39 were identical (except for the glycan) and that their amino termini were at or near the beginning of the S ORF came from an HBs subtype-correlated difference in the HBV DNA sequence. HBV isolates that contain the subtype allele *d* instead of *y* very often contain an insertion of 11 codons at the beginning of the S ORF (Figure 4). Thus, the products of the whole S ORF encoded by the *d*-subtype allele would be slightly larger than those encoded by the *y*-subtype allele.[61] This could be verified experimentally by immunoblotting using MA18/7.[30] Conclusive evidence supporting this contention was obtained by expression of the preS1 DNA sequence in *Escherichia coli* under the control of the *lacUV5* promoter. The resulting fusion protein of 17 kDa specifically bound MA18/7.[62] The complete primary structure of the three known HBs proteins could be delineated as shown in Figure 3. The presumed arrangement of the three HBs proteins in virions and HBs particles is shown in Figure 5.

Details of the preS1 structure were confirmed by other groups. Antisera against partial peptides of the preS1 domain reacted in immunoblots with the large HBs proteins.[63,64] Wong et al.[65] expressed the preS sequence (amino acids 27 to 134) as a hybrid protein in *E. coli*, purified it by affinity chromatography, and raised antibodies that reacted selectively with the larger HBs proteins.

VIII. COMPARISON OF HUMAN AND ANIMAL HBs PROTEINS

It was helpful for the analysis of the very limited amounts of large and middle HBs proteins to compare the DNA sequences of various hepadnaviruses. All three well-characterized mammalian hepadnaviruses have a tripartite S ORF comprising the S gene, preS2, and preS1 sequences. In the HBV of woodchucks (W) and ground squirrels (G), the S gene is slightly shorter than in human HBV, and the preS2 and preS1 regions are slightly longer (Figure 4). While the sequence homology between the rodent viruses and human virus is relatively low, the secondary structure and hydrophobicity of the predicted polypeptide chain is well maintained.[66] The protein composition of the ground squirrel HBsAg[21] and of woodchuck HBsAg[67] is indeed similar to that of human HBsAg, although it has not been as completely analyzed. The colinearity of the three HBs protein pairs in these animals was demonstrated by two-dimensional partial peptide maps.[67] The glycosylation pattern of WHBs and GHBs has not been experimentally studied, but the conservation of the glycosylation signals at the amino terminus of preS2 and in the hydrophilic antigen loop of the S gene suggests that the pattern is the same as in the human HBs. Detection of anti-preS antibodies confirmed that the preS sequence of GHBV is indeed expressed *in vivo*.[68]

The S ORF of avian HBV is significantly different from the mammalian HBV ORFs. While the division in a hydrophobic S gene and a more hydrophilic preS region is conserved, the division into preS1 and preS2 is obviously missing,[69] or at least unclear.[70,71] In contrast to the mammalian HBVs, there is no mRNA starting upstream of a potential preS2 protein start site.[72] Although some N-glycosylation signals are present in the avian S ORFs no glycosylation is detected experimentally.[72a] The HBs particles from infected ducks (D) are much larger than the mammalian HBs spheres, and filaments are missing. The DHBs contains easily

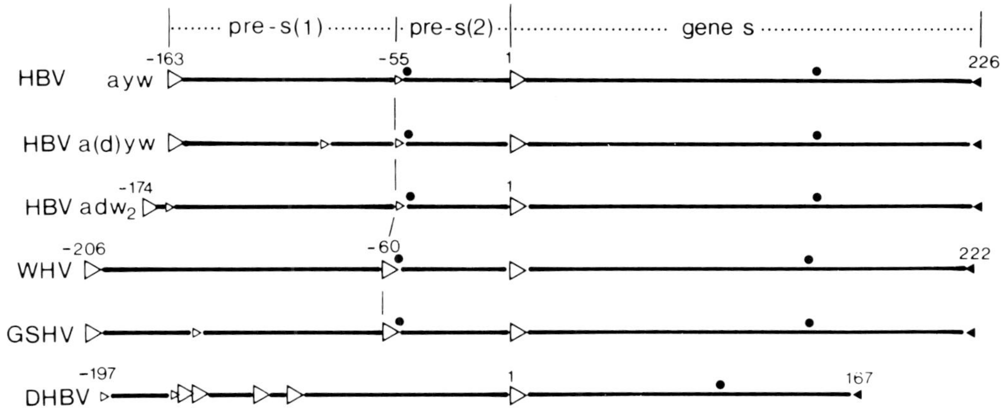

FIGURE 4. Genetic organization of S open reading frames (ORFs) in various hepadnaviruses. Δ, Strong initiation signals for protein synthesis according to the rules of M. Kozak;[178] Δ, weak initiation signals; ▲, stop signal; •, conserved glycosylation signal. For this figure the 5′ end of the S domain was taken as number 1. Sequences were taken from the following references: HBV *ayw*, Galibert et al.[45]; HBV *a(d)yw*, Pasek et al.[179]; HBV *adw₂*, Valenzuela et al.[180]; WHV, Galibert et al.[181]; GSHV, Seeger et al.[182]; DHBV, Mandart et al.[183] No obvious division of the preS domain into preS1 and preS2 is apparent for DHBV.

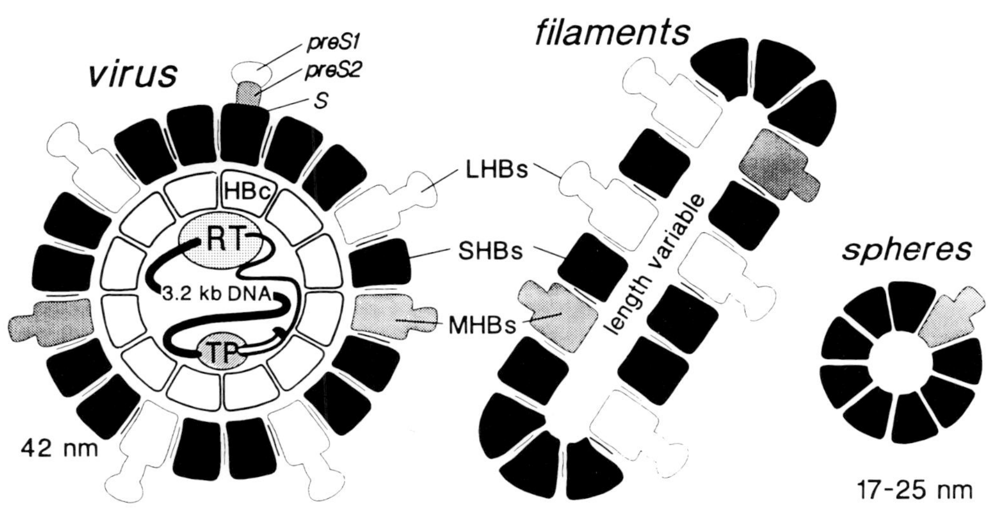

FIGURE 5. Schematic representation of the three HBs-containing particles. The model shows the approximate proportions, sizes, and spatial relations of the protein subunits, but their exact shape is not known. RT, reverse transcriptase; TP, terminal protein of the complete DNA strand. An exact ratio between HBc subunits and SHBs or S domains was not assumed, but cannot be excluded. The HBs spheres are shown without LHBs, but a proportion of HBs spheres contains small amounts of LHBs.

detectable levels of large HBs protein, but the smaller DHBs protein is much more abundant (Figure 2). A middle-sized DHBs protein has been found in infected duck liver[70,71,72b] and has been considered a preS2 protein.[70] However, there is no genetic or biochemical evidence that this protein corresponds to the middle-sized proteins of mammalian HBVs. The absence of the middle-sized protein in secreted DHBs particles appears contrary to this assumption. The absence of middle-sized DHBs, the different morphology of DHBs particles, plus the differences in the C ORF and the absence of an X ORF suggest that avian and mammalian HBVs may be divided into two genera within the *Hepadnaviridae*.[73,74]

IX. NOMENCLATURE AND NUMBERING

A clear nomenclature to distinguish the three (or two) HBs proteins has not yet been acknowledged. A group of scientists agreed at the first conference on the Molecular Biology of HBVs at the Cold Spring Harbor Laboratories (1985) that the three colinear proteins may be named large, middle, and small surface proteins. We suggest the abbreviations LHBs, MHBs, and SHBs. The letters HBs may, however, be omitted if distinction from other HBV proteins is not necessary. The often-used term preS1 or preS2 protein is somewhat misleading, because all three natural HBs proteins share the sequence of SHBs, and LHBs and MHBs also contain the preS2 sequence. Moreover, the prefix "pre" suggests to the uninformed reader that these preS proteins would be precursors. We suggest the three parts of the S ORF should be considered as the preS1, preS2, and S domains. PreS1 and preS2 sequences cannot be considered as genes because normally they cannot be expressed independently from the S gene. It appears, however, correct to consider the S ORF as composed of three colinear genes, *lhbs, mhbs,* and *shbs*. Numbering of the genes should start with their first base or codon, respectively. Thus, if MHBs is considered, numbering should start with the first codon of preS2. However, this is often handled differently in the literature. Numbering of the domains may be more practical, and is used here. Numbering of the preS1 domain is furthermore confused by the 5′ terminal insertion of 11 codons in the *d* subtype allele. Some authors prefer to include this insertion in their numbering, even if they refer to the *y* subtype allele, which does not have this insertion.

X. PROPERTIES OF SHBs

SHBs determines the morphogenesis and secretion of HBs spheres and has often been used synonymously with HBsAg. SHBs is absolutely required, but not sufficient, for secretion of HBV particles from cells that have been transfected with a replication-competent HBV genome.[75] The sequence of SHBs has three hydrophobic and two moderately hydrophilic regions. The amino-terminal hydrophobic sequence from amino acids 11 to 29 is predicted to form an α helix by the rules of Chou and Fasman. It has the ability to insert into the endoplasmic reticulum (ER) membrane and to cause translocation of the upstream sequences to the ER lumen without being cleaved by the signal peptidase (signal I).[76] The property of signal I to act as a noncleaved amino-terminal signal sequence has also been shown using β-globin gene fusion constructs.[77] The second very hydrophobic SHBs α helix from amino acids 80 to 98 is also inserted into the ER membrane (signal II), but in contrast to signal I it allows only translocation of downstream sequences while itself and upstream regions are not translocated.[78] The translocation stop signal present in signal II is possibly due to the two arginines at its amino terminus.

The orientations of the hydrophilic regions upstream of signal I and between signals I and II have not been experimentally studied. The translocation studies suggest that the amino terminus is at the surface of the particles, but refolding during maturation cannot be excluded. The orientation of signal II suggests that the sequence immediately upstream of this signal may face to the cytoplasmic side or, after maturation and secretion, to the interior of the particle. *In vitro* topogenesis studies using microsomes indicate that this sequence is sensitive to pronase treatment, while translocated regions are protected, supporting this model.[78]

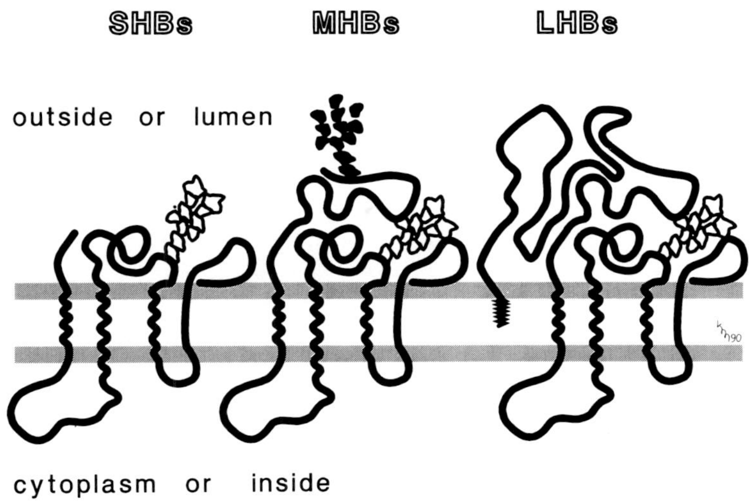

FIGURE 6. Schematic model of the three HBs proteins. The length of polypeptide chains is drawn to scale, and was adjusted to β-pleated chains except for α helices. Predictions of β turns were neglected but bends in the chain induced by prolines were adapted into the model. Many details of the model are speculative (see text). The gray lines represent the endoplasmic reticulum (ER) membrane.

If our schematic presentation in Figures 6 and 7 of the membrane orientation of HBs protein is correct, the peptide sequence between signals I and II may provide a hydrophilic coating for the SHBs molecule on the cytoplasmic side. A similar model was suggested by C. R. Howard et al.[79] This may facilitate attachment and enveloping of the HBV core particle or of the ribonucleoprotein from hepatitis D virus. This region carries five cysteines that may contribute to morphogenesis, but they may also interact with one or more cysteines of the HBc protein. It has in fact been suggested that HBs and HBc protein cannot be separated without cleavage of disulfide bonds,[80] but a functional and genetic analysis on the interaction between HBs and HBc protein has not yet been reported.

A model that suggests two additional membrane translocations between signals I and II and a partial surface exposure of these regions is problematic because there are no sufficiently hydrophobic stretches.[81] It appears possible that the mature particles may form hydrophilic channels. However, there are no clear data that suggest a superficial exposure of this sequence.

A. EPITOPE STRUCTURES

The region from amino acids 98 to 160 forms the surface of the SHBs spheres and partly those of HBs filaments and virions. The region contains clusters of cysteines that form inter- and intramolecular disulfide bridges. The exact cross linking of these many cysteines has not yet been unraveled, but it has been suggested that cysteine-121, -123, and -221 contribute to intermolecular cross linking.[81] The cross linking probably results in a very rigid structure of the polypeptide backbone that is very different from the structure with partially or completely cleaved cysteine bridges. However, not all conformational epitopes of this region depend on disulfide bonds. In a panel of 13 monoclonal antibodies against SHBs, 1 was sequential, 3 conformational but independent of disulfide bonds, and the remaining 9 were disulfide dependent.[81a] Sequential epitopes of SHBs are rare or not very immunogenic as shown by antisera against SHBs, which give a very high titer with HBs spheres as antigen, but do not react at all in immunoblots.[82] Only immunization with denaturated SHBs induces satisfactory levels of antibody against sequential epitopes.[30]

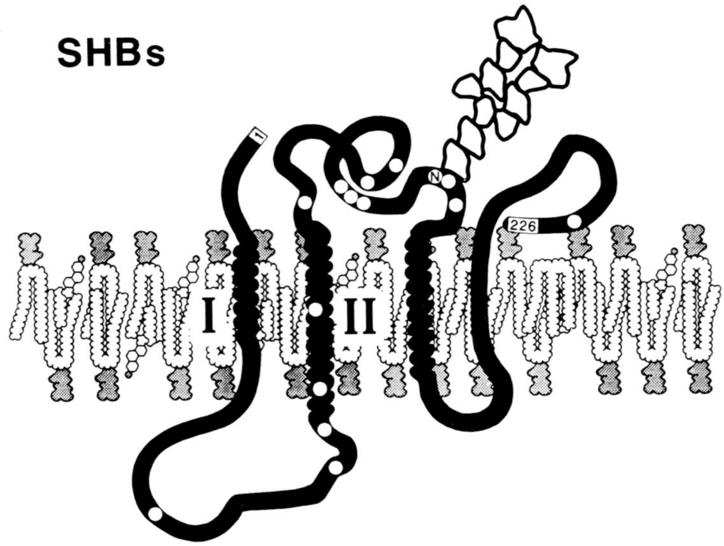

FIGURE 7. Details of SHBs. I, Signal peptide that induces translocation of upstream sequences; II, signal peptide that induces translocation of downstream sequences. White circles refer to cysteines. The background symbolizes lipids of the ER membrane. The monomers of the biantennary glycan linked to asparagine-146 are shown as irregularly shaped elements.

The SHBs antigen loop is an efficient immunogen for B cells in most but not all individuals. Genetic restriction of the anti-HBs production is mainly due to a limited ability to present T helper cell epitopes derived from SHBs (see Chapter 15). The immunogenicity of SHBs is probably increased by the assembly of repeating subunits (approximately 100 per HBs sphere). The assembly also creates epitopes that are formed at the interface of two subunits and are thus dependent on quarternary structure. At least one monoclonal antibody (RF1) is known that can react in immunoblots with dimers or oligomers.[83] Moreover, this monoclonal antibody has been shown to neutralize HBV in chimpanzee challenge experiments[84] and to terminate viremia and antigenemia in patients with immune globulin deficiency.[85]

The existence of a neutralizing antibody suggests that there are functions for SHBs antigen in the process of adhesion or penetration of HBV. However, no such function has yet been recognized. The binding of HBs particles to the monkey kidney cell line (Vero cells) has been reported but it is not clear if this has any importance in the biology of the HBV infection.[86] Binding of yeast-derived SHBs particles to various kinds of cells, including human hepatocytes, was also reported,[87] but it appears questionable if this type of binding reflects the natural situation. Incubation of recombinant SHBs spheres from mouse fibroblasts or of natural HBs spheres with human hepatoma cells did not generate SHBs-dependent binding.[87a]

Vaccination against HBV depends today on injection of several doses of SHBs and its success is measured by titers of anti-HBs. However, it is not precisely known how anti-HBs protects against infection. Recent experiments indicate that the temporal relationship between circulating virus and anti-HBs may be important in preventing infection. Postexposure protection was not possible by hepatitis B immune globulin (HBIG) in experimentally infected chimpanzees,[88] but active immunization after exposure protected the animals.[89] These remarkable experiments suggest that HBIG may not be able to inhibit spread of virus after infection has occurred, and that induction of anti-ShBs may not be the only mechanism by which active vaccination with SHBs particle protects. Nevertheless, the ability to protect before exposure with HBIG suggests that anti-SHBs is useful and can inhibit the primary events in HBV infection.

B. SUBTYPES

The high variability and the existence of subtype-specific epitopes in the HBs antigen loop suggest further that there is a selective pressure on this region by the immune system. The epitopes that are constantly present in all human HBs sample are considered by definition as determinant *a*. There are also mutually exclusive subtype determinants, *d* or *y*, and *w* or *r*. Exchange of lysine-122 in SHBs to arginine causes a subtype shift from *d* to *y*.[90] Exchange of arginine-160 to lysine causes the shift from *r* to *w*.[91] Replacement of these basic amino acids by other amino acids causes total loss of the subtype determinants.[92,93] However, other exchanges are also important for the formation of subtype specificity.[94] Moreover, "*a*" and subtype determinants cannot be clearly separated from each other. Subtyping by monoclonal antibodies[90] or by sequencing gives a much larger number of SHBs subtypes than the classical serology. There have been efforts to subdivide allele *w* into further subspecificities w_1 to w_4,[95,96] and determinants *g* or *h* have also been defined, but these have not reached the status of general recognition. Defining reactivities of monoclonals[90] or clusters of subtypical amino acids after sequencing[97] would be the optimal way of subtyping, if it is required.

For ethnological, epidemiological, or forensic reasons it may be interesting to assay subtypes of SHBs. The rare subtype of adw_4 is found in Western Amazonia and in Polynesia. This suggests a common origin of the two populations or at least an exchange of HBV between these two very distant regions.[98] For socioepidemiological reasons it would be very interesting to find out if the higher rate of HBV carriers in the black U.S. population is a consequence of their African heritage, or if HBV was introduced in America and spread more efficiently due to socioeconomical circumstances. In black Africa, subtype *ayw* is predominant, whereas in the U.S. and Northern Europe the subtype adw_2 is more abundant. Infections of patients by physicians, especially surgeons and dentists who are HBV carriers, have often been reported.[99-101] Subtyping of HBs may be helpful to establish the infectious chain in such cases, but for statistical reasons one needs more markers than just *d*, *y*, *w*, and *r*, unless the subtype is very rare in the affected population. Diagnostic sequencing of hypervariable regions in the S ORF generated by PCR may be helpful for this purpose.

It has been suggested that HBs subtypes are irrelevant for vaccines against HBV, because vaccination against one subtype protected against the challenge by another very different subtype.[102] However, experience with all infectious agents teaches that variability of exposed regions is one way in which the agents escape immunological elimination in a population. Thus it would be unwise to neglect these subtype-specific antibody responses, or to use a single subtype vaccine all over the world. An escape mutant has in fact been identified in successfully vaccinated babies from HBV carrier mothers with subtype *ayw* in southern Italy. The mutant virus has an arginine residue instead of a glycine residue at position 145.[103]

C. GLYCOSIDE OF SHBs

Asparagine-146 of SHBs is linked in a proportion of SHBs polypeptides to a biantennary complex glycan, the structure of which has been elucidated.[104,105] The glycosylation follows the typical biosynthetic pathway of other N-linked glycoproteins and is highly cell type dependent in its exact synthesis. In contrast to a previous model, there is no clear 1:1 stoichiometry between glycosylated and nonglycosylated SHBs.[48] Staining intensities by silver suggest that 60% of liver-derived SHBs is nonglycosylated, but the glycoside may possibly interact positively or negatively with this stain.

Nonhepatic animal cells that have been transfected with the S gene usually glycosylate

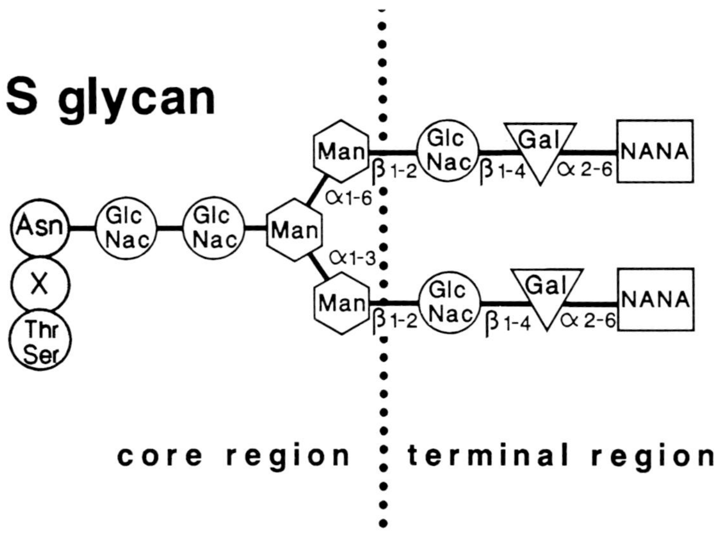

FIGURE 8. Structure of the glycan linked to SHBs or S domains (asparagine-146) in HBs spheres from HBV carriers. GlcNAc, *N*-acetylglucosamine; Man, mannose; Gal, galactosamine; NANA, *N*-acetylneuraminic acid. (Data compiled by X. Y. Lu.)

asparagine-146 less effectively than hepatic cells glycosylate SHBs.[106] Yeast cells do not glycosylate SHBs at all and still form HBs spheres of high antigenicity that contain protective epitopes.[107] However, yeast cells do not secrete HBsAg. The conservation of the glycosylation in the mammalian HBVs suggests, however, that the glycoside has an important role. Transfection of a mutant HBV genome without this glycosylation signal into a woodchuck or a ground squirrel would solve the question of whether the glycoside is necessary for the viability of the virus. As a component of the vaccine it is obviously unnecessary, or it may even be detrimental, since it represents a host component. Moreover, it masks a major "*a*" epitope between amino acids 137 and 149.[108] This epitope is essential for protection by SHBs vaccines[103] and by a certain monoclonal antibody.[109]

The linkage between asparagine-146 and the first *N*-acetyl glucosamine of the glycan side chain seems to be hidden or at least not easily accessible, because glycopeptidase F cannot remove the glycan without denaturation of SHBs spheres.[109a] The binding pattern of lectins is in agreement with the structure shown in Figure 8, which was derived from gas chromatography coupled with mass spectrometry.[104,105] Concanavalin A binds to the typical biantennary structure; *Sambucus nigra* agglutinin (SNA) binds to the terminal sialic acid; and *Ricinus communis* agglutinin (RCA) binds to the galactose residue after removal of the sialic acid.[109b]

D. ASSEMBLY OF HBs SPHERES

Formation of SHBs spheres occurs in animal cells in the absence of other viral genes. Secretion of the spheres may be slow or ineffective in invertebrate cells like *Xenopus laevis* eggs[110] or insect cells.[111] As mentioned before, yeast does not secrete HBs spheres, and it is even questionable if such HBs spheres are formed within the yeast cell. Apparently, the HBs spheres are formed only after disruption of the yeast cells. Generation of highly antigenic particles from yeast requires steps in the presence of denaturing reagents.[112] These observations suggest that cellular factors support conformational folding of SHBs and assembly, but they are not absolutely required. Natural maturation of HBs spheres within cells involves inter- and intramolecular cross linking of cysteines and budding into the ER, where addition and modification of the glycoside occur. Enzymes like disulfide isomerase and proline *cis-trans*-isomerase are probably useful for acquisition of the final conformation, since both amino acids are unusually abundant in SHBs. The role of heat shock proteins is not yet recognized.

Only signals I and II have been recognized as absolute requirements for topogenesis, while the topology of the hydrophobic carboxyl-terminal region remains obscure. Classical predictions of secondary structure suggest only β sheets for this region, but spectroscopic data suggest that there are more than the two α helices of signals I and II.[81] Reactivity of the carboxyl-terminal region with hydrophobic reagents suggests that this carboxyl-terminal sequence is buried within the lipid bilayer,[16] as schematically shown in Figures 6 and 7. There is no indication that assembly occurs via dimers as stable intermediates. Gel electrophoresis in SDS polyacrylamide gels without prior reduction does not show more than traces of monomers or oligomers. Partial reduction generates a ladder of oligomers with the dimer being just one of them.[112a]

The lipid of natural HBs spheres is reported to be similar to that of liver-derived high-density lipoprotein. The lipid composition is 60% phosphatidylcholine, 14% cholesteryl ester, 15% cholesterol, and 3% triglycerides and suggests that the lipid originates from the ER and not from the plasma membrane, which has more sphingomyelin and less phosphatidylcholine.[16] Lipid is apparently necessary for the formation of 20-nm spheres, because detergent-extracted SHBs proteins form large micelles after removal of the detergent.[113] The nature and origin of the lipid do not seem to be critical, because detergent-extracted SHBs incorporates yeast lipids and generates 20-nm spheres under these artificial conditions.[114]

E. INCORPORATION OF OTHER SEQUENCES

Although the analysis of elements necessary for topogenesis and morphogenesis of HBs spheres is still incomplete, it is clear that it is possible to express foreign DNA sequences fused to the S gene without losing the HBs polypeptide properties of morphogenesis and secretion.[115,116] Partial sequences of enteroviral VP1 or human immunodeficiency virus (HIV) GP120[117] have been introduced into the S gene and were found on the resulting hybrid particles. Sequences have been introduced within the antigenic loop, but the more logical approach is to introduce foreign sequences at the amino terminus, thus following the natural example of MHBs and LHBs. One must consider that introduction of foreign sequences at the antigenic loop will destroy many SHBs epitopes, while introduction of sufficiently large foreign sequences at the amino terminus will lead to masking of SHBs epitopes (see Section XI.A). Thus, such altered SHBs molecules may no longer be used as vaccines against HBV.

XI. PROPERTIES OF MHBs

A. TOPOLOGY

All available data suggest that the preS2 domain of MHBs is exposed at the surface of virions and HBs particles (Figure 9). The tyrosine of the preS2 domain is more efficiently surface labeled by ^{125}I than are tyrosines in the S domain of MHBs.[52] Antibodies against preS2 peptides bind to HBs spheres.[58] Moreover, the preS2 domain seems to cover large parts of the S domain in MHBs. Treatment of HBs spheres with trypsin cleaves MHBs at arginine-47[60] and increases the HBs reactivity of such particles.[30]

The proportion of MHBs in HBs spheres is variable from carrier to carrier. Viremic carriers with high concentrations of HBeAg usually have more MHBs than low-titer viremic carriers, but MHBs is always a minor component.[52,54] The relative proportion of MHBs in virions,

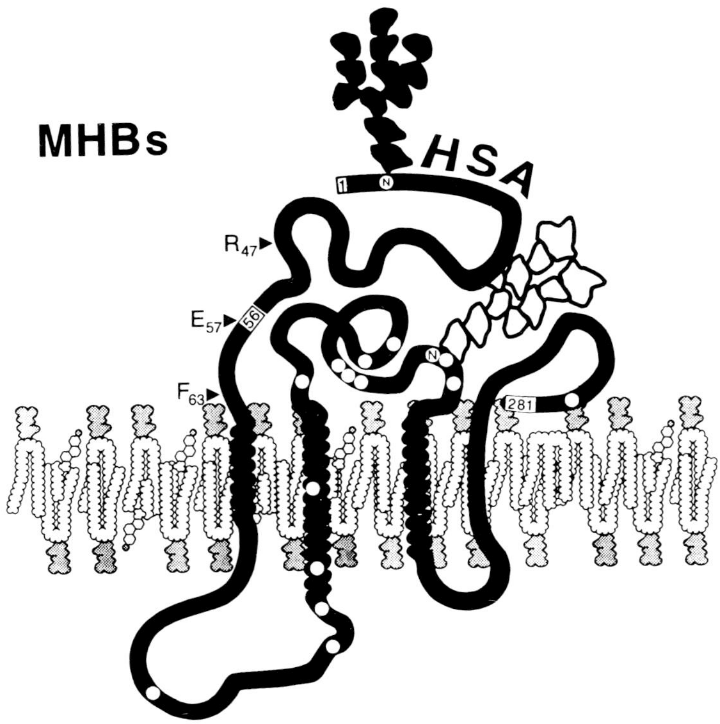

FIGURE 9. Details of MHBs. The S domain is assumed to be as in SHBs. The triantennary glycan at asparagine-4 of preS2 is shown in black. HSA, binding site for modified human serum albumin. The cleavage sites for trypsin (R_{47}), V8 protease (E_{57}), and chymotrypsin (probably F_{63}) in MHBs are also shown. The S domain begins at amino acid residue 56.

filaments, and spheres from one donor is almost constant.[30] MHBs is less abundant than LHBs in virions and more abundant than LHBs in HBs spheres. MHBs is not required for secretion of virions in cells that have been transfected by replication-competent HBV genomes.[75]

B. ALBUMIN BINDING

An intensely studied property of the preS2 domain is its binding of artificially polymerized human serum albumin (pHSA).[55] A correlation between high viremia and the presence of a so-called pHSA receptor was observed by many researchers. Only pHSA from human or chimpanzee serum, but not from other animals, bound to preS2.[118] Furthermore, pHSA binds to liver cells, albeit this binding is not species specific.[119] All of this evidence suggested that the preS2 domain would act via pHSA as a species- and organ-specific attachment site of HBV.[120] This view is corroborated by the fact that an antiserum against a preS2 peptide (amino acids 1 to 24) was able to neutralize a defined HBV inoculum used to experimentally infect chimpanzees.[121] Furthermore, immunization of chimpanzees with a conjugate of a preS2 peptide (amino acids 14 to 32) with keyhole limpet hemocyanin protected them against an HBV challenge.[122] Immunization of chimpanzees with the whole preS2 peptide was also protective.[123]

The importance of the pHSA binding was questioned by the observation that only pHSA *in vitro* cross linked by glutaraldehyde was able to bind the preS2 domain.[124] A more physiological way to modify HSA is cross linking with γ-glutamyltransferase, which also induces binding of HSA to preS2 domain.[125] Recent observations with more sensitive assays showed that unmodified monomeric HSA can also bind to preS2. However, the small proportion of HSA that binds to preS2 suggested that a naturally occurring modification or alteration of HSA is necessary for binding.[126] Possibly, cross linking by glutaraldehyde mimicks or increases this alteration. A lactosaminated human serum albumin was linked with the antiviral substance ara-AMP. The sharp decrease of viremia in patients who received this drug may have been due to binding of this artificially modified HSA to virions and subsequent hepatic uptake by the liver via the lactose receptors.[127]

The amount of preS2-binding HSA in normal human serum is so low that carriers with more than 10 μg HBs spheres per milliliter usually exhibit exposed binding sites for pHSA.[128-130] The presence of free preS2 domains in relation to viremia is explained both by the limited amount of modified HSA and the higher proportion of MHBs in HBs spheres from viremic carriers.[52,54] The masking of preS2 epitope Q19/10 in HBs particles from nonviremic HBsAg carriers was directly shown by immunoblotting using monoclonal antibody Q19/10.[128]

Nearly all preS-specific monoclonal antibodies inhibit binding of MHBs-containing HBs spheres to HSA that is coated to a solid phase, but this inhibition is usually less than 50%, while one monoclonal antibody inhibits more efficiently. This monoclonal antibody is most likely directed against an epitope between amino acids 19 and 28.[128] This site has a short homologous sequence, PAGGSS, that is also present in SHBs. It has been speculated that this homology also causes binding of HSA to SHBs.[131] In our experience such binding cannot be observed *in vitro*, but at high HSA concentrations such binding has been reported.[132] Affinity of natural HSA for MHBs or LHBs is so low that no binding can be observed in Western blotting analysis of HBs particles, but pHSA- or glutaraldehyde-treated monomeric HSA binds better.[126]

Membranes from fresh human liver are able to bind natural HBs spheres[133] or recombinant MHBs[134] when they are pretreated with pHSA. Although direct evidence is lacking, a naturally occurring pHSA may be present in low amounts at liver membranes and contribute to the attachment or even the endocytosis of HBs particles and virions. Such a natural pHSA might not circulate in the blood, because of its affinity for hepatocytes. Moreover, it might compete with circulating HSA for the preS2 attachment site.

The interaction of preS2 with natural HSA is reversible and may wane during storage in the cold[135] or during purification in high-salt solutions.[126] Possibly, the alteration of HSA that allows for binding is induced by heat, because reheating restored the complex of HBsAg and HSA.[135]

However, natural HBs spheres also contain some covalently bound HSA, which can be removed only by chemical cleavage of disulfide bonds. Since the preS domains do not contain cysteine residues, cross linking to SHBs or S domains is likely.[126] HSA has a free sulfhydryl group, while HBs spheres have only disulfide bridges. This suggests that cross linking between HSA and the S domain occurs during folding of the HSA and S domains.

It has been speculated that the presence of HSA in HBs particles contributes either to immune tolerance to HBsAg or to autoimmunity against HSA.[136] Although there is no clear evidence for this, one consequence of HSA binding to preS2 is obvious: HSA would cover the dominant B cell epitope in preS2 (amino acids 13 to 19).[137,138]

C. LIVER RECEPTOR FOR PreS2 GLYCAN

Natural HBs spheres from carrier plasma bind after a prolonged incubation time to the human hepatoma cell line HepG2. The binding is organ and species specific insofar as HeLa cells or mouse hepatocytes do not bind. The binding is not only independent of pHSA, but this substance even inhibits the binding. The binding between HBs spheres and HepG2 requires the specific glycoside structure that is linked to preS2. Removal of this glycan by peptidoglycanase F or pretreatment with certain glycan-dependent preS2 antibodies like F124 abolish the binding. Recombinant MHBs from mouse fibroblasts that contains a simple mannose-rich glycan at the preS2 domain also does not bind to HepG2 cells, while recombinant MHBs from Chinese hamster ovary (CHO) cells that contains a heterogeneous mixed-type glycan binds.[138a] The receptor for this HBs site is probably a novel lectin of liver cells that is specific for mixed-type N-linked glycan.

The absence or low activity of this lectin in fresh liver membranes suggests that it may not play an important role in the natural infection process.[133] Furthermore, infection of liver by HBV *in vivo* is highly efficient while HepG2 cells can barely be infected, although they produce HBV efficiently after transfection.[139] The question arises if the postulated lectin is related to a certain state of differentiation or dedifferentiation. Moreover, a speculative link to malignancy may be seen in the fact that a C-terminally truncated MHBs gene product from a human hepatoma was found to be a trans-activator.[140] Although the intracellular distribution of this truncated MHBs has not yet been studied, it might be a secreted or cell surface structural protein because it still contains translocation signal I of the S domain. Such a truncated MHBs would also be glycosylated and would bind to the novel lectin. It is known that regulation of cell growth depends on the glycosylation state of the cell membrane.

D. PROTEOLYSIS-INDUCED BINDING TO CELLS

Another type of MHBs-dependent binding is found after proteolytic cleavage of HBs spheres. In this case binding is no longer cell type specific. Besides HepG2 cells, HeLa cells and mouse fibroblasts also bind this modified MHBs. Moreover, the binding becomes acid resistant while the preS2 glycan-dependent binding is reverted by low pH. This protease-induced binding occurs only with MHBs containing HBs spheres, but not with spheres consisting only of SHBs. However, the source of the HBs spheres (natural or mouse fibroblasts) is irrelevant. Trypsin and V8 protease activate this type of binding, but low doses of chymotrypsin abolish it. The sizes of the MHBs fragments show that trypsin cleaves at arginine-47, V8 at glutamic acid-57, and chymotrypsin at a site downstream but close to glutamic acid-57 (Figure 9). The data suggest that MHBs could act like a viral fusion protein that can be activated by proteases. The fusion would require hydrophobic sequences like the signal I in the S domain. Chymotrypsin, with its affinity for aromatic amino acids, would, however, degrade the hypothetical fusion sequence.[140a] Further experiments are required to test these preliminary conclusions, because it is currently not understood why SHBs does not induce this type of binding without any protease treatment.

E. IMMUNOGENICITY OF PreS2

If only sequential epitopes are considered, preS2 is highly immunogenic or even immunodominant in HBs spheres.[58] However, if the conformational SHBs epitopes are compared with all preS2 epitopes, no immunodominance of preS2 can be observed.[60,82] Several subtype-specific epitopes are present in the preS2 domain.[141] The preS2 sequence is more conserved between the various subtypes in the N-terminal half than in the C-terminal half sequence, which is highly variable. Whether the higher variability of the C-terminal sequence reflects functional insignificance or, in contrast, high importance due to immunological pressure, remains unknown. Several subtype-specific monoclonal antibodies that bind this region have been identified. Assuming a role for the proteolytic cleavage at arginine-47, the second possibility would be favored. For a more detailed discussion on the immunogenicity of HBs proteins, see Chapter 15.

The high sensitivity of the preS2 domain to proteases has led to its loss in pepsin-treated

HBs spheres that have been used as hepatitis B vaccines. Expression of MHBs or the preS2 domains in yeast or bacteria[141a] leads to a large proportion of degraded products. Moreover, yeast also seems to glycosylate preS2 at serine or threonine residues in addition to N-glycosylation.[142,143] However, MHBs seems to be stable in transfected animals cells. A vaccine containing approximately 30% MHBs and 70% SHBs has been produced using transfected CHO cells[144] and successfully tested in human recipients.[145]

XII. LARGE HBs PROTEIN

LHBs was first clearly detected in silver-stained SDS gels of purified virions and filaments as proteins P39 and GP42.[60] A monoclonal antibody against preS1 (amino acids 30 to 36), MA18/7,[130] helped to distinguish this large HBs protein from dimers of SHBs that band around 44 to 48 kDa.

A. TOPOLOGY

The accessibility of the preS1 domain for MA18/7 in virions, filaments, and HBs spheres showed that at least parts of the preS1 domain are exposed on the surface of these particles.[30] Accessibility of a tryptic cleavage site at arginine-113 (numbering for HBV adw_2 isolate) shows that the hydrophilic region around this site is also surface accessible. However, tryptic cleavage sites of LHBs in the preS2 domain are not accessible[82] and surface iodination of tyrosines that are accessible in SHBs and MHBs is not possible in LHBs.[145a] Moreover, there is no glycan linked to the preS2 domain of LHBs. This allows distinction of preS2 domains in LHBs and MHBs without size determination, using monoclonal antibodies that react with the glycopeptide of MHBs, such as Q19/10.[129]

These findings suggest a model (Figure 10) in which preS1 faces the surface of HBs particles or virions, while S and preS2 are hidden. The orientation of the amino-terminal sequence before amino acid 20 and from amino acid 48 to 94 has not been experimentally established. There is much sequence variation in the moderately hydrophobic central region. The model in Figures 6 and 10 suggests a more hidden topology, but it appears possible that this region even reaches through the lipid bilayer to the cytoplasmic side.

The glycine-2 of preS1 at the amino terminus is known to be linked with myristic acid.[146] Therefore, we suggest that this C^{14} fatty acid attaches the amino terminus into the lipid bilayer. The orientation of the fatty acid is unknown, but in the absence of adjacent hydrophobic α helices a surface-directed position was assumed.

B. EFFECT OF LHBs ON SECRETION

The proportion of LHBs to the other HBs proteins has a strong effect on the secretion of HBs particles. This was most convincingly shown by expression of the S ORF in transgenic mice. In the absence of significant LHBs levels, HBs spheres are secreted into the blood and no HBs is stored in the liver. If expression of LHBs is put under the control of a strong inducible promoter like that of metallothionein, induction leads to a decrease of HBsAg in the serum, but the liver contains much LHBs.[147] A similar influence of increasing LHBs on the retention of HBsAg was found in microinjected *Xenopus laevis* oocytes[110] and in mouse fibroblasts with high-level expression of LHBs.[148] When expressed by a vaccinia vector, LHBs is still immunogenic, although it is not secreted.[149] The retention of LHBs in the ER is, surprisingly, not caused by the myristylation of LHBs but by a sequence located between amino acids 6 and 19.[150] Continuous high-level expression of LHBs in transgenic mice leads inevitably to storage of LHBs filaments in the ER, to storage disease,[151] and finally to hepatoma. Most interestingly, these LHBs-induced hepatomas no longer express LHBs.[152]

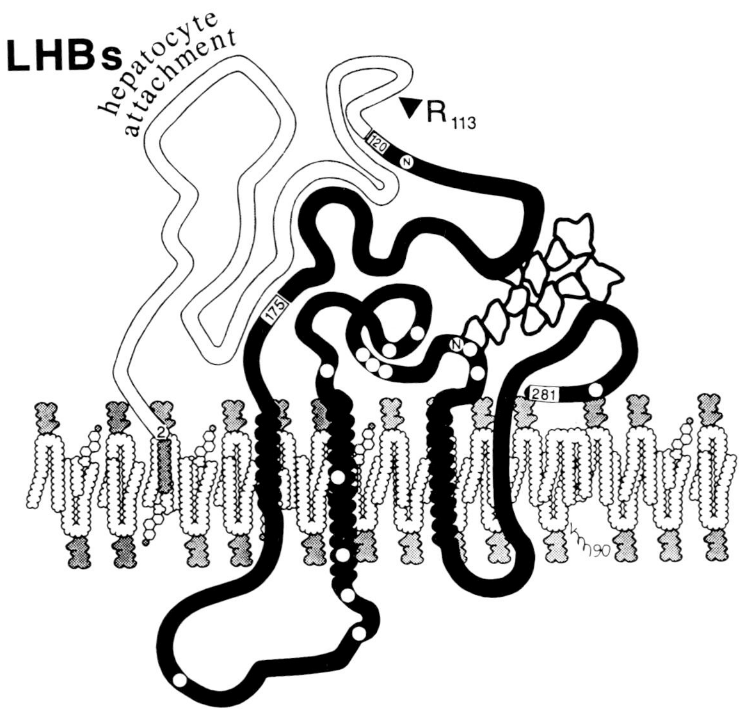

FIGURE 10. Details of LHBs. The preS1 domain is shown in white. The myristic acid at glycine-2 is shown in gray. Clostripain cleaves first at arginine-113 of LHBs. The cleavage sites of MHBs are not accessible in LHBs.

In human HBV carriers, no selective storage of LHBs is observed during the high-level replicative phase.[153] Infected hepatocytes stain equally well for LHBs, MHBs, and SHBs, and all three proteins are detected in high amounts in the serum.[154] The production and secretion of virions and viral proteins is schematically shown in Figure 11. In HBsAg carriers without high-level replication, a different pattern is found. Many hepatocytes are positive for LHBs, often in a submembranous location, but MHBs and SHBs are less intensely stained and less frequently found[155,155a] Electron microscopy shows filaments in the ER of the LHBs-positive hepatocytes.[155] Using an enzyme-linked immunosorbent assay (ELISA), low but easily detectable amounts of LHBs are found in serum 20-nm particles of low-viremic HBsAg carriers.[130,156] The selective intrahepatic storage of LHBs and low amounts of LHBs in serum of such carriers were confirmed by immunoblotting.[157] The presence of secreted HBs spheres with a low proportion of LHBs suggests that low-viremic HBsAg carriers have two types of hepatocytes, as shown in Figure 12A and B: one that is more numerous and stores LHBs, and another that expresses and secretes SHBs with small amounts of MHBs and traces of LHBs.

C. MORPHOGENESIS OF HBs PARTICLES AND LHBs

The morphological change from HBs spheres to HBs filaments resulting from an increased proportion of LHBs within the particles was found in HeLa cells that were transfected with an S ORF-expressing plasmid. These cells secreted both spheres that were devoid of LHBs and HBs filaments that contained approximately 30% LHBs. A mouse fibroblast cell line LTK that was transfected by the same plasmid secreted only HBs spheres containing trace amounts of LHBs (Figure 13).[31,106] Treatment of natural HBs filaments with clostripain cleaves LHBs at

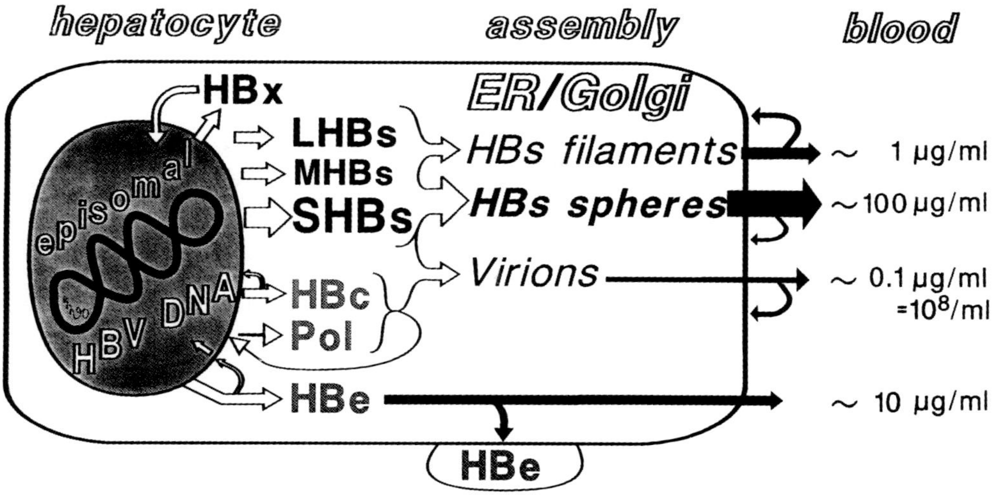

FIGURE 11. Expression of HBV proteins by infected hepatocytes and secretion into the blood of chronic carriers with high viremia. Messenger RNAs are transcribed from episomal HBV DNA. The largest mRNAs encode HBe, HBc, and the polymerase. HBe is distributed throughout the cell and is also secreted. HBc and Pol assemble with the pregenomic mRNA into core particles. These acquire an envelope that contains LHBs. Excessive HBs proteins assemble either into HBs spheres, or to HBs filaments if much LHBs is incorporated. The concentrations of HBV particles in blood are typical for the late incubation period of acute hepatitis B, or for persistent HBV infection with low antiviral immune reactions. HBx is probably continuously expressed and activates transcription of viral mRNAs.

arginine-113 and converts the filaments to strings of beads.[157a] These findings show that the proportion of LHBs is responsible for the morphology of HBs filaments. According to the composition of natural HBs particles from carrier serum, the limit between the morphology of spheres and filaments is probably between 5 and 10% LHBs. When the above-mentioned LHBs-expressing HeLa cells were grown as a nude mouse tumor, the expression of SHBs and MHBs dropped below 50% of the HBs polypeptides and HBsAg was no longer secreted.[106]

LHBs is required for the secretion of virions from cells that have been transfected by replication-competent HBV genomes.[75] Secretion of DHBV from infected hepatocytes also requires expression of L-DHBs.[158] The high proportion of LHBs in natural virions from carrier blood is in keeping with this finding. Furthermore, virions secreted by the transfected hepatoma cell line HepG2.2.15[139] contain a relatively high proportion of LHBs.[158a] It is controversial whether core particles alone can be secreted by viable transfected cells.[159,160] An interaction between LHBs and HBc must be postulated in a system where LHBs is required for viral maturation. The amino-terminal and central hydrophobic sequences of preS1 are the best candidates for a direct interaction, but an indirect interaction via a conformational change of the S domain, or SHBs, or via a cellular protein is also plausible. Myristylation of LHBs was obviously not required for LHBs-dependent secretion of virions in an HBV-replicating cell system.[75] However, an essential role of LHBs for virion formation might be questioned, because HBV-transfected LTK cells produce and secrete virus-like particles almost as efficiently as HepG2.2.15 cells, although they barely express LHBs.[161]

D. PreS1 ATTACHMENT SITE TO LIVER

Somewhat better understood than the role in virus maturation is the function of the preS1 domain as an attachment site of HBV to liver cells. In the seminal experiment of Neurath et al.,[162] suspended HepG2 cells were bound to immobilized natural HBsAg particles. This binding could be competed with peptide (amino acids 27 to 48) of preS1, and inhibited with

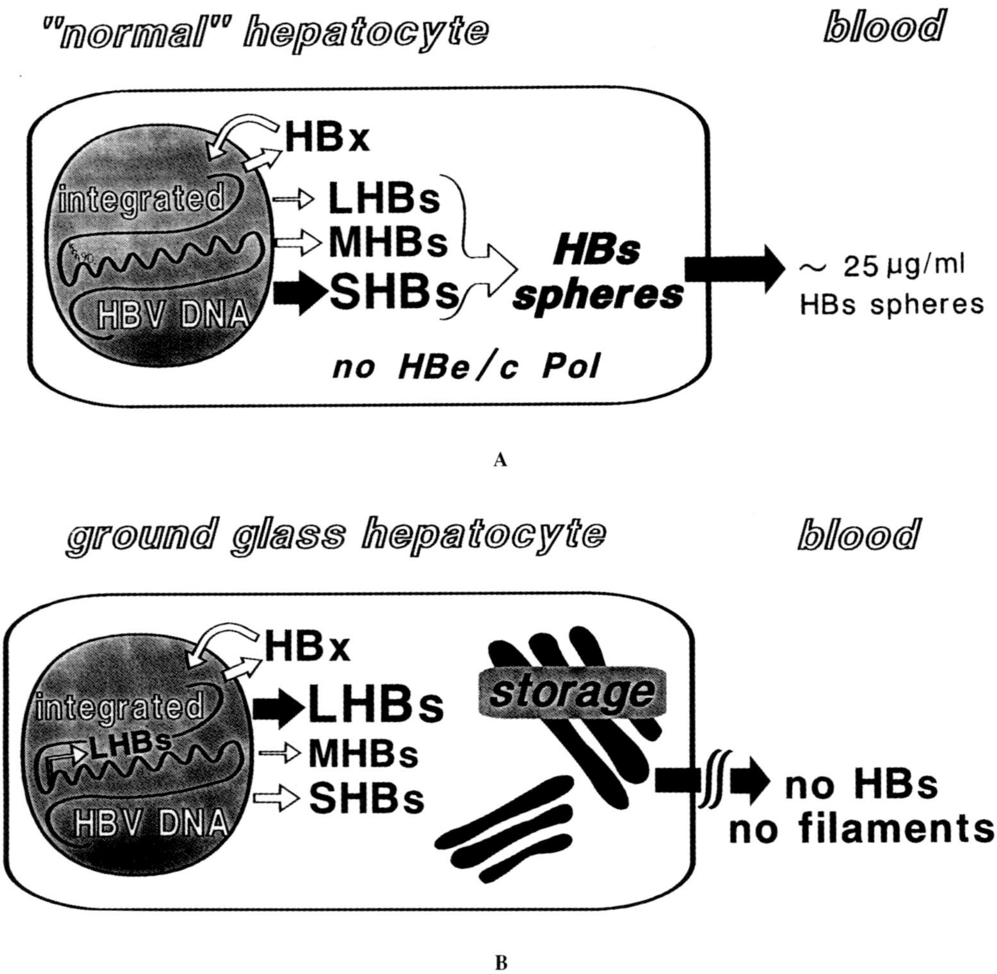

FIGURE 12. Expression of HBs spheres by hepatocytes that contain integrated fragments of the HBV genomes. Depending on the structure of the integrated viral DNA, the flanking cellular DNA, and the state of differentiation or regulation, hepatocytes may express predominantly SHBs (A) or LHBs (B). In the first case, HBs spheres with very little LHBs are secreted and reach levels in the blood as high as 25 µg/ml. In the second case, LHBs is stored in the ER and no HBs spheres or filaments are secreted. Immune histology suggests that the second case is more frequent in low-viremic carriers. HBx may or may not be expressed in the HBs-expressing cells, but episomal HBV DNA, HBe/cAg, and Pol would usually be absent.

antibodies against such peptides. This finding was confirmed using recombinant LHBs and immobilized plasma membranes from human liver.[134] A direct correlation was found between the degree of binding and the proportion of LHBs in natural HBs particles. Virions bound better than HBs filaments and these bound much better than spheres.[133] Binding could be inhibited by monoclonal antibodies that were directed against the preS1 attachment site,[133,163] but not by a monoclonal antibody against preS1 (amino acids 94 to 117) or by monoclonal antibodies against preS2.[133] The sequence around the binding site is highly homologous to a sequence of IgA[164,164a] and, possibly, it binds to an IgA receptor of the liver.[164a] The preS1 attachment site recognizes not only a receptor on liver cells, but also on monocytes and lymphocytes.[165] The interaction between preS1 and liver cells may possibly depend on a

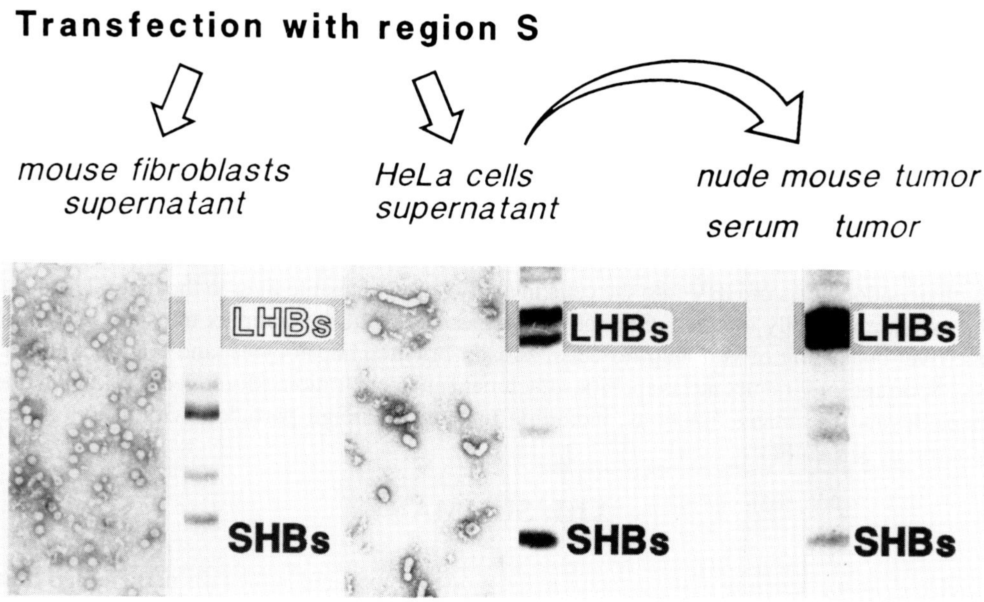

FIGURE 13. Cell type- and growth-dependent expression of S ORF (termed here region S) in cultured mouse fibroblasts, HeLa cells in culture, and HeLa cells in nude mice. After stable transfection, fibroblasts express HBs spheres that contain SHBs (P24, GP27) and MHBs (GP33, GP36), but no LHBs as shown by immune blotting. Stably transfected HeLa cells secrete HBs filaments that contain LHBs (P39, GP42) and SHBs. As nude mouse tumors, these cells do not secrete HBs particles, but LHBs accumulates within the tumor cells. The generation of the cells and tumors has been described.[31,106]

further mechanism. A 46-kDa protein from human liver cell membranes bound ^{125}I-labeled LHBs. This protein was much more abundant in serum and seems to be apolipoprotein H. The interaction between LHBs and apoH can be inhibited by antibodies against preS1 amino acids 27 to 48.[165a]

E. ROLE OF PreS DOMAINS IN INFECTIVITY

A disadvantage of the current studies on attachment of HBV is the fact that the systems used do not efficiently produce HBV after attachment of HBV. The reason for this is not clear, but in the case of HepG2 cells one would assume a block in penetration, uncoating, or transport of the genome to the nucleus, because the cells produce high-titer infectious virions after transfection.[139] Recently, cell systems and growth conditions have been developed that allow for weak replication of HBV after viral infection.[166-168] Data on neutralization of infectivity in these systems have not yet been reported. However, natural HBV inocula have been neutralized by rabbit antisera against preS1 (amino acids 12 to 47)[169] and chimpanzees have been directly immunized using a preS1 peptide (amino acids 12 to 47).[170] One might argue that these antibodies do not act by blocking an essential attachment or penetration function, but by aggregation and guiding the virus to monocytes and other Fc receptor-bearing cells. However, the suggestion that not all HBV antibodies and all HBs peptides are equally protective argues against this trivial protection mechanism. Infection of duck hepatocytes by duck HBV (DHBV) is highly efficient and can be blocked by monoclonal antibodies against the preS region of DHBV.[171] Since there is no homology between the preS region of DHBV and preS1 region of human HBV, no analogy can be postulated.

17. **Gerlich, W. and May, G.,** Hepatitis assoziiertes Antigen: Reinigung und Eigenschaften, *Zbl. Bakt. Hyg. I. Abt. Orig. A*, 224, 49, 1973.
18. **Gerlich, W. H. and Thomssen, R.,** Standardized detection of hepatitis B surface antigen: determination of its serum concentration in weight units per volume, *Dev. Biol. Stand.*, 30, 78, 1975.
18a. **Gerlich, W. H.,** unpublished observations, 1990.
19. **Summers, J., Smolec, J. M., and Snyder, R.,** A virus similar to human hepatitis B virus associated with hepatitis and hepatoma in woodchucks, *Proc. Natl. Acad. Sci. U.S.A.*, 75, 4533, 1978.
20. **Marion, P. L., Oshiro, L. S., Regnery, D. C., Scullard, G. H., and Robinson, W. S.,** A virus in Beechey ground squirrels that is related to hepatitis B virus of humans, *Proc. Natl. Acad. Sci. U.S.A.*, 77, 2941, 1980.
21. **Gerlich, W. H., Feitelson, M. A., Marion, P. L., and Robinson, W. S.,** Structural relationships between the surface antigens of ground squirrel hepatitis virus and human hepatitis B virus, *J. Virol.*, 36, 787, 1980.
22. **Mason, W. S., Seal, G., and Summers, J.,** Virus of Pekin ducks with structural and biological relatedness to human hepatitis B virus, *J. Virol.*, 36, 829, 1980.
22a. **Lottman, S., Heermann, K.-H., and Gerlich, W. H.,** unpublished observations,1990.
23. **Zyzik, E., Gerlich, W. H., Uy, A., Köchel, H., and Thomssen, R.,** Assay of hepatitis B virus genome titers in sera of infected subjects, *Eur. J. Clin. Microbiol.*, 5, 330, 1986.
24. **Davison, F., Alexander, G. J. M., Trowbridge, R., Fagan, E. A., and Williams, R.,** Detection of hepatitis B virus DNA in spermatozoa, urine, saliva, and leucocytes of chronic HBsAg carriers, *J. Hepatol.*, 4, 37, 1987.
25. **Ben-Aryeh, H., Ur, I., and Ben-Prath, E.,** The relationship between antigenaemia and excretion of hepatitis B surface antigen in human whole saliva and in gingival crevicular fluid, *Arch. Oral Biol.*, 30, 97, 1985.
26. **Linnemann, C. C,. Jr. and Goldberg, S.,** HBsAg in breast milk, *Lancet*, ii, 155, 1974.
27. **Karayiannis, P., Novick, D. M., Lok, A. S., Fowler, M. J., Monjardino, J., and Thomas, H. C.,** Hepatitis B virus DNA in saliva, urine, and seminal fluid of carriers of hepatitis B e antigen, *Br. Med. J.*, 290, 1853, 1985.
28. **Di Bisceglie, A. M., Dusheiko, G. M., and Kew, M. C.,** Detection of markers of hepatitis B virus infection in urine of chronic carriers, *J. Med. Virol.*, 16, 337, 1985.
29. **Stamm, B., Gerlich, W. H., and Thomssen, R.,** Experiments for the development of a hepatitis B vaccine: immunogenicity of HBsAg in guinea pigs, *Med. Microbiol. Immunol.*, 167, 83, 1979.
29a. **Lottman, S.,** unpublished observations, 1990.
30. **Heermann, K. H., Goldmann, U., Schartz, W., Seyffarth, T., Baumgarten, H., and Gerlich, W. H.,** Large surface proteins of hepatitis B virus containing the pre-s sequence, *J. Virol.*, 52, 396, 1984.
31. **Marquardt, O., Heermann, K.-H., Seifer, M., and Gerlich, W. H.,** Cell specific expression of preS1 antigen and secretion of hepatitis B virus surface antigen, *Arch. Virol.*, 96, 249, 1987.
32. **Thomssen, R., Gerlich, W. H., Böttcher, U., Legler, K., Ritter, S., Stibbe, W., Weinmann, W., Klinge, O., and Pfeifer, U.,** Safety and potency aspects in the preparation of an experimental HBsAg vaccine, *Dev. Biol. Stand.*, 54, 23, 1983.
33. **Fields, H. A., Bruguera, P., de-la-Torre, N., Puig, J., and Anderson, L. J.,** Purity, antigenicity and immunogenicity of the hepatitis B surface antigen purified by five different methods, *J. Virol. Methods*, 22, 283, 1988.
34. **Tabor, E., Buynak, E., Smallwood, L. A., Snoy, P., Hilleman, M., and Gerety, R.J.,** Inactivation of hepatitis B virus by three methods: treatment with pepsin, urea, or formalin, *J. Med. Virol.*, 11, 1, 1983.
34a. **Heermann, K.-H.,** manuscript in preparation, 1991.
35. **Howard, C. R. and Burell, C. J.,** Structure and nature of hepatitis B antigen, *Prog. Med. Virol.*, 22, 36, 1976.
36. **Chairez, R., Steiner, S., Melnik, J. L., and Dreesman, G. R.,** Glycoproteins associated with hepatitis B antigen, *Intervirology*, 1, 224, 1973.
37. **Peterson, D. L., Roberts, J. M., and Vyas, G. N.,** Partial amino acid sequence of two major component polypeptides of hepatitis surface antigen, *Proc. Natl. Acad. Sci. U.S.A.*, 74, 1530, 1977.
38. **Peterson, D. L.,** Isolation and characterization of the major protein and glycoprotein of hepatitis B surface antigen, *J. Biol. Chem.*, 256, 6975, 1981.
39. **Peterson, D. L., Nath, N., and Gavilanes, F.,** Structure of hepatitis B surface antigen: correlation of subtype with amino acid sequence and location of the carbohydrate moiety, *J. Biol. Chem.*, 257, 10414, 1981.
40. **Koistinen, V. U.,** Hepatitis B surface antigen polypeptides: artifactual bands in sodium dodecyl sulfate polyacrylamide gel electrophoresis caused by aggregation, *J. Virol.*, 35, 20, 1980.
41. **Shih, I. W.-K. and Gerin, J. L.,** Proteins of hepatitis B surface antigen, *J. Virol.*, 21, 347, 1977.
42. **Cabral, G. A., Cabral, F. M., Funk, G. A., Sanchez, Y., Hollinger, F. B., Melnick, J. L., and Dreesman, G. R.,** Cellular and humoral immunity in guinea pigs to two major polypeptides derived from hepatitis B surface antigen, *J. Gen. Virol.*, 38, 339, 1978.
43. **Sanchez, Y., Ionexcu-Matiu, I., Dreesman, G. R., Hollinger, F. B., and Melnick, J. L.,** Evidence for the presence of repeating antigenic determinants in the major and minor polypeptides derived from hepatitis B surface antigen, *Virology*, 114, 71, 1981.

44. **Valenzuela, P., Gray, P., Quiroga, M., Zaldivor, J., Goodman, H. M., and Rutter, W. J.,** Nucleotide sequence of the gene coding for the major protein of hepatitis B virus surface antigen, *Nature (London)*, 280, 815, 1979.
45. **Galibert, F., Mandart, E., Fitoussi, F., Tiollais, P., and Charnay, P.,** Nucleotide sequence of the hepatitis B virus genome (subtype *ayw*) cloned in *E. coli*, *Nature (London)*, 281, 646, 1979.
46. **Moriarty, A. M., Hoyer, B. H., Shih, J. W.-K., Gerin, J. L., and Hamer, D. H.,** Expression of the hepatitis B virus surface antigen gene in cell culture using a simian virus 40 vector, *Proc. Natl. Acad. Sci. U.S.A.*, 78, 2606, 1981.
47. **Liu, C. C., Yansura, D., and Levinson, A. D.,** Direct expression of hepatitis B surface-antigen in monkey cells from an SV40 vector, *DNA*, 1, 213, 1982.
48. **Tiollais, P., Charnay, P., and Vyas, G. N.,** Biology of hepatitis B virus, *Science*, 213, 406, 1981.
49. **Skelly, J., Howard, C. R., and Zuckerman, A. J.,** Analysis of hepatitis B surface antigen components solubilized with Triton X-100, *J. Gen. Virol.*, 44, 679, 1979.
50. **Gerin, J. L., Holland, P. V., and Purcell, R.H.,** Australia antigen: large scale purification from human serum and biochemical studies of its proteins, *J. Virol.*, 7, 569, 1971.
51. **Dreesman G. R., Chairez, R., Suarez, M., Hollinger, F. B., Courtney, R. J., and Melnick, J. L.,** Production of antibodies to individual polypeptides derived from purified hepatitis B surface antigen, *J. Virol.*, 16, 508, 1975.
52. **Stibbe, W. and Gerlich, W. H.,** Variable protein composition of hepatitis B surface antigen from different donors, *Virology*, 123, 436, 1982.

52a. **Heermann, K.-H.,** unpublished observations, 1990.

53. **Stibbe, W. and Gerlich, W. H.,** Structural relationships between minor and major proteins of hepatits B surface antigen, *J. Virol.*, 46, 626, 1983.
54. **Machida, A., Kishimoto, S., Ohnuma, H., Miyamoto, H., Baba, K., Oda, K., Nakamura, T., Miyakawa, M., and Mayumi, M.,** A hepatitis B surface antigen polypeptide (P31) with the receptor for polymerized human as well as chimpanzee albumins, *Gastroenterology*, 85, 268, 1983.
55. **Machida, A., Kishimoto, S., Ohnuma, H., Baba, K., Ito, Y., Miyamoto, H., Funatsu, G., Oda, K., Usuda, S., Togami, S., Nakamura, T., Miyakawa, Y., and Mayumi, M.,** A polypeptide containing 55 amino acid residues coded by the pre-S region of hepatitis B virus deoxyribonucleic acid bears the receptor for polymerized human as well as chimpanzee albumins, *Gastroenterology*, 86, 910, 1984.
56. **Persing, D. H., Varmus, H. E., and Ganem, D.,** A frameshift mutation in the pre-S region of the human hepatitis B virus genome allows production of surface antigen particles but eliminates binding to polymerized albumin, *Proc. Natl. Acad. Sci. U.S.A.*, 82, 3440, 1985.
57. **Michel, M. L., Pontisso, P., Sobczak, E., Malpiece, Y., Streeck, R. E., and Tiollais, P.,** Synthesis in animal cells of hepatitis B surface antigen particles carrying a receptor for polymerized human serum albumin, *Proc. Natl. Acad. Sci. U.S.A.*, 81, 7708, 1984.
58. **Neurath, A. R., Kent, S. B., and Strick, N.,** Location and chemical synthesis of a pre-S gene coded immunodominant epitope of hepatitis B virus, *Science*, 224, 392, 1984.
59. **Thomssen, R., Gerlich, W. H., Böttcher, U., Stibbe, W., and Legler, K.,** Herstellung und Erprobung eines Hepatitis B-Impfstoffes, *Dtsch. Med. Wschr.*, 179, 125, 1982.
60. **Stibbe, W. and Gerlich, W.H.,** Characterization of pre-S gene products in hepatitis B surface antigen, *Dev. Biol. Stand.*, 54, 33, 1983.
61. **Tiollais, P., Dejean, A., Brechot, C., Michel, M. L., Sonigo, P., and Wain-Hobson, S.,** Structure of hepatitis B virus DNA, in *Viral Hepatitis and Liver Disease*, Vyas, G.N., Dienstag, J.L., Hoofnagle, J.H., Eds., Grune & Stratton, Orlando, 1984, 49.
62. **Uy, A.,** Expression der Prä-Core und der Prä-S Sequenz aus dem Hepatitis B Virus Genom in Escherichia Coli und immunologische Charakterisierung der Produkte, *Thesis*, Göttingen, 1984.
63. **Neurath, A. R., Kent, S. B. H., Strick, N., Taylor, P., and Stevens, C. E.,** Hepatitis B virus contains pre-S gene-encoded domains, *Nature (London)*, 315, 154, 1985.
64. **Takahashi, K., Kishimoto, S., Ohnuma, H., Machida, A., Takai, E., Tsuda, F., Miyamoto, H., Tanaka, T., Matsushita, K., Oda, K., Miyakawa, Y., and Mayumi, M.,** Polypeptides coded for by region preS and gene S of hepatitis B virus DNA with the receptor for polymerized human serum albumin: expression on hepatitis B particles produced in the HBeAg or anti-HBe phase of hepatitis B virus infection, *J. Immunol.*, 136, 3467, 1986.
65. **Wong, D. T., Nath, N., and Sninsky, J. J.,** Identification of hepatitis B virus polypeptides encoded by the entire pre-s open reading frame, *J. Virol.*, 55, 223, 1985.
66. **Schaeffer, E. and Sninsky, J. J.,** Predicted secondary structure similarity in the absence of primary amino acid sequence homology: hepatitis B virus open reading frames. *Proc. Natl. Acad. Sci. U.S.A.*, 81, 2902, 1984.

67. **Feitelson, M., Marion, P. L., and Robinson, W. S.,** The nature of polypeptides larger in size than the major surface-antigen components of hepatitis B and like viruses in ground squirrels, woodchucks and ducks, *Virology*, 130, 76, 1983.
68. **Persing, D. H., Varmus, H. E., and Ganem, D.,** Antibodies to pre-S and X determinants arise during natural infection with ground squirrel hepatitis virus, *J. Virol.*, 60, 177, 1986.
69. **Schlicht, H. J., Kuhn, C., Guhr, B., Mattaliano, R. J., and Schaller, H.,** Biochemical and immunological characterization of the duck hepatitis B virus envelope proteins, *J. Virol.*, 61, 2280, 1987.
70. **Yokosuka, O., Omata, M., and Ito, Y.,** Expression of pre-S1, pre-S2, and C proteins in duck hepatitis B virus infection, *Virology*, 167, 82, 1988.
71. **Li, J.-S., Cova, L., Buckland, R., Lambert, V., Deléage, G., and Trepo, C.,** Duck hepatitis B virus can tolerate insertion, deletion, and partial frameshift mutation in the distal pre-S region, *J. Virol.*, 63, 4965, 1989.
72. **Büscher, M., Reiser, W., Will, H., and Schaller, H.,** Transcripts and the putative RNA pregenome of duck hepatitis B virus: implications for reverse transcription, *Cell*, 40, 717, 1985.
72a. **Lottmann, S., Heermann, K.-H., and Gerlich, W.H.,** unpublished observations, 1990.
72b. **Lottman, S., et al.,** unpublished observations, 1990.
73. **Sprengel, R., Kuhn, C., Will, H., and Schaller, H.,** Comparative sequence analysis of duck and human hepatitis B virus genomes, *J. Med. Virol.*, 15, 323, 1985.
74. **Mandart, E., Kay, A., and Galibert, F.,** Nucleotide sequence of a cloned duck hepatitis B virus genome: comparison with woodchuck and human hepatitis B virus sequences, *J. Virol.*, 49, 782, 1984.
75. **Bruss, V. and Ganem, D.,** The role of envelope proteins in hepatitis B virus assembly, *Proc. Natl. Acad. Sci. U.S.A.*, in press.
76. **Eble, B. E., Lingappa, V. R., and Ganem, D.,** The N-terminal (pre-S2) domain of a hepatitis B virus surface glycoprotein is translocated across membranes by downstream signal sequences, *J. Virol.*, 64, 1414, 1990.
77. **Eble, B. E., Lingappa, V. R., and Ganem, D.,** Hepatitis B surface antigen: an unusual secreted protein initially synthesized as a transmembrane polypeptide, *Mol. Cell. Biol.*, 6, 1454, 1986.
78. **Eble, B. E., MacRae, D. R., Lingappa, V. R., and Ganem, D.,** Multiple topogenic sequences determine the transmembrane orientation of the hepatitis B surface antigen, *Mol. Cell. Biol.*, 7, 3591, 1987.
79. **Howard, C. R., Stirk, H. J., Brown, S. E., and Steward, M. W.,** Towards the development of synthetic hepatitis B vaccines, in *Viral Hepatitis and Liver Disease*, Zuckerman, A.J., Ed., Alan R. Liss, New York, 1988, 1094.
80. **Petit, M. A., Capel, F., and Pillot, J.,** Demonstration of a firm association between hepatitis B surface antigen proteins bearing polymerized human albumin binding sites and core-specific determinants in serum hepatitis B viral particles, *Mol. Immunol.*, 22, 1279, 1985.
81. **Guerrero, E., Gavilanes, F., and Peterson, D. L.,** Model for the protein arrangement in HBsAg particles based on physical and chemical studies, in *Viral Hepatitis and Liver Disease*, Zuckerman, A. J., Ed., Alan R. Liss, New York, 1988, 606.
81a. **Deepen, R., Heermann, K.-H., and Gerlich, W.H.,** unpublished observations, 1990.
82. **Heermann, K.-H., Kruse, F., Seifer, M., and Gerlich, W. H.,** Immunogenicity of the gene S and pre-S domains in hepatitis B virions and HBsAg filaments, *Intervirology*, 28, 14, 1987.
83. **Hauser, P., Thomas, H. C., Waters, J., Simoen, E., Voet, P., De Wilde, M., Stephenne, J., and Pétre, J.,** Induction of neutralizing antibodies in chimpanzees and humans by a recombinant yeast-derived hepatitis B surface antigen particle, in *Viral Hepatitis and Liver Disease*, Zuckerman, A. J., Ed., Alan R. Liss, New York, 1988, 1031.
84. **Iwarson, S., Tabor, E., Thomas, H. C., Goodall, A., Waters, J., Snoy, P., Shih, J. W., and Gerety, R. J.,** Neutralization of hepatitis B virus infectivity by a murine monoclonal antibody: an experimental study in the chimpanzee, *J. Med. Virol.*, 16, 89, 1985.
85. **Lever, A. M. L., Waters, J., Brook, M. G., Karayiannis, P., and Thomas, H. C.,** Treatment of chronic hepatitis B virus infection with monoclonal antibody to the hepatitis B virus surface antigen in two patients with hypogammaglobulinaemia, in *Viral Hepatitis and Liver Disease*, Zuckerman, A. J., Alan R. Liss, New York, 1988, 961.
86. **Peeples, M. E., Komai, K., Radek, R., and Bankowski, M. J.,** A cultured cell receptor for the small S protein of hepatitis B virus, *Virology*, 160, 135, 1987.
87. **Leenders, W. J., Glansbeek, H. L., Bruin, W. C. C., and Yap, S.-H.,** Binding of the major and large HBsAg to human hepatocytes and liver plasma membranes: putative external and internal receptors for infection and secretion of hepatitis B virus, *Hepatology*, 12, 141, 1990.
87a. **Lu, X., Heerman, K.-H., and Gerlich, W.H.,** in preparation.
88. **Wahl, M., Iwarson, S., Snoy, P., and Gerety, R. J.,** Failure of hepatitis B immune globulin to protect against experimental infection in chimpanzees, *J. Hepatol.*, 9, 198, 1989.

89. **Iwarson, S., Wahl, M., Ruttimann, E., Snoy, P., Seto, B., and Gerety, R. J.,** Successful postexposure vaccination against hepatitis B in chimpanzees, *J. Med. Virol.*, 25, 433, 1988.
90. **Peterson, D. L., Paul, D. A., Lam, J., Tribby, I. I., and Achord, D. T.,** Antigenic structure of hepatitis B surface antigen: identification of the "d" subtype determinant by chemical modification and use of monoclonal antibodies, *J. Immunol.*, 132, 920, 1984.
91. **Okamoto, H., Imai, M., Miyakaw, Y., and Mayumi, M.,** Site-directed mutagenesis of hepatitis B surface antigen sequence at codon 160 from arginine to lysine for conversion of subtypic determinant from r to w, *Biochem. Biophys. Res. Commun.*, 148, 500, 1987.
92. **Okamoto, H., Omi, S., Wang, Y., Itoh, Y., Tsuda, F., Tsuda, F., Tanaka, T., Akahane, Y., Miyakawa, Y., and Mayumi, M.,** The loss of subtypic determinants in alleles, d/y or w/r, on hepatitis B surface antigen, *Mol. Immunol.*, 26, 197, 1989.
93. **Ashton-Rickardt, P. G. and Murray, K.,** Mutations that change the immunological subtype of hepatitis B virus surface antigen and distinguish between antigenic and immunogenic determination, *J. Med. Virol.*, 29, 204, 1989.
94. **Gerin, J. L., Alexander, H., Shih, J. W.-K., Purcell, R. H., Dapolito, G., Engle, R., Green, N., Sutcliffe, J. G., Shinnick, T. M., and Lerner, R. A.,** Chemically synthesized peptides of hepatitis B surface antigen duplicate the d/y specificities and induce subtype-specific antibodies in chimpanzees, *Proc. Natl. Acad. Sci. U.S.A.*, 80, 2365, 1983.
95. **Courouce-Pauty, A. M., Plancon, A., and Soulier, J. P.,** Distribution of HBsAg subtypes in the world, *Vox Sanguinis*, 44, 197, 1983.
96. **Courouce, A. M., Holland, P. V., Muller, J. Y., and Soulie, J. P.,** HBs-antigen subtypes, *Bibl. Haematol.*, 42, 20, 1976.
97. **Okamoto, H., Tsuda, F., Sakugawa, H., Sastrosoewignijo, R. I., Imai, M., Miyakawa, Y., and Mayumi, M.,** Typing hepatitis B virus by homology in nucleotide sequence: comparison of surface antigen subtypes, *J. Gen. Virol.*, 69, 2575, 1988.
98. **Gaspar, A. M. and Yoshida, C. F.,** Geographic distribution of HBsAg subtypes in Brazil, *Mem. Inst. Oswaldo Cruz*, 82, 253, 1987.
99. **Communicable Disease Surveillance Centre and the Epidemiological Research Laboratory of the Public Health Laboratory Service, together with a district control-of-infection service,** Acute hepatitis B associated with gynaecological surgery, *Lancet*, i, 1, 1980.
100. **Rimland, D., Parkin, W. E., Miller, G. B., Jr., and Schrack, W. D.,** Hepatitis B outbreak traced to an oral surgeon, *N. Engl. J. Med.*, 296, 953, 1977.
101. **Snydman, D. R., Hindman, S. H., Wineland, M. D., Bryan, J. A., and Maynard, J. E.,** Nosocomial viral hepatitis B: a cluster among staff with subsequent transmission to patients, *Ann. Intern. Med.*, 85, 573, 1976.
102. **Purcell, R. H. and Gerin, J. L.,** Hepatitis B subunit vaccine. A preliminary report of safety and efficacy test in chimpanzees, *Am. J. Med. Sci.*, 270, 395, 1975.
103. **Carman, W. F., Zanetti, A. R., Karayiannis, P., Waters, J., Manzillo, G., Tanzi, E., Zuckerman, A. J., and Thomas, H. C.,** Vaccine-induced escape mutant of hepatitis B virus, *Lancet*, 336, 325, 1990.
104. **Hanaoka, K., Takasaki, S., Kobata, A., Miyamoto, H., Nakamura, T., and Mayumi, M.,** Structure of asparagine-linked sugar chains in the major polypeptide of hepatitis B surface antigen, *J. Biochem.*, 99, 1273, 1986.
105. **Gillece-Castro, B. L., Fisher, S. J., Tarentino, A. L., Peterson, D. L., and Burlingame, A. L.,** Structure of the oligosaccharide portion of human hepatitis B surface antigen, *Arch. Biochem. Biophys.*, 256, 194, 1987.
106. **Marquardt, O., Heermann, K.-H., Seifer, M., and Gerlich, W. H.,** Cell type specific expression of preS1 antigen and secretion of hepatitis B surface antigen, *Postgrad. Med. J.*, 63, 41, 1987.
107. **Valenzuela, P., Medina, A., Rutter, W. J., Ammerer, G., and Hall, B. D.,** Synthesis and assembly of hepatitis B virus surface antigen particles in yeast, *Nature (London)*, 298, 347, 1982.
108. **Howard, C. R., Brown, S. E., Hogben, D. N., Zuckerman, A. J., Murray-Lyon, I. M., and Steward, M. W.,** Analysis of antibody responses to hepatitis B surface antigen, in *Viral Hepatitis and Liver Disease*, Vyas, G.N., Dienstag, J.L., and Hoofnagle, J.H., Eds., Grune & Stratton, New York, 1984, 561.
109. **McMahon, G., McCarthy, L. A., Dottavio, D., and Ostberg, L.,** Surface antigen and polymerase gene variation in hepatitis B virus isolates from a monoclonal antibody treated liver transplant patient, Abstract 82, in Int. Symp. Viral Hepatitis and Liver Disease, Houston, 1990.

109a. **Lu, X. Y., Heermann, K.-H., and Gerlich, W. H.,** unpublished observations, 1990.

109b. **Lu, X. Y.,** unpublished observations, 1990.

110. **Standring, D. N., Ou, J. H., and Rutter, W. J.,** Assembly of viral particles in *Xenopus* oocytes: pre-surface-antigens regulate secretion of the hepatitis B viral surface envelope particle, *Proc. Natl. Acad. Sci. U.S.A.*, 83, 9338, 1986.

111. **Lanford, R. E., Luckow, V., Kennedy, R. C., Dreesman, G. R., Notvall, L., and Summers, M. D.,** Expression and characterization of hepatitis B surface antigen polypeptides in insect cells with a baculovirus expression system, *J. Virol.*, 63, 1549, 1989.
112. **Wampler, D. E., Lehman, E. D., Boger, J., McAleer, W. J., and Scolnick, E. M.,** Multiple chemical forms of hepatitis B surface antigen produced in yeast, *Proc. Natl. Acad. Sci. U.S.A.*, 82, 6830, 1985.
112a. **Heermann, K.-H.,** unpublished observations, 1990.
113. **Howard, C. R., Young, P. R., Lee, S., Dixon, J., Zuckerman, A. J., McAleer, W. J., and Lehman, E. D.,** Hepatitis B surface antigen polypeptide micelles from antigen expressed in *Saccharomyces cerevisiae*, *J. Virol. Methods*, 14, 25, 1986.
114. **Petre, J., Von Wijnendaele, F., De Neys, B., Conrath, K., and Van Opstal, O.,** Development of a hepatitis B vaccine from transformed yeast cells, *Postgrad. Med. J.*, 63, 73, 1987.
115. **Delpeyroux, F., Chenciner, N., Lim, A., Lambert, M., Malpiece, Y., and Streeck, R. E.,** Insertions in the hepatitis B surface antigen. Effect on assembly and secretion of 22-nm particles from mammalian cells, *J. Mol. Biol.*, 195, 343, 1987.
116. **Delpeyroux, F., Peillon, N., Blondel, B., Crainic, R., and Streeck, R. E.,** Presentation and immunogenicity of the hepatitis B surface antigen and a poliovirus neutralization antigen on mixed empty envelope particles, *J. Virol.*, 62, 1836, 1988.
117. **Michel, M. L., Mancini, M., Sobczak, E., Favier, V., Guetard, D., Bahraoui, E. M., and Tiollais, P.,** Induction of anti-human immunodeficiency virus (HIV) neutralizing antibodies in rabbits immunized with recombinant HIV-hepatitis B surface antigen particles, *Proc. Natl. Acad. Sci. U.S.A.*, 85, 7957, 1988.
118. **Neurath, A. R. and Kent, S. B. H.,** The pre-S region of hepadnavirus envelope proteins, *Adv. Virus Res.*, 34, 65, 1988.
119. **Trevisan, A., Gudat, F., Guggenheim, R., Krey, G., Dürmüller, V., Luond, G., Duggelin, M., Landmann, J., Rondelli, P., and Bianchi, L.,** Demonstration of albumin receptors on isolated human hepatocytes by light and scanning electron microscopy, *Hepatology*, 2, 832, 1982.
120. **Thung, S. N. and Gerber, M. A.,** Polyalbumin receptors: their role in the attachment of hepatitis B virus to hepatocytes, *Semin. Liver Dis.*, 4(1), 69, 1984.
121. **Neurath, A. R., Kent, S. B. H., Parker, K., Prince, A. M., Strick, N., Brotman, B., and Sproul, P.,** Antibodies to a synthetic peptide from the preS 120-145 region of the hepatitis B virus envelope are virus-neutralizing, *Vaccine*, 4, 35, 1986.
122. **Itoh, Y., Takai, E., Ohnuma, H., Kitajima, K., Tsuda, F., Machida, A., Mishiro, S., Nakamura, T., Miyakawa, Y., and Mayumi, M.,** A synthetic peptide vaccine involving the product of the pre-S(2) region of hepatitis B virus DNA: protective efficacy in chimpanzees, *Proc. Natl. Acad. Sci. U.S.A.*, 83, 9174, 1986.
123. **Emini, E. A., Larson, V., Eichberg, J., Conard, P., Garsky, V. M., Lee, D. R., Ellis, R. W., Miller, W. J., Anderson, C. A., and Gerety, R. J.,** Protective effect of a synthetic peptide comprising the complete preS2 region of the hepatitis B virus surface protein, *J. Med. Virol.*, 28, 7, 1989.
124. **Yu, M. W., Finlayson, J. S., and Shih, J. W.,** Interaction between various polymerized human albumins and hepatitis B surface antigen, *J. Virol.*, 55, 736, 1985.
125. **Thung, S. N., Wang, D. F., Fasy, T. M., Hood, A., and Gerber, M. A.,** Hepatitis B surface antigen binds to human serum albumin cross-linked by transglutaminase, *Hepatology*, 9, 726, 1989.
126. **Krone, B., Lenz, A., Heermann, K.-H., Seifer, M., Lu, X., and Gerlich, W. H.,** Interaction between hepatitis B surface proteins and monomeric human serum albumin, *Hepatology*, 11, 1050, 1990.
127. **Fiume, L., Cerenzia, M. R. T., Bonino, F., Busi, C., Mattioli, A., Brunetto, M. R., Chiaberge, E., and Verme, G.,** Inhibition of hepatitis B virus replication by vidarabine monophosphate conjugated with lactosaminated serum albumin, *Lancet*, ii, 13, 1988.
128. **Krone, B., Gerlich, W. H., Heermann, K. H., Lenz, A., Lu, X., Seifer, M., and Thomssen, R.,** Significance of pre-S antigens for hepatitis B vaccines, in *Progress in Hepatitis B Immunization*, Vol. 194, Coursaget, P. and Tong, M. J., Eds., Colloque INSERM/John Libbey Eurotext, 1990, 21.
129. **Heermann, K.-H., Waldeck, F., and Gerlich, W. H.,** Interaction between native human serum and the preS2 domain of hepatitis B virus surface antigen, in *Viral Hepatitis and Liver Disease*, Zuckerman, A.J., Ed., Alan R. Liss, New York, 1988, 697.
130. **Deepen, R., Heermann, K.-H., Uy, A., Thomssen, R., and Gerlich, W. H.,** Assay of preS epitopes and preS1 antibody in hepatitis B virus carriers and immune persons, *Med. Microbiol. Immunol.*, 179, 49, 1990.
131. **Mizokami, M., Orito, E., Sugiura, S., Kameshima, N., and Yamamoto, M.,** Polymerized human serum albumin receptor in the S gene of the hepatitis B virus and determination of its amino acid sequence, in *Viral Hepatitis and Liver Disease*, Zuckerman, A.J., Ed., Alan R. Liss, New York, 1988, 701.
132. **Ishihara, K., Waters, J. A., Pignatelli, M., and Thomas, H. C.,** Characterisation of the polymerized and monomeric human serum albumin binding sites on hepatitis B surface antigen, *J. Med. Virol.*, 21, 89, 1987.

133. **Pontisso, P., Ruvoletto, M. G., Gerlich, W. H., Heermann, K.-H., Bardini, R., and Alberti, A.,** Identification of an attachment site for human liver plasma membranes on hepatitis B virus particles, *Virology*, 173, 522, 1989.
134. **Pontisso, P., Petit, M. A., Bankowski, M. J., and Peeples, M. E.,** Human liver plasma membranes contain receptors for the hepatitis B virus pre-S1 region and, via polymerized human serum albumin, for the pre-S2 region, *J. Virol.*, 63, 1981, 1989.
135. **Lu, X., Yao, G., and Tian, Y.,** The interaction between native serum albumin and hepatitis B virus, *Arch. Virol.*, 98, 163, 1988.
136. **Hellström, U. and Sylvan, S.,** Significance of pre-S2 peptide of hepatitis B virus: Should it be in the vaccine?, in *Prog. Med. Virol.*, Vol. 35, Melnick, J. L., Ed., S. Karger, Basel, 1988, 76.
137. **Milich, D. R., McLachlan, A., Chisari, F. V., Nakamura, T., and Thornton, G. B.,** Two distinct but overlapping antibody binding sites in the pre-S(2) region of HBsAg localized within 11 continuous residues, *J. Immunol.*, 137, 2703, 1986.
138. **Budkowska, A., Riottot, M. M., Dubreuil, P., Lazizi, Y., Brechot, C., Sobczak, E., Petit, M. A., and Pillot, J.,** Monoclonal antibody recognizing pre-S(2) epitope of hepatitis B virus: characterization of pre-S(2) epitope and anti-pre-S(2) antibody, *J. Med. Virol.*, 20, 111, 1986.

138a. **Lu, X. Y., Heermann, K.-H., and Gerlich, W. H.,** manuscript in preparation.

139. **Sells, M. A., Chen, M.-L., and Acs, G.,** Production of hepatitis B virus particles in Hep G2 cells transfected with cloned hepatitis B virus DNA, *Proc. Natl. Acad. Sci. U.S.A.*, 84, 1005, 1987.
140. **Kekulé, A. S., Lauer, U., Meyer, M., Caselmann, W. H., Hofschneider, P. H., and Koshy, R.,** The preS2/S region of integrated hepatitis B virus DNA encodes a transcriptional transactivator, *Nature (London)*, 343, 457, 1990.

140a. **Lu, X. Y.,** unpublished observations, 1990.

141. **Neurath, A. R., Kent, S. B. H., Strick, N., and the Hepatitis B Vaccine Trial Study Groups,** Detection of antiviral antibodies with predetermined specificity using peptide-β-lactamase conjugates: application to antibodies specific for the preS regions of the hepatitis B virus envelope proteins, *J. Gen. Virol.*, 67, 453, 1986.

141a. **Teepen, R.,** unpublished observations, 1990.

142. **Imamura, T., Araki, M., Miyanohara, A., Nakao, J., Yonemura, H., Ohtomo, N., and Matsubara, K.,** Expression of hepatitis B virus middle and large surface antigen genes in *Saccharomyces cerevisiae*, *J. Virol.*, 61, 3543, 1987.
143. **Rutgers, T., Cabezon, R., Harford, N., Vanderbrugge, D., Descurieux, M., Van Opstal, O., Van Wijnendaele, F., Hauser, P., Voet, P., and De Wilde, M.,** Expression of different forms of hepatitis B virus envelope proteins in yeast, in *Viral Hepatitis and Liver Disease*, Zuckerman, A.J., Alan R. Liss, New York, 1988, 304.
144. **Adamowicz, P., Tron, F., Vinas, R., Mevelec, M. N., Diaz, I., Courouce, A. M., Mazert, M. C., Lagarde, D., and Girard, M.,** Hepatitis B vaccine containing the S and preS-2 antigens produced in Chinese hamster ovary cells, in *Viral Hepatitis and Liver Disease*, Zuckerman, A.J., Alan R. Liss, New York, 1988, 1087.
145. **Tron, F., Degos, F., Bréchot, C., Courouce, A.-M., Goudeau, A., Marie, F.-N., Adamowicz, P., Saliou, P., Laplanche, A., Benhamou, J.-P., and Girard, M.,** Randomized dose range study of a recombinant hepatitis B vaccine produced in mammalian cells and containing the S and preS2 sequences, *J. Infect. Dis.*, 160, 199, 1989.

145a. **Heermann, K.-H. and Gerlich, W.H.,** unpublished observations, 1990.

146. **Persing, D. H., Varmus, H. E., and Ganem, D.,** The preS1 protein of hepatitis B virus is acylated at its amino terminus with myristic acid, *J. Virol.*, 61, 1672, 1987.
147. **Chisari, F. V., Filippi, P., McLachlan, A., Milich, D. R., Riggs, M., Lee, S., Palmiter, R. D., Pinkert, C. A., and Brinster, R. L.,** Expression of hepatitis B virus large envelope polypeptide inhibits hepatitis B surface antigen secretion in transgenic mice, *J. Virol.*, 60, 880, 1986.
148. **Persing, D. H., Varmus, H. E., and Ganem, D.,** Inhibition of secretion of hepatitis B surface antigen by a related presurface polypeptide, *Science*, 234, 1388, 1986.
149. **Cheng, K. C., Smith, G. L., and Moss, B.,** Hepatitis B large surface protein is not secreted but is immunogenic when selectively expressed by recombinant vaccinia virus, *J. Virol.*, 60, 337, 1986.
150. **Kuroki, K., Russnak, R., and Ganem, D.,** Novel N-terminal amino acid sequence required for retention of a hepatitis B virus glycoprotein in the endoplasmic reticulum, *Mol. Cell. Biol.*, 9, 4459, 1989.
151. **Chisari, F. V., Filippi, P., Buras, J., McLachlan, A., Popper, H., Pinkert, C. A., Palmiter, R. D., and Brinster, R. L.,** Structural and pathological effects of synthesis of hepatitis B virus large envelope polypeptide in transgenic mice, *Proc. Natl. Acad. Sci. U.S.A.*, 84, 6909, 1987.
152. **Chisari, F. V., Klopchin, K., Moriyama, T., Pasquinelli, C., Dunsford, H. A., Sell, S., Pinkert, C. A., Brinster, R. L., and Palmiter, R. D.,** Molecular pathogenesis of hepatocellular carcinoma in hepatitis B virus transgenic mice, *Cell*, 59, 1145, 1989.

153. **Hadziyannis, S., Raimondo, G., Papaioannou, C., Anastassakos, C., Wong, D., Sninsky, J., and Shafritz, D.,** Expression of pre-S gene-encoded proteins in liver and serum during chronic hepatitis B virus infection in comparison to other markers of active virus replication, *J. Hepatol.*, 5, 253, 1987.
154. **Thung, S. N., Gerber, M. A., Kasambalides, E. J., Gilja, B. K., Keh, W., and Gerlich, W. H.,** Demonstration of pre-S polypeptides of hepatitis B virus in infected livers, *Hepatology*, 6, 1315, 1986.
155. **Dienes, H. P., Gerlich, W. H., Wörsdörfer, M., Gerken, G., Bianchi, L., Hess, G., Meyer zum Büschenfelde, K.-H.,** Hepatic expression patterns of the large and middle hepatitis B virus surface proteins in viremic and nonviremic chronic hepatitis B, *Gastroenterology*, 88, 1, 1990.
155a. **Hsu, H. C.,** personal communication, 1990.
156. **Hu, P. S., Carithers, R. L., Jr., Fiorenza, V., and Peterson, D. L.,** Quantitative studies of the hepatitis B viral pre-S proteins: lack of correlation with the HBeAg status, *J. Virol. Methods*, 16, 97, 1987.
157. **Gerken, G., Manns, M., Gerlich, W. H., Hess, G., and Meyer zum Büschenfelde, K.-H.,** Immune blot analysis of viral surface proteins in serum and liver of patients with chronic hepatitis B virus infection, *J. Med. Virol.*, 29, 261, 1988.
157a. **Heermann, K.-H.,** unpublished observations, 1990.
158. **Horwich, A. L., Furtak, K., Pugh, J., and Summers, J.**, Synthesis of hepadnavirus particles that contain replication-defective duck hepatitis B virus genomes in cultured HuH7 cells, *J. Virol.*, 64, 642, 1990.
158a. **Heerman, K.-H. and Gerlich, W. H.,** unpublished observations, 1990.
159. **Jean-Jean, O., Levrero, M., Will, H., Perricaudet, M., and Rossignol, J. M.,** Expression mechanism of the hepatitis B virus (HBV) C gene and biosynthesis of HBe antigen, *Virology*, 170, 99, 1989.
160. **Ohori, H. and Matsuda, K.,** Intra- and extracellular distribution and immunochemical characterization of hepatitis B virus nucleocapsid proteins produced by a human hepatoma cell line transfected with cloned viral DNA, *Virology*, 168, 40, 1989.
161. **Seifer, M., Heermann, K. H., and Gerlich, W. H.,** Expression pattern of the hepatitis B virus genome in transfected mouse fibroblasts; submitted.
162. **Neurath, A. R., Kent, S. B. H., Strick, N., and Parker, K.,** Identification and chemical synthesis of a host cell receptor binding site on hepatitis B virus, *Cell*, 46, 429, 1986.
163. **Petit, M. A., Dubanchet, S., and Capel, F.,** A monoclonal antibody specific for the hepatocyte receptor binding site on hepatitis B virus, *Mol. Immunol.*, 26, 531 1989.
164. **Neurath, A R. and Strick, N.**, Antigenic mimicry of an immunoglobulin A epitope by a hepatitis B virus cell attachment site, *Virology*, 178 631, 1990.
164a. **Howard, C. R.** in press.[12]
164b. **Pontisso, P.,** in press.[12]
165. **Pontisso, P., Morsica, G., Zambello, R., Colletta, C., Chemello, L., and Alberti, A.,** Hepatitis B virus binds to peripheral blood mononuclear cells via the preS1 protein, *J. Hepatol.*, in press.
165a. **Peeples, M.,** personal communication, 1990.
166. **Bchini, R., Capel, F., Danguet, C., Dubanchet, S., and Petit, M. A.,** In vitro infection of human hepatoma (HepG2) cells with hepatitis B virus, *J. Virol.*, 64, 3025, 1990.
167. **Gripon, P., Diot, C., Theze, N., Fourcel, I., Loreal, O., Brechot, C., and Guguen-Guillouzo, C.,** Hepatitis B virus infection of adult human hepatocytes cultured in the presence of dimethyl sulfoxide, *J. Virol.*, 62, 4136, 1988.
168. **Ochiya, T., Tsurimoto, R., Ueda, K., Okubo, K., Shiozawa, M., and Matsubara, K.,** An *in vitro* system for infection with hepatitis B virus that uses primary human fetal hepatocytes, *Proc. Natl. Acad. Sci. U.S.A.*, 86, 1875, 1989.
169. **Neurath, A. R., Strick, N., Kent, S. B. H., Parker, K., Seto, B., and Girard, M.,** Design of synthetic peptides mimicking the immunologic and biologic functions of the pre-S1 sequence of the hepatitis B virus envelope protein, in *Vaccine 88*, Ginsberg, H., Brown, F., Lerner, R. A., and Chanock, R. M., Eds., Cold Spring Harbor Laboratory, Cold Spring Harbor, New York, 1988, 229.
170. **Thornton, G. B., Moriarty, A. M., Milich, D. R., Eichberg, J. W., Purcell, R.H., and Gerin, J.L.,** Protection of chimpanzees from hepatitis-B virus infection after immunization with synthetic peptides: identification of protective epitopes in the pre-S region, in *Vaccines 89: Modern Approaches to New Vaccines Including Prevention of AIDS*, Lerner, R. A., Ginsberg, H., Chanock, R. M., and Brown, F., Eds., Cold Spring Harbor Laboratory, Cold Spring Harbor, New York, 1989, 467.
171. **Pugh, J. C. and Summers, J. W.,** Infection and uptake of duck hepatitis B virus by duck hepatocytes maintained in the presence of dimethyl sulfoxide, *Virology*, 172, 564, 1989.
172. **Bonino, F., Heermann, K. H., Rizzetto, M., and Gerlich, W. H.,** Hepatitis delta virus: protein composition of delta antigen and its hepatitis B virus-derived envelope, *J. Virol.*, 58, 945, 1986.
173. **Ponzetto, A., Hoyer, B. H., Popper, H., Engle, R., Purcell, R. H., and Gerin, J. L.,** Titration of the infectivity of hepatitis D virus in chimpanzees, *J. Infect. Dis.*, 155, 72, 1987.

174. **Ponzetto, A., Cote, P. J., Popper, H., Hoyer, B. H., London, W. T., Ford, E. C., Bonino, F., Purcell, R. H., and Gerin, J. L.,** Transmission of the hepatitis B virus-associated δ agent to the eastern woodchuck, *Proc. Natl. Acad. Sci. U.S.A.*, 81, 2208, 1984.

175. **Budkowska, A., Dubreuil, P., Gerlich, W. H., Lazizi, Y., and Pillot, J.,** Occurrence of pre-S1 antigen in viremic and nonviremic carriers of hepatitis B surface antigen, *J. Med. Virol.*, 26, 217, 1988.

176. **Gerken, G., Manns, M., Gerlich, W. H., Hess, G., and Meyer zum Bueschenfelde, K. H.,** Pre-S encoded surface proteins in relation to the major viral surface antigen in acute hepatitis B virus infection. *Gastroenterology*, 92, 1864, 1987.

177. **Thoma, H. A., Hemmerling, A. E., and Hötzinger, H.,** Evaluation of immune response in a third generation hepatitis B vaccine containing pre-S proteins in comparative trials, in *Progress in Hepatitis B Immunization*, Vol. 194, Coursaget, P. and Tong, M.J., Eds., Colloque INSERM/John Libbey Eurotext, 1990, 35.

178. **Kozak, M.,** Comparison of initiation of protein synthesis in procaryotes, eucaryotes and organelles, *Microbiol. Rev.*, 47, 1, 1983.

179. **Pasek, M., Goto, T., Gilbert, W., Zink, B., Schaller, H., McKay, P., Leadbetter, G., and Murray, K.,** Hepatitis B virus genes and their expression in *E. coli*, *Nature (London)*, 282, 575, 1979.

180. **Valenzuela, P., Quiroga, M., Zaldivar, J., Gray, P., and Rutter, W. J.,** The nucleotide sequence of the hepatitis B genome and the identification of the major viral genes, in *Animal Virus Genetics*, Fields, B., Jaenisch, R., and Fox, C. F., Eds., Academic Press, New York, 1980, 57.

181. **Galibert, F., Chen, T. N., and Mandart, E.,** Nucleotide sequence of a cloned woodchuck hepatitis virus genome: comparison with the hepatitis B virus sequence, *J. Virol.*, 41, 51, 1982.

182. **Seeger, C., Ganem, D., and Varmus, H. E.,** Nucleotide sequence of an infectious molecularly cloned genome of ground squirrel hepatitis virus, *J. Virol.*, 51, 367, 1984.

183. **Mandart, E., Kay, A., and Galibert, F.,** Nucleotide sequence of a cloned duck hepatitis B virus genome: comparison with woodchuck and human hepatitis B virus sequences, *J. Virol.*, 49, 782, 1984.

Chapter 7

THE MOLECULAR BIOLOGY OF THE HEPATITIS B VIRUS CORE PROTEIN

David N. Standring

TABLE OF CONTENTS

I. PERSPECTIVE

The icosahedral hepatitis B virus nucleocapsid or core particle plays a central role in the production of an infectious viral Dane particle. During its formation, the core particle must recruit into its structure the essential ingredients of viral replication, the pregenomic RNA and the viral reverse transcriptase. On its completion, the core particle must migrate to the endoplasmic reticulum, where the viral surface antigens and lipid are being continuously assembled into 22-nm envelope structures and delivered to the extracellular environment via the secretory pathway. Formation of the infectious virion is a rare event (the vast majority of serum hepatitis B particles are empty envelope structures) that requires the nucleocapsid to "capture" the assembling surface proteins and thus share their passage through the secretory pathway. The core particle presumably provides the template on which the surface envelope components assemble, drawn by direct molecular interactions. While many viral icosahedral nucleocapsids comprise at least three different structural proteins, and enveloped viruses may use an additional protein(s) to link the nucleocapsid and the envelope, the hepatitis B virus capsid contains only a single protein, the viral core protein, with which to accomplish these various tasks. The properties of this remarkable protein are reviewed here.

II. GENERAL PROPERTIES OF CORE PARTICLES AND CORE PROTEIN

A. THE CORE PARTICLE

The 42-nm infectious hepatitis B virus (HBV) particle, found in the serum of infected patients, has a double-shelled structure in which an outer envelope (containing three related surface antigens plus lipid) coats the 28-nm viral nucleocapsid. Nonionic detergents disrupt the outer envelope,[1,2] releasing intact HBV nucleocapsids (commonly referred to as virion "cores" or core particles) that contain the partially double-stranded HBV DNA genome. Nucleocapsids that have not yet been coated by the surface envelope occur free in hepatocytes. These immature cores contain a spectrum of viral RNA or DNA molecules representing intermediates in the reverse transcription pathway for HBV replication.[3,4] Nucleocapsids isolated from either source are particles with a buoyant density of 1.30 to 1.37 g/ml on equilibrium gradient density centrifugation.[3-7] The low-density particles have been reported to be polymerase-negative "empty" cores, devoid of viral nucleic acid, while the denser particles are replication competent.[6-8] It is not clear whether the empty particles represent defective species found *in vivo* or whether they result from the isolation protocols; cesium chloride, in particular, has been reported to cause morphological changes in HBV capsids.[5]

B. CORE PROTEIN

By mass, the two major components of the HBV core particle are the viral nucleic acid and a single structural core polypeptide of 21 kDa. The sera of HBV-infected patients contain a second core-related species of approximately 17 kDa that represents a nonparticulate, secreted product of the precore locus (see Section VIII). The two proteins carry serologically distinguishable antigenic determinants; the intact nucleocapsid displays the core antigen determinant (HBcAg or HBc), while the serum protein carries e antigen or HBeAg. Unfortunately, the proteins are frequently identified by their antigenic names in the literature; core protein is often called core antigen, while secreted precore products are referred to as e antigen or HBeAg. This terminology is at best confusing and at worst misleading; the antigenic identities vary with the physical state of the molecule and, in any case, are not as independent as was originally thought. A more appropriate nomenclature is one that identifies the protein by size and by the locus from which it originated; e.g., a 22.5-kDa precore protein, P22.5preC, or P22.5pc. The antigenic names should be retained for specific reference to immunologically characterized molecules.

C. SUBCELLULAR LOCALIZATION OF CORE PROTEIN

The simple question of where core protein and nucleocapsids are found within the cell has yielded no simple answer. Early histochemical studies, employing γ-globulin fractions from HBV carriers and paraffin sections of liver tissue, found core protein predominantly in the nucleus. However, cytoplasmic core staining is readily destroyed by formalin fixation and paraffin embedding.[9,10] Recent papers, mostly utilizing commercial antisera and frozen sections, have reported purely nuclear, mixed nuclear and cytoplasmic, or purely cytoplasmic core staining.[10-14] Electron microscopy (EM) studies[12,14,15] confirm this broad subcellular distribution and reportedly show that cytoplasmic P21.5 is largely nonparticulate while nuclear core is found in capsid-like structures (although at least one EM study has detected packed arrays of cytoplasmic core particles[16]), implying that P21.5 is synthesized in the cytoplasm, then migrates into the nucleus where it assembles into particles.[12,14,15] In contrast, *in situ* hybridization studies[9,17-19] revealed HBV replicative intermediates, and thus presumably core particles, in the hepatocyte cytoplasm and not in the nucleus; indeed, cells with nuclear core did not exhibit nuclear viral DNA.[9] According to these conflicting views, nuclear core could be an essential component of virus assembly, or merely "excess antigen accumulation unrelated to the virus replication strategy."[9] The core protamine tail contains the kind of short, positively charged amino acid stretches that typically provide nuclear transport signals.[20] We examined this possibility by injecting *Xenopus* oocytes with HBV C mRNA or with an RNA encoding the human immunodeficiency virus (HIV) Tat protein (a known nuclear protein). When the giant oocyte nuclei (germinal vesicles) were subsequently manually separated from the cytoplasm, the bulk of the P21.5 was found in the cytoplasm while Tat accumulated in the nucleus.[21] Thus, we can find no evidence for a functional nuclear transport signal in core protein and the mechanism of core nuclear localization remains a mystery.

III. ORGANIZATION AND TRANSCRIPTION OF THE CORE GENE

A. THE C GENE AND PreC REGION

In hepadnavirus genomes a single contiguous open reading frame (ORF) encodes the capsid protein. The 5′ end of the C ORF contains two conserved in-frame AUG codons, separated by 28 amino acid codons in the case of HBV. For the avian and mammalian viruses, translation initiating at the second AUG codon would produce a protein of the correct molecular weight for core. Expression of the C ORF in bacteria,[22,23] yeast, [24,25] and mammalian cells[26-28] has provided unequivocal proof that the second AUG directs the synthesis of authentic core protein. Interestingly, efforts to sequence the amino terminus of core protein directly by the Edman procedure have failed, suggesting that the the amino terminus of core may be blocked by an unidentified modification.[29] The region between the first and second AUGs of the ORF is termed the precore region. Expression of the entire precore/core gene is now known to result in the rerouting of core-related proteins into the secretory pathway (via a hydrophobic signal peptide located at the start of precore), with concomitant secretion of e antigen (see Section VIII). For the sake of convenience, proteins generated from the precore AUG will be referred to here as precore proteins, even though they may not contain the bulk of the precore region.

B. TRANSCRIPTION OF THE CORE GENE

The greater than full-length 3.5-kb pregenomic RNA is the C ORF mRNA. Expression of both precore and core proteins during the viral life cycle is accomplished via heterogeneous transcriptional initiation; three distinct 3.5-kb RNAs initiate at sites that are only ~30 bp apart yet straddle the precore AUG.[30] The two longer 3.5-kb species contain the precore AUG; microinjection of a synthetic precore mRNA into *Xenopus* oocytes results in the translation of only precore species with little or no core protein.[31] The shortest RNA lacks the precore AUG and yields only core protein.

C. CORE-RELATED PROTEINS

Aside from the conventional core and precore proteins, there are two reported sources of additional, core-related species; spliced pregenomic mRNAs and core–Pol fusion proteins. A novel class of transcripts, representing spliced versions of the pregenomic RNAs, has recently been described in mammalian cells transfected with HBV DNA.[32,33] One such spliced RNA contains a splice donor site at the 3′ end of ORF C and an acceptor site within ORF S. The predicted core protein translated from this RNA is identical to authentic core except that the final amino acid (Cys) is deleted. Although this protein has not yet been characterized *in vivo*, it is of interest in view of the proposed role of this terminal cysteine in particle stabilization via disulfide bridges (see Section V). Another potential core gene product is a core-Pol fusion protein, the result of ribosomal frameshifting during translation, which has so far been reported only as an infrequent 35-kDa product in liver tissue samples from patients with HBV-related hepatocellular carcinoma or cirrhosis.[34,35] Core-Pol fusion products are not precursors for hepadnaviral reverse transcriptases,[36,37] unlike retroviral Gag-Pol fusion proteins, but a possible function for such molecules cannot be ruled out at this time.

IV. THE PRIMARY STRUCTURE OF CORE PROTEIN AND COMPARISON WITH OTHER CAPSID PROTEINS

A. THE PRIMARY SEQUENCE OF CORE PROTEIN

The HBV C ORF encodes a protein of 183 (all subtypes except *adw*) or 185 (*adw*) amino acids with a molecular mass of 21,000 Da (*ayw*) or 21,399 Da (*adw*); thus the core protein is termed P21 or P21.5. The precore region provides an additional 29 amino-terminal residues so that the complete precore/core ORF (*adw* subtype) encodes a predicted precore protein of 24,647 Da (P25 or P25prec). Analysis of the P21.5 core protein sequence reveals that the polypeptide divides into two distinct regions. The first 149 residues (the "β-barrel" domain, see Section V) display an unusually high content of hydrophobic amino acids compared to typical soluble proteins. The final 34 (*ayw*) or 36 (*adw*) residues compose an extremely basic carboxyl terminal tail region; 47% (17/36) of the tail residues are arginines that are organized into four clusters of three or more residues flanked by Ser-Pro or Ser-Gln pairs. The tail region bears a striking sequence resemblance to protamine-like molecules. The precore region is highly hydrophobic; its most remarkable feature is an unusually high content of four Cys residues.

B. SIMILARITY BETWEEN THE HEPADNAVIRAL CORE PROTEINS

The core proteins of ground squirrel hepatitis B virus (GSHV) and woodchuck hepatitis B virus (WHV) are closely related to the HBV protein by both size and sequence. The avian virus core proteins are much larger (the 32 to 35-kDa duck core protein contains 262 amino acids) and much less related, but the main features, such as the precore region, the hydrophobic body, and the arginine-rich tail, are retained. The sequences of the hepadnaviral core proteins are shown aligned in Figure 1. The HBV and avian core proteins share limited homology concentrated mainly at the amino and carboxyl termini. When the first 39 amino acids of the HBV protein (starting at the C ORF methionine) are aligned with analogous residues of the duck and heron viruses, 41% show identity and a further 23% represent conservative substitutions.[38] The last ~90 amino acids of the proteins constitute a homologous segment that includes the arginine-rich tail. However, the precise structure of the avian tail domain differs from that of the HBV tail. Clustered basic residues punctuate the last 48 amino acids of duck core, but are neither flanked by Ser-Pro dipeptides nor organized into clearly repeating units. The final 13 amino acids of the duck hepatitis B virus (DHBV) tail contain six Ser residues and no Arg clusters. It has been argued[38] that the central portion of the duck core protein, which displays no homology to P21.5, is related to the HBV X trans-activator protein. However, no functional trans-activation capability has been reported for the DHBV core protein and the weak homology extends over very short regions, mainly toward the carboxyl terminus of the HBV X protein (residues 117 to 132). Thus, this hypothesis remains largely untested. The 43-amino acid avian precore region is longer than the mammalian counterpart and not very related, but the hydrophobicity and the four Cys residues are conserved.

C. SIMILARITY WITH OTHER VIRAL CAPSID PROTEINS

The HBV core protein and retroviral nucleocapsid (Gag) proteins share modest but significant amino acid similarity. The 30-kDa Moloney murine leukemia virus Gag protein displays 41% similarity with P21.5 over 98 amino acids located at the carboxyl termini of the respective proteins.[39] Both proteins contain an arginine-rich tail.

Argos and Fuller[40] noted low but significant similarity between the hepadnavirus and picornavirus capsid proteins, particularly those of mengo virus, poliovirus, and human rhinovirus. The least homologous region is the core tail domain. The most analogous basic segment of picornavirus capsids occurs at the amino terminus. Excluding the core tail, the mengo virus vp3 and WHV capsid proteins share 23% identity over 160 aligned residues with a further 42% representing conservative changes. The similarity is distributed fairly evenly throughout the two proteins, leading to the suggestion[40] that it may be the basis of similar three-dimensional structures (see Section V).

```
adw HBV   MDIDPYKEFGATVEL LSFLPSDFFPSVRDL LDTA-SALYREALES PEH-----CSPHHTA LRQAILCWGELM--- ----TLATWVGNNLE
ayw HBV   MDIDPYKEFGATVEL LSFLPSDFFPSVRDL LDNA-SALYREALES PEH-----CSPHHTA LRQAILCWGELM--- ----TLATWVGGNLE
GSHV      MDIDPYKEFGSSYQL LNFLPLDFFPDLNAL VDTA-AALYEEELTG REH-----CSPHHTA IRQALVCWEELT--- ----RLITWMSENTT
WHV       MDIDPYKEFGSSYQL LNFLPLDFFPDLNAL VDTA-TALYEEELTG REH-----CSPHHTA IRQALVCWDELT--- ----KLIAWMSSNIT
DHBV      MDINASR---ALANV YD-LPDDFFPKIDDL VRDAKDALE-PYWKS DSIKKHVLIATHFVD LIED--FWQTTQGMH EIAESLRAVIPPTTT

adw HBV   D--P----------- --------------- -------------AS RDLVVNYVNTNMGLK IRQLLWFHISCLTFG RETVLEYLVSFGVWI
ayw HBV   D--P----------- --------------- -------------IS RDLVVSYVNTNMGLK FRQLLWFHISCLTFG RETVIEYLVSFGVWI
GSHV      ---E----------- --------------- -------------EV RRIIVDHVNNTWGLK VRQTLWFHLSCLTFG QHTVQEFLVSFGVWI
WHV       S--E----------- --------------- -------------QV RTIIVNHVNDTWGLK VRQSLWFHLSCLTFG QHTVQEFLVSFGVWI
DHBV      PVPPGYLIQHEEAEE IPLGDLFKHQEERIV SFQPDYPITARIHAH LKAYAKINEESLD-R ARRLLWWHYNCLLWG EAQVTNYISRLRTWL

adw HBV   RTPPAYRPPNAPILS TLPETTVVRRRDR-- -GRSPRRRTPSPRRR RSQS---PRRRRSQS RESQC
ayw HBV   RTPPAYRPPNAPILS TLPETTVVRRR---- -GRSPRRRTPSPRRR RSQS---PRRRRSQS RESQC
GSHV      RTPAPYRPPNAPILS TLPEHTVIRRRGGSR AARSPRRRTPSPRRR RSQS---PRRRRSQS PASNC
WHV       RTPAPYRPPNAPILS TLPEHTVIRRRGGAR ASRSPRRRTPSPRRR RSQS---PRRRRSQS PSANC
DHBV      STPEKYRGRDAPTIE AITRPIQVAQGGRKT TTGTRKPRGLEPRRR KVKTTVVYGRRRSKS RERRAPTPQRAGSPL PRSSSSHHRSPSPRK
```

FIGURE 1. Comparison of the hepadnavirus core protein sequences. The core protein sequences were drawn from the PIR protein sequence database of the National Biomedical Research Foundation and aligned by computer, then visually adjusted.

V. NUCLEOCAPSID ARCHITECTURE AND ASSEMBLY

A. THE STRUCTURE OF THE CORE PARTICLE

Electron micrographs of negatively stained HBV nucleocapsids depict uniform spherical structures with a 28.3-nm average diameter.[41] Positively stained particles display a 21-nm diameter electron-dense nucleoid center that presumably represents the viral nucleic acid.[41] More recent cryoelectron microscopy and conventional negative-staining studies provide similar results.[42] The apparent size of the core particle varies considerably with the method of analysis; cryoelectron microscopy and negative staining give particle sizes of 28.0 ± 1.9 and 31.3 ± 1.9 nm, respectively, while ultrathin sectioning gives a 25% smaller size (21.1 ± 1.4 nm), presumably reflecting shrinkage during sample preparation.[42] High-resolution micrographs reveal that negatively stained cores display a uniform peripheral pattern of spikes that appear to represent individual core protein molecules.[41] Analysis of the micrographs suggests that ~180 core polypeptides are regularly arranged on an icosahedral surface lattice with a $T = 3$ symmetry. Although the capsid proteins of many icosahedral viruses are organized into distinctive pentameric or hexameric capsomere substructures, core molecules are only infrequently seen in small capsomere structures; generally no clustering is observed.[41,42]

B. LOCATION OF THE VIRAL NUCLEIC ACID

The above EM studies are consistent with the viral nucleic acid being localized inside core particles. Further corroborative evidence is provided by biochemical studies demonstrating the inaccessibility of the encapsidated viral nucleic acid to ribonuclease A [3,4] or, in the case of DHBV cores, to DNase I or restriction enzymes unless the core particles are first treated so as to disrupt them.[3,43]

C. CORE POLYPEPTIDE TOPOGRAPHY WITHIN THE PARTICLE

A topographic model for core protein within the core particle has been predicted[40] based on the known three-dimensional structure of the mengo virus capsid protein vp3. To generate the model, the WHV core polypeptide was superimposed onto the vp3 structure using short segments of highest amino acid similarity to direct the alignment. The model predicts a conserved β-sheet framework for both proteins and a common eight-stranded, anti-parallel β-barrel structure. This superstructural motif is utilized by a wide variety of simple enveloped and nonenveloped viruses as a means to encapsidate RNA in a simple icosahedral shell.[40] Clearly, this theoretical core-folding model is exciting, since it implies that studies on core protein will benefit from the large body of work on picornavirus and retrovirus nucleocapsids, but its accuracy is still uncertain. The low sequence conservation between the core and vp3 proteins means that the alignment process is fairly arbitrary, thus lowering the resolution of the model. The model is consistent with the limited experimental data (for instance, it places exposed antigenic determinants in exposed locations), but it is not clear if it will provide important insights into, say, interactions mediating core dimer formation or the linkage between core and envelope components.

Since vp3 lacks a comparable carboxyl-terminal tail domain, the core-folding model specifically excludes the arginine-rich tail. Actually, the amino terminus of the mengo virus vp3 protein (also believed to be involved in nucleic acid binding) may provide an excellent structural model for the core tail. The X-ray structure of vp3 reveals a protein organized into two very distinct domains: a tightly folded, eight-stranded β-barrel domain from which ~40 amino-terminal residues protrude as an open, elongated β sheet.[44] Such a structure is quite consistent with the known properties of core protein; the putative β-barrel domain of core polypeptide appears to be very tightly folded, as evidenced by its extreme resistance to proteolysis [21,31,45] and to radiolabeling by chemical reagents,[46] while the protamine-like tail is exquisitely sensitive to proteases,[21] suggesting an open structure. In vp3, the carboxyl terminus lies in very close spatial proximity to the point where the amino-terminal domain emerges from the β barrel. Simply excising the vp3 amino-terminal domain and pasting it onto the end of the carboxyl terminus would convert the vp3 structure into a potential model structure for core. In vp3, the part of the amino tail adjacent to the β barrel faces the inside of the particle; the portion further from the structural domain appears to angle away from the interior. It is interesting to note that the amino-terminal half of the HBV core tail is inaccessible to antibodies and thus probably lies on the interior of the particle,[47] while the last 13 amino acids of duck core lie on the particle exterior accessible to a specific antibody.[48]

D. CORE PARTICLE ASSEMBLY

In several heterologous systems, core protein assembles into RNA-containing core particles in the absence of other viral proteins or the correct pregenomic RNA (see Section VI). The RNA requirement for assembly (if any) has not been elucidated. Deletion studies[46] suggest that a minimal segment of ~130 core polypeptide residues (~70% of the protein) from the β-barrel domain is required for particle assembly. At the amino terminus, about 10 residues can be deleted[46] and at least 18 extra amino acids of foreign sequence added[49] without compromising particle integrity. Interestingly, the foreign sequence is effective at eliciting an immune response.[49] Pure precore polypeptide (P25), with its 29 extra amino-terminal residues, also assembles into cores in yeast, although these are less stable to high salt than P21 cores.[24] At the carboxyl terminus, deletions as far as residue 144 support assembly, but deletion down to residue 139 aborts particle formation[50]; thus the entire protamine tail domain can be eliminated without loss of particle integrity. In *Escherichia coli*, recombinant tail-minus core (r-P17) is assembled into particles that appear very similar to authentic P21.5 cores by EM. The r-P17 particles appeared slightly larger than wild-type cores (30 vs. 28 nm, respectively) in one study,[51] but ~10% smaller in a prior study.[52] These differences may reflect minor variations in the structure of the tail-minus mutants used in the two cases. The effect of additions to the tail region has not been reported so far.

E. THE INTERACTIONS BETWEEN CORE PROTEINS

The arginine-rich tail, although dispensable for particle assembly, plays a structural role in the wild-type nucleocapsid and contributes to its stability. Compared to r-P21.5 cores, r-P17 particles are more easily dissociated by urea[51] or by high pH.[52] The interactions between core protein molecules in these two kinds of particles have been examined by sodium dodecyl sulfate polyacrylamide gel electrophoresis (SDS-PAGE) under reducing versus nonreducing conditions.[51] When r-P17 particles are disrupted by SDS in the absence of reduction, the tail-minus core protein migrates as a 34-kDa dimer. Under strongly reducing conditions the dimers dissociate (with some difficulty) into monomers. These data suggest that a disulfide-linked dimer of r-P17 is the basic nucleocapsid building block. Noncovalent, detergent-sensitive interactions between dimers link them into particles. In contrast, r-P21.5 forms much higher molecular weight, SDS-resistant aggregates that fail to enter acrylamide gels under nonreducing conditions. In the presence of high concentrations of SDS and 2-mercaptoethanol, r-P21.5 migrates largely as a monomer, although 42-kDa dimers are always seen. This suggests that core dimers remain the basic building block, but that additional disulfide bridges between dimers must occur in the wild-type core particle. HBV core protein contains three cysteines in the β-barrel domain of the molecule (residues 48, 61, and 107) and one at the end of the tail (residue 183 or 185). These residues are highly conserved in the HBV subtypes and in the mammalian hepadnaviruses. Thus we can envisage a model in which (1) core β-barrel domains form dimers via one or two disulfide bridges; (2) dimers then associate by noncovalent, presumably hydrophobic, interactions; and (3) dimers covalently link via a disulfide bridge between a tail cysteine in one dimer and either a free β-barrel domain cysteine or another tail cysteine in an adjacent dimer. This would ultimately generate a network of disulfide-bridged dimers within a covalently closed particle. Such a model provides an attractive explanation for the superior stability of r-P21.5 vs. r-P17 cores, although the nucleic acid content of the particle, which is probably reduced in the case of r-P17 (see Section VI), could also determine particle stability. However, the model cannot apply to the duck core protein, which contains only one cysteine!

VI. CORE-NUCLEIC ACID INTERACTIONS

A crucial event in the viral life cycle is the recruitment of the RNA pregenome, the viral reverse transcriptase (Pol), the terminal protein (now known to be derived from the amino terminus of Pol), and quite possibly other products of viral or even cellular origin, into the assembling core particle. Inside this shell, suitably isolated from the cellular environment, reverse transcription of the RNA pregenome yields the familiar partially double-stranded viral DNA. The formation of a replication-competent nucleocapsid demands very selective packaging of the viral pregenome RNA, which presumably depends on very specific interactions between core protein and a cis-acting packaging signal located within the pregenome RNA. Recent data suggest that this process also requires a trans-acting factor that is a product of the *pol* gene.[53] The heirarchy of the interactions between the core, Pol, and pregenome components has not been established. In addition to the selective packaging event, core protein presumably mediates a conceptually distinct, more general level of interaction with the viral genome by binding the RNA sequences outside of the packaging signal and neutralizing their charge in such a way as to stabilize the tightly wound genome within the core particle.

A. PACKAGING DISCRIMINATION

Selective RNA packaging has been clearly established for GSHV,[4] which produces multiple 2.3-kb subgenomic envelope RNAs as well as three 3.5-kb pregenomic RNAs with 5′ ends clustered at nucleotides –25, –16, and +6, relative to the initiating methionine codon of the C ORF. Only the +6 RNA species was found in immature core particles; the discrimination between this and other viral RNAs within the particle is at least 100 to 1.[4] Packaging discrimination between genomic and subgenomic RNA species can easily be explained by a cis-acting packaging signal that is exclusive to the pregenomic RNA. However, the sharp discrimination between the GSHV 3.5-kb RNAs is much more surprising; as few as 21 additional nucleotides at the 5′ end of otherwise identical 3.5-kb mRNAs, which should all contain the presumptive packaging signal, is sufficient to derail packaging. The way in which this degree of selectivity is achieved is unknown but four possible mechanisms, falling into two broad classes, have been proposed.[4] The first class evokes the specific nucleotides at the 5′ end of the mRNA; either the extra 5′ nucleotides in the longer RNAs alter the local RNA secondary structure so as to prevent packaging signal recognition, or the packaging machinery recognizes the packaging signal in conjunction with another feature, such as the RNA cap structure, in a spatially defined manner. In the second class of model, packaging discrimination is achieved essentially as a result of differential protein production by the three RNAs. The two longer mRNAs encode the precore rather than core protein. Positioning the translational machinery for the synthesis of the precore protein might obstruct access to the packaging signal. Alternatively, the targeting of the precore signal peptide to the endoplasmic reticulum (see below) directs the mRNA to a cellular location where packaging is sterically

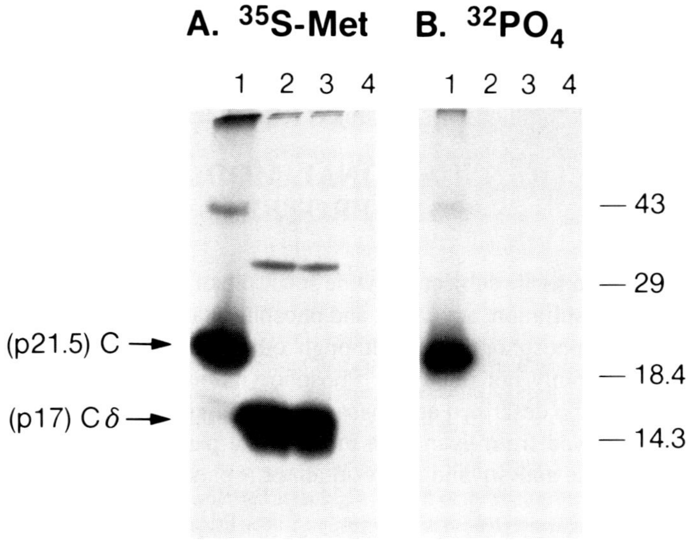

FIGURE 4. *In vivo* phosphorylation of core protein occurs in the protamine tail domain. Oocytes were metabolically labeled with (A) [^{35}S]methionine or (B) $^{32}P_i$ after being injected with C mRNA (lane 1), Cδ mRNA (lanes 2 and 3), or no RNA (lane 4). The positions of the C and Cδ proteins are indicated to the left of the figure; the positions of marker proteins (sizes in kilodaltons) are shown at the right.

changes in DHBV core protein reflect conformational changes rather than simple molecular weight increases.

C. ORIGIN OF THE KINASE ACTIVITY

Is the core kinase an intrinsic activity of core protein, a separate viral function, or a cellular enzyme? Current evidence favors the latter. As described above, duck core expressed in *E. coli* does not appear to be phosphorylated.[57] Recombinant HBV P21.5 core isolated from *E. coli* displays no *in vitro* autophosphorylation.[55,58] Thus, the kinase activity is not intrinsic to core protein. However, P21.5 is phosphorylated on expression in higher eucaryotic cells in the absence of other viral proteins[61] (Figure 4), implying that a cellular kinase is responsible for the phosphorylation. Recent studies by Laub's group suggest that a 90-kDa Mg^{2+}-dependent kinase is associated with HBV core particles.[63]

D. BIOLOGICAL SIGNIFICANCE OF PHOSPHORYLATION

There is no documented role for the kinase or phosphocore in the HBV life cycle and no certainty that there even is such a role. However, there are some clues that link core phosphorylation with viral replication. Low-density, polymerase-negative cores display more kinase activity than the higher density, polymerase-positive cores,[6-8] suggesting a correlation between core phosphorylation and the replicative state of HBV. Additional support comes from experiments[57] in which duck core was mutationally truncated so as to remove the phosphorylation target residues within the tail domain without altering the amino acid sequence encoded by the overlapping *pol* reading frame. A mutant lacking the final 36 amino acids of duck core (containing nine potential Ser phosphorylation sites) displayed only a single core species on SDS-PAGE, implying that phosphorylation had not occurred. In HepG2 cells, which are permissive for DHBV replication, the corresponding virus assembled correctly and domains form dimers via one or two disulfide bridges; (2) dimers then associate by noncovalent, presumably hydrophobic, interactions; and (3) dimers covalently link via a disulfide bridge between a tail cysteine in one dimer and either a free β-barrel domain cysteine or another tail cysteine in an adjacent dimer. This would ultimately generate a network of disulfide-bridged dimers within a covalently closed particle. Such a model provides an attractive explanation for the superior stability of r-P21.5 vs. r-P17 cores, although the nucleic acid content of the particle, which is probably reduced in the case of r-P17 (see Section VI), could also determine particle stability. However, the model cannot apply to the duck core protein, which contains only one cysteine!

VI. CORE-NUCLEIC ACID INTERACTIONS

A crucial event in the viral life cycle is the recruitment of the RNA pregenome, the viral reverse transcriptase (Pol), the terminal protein (now known to be derived from the amino terminus of Pol), and quite possibly other products of viral or even cellular origin, into the assembling core particle. Inside this shell, suitably isolated from the cellular environment, reverse transcription of the RNA pregenome yields the familiar partially double-stranded viral DNA. The formation of a replication-competent nucleocapsid demands very selective packaging of the viral pregenome RNA, which presumably depends on very specific interactions between core protein and a cis-acting packaging signal located within the pregenome RNA. Recent data suggest that this process also requires a trans-acting factor that is a product of the *pol* gene.[53] The heirarchy of the interactions between the core, Pol, and pregenome components has not been established. In addition to the selective packaging event, core protein presumably mediates a conceptually distinct, more general level of interaction with the viral genome by binding the RNA sequences outside of the packaging signal and neutralizing their charge in such a way as to stabilize the tightly wound genome within the core particle.

A. PACKAGING DISCRIMINATION

Selective RNA packaging has been clearly established for GSHV,[4] which produces multiple 2.3-kb subgenomic envelope RNAs as well as three 3.5-kb pregenomic RNAs with 5′ ends clustered at nucleotides −25, −16, and +6, relative to the initiating methionine codon of the C ORF. Only the +6 RNA species was found in immature core particles; the discrimination between this and other viral RNAs within the particle is at least 100 to 1.[4] Packaging discrimination between genomic and subgenomic RNA species can easily be explained by a cis-acting packaging signal that is exclusive to the pregenomic RNA. However, the sharp discrimination between the GSHV 3.5-kb RNAs is much more surprising; as few as 21 additional nucleotides at the 5′ end of otherwise identical 3.5-kb mRNAs, which should all contain the presumptive packaging signal, is sufficient to derail packaging. The way in which this degree of selectivity is achieved is unknown but four possible mechanisms, falling into two broad classes, have been proposed.[4] The first class evokes the specific nucleotides at the 5′ end of the mRNA; either the extra 5′ nucleotides in the longer RNAs alter the local RNA secondary structure so as to prevent packaging signal recognition, or the packaging machinery recognizes the packaging signal in conjunction with another feature, such as the RNA cap structure, in a spatially defined manner. In the second class of model, packaging discrimination is achieved essentially as a result of differential protein production by the three RNAs. The two longer mRNAs encode the precore rather than core protein. Positioning the translational machinery for the synthesis of the precore protein might obstruct access to the packaging signal. Alternatively, the targeting of the precore signal peptide to the endoplasmic reticulum (see below) directs the mRNA to a cellular location where packaging is sterically hindered or in some way disfavored; it also ensures that the mRNA and its protein product are rapidly segregated. In contrast, the shorter RNA directs the production of core protein that is free to interact with RNA in the cytoplasm. In the cis-packaging model a nascent core protein would bind directly to the same RNA molecule from which it was translated, thus tightly coupling core synthesis and packaging of the short RNA.

B. THE PACKAGING SIGNAL

By analogy with retroviruses the hepadnaviral packaging signal is probably a cis-acting sequence located within the pregenomic RNA. However, specific packaging of hepadnavirus genomes, unlike retroviral genomes, is conditional on the presence of a functional *pol* gene product,[53] so that attempts to map the packaging signal can only be expected to succeed if Pol is supplied in trans. Deletion analysis has revealed that sequences within or close to the precore region are essential for selective packaging.[54] An 85-nucleotide stretch of HBV RNA sequence starting downstream from direct repeat 2 (DR2) is apparently sufficient, when appended to the 5′ end of a heterologous β-galactosidase mRNA, to direct this chimera to be packaged into core particles,[54] as long as Pol is supplied in trans. Since the efficiency of this chimeric packaging is unclear this result does not necessarily rule out a role for distal signals and it remains possible that downstream sequences contribute to the efficiency and/or selectivity of packaging.

C. PROTEIN SEQUENCES THAT INTERACT WITH NUCLEIC ACID

Both Southwestern blotting and the direct analysis of RNA packaging into particles have been employed to elucidate which regions of core protein interact with nucleic acid. Recently, binding of [^{32}P]DNA to immobilized core protein was demonstrated using Southwestern blots.[7] Subsequently, a rather weaker binding activity has been demonstrated for [^{32}P]RNA.[51,55] Neither binding reaction so far displays significant specificity with regard to nucleotide sequence; vector RNA or DNA probes give signals comparable to those obtained with HBV sequences of similar size.[51,55] Packaging of RNA into core particles in solution has so far been studied in *E. coli* and in the HepG2 cell tissue culture system for viral replication.

Perusal of the HBV literature suggests that the protamine-like tail of core (shown in detail in Figure 2) is the almost universally accepted nucleic acid-binding domain. While this idea appears reasonable (and derives precedent from the study of RNA binding by other viruses) there is little direct evidence to support it, and two recent studies provide apparently contradictory evidence. The initial challenge came from the observation[56] that purified serum e antigen (P17e and P18e), thought to lack the protamine-like tail region, binds DNA on a Southwestern blot. Schlicht et al.[57] has shown that deletion of the last 36 DHBV core residues does not impair the ability of the resultant truncated core to package RNA with close to wild-type efficiency in cultured human liver cells. These two studies certainly indicate that most or perhaps all of the protamine-like tail can be deleted without eliminating the ability of core protein to bind nucleic acid. However, in the latter study a small portion of the tail region was not deleted (one Arg cluster lies 48 residues from the end of duck core), while in the former study the tail was presumed rather than proved to be completely absent. Since e antigen derived from HBV subtype *adw* is incompletely processed at the carboxyl terminus (see Section VIII) this assumption may be dangerous; indeed, the e antigen used in this work displays the telltale heterogeneity generally associated with subtype *adw*.

To clarify the contribution of the protamine domain to nucleic acid binding we constructed *E. coli* expression vectors and expressed both wild-type core (r-P21.5) and the Cδ mutant, which encodes a 17-kDa core protein (r-P17) that ends at Val-149 and cleanly lacks the entire 36-amino acid tail region. We found that immobilized r-P21.5 bound [^{32}P]DNA probes as

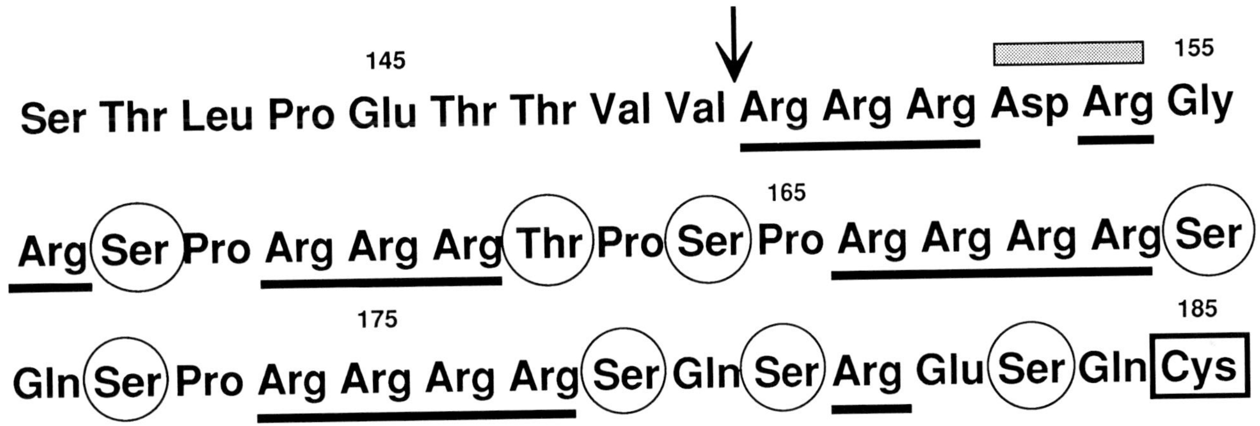

FIGURE 2. The structure of the core tail domain. The last 45 amino acids of the HBV core protein (subtype *adw*) are shown (residues 141 to 185). Residues that may have a particular functional significance are highlighted as follows: the carboxyl-terminus/putative cleavage site of HBeAg (vertical arrow next to Val-149); the Asp-Arg (or Arg-Asp) insertion in the *adw* tail that appears to interfere with the complete processing of the carboxyl terminus of HBeAg (shaded bar above residues); the seven possible Ser targets and one Thr target for phosphorylation (open circles); Arg blocks and single residues, potentially involved in nucleic acid binding (heavy underlining); Cys-185, which may be involved in multimerization of core dimers within the particle (boxed).

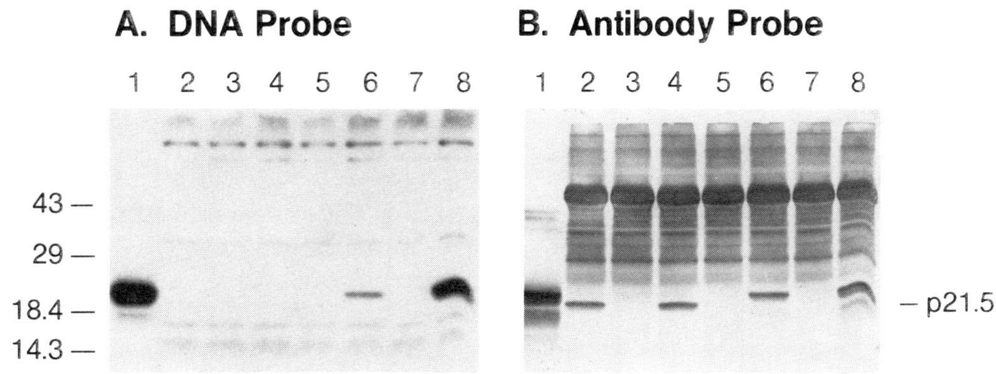

FIGURE 3. DNA binding by core protein generated in *Xenopus* oocytes or *E. coli*. (A) Southwestern analysis of DNA binding; (B) Western analysis of core protein. Core protein was immunoprecipitated from *Xenopus* oocytes that were uninjected (lane 7) or injected with synthetic C RNA (lanes 2 and 4), a control RNA (lanes 3 and 5), or recombinant *E. coli* core protein (lane 6). Also shown is *E. coli* core that was untreated (lane 1) or added to homogenized oocytes then recovered by immunoprecipitation (lane 8). The positions of authentic core protein (P21.5) and marker proteins (in kilodaltons) are indicated at the right and left of the figure, respectively. Note that the *E. coli* protein is slightly larger in size due to a small addition of sequence at the amino terminus. We have verified that this addition does not affect binding. For general oocyte methodology and microinjection methodology, see Reference 31. Proteins were resolved by 15% SDS-PAGE, then electrophoretically transferred onto Immobilon (Millipore, Bedford, MA). The membrane was first probed with a ^{32}P-labeled DNA sequence and autoradiographed to measure DNA binding, then reprobed with anti-core antibody to visualize the relative amount of core in each lane.

expected, but r-P17 did not.[55] Independent studies have confirmed the binding inactivity of similar tail-minus mutants.[50,51] We conclude that a portion of the tail domain is in fact essential for nucleic acid binding and are currently using directed mutagenesis to determine the nature of the essential residues.

As a simple working model for the more general form of core–RNA interaction, we propose that the N-terminal half of the tail faces the interior of the particle and binds RNA, while the C-terminus of the tail, which may lie more to the outer surface of the particle, plays a lesser part in RNA binding. This model is consistent with both the studies summarized above and with the structural information available for core (see Section V). It does not presume to explain the potentially complex interactions involved in the specific packaging event and should not be taken to imply that the N-terminal half of the tail domain is sufficient for interaction with RNA (particularly for selective packaging). Interior-facing residues from the β-barrel domain may very well contribute binding energy but, in the absence of the tail region, their affinity for nucleic acid is too weak to allow detection.

Since the above model is largely derived using *E. coli* core protein, we have immunoprecipitated P21.5 from C mRNA-injected *Xenopus* oocytes and analyzed its ability to bind nucleic acid by Southwestern blotting. Surprisingly, we found that the *Xenopus*-derived core does not bind DNA under conditions where *E. coli* P21.5 gives a strong signal (Figure 3A, lanes 2 and 4 versus lane 1). Extensive control experiments have assured us that there is no trivial explanation for this difference. Thus, immunoprecipitation per se does not render core unable to bind DNA (Figure 3A, lane 8); we have also injected recombinant *E. coli* core particles into oocytes, recovered them, and shown that they can still bind nucleic acid (Figure 3A, lane 6). We have also performed a Western analysis to confirm that the lanes contain comparable amounts of immobilized core protein (Figure 3B). Attempts to demonstrate nucleic acid binding by immobilized HBV core protein derived from virions or immature cores have also failed.[55,58] We believe that a specific core protein modification is the most probable cause of this loss of binding activity. Since P21.5 is phosphorylated (see Section VII) in oocytes and mammalian cells, but not in *E. coli*, we are attempting to determine whether phosphorylation can modulate the nucleic acid-binding ability of core protein. We do not believe that these results negate the model for core–nucleic acid interactions presented above, but they do remind us that binding may be subject to complex regulation in eukaryotic cells.

VII. POSTTRANSLATIONAL MODIFICATION OF CORE PROTEIN

Retroviral capsid proteins are subject to a wide spectrum of posttranslational modifications, including myristylation, sulfation, acylation, and phosphorylation. The only known modification of core protein is phosphorylation, although other kinds of modification have been suggested[45] and are certainly not ruled out. Phosphorylation was first reported in 1980 by Albin and Robinson,[59] who described an *in vitro* kinase activity present in purified immature HBV core particles derived from liver or in mature core particles released from virions by detergent. Both the kinase activity and the resultant phosphocore product have subsequently been identified *in vivo*.

A. *IN VITRO* KINASE ACTIVITY

When core particles are treated with [γ-^{32}P]ATP *in vitro*, the 21-kDa core protein is phosphorylated. This phosphorylation, which is common to all the *Hepadnaviridae* core proteins, occurs on one or more Ser residues[6,8] or, to a lesser extent, on Thr residues.[8] Tyrosine phosphorylation has not been implicated to date. Relatively few of the 180 core protein subunits within the capsid particle appear to be targets for the *in vitro* activity.[6] The original studies did not define the exact sites of phosphorylation (the sequence of core protein was unknown at that time), but it was noted that the labeled tryptic phosphopeptides generated from either GSHV or HBV ^{32}P-labeled core protein have identical mobilities.[8,60] Argos and Fuller[40] recently deduced, from the known sequence of the two proteins, that these phosphopeptides could correspond only to HBV core residues 155 to 157, 160 to 164, and 168 to 172. Thus, the carboxyl-terminal tail of core (with six Ser and one Thr residue) is the phosphorylation target (Figure 2).

B. PHOSPHORYLATION *IN VIVO*

Recombinant HBV P21.5 expressed in *Xenopus* oocytes can be metabolically labeled with $^{32}P_i$ (Figure 4B, lane 1) but the Cδ core mutant (truncated at Val-149) cannot be phosphorylated (Figure 4B, lanes 2 and 3), providing direct experimental evidence that phosphorylation occurs *in vivo* and that the tail is again the target.[21] Phosphocore has also been reported in mammalian cells transfected with the HBV core gene.[61] DHBV core protein, isolated from liver core particles, migrates as at least four different species with sizes between 32 and 34 kDa on SDS-PAGE.[57,62] Alkaline[62] or acid[57] phosphatase can convert these species to the highest mobility 32-kDa product, suggesting that the higher molecular weight bands represent phosphocore proteins. Again this phosphorylation is associated with the tail region since a duck core mutant missing the final 36 tail residues migrates as a single (presumably unphosphorylated) 32-kDa species on SDS gels.[57] Curiously, no such multiple species have been reported for HBV phosphocore; phosphorylated (as synthesized in mammalian cells or frog oocytes) or unphosphorylated (as produced in *E. coli*, see Section VII.C) P21.5 have identical mobilities on SDS gels.[55] Presumably the phosphorylation-associated mobility

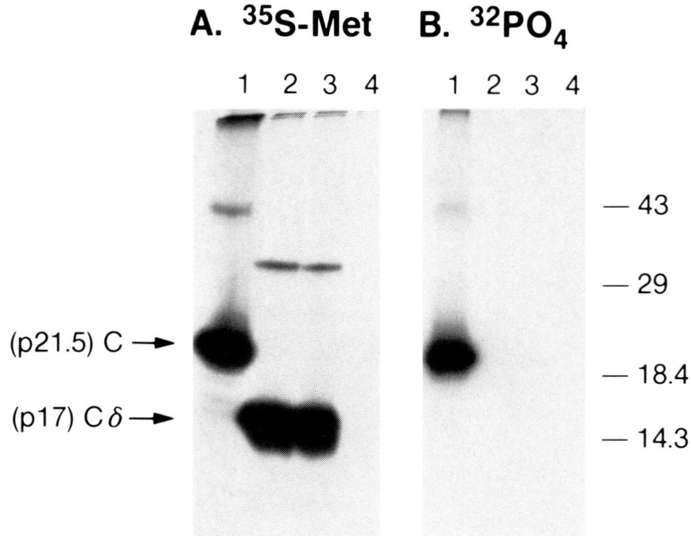

FIGURE 4. *In vivo* phosphorylation of core protein occurs in the protamine tail domain. Oocytes were metabolically labeled with (A) [^{35}S]methionine or (B) $^{32}P_i$ after being injected with C mRNA (lane 1), Cδ mRNA (lanes 2 and 3), or no RNA (lane 4). The positions of the C and Cδ proteins are indicated to the left of the figure; the positions of marker proteins (sizes in kilodaltons) are shown at the right.

changes in DHBV core protein reflect conformational changes rather than simple molecular weight increases.

C. ORIGIN OF THE KINASE ACTIVITY

Is the core kinase an intrinsic activity of core protein, a separate viral function, or a cellular enzyme? Current evidence favors the latter. As described above, duck core expressed in *E. coli* does not appear to be phosphorylated.[57] Recombinant HBV P21.5 core isolated from *E. coli* displays no *in vitro* autophosphorylation.[55,58] Thus, the kinase activity is not intrinsic to core protein. However, P21.5 is phosphorylated on expression in higher eucaryotic cells in the absence of other viral proteins[61] (Figure 4), implying that a cellular kinase is responsible for the phosphorylation. Recent studies by Laub's group suggest that a 90-kDa Mg^{2+}-dependent kinase is associated with HBV core particles.[63]

D. BIOLOGICAL SIGNIFICANCE OF PHOSPHORYLATION

There is no documented role for the kinase or phosphocore in the HBV life cycle and no certainty that there even is such a role. However, there are some clues that link core phosphorylation with viral replication. Low-density, polymerase-negative cores display more kinase activity than the higher density, polymerase-positive cores,[6-8] suggesting a correlation between core phosphorylation and the replicative state of HBV. Additional support comes from experiments[57] in which duck core was mutationally truncated so as to remove the phosphorylation target residues within the tail domain without altering the amino acid sequence encoded by the overlapping *pol* reading frame. A mutant lacking the final 36 amino acids of duck core (containing nine potential Ser phosphorylation sites) displayed only a single core species on SDS-PAGE, implying that phosphorylation had not occurred. In HepG2 cells, which are permissive for DHBV replication, the corresponding virus assembled correctly and packaged pregenomic RNA but failed to replicate its genome, suggesting that phosphorylation is required for replication. However, the authors rightly point out that specific residues from the deleted tail segment, rather than phosphorylation per se, could cause the defect. A deletion removing only the last 12 amino acids of duck core (including five Ser residues) was wild type with respect to viral replication and assembly, but the resultant virion was not infectious in ducks. Again it is difficult to be certain if this effect is truly related to phosphorylation. As described in the previous section there is evidence (albeit indirect) that phosphorylation reduces the nucleic acid-binding ability of P21.5. This could in turn affect replication since the reverse transcription reaction essentially doubles the nucleic acid content of the particle. Finally, Pugh et al.,[62] noting that duck core protein is phosphorylated in immature cores but appears largely unphosphorylated (P32) in virions, speculate that dephosphorylation supplies either the signal or the driving force that causes the packaging of cores within the lipid envelope. An alternative interpretation of their data is that core dephosphorylation signals or even drives the replication process that accompanies virion maturation.

VIII. THE BIOSYNTHESIS OF SERUM e ANTIGEN

A. HISTORICAL PERSPECTIVE

One of the first known HBV markers was e antigen or HBeAg, described by Magnius and Espmark in 1972.[64] The clinical importance of HBeAg lies in its wide acceptance as a serological marker of HBV infection (particularly acute infection), although its reliability has been questioned.[35] HBeAg is immunologically related to core protein[5,65-68]; its carboxyl-terminal protein sequence suggests that HBeAg is a truncated core polypeptide which ends at Val-149 and lacks the final 34 amino acids of P21c.[68] HBeAg is a dimeric, nonparticulate protein or proteins of ~17 to 20 kDa that is often associated with antibodies.[69-72] HBeAg activity is present cryptically in core particles and is revealed when the particle is disrupted by sonication, by treatment with detergent plus 2-mercaptoethanol, or by proteases.[5,66,68,72] Limited pronase digestion of core particles releases a protein with HBeAg size and antigenicity that, by sequence analysis, terminates at Thr-147, only two residues short of authentic e.[68] Thus, it is hardly surprising that e antigen was first thought to be a breakdown product of intact HBV cores, although it remains difficult to see how such a process could generate substantial amounts of free serum HBeAg.

B. ROLE OF THE PRECORE SEQUENCES

The above dilemma was resolved by the recent discovery that the biogenesis of HBeAg is actually dependent on the precore region of HBV.[26-28] Recent studies on the expression of duck precore/core sequences have established that duck precore directs the secretion of heterogeneous duck e antigen species; introduction of a stop codon into the precore region of DHBV abrogates secretion of duck e antigen in tissue culture cells.[48] We have studied the production of HBeAg in detail using *Xenopus* oocytes injected with precore mRNA.[31] Figure 5 shows the complex pattern of accumulation of both intracellular and secreted precore proteins synthesized in oocytes and contrasts it with the simple pattern of core protein accumulation. The picture of HBeAg production emerging from these and other studies is as follows. Upon translation of the precore/core locus a cleavable 19-amino acid signal peptide, located at the start of the 29-amino acid HBV precore region, targets the HBeAg precursor to the secretory pathway. The precise removal of this signal peptide has been verified by protein-sequencing studies performed *in vitro*,[45] in *Xenopus* oocytes,[31] and in mammalian cells.[73] Signal peptide cleavage *in vitro* is dependent on the presence of microsomal membranes and occurs concomitantly with translocation across the endoplasmic reticulum (ER) membrane.[45] The initial product of translocation is a 22.5-kDa precore protein (P22.5) that contains the last 10 precore residues followed by the intact core protein (P21). Very little P22.5 is secreted as such.[31] Instead, some or all of the protamine tail is cleaved, depending on subtype (see Section VIII.C). The resultant ~17-kDa protein or group of proteins is secreted into the medium (presumably in dimeric form) and displays e antigen activity.[31]

C. THE CARBOXYL-TERMINAL PROCESSING OF PRECORE PROTEINS

Protein sequencing[68] first established that serum HBeAg terminates with the sequence 146-Thr-Thr-Val-Val-149 and lacks the last 34 tail residues of *ayw* core (see Figure 2). Recent work has confirmed that cleavage of carboxyl-terminal residues accompanies e antigen secretion. Schlicht et al.[48] used a specific antibody to demonstrate that secreted duck e antigen lacks the carboxyl terminus of duck core. We have shown that the secreted HBeAg proteins and intracellular P22.5 have the same amino-terminal protein sequence,[21,31] confirming that the precore signal peptide has been removed. Yet the smallest secreted proteins are almost 6 kDa smaller than P22.5, suggesting that protein sequence must have been removed from the carboxyl terminus. Using a precore mutant that terminates at Val-149, we further showed that the resultant secreted "tailless" e antigen is a single protein of the same molecular mass (predicted molecular mass of 17 kDa) as the smallest species derived from the wild-type precore ORF.[31]

Two puzzles remain concerning the processing of the precore C-terminal residues. First, what activity is responsible for the processing? Second, why do some studies report a single secreted HBeAg species while others find five or more species? For the processing activity, Miller[74] suggested that e antigen is derived from core (or precore) protein by self-proteolysis, based on a prediction of an aspartic acid-type protease activity located within the first ~50 amino acids of core. This region bears scattered homology to the protease regions of other retroviruses and includes a particularly conserved motif (LLDT, HBV core residues 30 to 33) in which the aspartate residue is invariant. The ubiquity of aspartic acid proteases and self-proteolysis (particularly of retroviral Gag-Pol precursors) in the viral world makes this idea intriguing. However, given the involvement of precore sequences in the generation of HBeAg and the intrinsic stability of core antigen synthesized in yeast, *Xenopus* oocytes, or mammalian cells, it is unlikely that self-proteolysis of core protein provides a significant mechanism for e antigen biogenesis. The final nail in the coffin for this hypothesis came when the putative protease active site residues within the C ORF (the invariant D and selected adjacent residues) were mutated without any effect on HBeAg synthesis or secretion (or on reverse transcription of the HBV genome).[75,76] Studies on the time course of translation and processing of precore proteins in frog oocytes indicate that the trimming of the carboxyl terminus is a late event in the secretory pathway for HBeAg, consistent with a cellular protease acting on the precore proteins inside the Golgi complex.[31] A protease that cleaves Arg-Arg dipeptides would provide a simple means to trim e antigen down to Val-149. Alternatively, a cellular aspartyl protease may be responsible for P22.5 trimming.[76] Viral aspartyl proteases frequently cleave X-Pro peptide bonds in Gag-Pol precursor proteins. The precore tail domain contains several Ser-Pro dipeptides (including residues 157 to 158, Figure 2) and cleavage at Ser-Pro, followed by trimming from a cellular carboxypeptidase activity, could generate the authentic e antigen terminus.

HBeAg heterogeneity has been shown to be strictly associated with the *adw* subtype.[46] In sera from patients infected with *adw* HBV, the ~17-kDa form of HBeAg is reduced or eliminated and replaced by multiple 18- to 21-kDa species.[46] Compared to other HBV subtypes, *adw* contains an Asp-Arg (or Arg-Asp) insertion (indicated by a shaded bar in Figure 2) just downstream from Val-149. This insertion apparently interferes with normal processing,

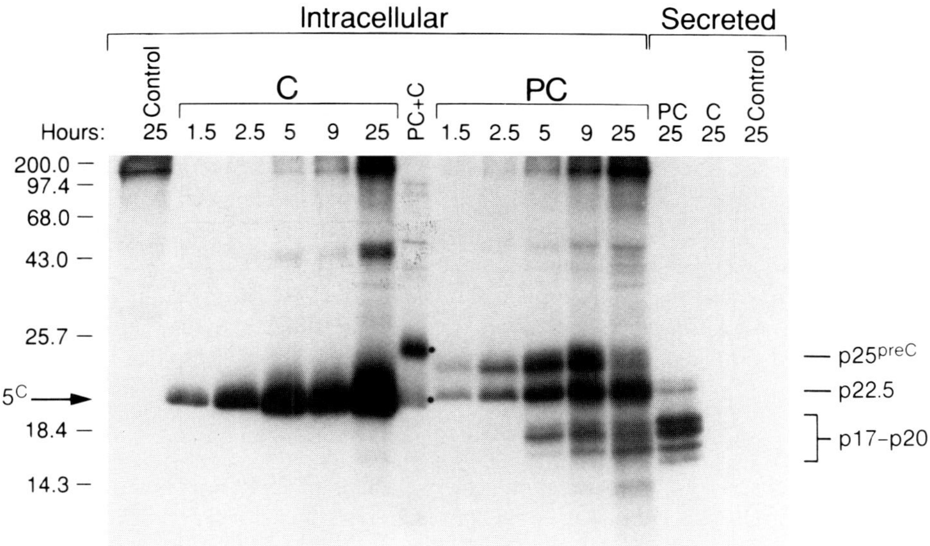

FIGURE 5. Expression of core and precore proteins in *Xenopus* oocytes. The figure is modified from Reference 31, where further details can be found. In brief, oocytes were injected with either C mRNA (C lanes) or PC mRNA (PC lanes), then labeled with [^{35}S]methionine for the times (in hours) indicated above each lane. Uninjected control oocytes labeled for 25 h are included in the left-hand lane. Core proteins were recovered from the medium (secreted proteins, shown in the three right-hand lanes) or from the oocyte (intracellular proteins, remaining lanes) by immunoprecipitation with anti-core antibody (Dako), resolved by 15% SDS-PAGE, and visualized by autoradiography. The position of core protein (p21.5^{C}) is indicated at the left of the figure, along with the sizes (in kilodaltons) of protein molecular weight markers. The positions and sizes of some of the precore species are shown at the right of the figure.

presumably via charge or steric factors, resulting in partial cleavages at downstream sites and a heterogeneous array of e antigens. In the case of duck e antigen, both multiple cleavages and one or two glycosylation events probably accompany secretion.[48] Mammalian HBeAgs have no glycosylation sites and to date there is no evidence of other modification steps *in vivo*.

D. UNUSUAL ASPECTS OF HBeAg SECRETION

One indication that the secretion process for precore proteins is complex is seen in the precore protein profile shown in Figure 5. Proteins of 23, 24, and 25 kDa, all larger than the 22.5-kDa precore translocation product, are typically seen in this and other experiments. Conventional models of protein secretion do not account for these species, nor for several reports that the precore region sends precore proteins to unexpected cellular locations. The precore region has recently been shown to direct uncharacterized precore species to the cellular plasma membrane, [77] where they may provide targets for the immune response to HBV. Previous histological[14,78] and immunological[79,80] studies have documented core-related proteins in hepatocyte plasma membranes. Two groups examined the translocation of the signal peptide-minus P22.5 *in vitro*, by translating preC mRNA in the presence of microsomal membranes, but reached rather different conclusions. One report[81] found mostly transmembraneous (i.e., only partially translocated) P22.5, but this was not substantiated by Garcia et al.,[45] who found that 20 to 30% of the P22.5 was fully translocated, while 70 to 80% was apparently released back into the cytoplasm (after removal of the signal peptide) and subsequently modified! In oocytes we have detected cytoplasmic P22.5, although only as a very minor species; most P22.5 appears to be translocated normally.[21] In cultured mammalian cells most precore proteins go through the secretory pathway, but a recent report[73] claims that ~15% of the intracellular p22.5 is released from the ER into the cytosol, from which it migrates to the cell nucleus, the 10-residue precore segment apparently functioning as a nuclear localization signal. However, it is not clear that the controls performed in this study rigorously rule out the possibility of contamination of nuclear subcellular fractions by ER components. In our oocyte studies we have consistently observed the expression of a 25-kDa precore protein that accumulates in the oocyte cytoplasm (often one of the major precore products); to our surprise, protein sequencing has verified that this species contains an intact signal peptide and is indeed P25![21] An uncharacterized 25-kDa species (presumably P25) has been reported in studies in which mammalian cells were transfected with precore sequences.[76,82]

The above observations are generally bizarre, sometimes confusing, largely inexplicable in terms of what we know about eukaryotic protein secretion, and of unknown biological significance. Nevertheless, the sheer weight of evidence makes it clear that the secretion of precore proteins is an interesting and unusual process that merits further study.

E. BIOLOGICAL FUNCTION OF PRECORE

The above observations could fuel unbridled speculation about the pivotal role of precore in almost all aspects of the viral life cycle. Such speculation is sharply constrained by elegant studies in the duck animal model that demonstrate beyond doubt that when the preC region is removed by selective mutation, the ability of the mutant virus to infect ducklings is essentially unaltered[48,83] and the course of viremia appears normal! Thus, it is impossible for precore to play a crucial role in the nucleic acid packaging, assembly, replication, export, or infectivity functions of the DHBV virion. Similarly, removal of the HBV precore region appears to have no effect on the formation or replication of HBV virions in HepG2 cells.[82] Biochemical data indicate that precore proteins cannot be detected in core particles derived from infected ducks,[62] making it unlikely that they play any crucial structural role in DHBV. No evidence of the presence of P25 or even P22.5 in authentic HBV cores has been presented.

Despite the above there is strong circumstantial evidence that the precore region must have a function, albeit a subtle one. First, the precore region is conserved throughout hepadnavirus evolution, although a simple point mutation could eliminate it. Second, the functional ability to direct HBeAg secretion is conserved in HBV and DHBV. Finally, viral capsid protein secretion is conserved through evolution as far as retroviruses, suggesting that it must confer some selective advantage. Secreted[84,85] and cell surface-associated[86] forms of the Moloney murine leukemia virus Gag protein have been reported. Expression of the secreted Gag proteins can be eliminated by mutations just prior to Gag without apparent detriment to the viral life cycle in cultured cells.[84,85] What kind of subtle role might these secreted capsid proteins play? Modulation of the host immune response during the course of viral infection seems the most probable answer for precore proteins. Cell surface precore proteins may provide the target for T cell-mediated clearance of virus-infected cells[79,80]; circulating HBeAg could modulate this response Alternatively, precore proteins could still play a minor structural role in the life cycle of the virus, perhaps facilitating its assembly or increasing its stability; minor alterations in physical structure and replication have been noted[84] for viral particles generated from a Moloney murine leukemia virus mutant deficient in secretion of glycosylated Gag proteins.

IX. THE ANTIGENIC PROPERTIES OF CORE AND PRECORE PROTEINS

The two serologically defined, core-related antigenic determinants are HBc (core antigen, present on the intact nucleocapsid) and HBe (e antigen, displayed by the secreted precore proteins or disrupted nucleocapsids). A much less well-characterized antigenic site has been localized to the amino terminus of the protamine tail.[47] The immune response to HBc/HBe may play an important role in the progress of hepatitis infection. IgM antibodies against HBc are the first anti-HBV antibodies to appear during acute viral infection. Virtually 100% of infected individuals produce high-titer IgM and IgG anti-HBc responses that persist for many years, whereas individual immune responses to the HBV envelope proteins vary widely (including no response). Seroconversion from HBeAg to anti-HBeAg is a late event in the immune response and generally correlates well with viral clearance.[87] Both the antigens and antibodies make valuable and sensitive markers of current and prior HBV infection and can be used to diagnose the status of infection.

The magnitude of the immune response to HBc and HBe is quite different; particulate HBc elicits a very powerful response compared to nonparticulate HBe.[52] Recombinant tail-minus core (r-P17) provides a particulate source of both HBc and HBe, yet only the former elicits a strong immune response in mice.[52] Thus the high-level response to HBcAg is intrinsic to the antigen itself. The immune response of inbred mouse strains to core particles is at least 80-fold greater than that against surface envelope particles.[88,89] The high immunogenicity of core particles makes them an attractive vehicle for presenting foreign epitopes to the immune system.[49,90]

Early immunologic studies of HBc/HBe often employed complex, poorly characterized patient sera to probe limited amounts of partially purified core particles or HBeAg; the development of pure recombinant core/e and high-quality rabbit and monoclonal antibody reagents has overcome these limitations. Monoclonal antibodies have been used to identify two HBe epitopes,[52,68,91-94] and either a single immunodominant HBc epitope[95,96] or two epitopes.[68] These monoclonals, in conjunction with recombinant core/e and sandwich-type solid-phase immunoassays, have provided evidence that the two antigenic activities are not as independent as was historically thought[52,68]; a minor core determinant has been found on serum e polypeptide and one e epitope (but not the second) occurs on core particles.[68] Since core and e proteins contain almost identical linear polypeptide chains it is no surprise that they share common antigenic determinants. The issue of how they present serologically distinct antigenic identities can be addressed only when the epitopes have been identified.

A. MAPPING OF B CELL EPITOPES

The fine structure of the c and e epitopes has been investigated by using defined antibody reagents to probe either synthetic peptides or purified fusion proteins containing subregions of core polypeptide. Using an extensive set of fusion proteins, Salfeld et al.[46] mapped the HBe1 and HBe2 epitopes to around core residues 85 and 138, respectively, and a single immunodominant HBc epitope was localized between residues 74 and 83, overlapping HBe1. Colucci et al.[97] used synthetic peptides corresponding to the four most hydrophilic regions of core to find a single immunodominant core epitope, but mapped it to residues 107 to 118. This peptide elicited a strong anti-HBc response in BALB/c mice. Peptide 73–85, corresponding to the epitope mapped by Salfeld et al.,[46] was recognized by sera from anti-HBc positive patients but not by the monoclonals, and was characterized as a minor core epitope. Unfortunately, the different strategies and reagents used by the two groups make it hard to evaluate their results, although the mapping studies of Salfeld et al. seem particularly exhaustive. Clearly, more studies will be required to confirm the identity of these antigenic determinants.

The above epitopes were mapped using minimal segments of core protein and are therefore linear rather than conformational. Comparison of the ability of a panel of antibody reagents to recognize short segments of core polypeptide (i.e., linear epitopes) vs. fusion proteins containing the majority of core protein (presumably retaining some conformation) reveals that some antibodies recognize their epitope only in the context of adequate conformation.[46] In particular, HBe2 and HBc appear to be highly conformational, while HBe1 is a relatively conformation-insensitive, linear determinant.[46] The core protein-folding model shows HBe1 (residue ~85) on the surface of the core particle.[40] This location and the linear nature of HBe1 are consistent with this antigen being presented by both core and e structures. HBe2 lies buried and inaccessible in the core particle but removal of the adjacent protamine tail in e antigen could be envisaged to open up this region and unmask the epitope. One puzzle that remains concerns the simultaneous presentation of the linear HBe1 and conformational HBc epitopes by core particles even though they colocalize to the same segment of core sequence (about amino acid 80).[46] This can be explained if the core protein molecules within the particle exist in two slightly different, or quasi-equivalent, structural states.[46] Quasi-equivalence of capsid protein structures is a common feature of icosahedral nucleocapsids.

B. MAPPING OF T CELL EPITOPES

Although HBV nucleocapsid particles (HBc) function as T cell-independent antigens when injected into athymic mice, the B cell response is dramatically elevated in euthymic mice, suggesting that T helper cell function is required for maximal antibody response.[88] The immune response to denatured core protein (i.e., HBeAg) is in fact dependent on this T helper function.[52,88] HBeAg and native core antigen are highly cross reactive and virtually indistinguishable at the T cell level,[88] despite the serological distinction between the two antigens, and T cells that have been primed with either HBeAg or HBcAg can be activated *in vitro* by the reciprocal antigen. Thus, the difference in immune response to HBcAg vs. HBeAg must be explained at the B cell level.

The sites involved in the T cell response to HBV core protein have been mapped using synthetic peptides. A 21-residue synthetic peptide corresponding to residues 100 to 120 was found to activate HBcAg-primed T cells *in vitro*.[88,98] T-cells primed by this peptide can be activated by HBcAg, further emphasizing the cross reactivity between HBeAg and HBcAg at the T cell level. Five additional peptides (1 to 20, 120 to 140, 70 to 94, 85 to 100 and 28 to 52) that induce HBcAg-specific T cell proliferation *in vitro* have since been identified from a panel of synthetic peptides that covers the entire core sequence; these peptides also elicited HBcAg-specific T cells in mice.[99] Different *H-2* congenic strains of mice responded differently, with some strains recognizing only a single peptide while others recognized multiple peptides with the response to one being predominant. In each case, a single peptide gave the predominant T cell response but both the dominant peptide and the magnitude of the response differed among strains. These experiments show that the T cell response to HBcAg depends on a limited number of discrete determinants on core and on the *H-2* haplotype of the host.

When a complex multiantigenic molecule such as HBV is bound and internalized by a B cell, it is degraded and then reexpressed on the cell surface in the form of multiple viral peptides. Thus, the repertoire of viral B cell and T cell determinants could potentially be presented simultaneously to the immune system, raising the possibility that T helper cells recognizing a core determinant may activate B cell clones directed against the surface envelope. The ability of HBcAg-specific T cells to prime antibody (IgG) production to HBsAg determinants has been examined in mice.[89,98] *In vivo* priming by either HBcAg or the synthetic T cell recognition peptide P120–140 followed by challenge with whole virus resulted in substantial antibody production to all three S antigens. If priming was omitted, or if HBsAg and HBcAg were supplied as a mixture rather than within the same particle, there was no anti-S response. Thus, HBcAg-primed T cells are competent to help activate B cell clones specific for preS1, preS2, and S provided the two antigens are present in the same particle. These observations may explain why vaccination with core particles has been reported to confer protection against challenge by HBV[100] or WHV.[101]

X. CONCLUSIONS AND PROSPECTS

The preceding review provides a testament to the remarkable properties and multifunctional nature of the hepatitis core protein. Progress has accelerated in the past few years. Until 1986 the biogenesis of e antigen was essentially a mystery. Today the entire HBeAg secretion pathway and the nature of the various processing steps involved are fairly well characterized. Since 1985, studies have begun to illuminate the interactions between core polypeptide and RNA sequences, leading to recent revelations about the role of Pol in the process of selectively packaging the pregenome RNA. Many features of the immune response to core, at both the B and T cell level, have been characterized and the nature of the antigenic sites is being clarified. Thus the description of these and most other properties of core is almost complete. Only the nature of the interactions between core and the viral envelope proteins or the viral polymerase has so far evaded descriptive analysis. However, there has been much less overall progress on the molecular mechanism or the biological significance behind many of these properties. For example, we have insufficient data on which to base a detailed model of the interactions between core, Pol, and the RNA-packaging signal. Moreover, although we know some details about the nature of HBeAg or of phosphocore, our attempts to explain the biological role of these species are little more than speculation. The challenge that faces the field is to move from the descriptive phase to tackle these important, although experimentally more difficult, issues.

ACKNOWLEDGMENTS

I would like to thank my present and former collaborators, Tom Hatton, Siliang Zhou, and Maureen Walter, for their help, enthusiasm, and contributions to the work cited in this chapter. I also thank Tom Hatton for providing Figure 1, and W. J. Rutter for his encouragement. I thank many colleagues in the field for gifts of reagents, for valuable discussions, and for communicating their data prior to publication. This work was supported by National Institutes of Health Grant AI25056.

REFERENCES

1. **Almeida, J. D., Rubenstein, D., and Stott, E. J.,** New antigen-antibody system in Australia antigen-positive hepatitis, *Lancet*, ii, 1225, 1971.
2. **Fauvel, M., Babiuk, L., Sheaff, E. T., and Spence, L.,** Preparation of a Dane core (hepatitis B) antigen from human plasma, *Can. J. Microbiol.*, 21, 905, 1975.
3. **Summers, J. and Mason, W. S.,** Replication of the genome of a hepatitis B-like virus by reverse transcription of an RNA intermediate, *Cell*, 29, 403, 1982.
4. **Enders, G. H., Ganem, D., and Varmus, H. E.,** 5′-Terminal sequences influence the segregation of ground squirrel hepatitis virus RNAs into polyribosomes and viral core particles, *J. Virol.*, 61, 35, 1987.
5. **Ohori, H., Yamaki, M., Onodera, S., Yamada, E., and Ishida, N.,** Antigenic conversion from HBcAg to HBeAg by degradation of hepatitis B core particles, *Intervirology*, 13, 74, 1980.
6. **Gerlich, W. H., Goldmann, U., Muller, R., Stibbe, W., and Wolff, W.,** Specificity and localization of the hepatitis B virus-associated protein kinase, *J. Virol.*, 42, 761, 1982.
7. **Petit, M.-A. and Pillot, J.,** HBc and HBe antigenicity and DNA-binding activity of major core protein P22 in hepatitis B virus core particles isolated from the cytoplasm of human liver cells, *J. Virol.*, 53, 543, 1985.
8. **Feitelson, M. A., Marion, P. L., and Robinson, W. S.,** Core particles of hepatitis B virus and ground squirrel hepatitis virus. II. Characterization of the protein kinase reaction associated with ground squirrel hepatitis virus and hepatitis B virus, *J. Virol.*, 43, 741, 1982.

9. **Burrell, C. J., Gowans, E. J., and Marmion, B. P.,** High levels of cytoplasmic hepatitis B core antigen as reliable marker of HBV DNA replication, *Lancet*, i, 454, 1985 (letter).
10. **Sansonno, D. E., Fiore, G., Bufano, G., and Manghisi, O. G.,** Cytoplasmic localization of hepatitis B core antigen in hepatitis B virus infected livers, *J. Immunol. Methods*, 109, 245, 1988.
11. **Gudat, F. and Bianchi, L.,** Evidence for phasic sequences in nuclear HBcAg formation and cell membrane-directed flow of core particles in chronic hepatitis B, *Gastroenterology*, 73, 1194, 1977.
12. **Yamada, G. and Nakane, P. K.,** Hepatitis B core and surface antigens in liver tissue. Light and electron microscopic localization by the peroxidase-labeled antibody method, *Lab. Invest.*, 36, 649, 1977.
13. **Ulich, T. R., Anders, K., Layfield, L., Cheng, L., and Lewin, K. J.,** Chronic active hepatitis of hepatitis B and non-A, non-B etiology. Immunohistochemical localization of hepatitis B core antigen in a series of needle biopsies, *Arch. Pathol. Lab. Med.*, 109, 403, 1985.
14. **Kojima, T., Bloemen, J., and Desmet, V. J.,** Immune electron microscopic demonstration of hepatitis B core antigen (HBcAg) in liver cell plasma membranes, *Liver*, 7, 191, 1987.
15. **Huang, S. N. and Neurath, A. R.,** Immunohistologic demonstration of hepatitis B virus antigens in liver with reference to its significance in liver injury, *Lab. Invest.*, 40, 1, 1979.
16. **Michalak, T. and Nowosawski, A.,** Crystalline aggregates of hepatitis B core particles in cytoplasm of hepatocytes, *Intervirology*, 17, 247, 1982.
17. **Burrell, C. J., Gowans, E. J., Jilbert, A. R., Lake, J. R., and Marmion, B. P.,** Hepatitis B virus DNA detection by *in situ* cytohybridisation: implications for viral replication strategy and pathogenesis of chronic hepatitis, *Hepatology*, 2, 85, 1982.
18. **Blum, H. E., Haase, A. T., Harris, J. D., Walker, D., and Vyas, G. N.,** Asymmetric replication of hepatitis B virus DNA in human liver: demonstration of cytoplasmic minus-strand DNA by blot analyses and *in situ* hybridization, *Virology*, 139, 87, 1984.
19. **Blum, H. E., Haase, A. T., and Vyas, G. N.,** Molecular pathogenesis of hepatitis B virus infection: simultaneous detection of viral DNA and antigens in paraffin-embedded liver-sections, *Lancet*, ii, 771, 1984.
20. **Dingwall, D. and Laskey, R.A.,** Protein import into the cell nucleus, *Annu. Rev. Cell Biol.*, 2, 367, 1986.
21. **Standring, D. N.,** unpublished observations, 1990.
22. **Edman, J. C., Hallewell, R. A., Valenzuela, P., Goodman, H. M., and Rutter, W. J.,** Synthesis of hepatitis B surface and core antigens in *E. coli*, *Nature (London)*, 291, 503, 1981.
23. **Stahl, S., Mackay, P., Magazin, M., Bruce, S. A., and Murray, K.,** Hepatitis B virus core antigen: synthesis in *E. coli* and application in diagnosis, *Proc. Nat. Acad. Sci. U.S.A.*, 79, 1606, 1982.
24. **Miyanohara, A., Imamura, T., Araki, M., Sugawara, K., Ohtomo, N., and Matsubara, K.,** Expression of hepatitis B virus core antigen gene in *Saccharomyces cerevisiae:* synthesis of two polypeptides translated from different initiation codons, *J. Virol.*, 59, 176, 1986.
25. **Kniskern, P. J., Hagopian, A., Montgomery, D. L., Burke, P., Dunn, N. R., Hofmann, K. J., Miller, W. J., and Ellis, R. W.,** Unusually high-level expression of a foreign gene (hepatitis B virus core antigen) in *Saccharomyces cerevisiae*, *Gene*, 46, 135, 1986.
26. **Ou, J.-H., Laub, O., and Rutter, W. J.,** Hepatitis B virus gene function: the precore region targets the core antigen to cellular membranes and causes the secretion of the e antigen, *Proc. Natl. Acad. Sci. U.S.A.*, 83, 1578, 1986.
27. **Roossinck, M. J., Jameel, S., Loukin, S. H., and Siddiqui, A.,** Expression of hepatitis B viral core region in mammalian cells, *Mol. Cell. Biol.*, 6, 1393, 1986.
28. **McLachlan, A., Milich, D. R., Raney, A. K., Riggs, M. G., Hughes, J. L., Sorge, J., and Chisari, F. V.,** Expression of hepatitis B virus surface and core antigens: influences of pre-S and precore sequences, *J. Virol.*, 61, 683, 1987.
29. **Yamaki, M., Ohori, H., Onodera, S., Ishida, N., and Maeda, H.,** Circular dichroism and biochemical properties of the hepatitis B virus core antigen, *Biochim. Biophys. Acta*, 706, 165, 1982.
30. **Enders, G. H., Ganem, D., and Varmus, H.,** Mapping the major transcripts of ground squirrel hepatitis virus: the presumptive template for reverse transcriptase is terminally redundant, *Cell*, 42, 297, 1985.
31. **Standring, D. N., Ou, J.-H., Masiarz, F. R., and Rutter, W. J.,** A signal peptide encoded within the precore region of hepatitis B virus directs the secretion of a heterogeneous population of e antigens in *Xenopus* oocytes, *Proc. Natl. Acad. Sci. U.S.A.*, 85, 8405, 1988.
32. **Suzuki, T., Masui, N., Kajino, K., Saito, I., and Miyamura, T.,** Detection and mapping of spliced RNA from a human hepatoma cell line transfected with the hepatitis B virus genome, *Proc. Natl. Acad. Sci. U.S.A.*, 86, 8422, 1989.
33. **Su, T. S., Lai, C.-J., Huang, J.-L., Lin, L.-H., Yauk, Y.-K., Chang, C., Lo, S. J., and Han, S.-H.,** Hepatitis B virus transcript produced by RNA splicing, *J. Virol.*, 63, 4011, 1989.
34. **Will, H., Salfeld, J., Pfaff, E., Manso, C., Theilmann, L., and Schaller, H.,** Putative reverse transcriptase intermediates of human hepatitis B virus in primary liver carcinomas, *Science*, 231, 594, 1986.

35. **Theilmann, L., Pfaff, E., Kommerell, B., Gmelin, K., Schaller, H., and Salfeld, J.,** Detection of hepatitis B virus core gene products in sera and liver of HBV-infected individuals, *J. Hepatol.*, 8, 77, 1989.
36. **Schlicht, H. J., Radziwill, G., and Schaller, H.,** Synthesis and encapsidation of duck hepatitis B virus reverse transcriptase do not require formation of core-polymerase fusion proteins, *Cell*, 56, 85, 1989.
37. **Chang, L.-J., Pryciak, P., Ganem, D., and Varmus, H. E.,** Biosynthesis of the reverse transcriptase of hepatitis B viruses involves *de novo* translational initiation not ribosomal frameshifting, *Nature (London)*, 337, 364, 1989.
38. **Feitelson, M. A. and Miller, R. H.,** X gene-related sequences in the core gene of duck and heron hepatitis B viruses, *Proc. Natl. Acad. Sci. U.S.A.*, 85, 6162, 1988.
39. **Miller, R. H. and Robinson, W. S.,** Common evolutionary origin of hepatitis B virus and retroviruses, *Proc. Natl. Acad. Sci. U.S.A.*, 83, 2531, 1986.
40. **Argos, P. and Fuller, S. D.,** A model for the hepatitis B virus core protein: prediction of antigenic sites and relationship to RNA virus capsid proteins, *EMBO J.*, 7, 819, 1988.
41. **Onodera, S., Ohori, H., Yamaki, M., and Ishida, N.,** Electron microscopy of human hepatitis B virus cores by negative staining-carbon film technique, *J. Med. Virol.*, 10, 147, 1982.
42. **Yamaguchi, M., Hirano, T., Hirokawa, H., Sugahara, K., Mizokami, H., and Matsubara, K.,** Cryo-electron microscopy of hepatitis B virus core particles produced by transformed yeast: comparison with negative staining and ultrathin sectioning, *J. Electron Microsc.*, 37, 337, 1988.
43. **Radziwill, G., Zentgraf, H., Schaller, H., and Bosch, V.,** The duck hepatitis B virus DNA polymerase is tightly associated with the viral core structure and unable to switch to an exogenous template, *Virology*, 163, 123, 1988.
44. **Luo, M., Vriend, G., Kamer, G., Minor, I., Arnold, E., Rossmann, M. G., Boege, U., Scraba, D. G., Duke, G. M., and Palmenberg, A. C.,** The atomic structure of Mengo virus at 3.0 Å resolution, *Science*, 235, 182, 1987.
45. **Garcia, P. D., Ou, J.-H., Rutter, W. J., and Walter, P.,** Targeting of the hepatitis B virus precore protein to the endoplasmic reticulum membrane: after signal peptide cleavage translocation can be aborted and the product released into the cytoplasm, *J. Cell Biol.*, 106, 1093, 1988.
46. **Salfeld, J., Pfaff, E., Noah, M., and Schaller, H.,** Antigenic determinants and functional domains in core antigen and e antigen from hepatitis B virus, *J. Virol.*, 63, 798, 1989.
47. **Machida, A., Ohnuma, H., Takai, E., Tsuda, F., Tanaka, T., Naito, M., Munekata, E., Miyakawa, Y., and Mayumi, M.,** Antigenic sites on the arginine-rich carboxyl-terminal domain of the capsid protein of hepatitis B virus distinct from hepatitis B core or e antigen, *Mol. Immunol.*, 26, 413, 1989.
48. **Schlicht, H. J., Salfeld, J., and Schaller, H.,** The duck hepatitis B virus pre-C region encodes a signal sequence which is essential for synthesis and secretion of processed core proteins but not for virus formation, *J. Virol.*, 61, 3701, 1987.
49. **Clarke, B. E., Newton, S. E., Carroll, A. R., Francis, M. J., Appleyard, G., Syred, A. D., Highfield, P. E., Rowlands, D. J., and Brown, F.,** Improved immunogenicity of a peptide epitope after fusion to hepatitis B core protein, *Nature (London)*, 330, 381, 1987.
50. **Birnbaum, F. and Nassal, M.,** Hepatitis B virus nucleocapsid assembly: primary structure requirements in the core protein, *J. Virol.*, 64, 3319, 1990.
51. **Gallina, A., Bonelli, F., Zentilin, L., Rindi, G., Muttini, M., and Milanesi, G.,** A recombinant hepatitis B core antigen polypeptide with the protamine-like domain deleted self-assembles into capsid particles but fails to bind nucleic acids, *J. Virol.*, 63, 4645, 1989.
52. **Milich, D. R., McLachlan, A., Stahl, S., Wingfield, P., Thornton, G. B., Hughes, J. L., and Jones, J. E.,** Comparative immunogenicity of hepatitis B virus core and e antigens, *J. Immunol.*, 141, 3617, 1988.
53. **Hirsch, R. C., Lavine, J. E., Chang, L. J., Varmus, H. E., and Ganem, D.,** Polymerase gene products of hepatitis B viruses are required for genomic RNA packaging as well as for reverse transcription, *Nature (London)*, 344, 552, 1990.
54. **Junker-Niepmann, M., Bartenschlager, R., and Schaller, H.,** A short *cis*-acting sequence is required for hepatitis B virus pregenome encapsidation and sufficient for packaging of foreign RNA, *EMBO J.*, 9, 3389, 1990.
55. **Hatton, T. and Standring, D. N.,** unpublished observations, 1990.
56. **Matsuda, K., Satoh, S., and Ohori, H.,** DNA-binding activity of hepatitis B e antigen polypeptide lacking the protaminelike sequence of nucleocapsid protein of human hepatitis B virus, *J. Virol.*, 62, 3517, 1988.
57. **Schlicht, H. J., Bartenschlager, R., and Schaller, H.,** The duck hepatitis B virus core protein contains a highly phosphorylated C terminus that is essential for replication but not for RNA packaging, *J. Virol.*, 63, 2995, 1989.
58. **Bruss, V. and Gerlich, W.,** personal communication, 1989.
59. **Albin, C. and Robinson, W. S.,** Protein kinase activity in hepatitis B virus, *J. Virol.*, 34, 297, 1980.

60. **Feitelson, M. A., Marion, P. L., and Robinson, W. S.,** Core particles of hepatitis B virus and ground squirrel hepatitis virus. I. Relationship between hepatitis B core antigen- and ground squirrel hepatitis core antigen-associated polypeptides by sodium dodecyl sulfate-polyacrylamide gel electrophoresis and tryptic peptide mapping, *J. Virol.*, 43, 687, 1982.
61. **Roossinck, M. J. and Siddiqui, A.,** *In vivo* phosphorylation and protein analysis of hepatitis B virus core antigen, *J. Virol.*, 61, 955, 1987.
62. **Pugh, J., Zweidler, A., and Summers, J.,** Characterization of the major duck hepatitis B virus core particle protein, *J. Virol.*, 63, 1371, 1989.
63. **Laub, O.,** personal communication, 1989.
64. **Magnius, L. O. and Espmark, J. A.,** New specificities in Australia antigen positive sera distinct from Le Bouvier determinants, *J. Immunol.*, 109, 1017, 1972.
65. **Takahashi, K., Akahane, Y., Gotanda, T., Mishiro, T., Imai, M., Miyakawa, Y., and Mayumi, M.,** Demonstration of hepatitis B e antigen in the core of Dane particles, *J. Immunol.*, 122, 275, 1979.
66. **MacKay, P., Lees, J., and Murray, K.,** The conversion of hepatitis B core antigen synthesized in *E. coli* into e antigen, *J. Med. Virol.*, 8, 237, 1981.
67. **Takahashi, K., Imai, M., Nomura, M., Oinuma, A., Machida, A., Funatsu, G., Miyakawa, Y., and Mayumi, M.,** Demonstration of the immunogenicity of hepatitis B core antigen in a hepatitis B e antigen polypeptide (P19), *J. Gen. Virol.*, 1981.
68. **Takahashi, K., Machida, A., Funatsu, G., Nomura, M., Usuda, S., Aoyagi, S., Tachibana, K., Miyakawa, Y., and Mayumi, M.,** Immunochemical structure of hepatitis B antigen in the serum, *J. Immunol.*, 130, 2903, 1983.
69. **Takahashi, K., Imai, M., Miyakawa, Y., Iwakiri, S., and Mayumi, M.,** Duality of hepatitis B e antigen in serum of persons infected with hepatitis B virus: evidence for the nonidentity of e antigen with immunoglobulins, *Proc. Natl. Acad. Sci. U.S.A.*, 75, 1952, 1978.
70. **Neurath, A. R., Strick, N., Szmuness, W., Stevens, C. E., and Harley, E. J.,** Radioimmunoassay of hepatitis B e-antigen (HBeAg): identification of HBeAg not associated with immunoglobulins, *J. Gen. Virol.*, 42, 493, 1979.
71. **Tedder, R. S. and Bull, F. G.,** Characterization of "e" antigen associated with hepatitis B, *Clin. Exp. Immunol.*, 35, 380, 1979.
72. **Takahashi, K., Imai, M., Gotanda, T., Sano, T., Oinuma, A., Mishiro, S., Miyakawa, Y., and Mayumi, M.,** Hepatitis B e antigen polypeptides isolated from sera of individuals infected with hepatitis B virus: comparison with HBeAg polypeptide derived from Dane particles, *J. Gen. Virol.*, 50, 49, 1980.
73. **Ou, J.-H., Yeh, C.-T., and Yen, T. S. B.,** Transport of hepatitis B virus precore protein into the nucleus after cleavage of its signal peptide, *J. Virol.*, 63, 5238, 1989.
74. **Miller, R. H.,** Proteolytic self-cleavage of hepatitis B virus core protein may generate serum e antigen, *Science*, 236, 722, 1987.
75. **Nassal, M., Galle, P. R., and Schaller, H.,** Proteaselike sequence in hepatitis B virus core antigen is not required for e antigen generation and may not be part of an aspartic acid-type protease, *J. Virol.*, 63, 2598, 1989.
76. **Jean-Jean, O., Salhi, S., Carlier, D., Elie, C., de Recondo, A.-M., and Rossignol, J.-M.,** Biosynthesis of hepatitis B virus e antigen: directed mutagenesis of the putative aspartyl protease site, *J. Virol.*, 63, 5497, 1989.
77. **Schlicht, H.-J., and Schaller, H.,** The secretory core protein of human hepatitis B virus is expressed on the cell surface, *J. Virol.*, 63, 5399, 1989.
78. **Chu, C. M. and Liaw, Y. F.,** Intrahepatic distribution of hepatitis B surface and core antigens in chronic hepatitis B virus infection. Hepatocyte with cytoplasmic/membranous hepatitis B core antigen as a possible target for immune hepatocytolysis, *Gastroenterology*, 92, 220, 1987.
79. **Mondelli, M., Vergani, G. M., Alberti, A., Vergani, D., Portmann, B., Eddleston, A. L., and Williams, R.,** Specificity of T lymphocyte cytotoxicity to autologous hepatocytes in chronic hepatitis B virus infection: Evidence that T cells are directed against HBV core antigen expressed on hepatocytes, *J. Immunol.*, 129, 2773, 1982.
80. **Pignatelli, M., Waters, J., Lever, A., Iwarson, S., Gerety, R., and Thomas, H. C.,** Cytotoxic T-cell responses to the nucleocapsid proteins of HBV in chronic hepatitis. Evidence that antibody modulation may cause protracted infection, *J. Hepatol.*, 4, 15, 1987.
81. **Bruss, V. and Gerlich, W. H.,** Formation of transmembraneous hepatitis B e-antigen by cotranslational *in vitro* processing of the viral precore protein, *Virology*, 163, 268, 1988.
82. **Junker, M., Galle, P., and Schaller, H.,** Expression and replication of the hepatitis B virus genome under foreign promoter control, *Nucleic Acids Res.*, 15, 10117, 1987.
83. **Chang, C., Enders, G., Sprengel, R., Peters, N., Varmus, H. E., and Ganem, D.,** Expression of the precore region of an avian hepatitis B virus is not required for viral replication, *J. Virol.*, 61, 3322, 1987.

84. **Fan, H., Chute, H., Chao, E., and Feuerman, M.,** Construction and characterization of Moloney murine leukemia virus mutants unable to synthesize glycosylated gag polyprotein, *Proc. Natl. Acad. Sci. U.S.A.*, 80, 5965, 1983.
85. **Schwartzberg, P., Colicelli, J., and Goff, S. P.,** Deletion mutants of Moloney leukemia virus which lack glycosylated gag protein are replication competent, *J. Virol.*, 46, 538, 1983.
86. **Pillemer, E., Koistra, D., Witte, O., and Weissman, I.,** Monoclonal antibody to the amino-terminal L sequence of murine leukemia virus glycosylated gag polyproteins demonstrates their unusual orientation in the cell membrane, *J. Virol.*, 57, 413, 1986.
87. **Hoofnagle, J., Shafritz, D., and Popper, H.,** Chronic type B hepatitis and the "healthy" HBsAg carrier state, *Hepatology*, 7, 758, 1987.
88. **Milich, D. R. and McLachlan, A.,** The nucleocapsid of hepatitis B virus is both a T-cell-independent and a T-cell-dependent antigen, *Science*, 234, 1398, 1986.
89. **Milich, D. R.,** Genetic and molecular basis for T- and B-cell recognition of hepatitis B viral antigens, *Immunol. Rev.*, 99, 71, 1987.
90. **Stahl, S. J. and Murray, K.,** Immunogenicity of peptide fusions to hepatitis B virus core antigen, *Proc. Natl. Acad. Sci. U.S.A.*, 86, 6283, 1989.
91. **Ferns, R. B. and Tedder, R. S.,** Monoclonal antibodies to hepatitis B e antigen (HBeAg) derived from hepatitis B core antigen (HBcAg): their use in characterization and detection of HBeAg, *J. Gen. Virol.*, 65, 899, 1984.
92. **Ohori, H., Shimizu, N., Yamada, E., Onodera, S., and Ishida, N.,** Immunological and morphological properties of HBeAg subtypes (HBeAg/1 and HBeAg/2) in hepatitis B virus core particles, *J. Gen. Virol.*, 65, 405, 1984.
93. **Matsuda, K., Kanno, A., and Ohori, H.,** Immunochemical characterization of hepatitis B e antigen subtypes, HBeAg/1 and HBeAg/2, in sera of hepatitis B virus carriers, *J. Med. Virol.*, 20, 219, 1986.
94. **Matsuda, K. and Ohori, H.,** Immunochemical characteristics of hepatitis B e antigen subspecificities, HBeAg/1 and HBeAg/2, *J. Immunol.*, 141, 1709, 1988.
95. **Waters, J. A., Jowett, T. P., and Thomas, H. C.,** Identification of a dominant immunogenic epitope of the nucleocapsid (HBc) of the hepatitis B virus, *J. Med. Virol.*, 19, 79, 1986.
96. **Ferns, R. B. and Tedder, R. S.,** Human and monoclonal antibodies to hepatitis B core antigen recognize a single immunodominant epitope, *J. Med. Virol.*, 19, 193, 1986.
97. **Colucci, G., Beazer, Y., Cantaluppi, C., and Tackney, C.,** Identification of a major hepatitis B core antigen (HBcAg) determinant by using synthetic peptides and monoclonal antibodies, *J. Immunol.*, 141, 4376, 1988.
98. **Milich, D. R., McLachlan, A., Thornton, G. B., and Hughes, J. L.,** Antibody production to the nucleocapsid and envelope of the hepatitis B virus primed by a single synthetic T cell site, *Nature (London)*, 329, 547, 1987.
99. **Milich, D. R., McLachlan, A., Moriarty, A., and Thornton, G. B.,** Immune response to hepatitis B virus core antigen (HBcAg): localization of T cell recognition sites within HBcAg/HBeAg, *J. Immunol.*, 139, 1223, 1987.
100. **Iwarson, S., Tabor, E., Thomas, H. C., Snoy, P., and Gerety, R. J.,** Protection against hepatitis B virus infection by immunization with hepatitis B core antigen, *Gastroenterology*, 88, 763, 1985.
101. **Roos, S., Fuchs, K., and Roggendorf, M.,** Protection of woodchucks from infection with woodchuck hepatitis virus by immunization with recombinant core protein, *J. Gen. Virol.*, 70, 2087, 1989.

Chapter 8

BIOSYNTHESIS AND ENZYMATIC FUNCTIONS OF THE HEPADNAVIRAL REVERSE TRANSCRIPTASE

Hans-Jürgen Schlicht, Ralf Bartenschlager, and Heinz Schaller

TABLE OF CONTENTS

I. INTRODUCTION

In 1973, Kaplan et al.[1] reported that hepatitis B virus particles purified from serum contain an enzymatic activity that incorporates radiolabeled nucleotides into the viral genome *in vitro*. Nine years later, Summers and Mason,[2] using the duck hepatitis B virus, demonstrated that this activity was due to the virus-specific polymerase that, astonishingly, had the features of a reverse transcriptase. Since that time, many attempts have been made to elucidate the biosynthesis and enzymatic activities of this enzyme, first by biochemical methods and later, after systems for the *in vitro* expression of viral proteins had been established, also by genetic analyses. However, only recently has major progress been made that led to the definition of the expression strategy of this protein as well as to the characterization of its major enzymatic activities.

In this chapter we will summarize recent data on the biosynthesis and enzymatic functions of the hepadnaviral reverse transcriptase. In the first section we will describe the structure of the polymerase gene and its expression strategy. In the following two parts we will discuss the different enzymatic and structural functions of the P gene product that have been defined to date, and finally we will briefly list some of the questions that are still unanswered.

II. STRUCTURE AND EXPRESSION STRATEGY OF THE HEPADNAVIRAL POLYMERASE GENE

As is shown in the upper part of Figure 1, the genome of the human hepatitis B virus (HBV), the prototype member of the mammalian hepadnaviruses, contains four open reading frames that can code for protein.[3] The genome organization of the avian viruses, as evidenced by the genome of the duck hepatitis B virus (DHBV; Figure 1, lower panel), is quite similar, the only major difference being that the X gene is missing. Biochemical studies employing antisera raised against viral gene products expressed in bacterial cells revealed that the major protein encoded by the C gene represents the viral capsid protein whereas the preS/S gene codes for the surface proteins. No protein products encoded by the X or P gene have been unambiguously identified in infected liver tissue or purified virus particles to date. Since the X gene is missing in the avian viruses, it was generally assumed that the P gene, which is also by far the largest open reading frame of the genome, encodes the viral polymerase. This assumption was later supported by the finding that the putative P gene product contains sequences that are also present in retroviral reverse transcriptases.[4]

How is the P gene expressed? As is shown in Figure 1, mapping of the viral transcripts did not reveal an mRNA that could directly encode a P gene product. The only mRNA that covers the whole P gene is the C mRNA or pregenomic RNA. The most 5′ AUG on this mRNA, however, is a C gene AUG. Consequently, this mRNA primarily encodes C gene products and, if at all, can give rise to P proteins only by an unusual expression mechanism. The subgenomic mRNAs cover only part of the P gene, which, as discussed below, is not sufficient for the production of a functional enzyme.

A possible explanation for this initially surprising finding resulted from a comparison of the hepadnavirus genomes with other genetic elements that replicate by reverse transcription and that also do not produce a polymerase mRNA. If the structures of these genetic elements are compared, a characteristic overlap between the 3′ end of the C or *gag* gene and the 5′ end of the P gene can be found in most cases.[5] The significance of this peculiar genomic arrangement became evident only recently when the molecular details of the biosynthetic pathway leading to polymerase production were elucidated for several retroviruses.[6-9] These

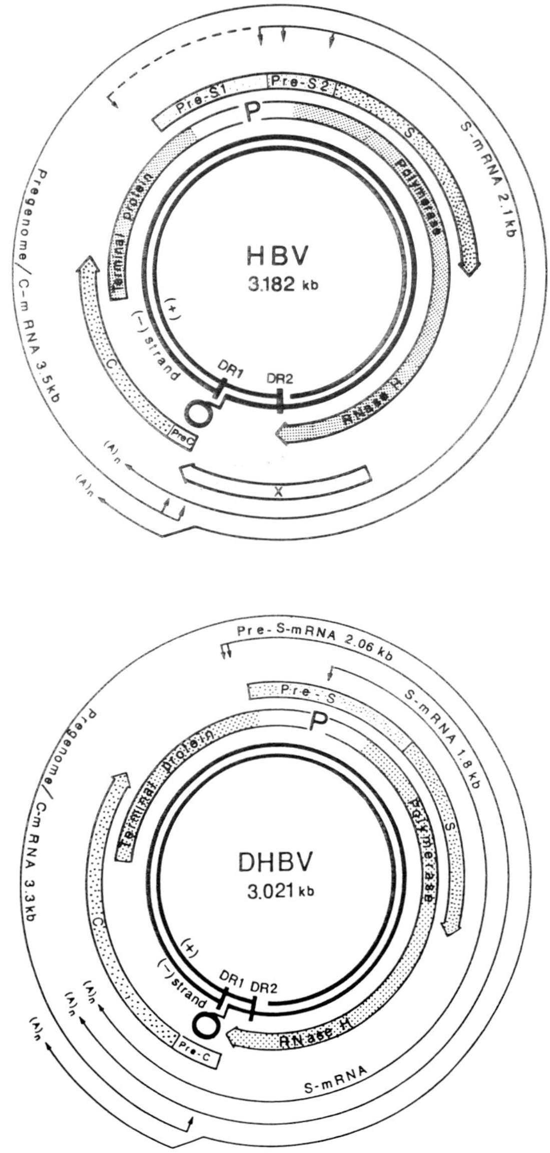

FIGURE 1. Genetic, physical, and transcriptional maps of the HBV and the DHBV genome. The inner circles represent the viral DNA. At the minus strand, a protein is covalently bound (open circle). The different open reading frames are depicted as arrows. The outer lines represent the major viral transcripts.

studies showed that retroviral polymerase is synthesized as a *gag/pol* fusion protein intermediate that is later proteolytically processed. Most interestingly, with the exception of murine and feline leukemia viruses, where the *pol* gene is expressed through suppression of a termination codon,[10,11] fusion was due to ribosomal frameshifting that occurs in the *gag/pol* overlap region. Since a C/P overlap was also found in all hepadnavirus genomes, it appeared very likely that the mechanism of polymerase production would be the same as for retroviruses, an assumption supported by the demonstration of C/Pol fusion proteins in some liver tumors.[12] However, one major difference is that in hepadnaviruses the P gene is in the +1 frame relative to the C gene, and not in the –1 frame as is the case for retroviruses. Thus, it also appeared possible that hepadnaviruses might express the polymerase by another biosynthetic pathway.

To clarify this point, genetic analyses were carried out immediately after a system had been established that allowed the analysis of virus mutants in tissue culture and *in vivo*.[13] In these experiments, which were first carried out with the duck hepatitis B virus, point mutations were introduced into the virus genome and the mutants either transiently expressed in tissue culture cells or transfected into experimental animals. These studies conclusively showed that the P gene is expressed independently from the C gene since C^-P^+ mutants could transcomplement C^+P^- mutants. Furthermore, mutations changing the P gene start codon, or introducing stop codons downstream of the P gene start codon, strongly reduced or abolished polymerase activity. Thus, the hepadnaviral polymerase is not synthesized as a core/polymerase fusion protein but rather is produced by translation initiation at a P gene AUG.[14,15] The nature of this initiation event, however, is still unclear. As has been mentioned above, no mRNA that could directly encode a P gene product has been detected for any hepatitis B virus. Whereas this finding could be ascribed to the rarity of this mRNA, another possibility is that the polymerase is expressed from the pregenomic mRNA, which also encodes the core protein, by internal initiation. In fact, very recently evidence has been obtained suggesting that synthesis of the HBV P gene product might be initiated by ribosomes that enter the mRNA at an internal sequence located slightly upstream of the P gene start AUG.[16,17]

III. ENZYMATIC FUNCTIONS OF THE HEPADNAVIRAL POLYMERASE

In principle, each reverse transcriptase must be endowed with three different enzymatic activities: an RNA-dependent DNA polymerase, a DNA-dependent DNA polymerase, and an RNase H activity. One part of the protein implicated in DNA polymerase activity is the region located around the highly conserved amino acid motif YMDD, which, for example, has been shown to be essential for the enzymatic activity of the human immunodeficiency virus (HIV) reverse transcriptase[18] and which is also present in the putative P gene product of the hepadnaviral polymerase (Figure 2C). Moreover, sequence comparisons with proteins with known RNase H activities revealed a weak homology at a position located downstream of the YMDD sequence (Figure 2D).[19,20] To analyze the significance of these sequence similarities for polymerase function, point mutations were introduced into a cloned HBV genome (Figure 2), and polymerase activity assayed after transient expression of the mutants in tissue culture cells. These studies demonstrated (1) that the YMDD motif is essential for the activity of the hepadnaviral reverse transcriptase and (2) that mutations in the RNase H homology region can result in the production of a P gene product that still produces a DNA minus strand but is devoid of plus strand synthesis.[20] These data demonstrate that the sequence similarities observed are indeed of functional significance. Another consequence of these findings is that, as in retroviruses, the RNase H domain of hepadnaviral reverse transcriptase is located carboxyterminal to the DNA polymerase part of the enzyme.

In Figure 2B, the functional organization of the hepadnaviral P gene as derived from these studies is summarized and compared to that of the retroviral *pol* gene. Accordingly, and in contrast to retroviruses, there is neither evidence for a P gene-encoded hepadnaviral protease, nor for a viral endonuclease. As for the latter, integration of hepatitis B virus genomes into host cell chromosomes is by no means an essential part of the viral life cycle, as it is for

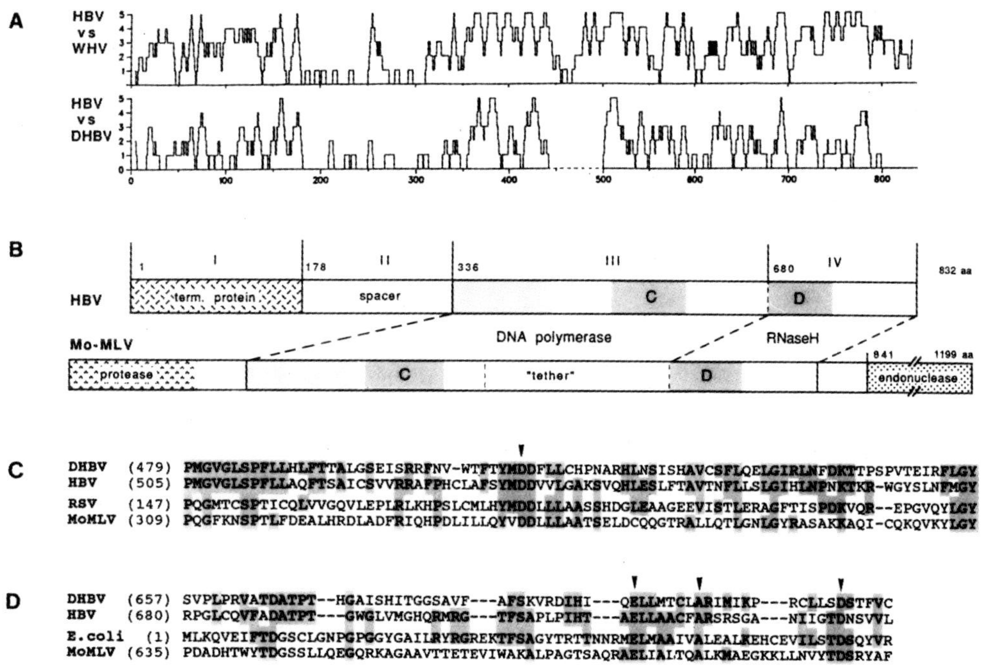

FIGURE 2. Structural and functional organization of the HBV P gene as derived from sequence comparisons between hepadnaviral and retroviral *pol* genes and by mutational analysis of the HBV P gene.[20] (A) Sequence similarity between the putative P gene products of two mammalian hepadnaviruses (HBV and WHV) and between a mammalian (HBV) and an avian (DHBV) hepadnavirus. A low y-axis value corresponds to a region of low homology. (B) Segmentation of the P gene into four domains corresponding to regions of high or low amino acid identity between different hepadnaviral P gene products, or sequence similarities with reverse transcriptase and RNase H domains of retroviruses. Region I is highly conserved between hepadnaviruses but has no counterpart in retroviral genomes. It most likely encodes the DNA terminal protein. Region II is poorly conserved between hepadnaviruses and can be partially deleted without affecting polymerase function. Regions III and IV are well conserved among hepadnaviruses and contain sequences similar to reverse transcriptase (region III) or RNase H (region IV) domains of retroviruses. (C) Alignment of the amino acid sequence of the HBV P gene segment labeled "C" in (B) (amino acids [aa] 505 to 587) with the corresponding DHBV sequence (aa 479 to 561) and a region of the P gene product of Rous sarcoma virus (RSV; aa 147 to 228) and Moloney murine leukemia virus (Mo-MuLV; aa 309 to 391). Amino acids identical between HBV and DHBV, or between one of the hepadnaviruses and RSV and/or Mo-MuLV, are shaded. (D) Alignment of the amino acid sequence of the HBV P gene segment labeled "D" (B) (aa 680 to 741) with the corresponding DHBV sequence (aa 657 to 719), the N-terminal part of *E. coli* RNase H (aa 1 to 74), and a region of the RNase H of Mo-MuLV (aa 635 to 705). Amino acids identical between HBV and DHBV or between one of the hepadnaviruses and *E. coli* and/or Mo-MuLV sequences are shaded. Amino acids changed during a mutational analysis[20] are indicated by arrowheads in (C) and (D).

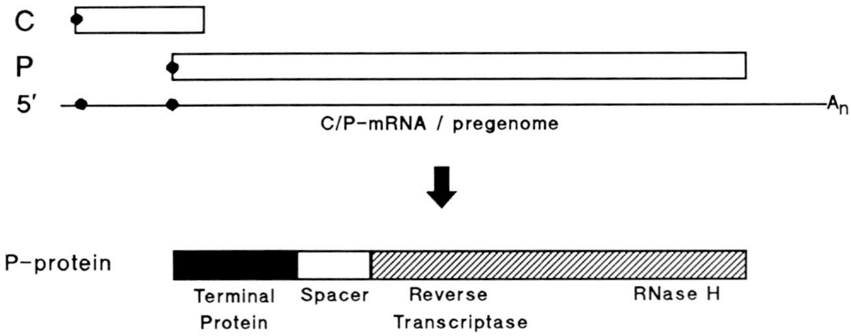

FIGURE 3. Expression strategy of the hepadnaviral P gene product. Synthesis of this protein is most likely due to an internal translation initiation within the C/P mRNA. The resulting protein consists of two domains connected by a spacer region that is not needed for polymerase function.

retroviruses, although it can occur in a nonspecific manner. Due to the circular nature of the viral genome and its complete coverage by viral genes (Figure 1), all integrated hepatitis B virus genomes are necessarily replication deficient. Thus, it is not surprising that hepatitis B viruses lack an endonuclease.

Similarly, there is no essential need for a hepadnaviral protease since the P gene product is synthesized unfused to the core protein and can be envisaged to function without proteolytic processing (Figures 3 and 5). However, instead of a protease, the hepadnaviral P gene product does contain another highly interesting function. As is shown in Figure 2A, sequence comparisons between different HBV or DHBV isolates revealed that the P gene product appears to be split into N-terminal and C-terminal domains with conserved amino acid sequences that are connected by a highly variable spacer region.[20,21] Mutational analysis showed that inframe insertions as well as certain point mutations introduced into either domain abolish polymerase activity,[14,20] whereas most of the spacer region can be deleted.[20,22] As has been described above, the C-terminal domain of the P gene product contains the polymerase and RNase H activities and therefore most likely represents the reverse transcriptase. What is the function of the N-terminal domain? In the case of the hepadnaviruses, initiation of reverse transcription is not primed by a tRNA molecule but most likely by a protein that remains covalently bound to the newly synthesized DNA minus strand (Chapter 11, this volume). The origin of this primer protein remained obscure for a long time. Recently, analyses carried out with antisera specific for individual domains of the DHBV P gene product demonstrated that this genome-linked protein, or DNA terminal protein, is derived from the N-terminal domain of the polymerase protein.[22,23] Thus, it appears that the hepadnaviral reverse transcriptase is a highly specialized, multifunctional enzyme that brings along its own primer for DNA synthesis. This might explain the extraordinary template specificity of this enzyme that, under native conditions, is virtually unable to switch to an exogenous template.[24] Some activity on a foreign template has been observed in activity gels only after partial denaturation of a nucleocapsid/virus fraction from HBV DNA-transfected human hepatoma cells.[25,26]

IV. A NONENZYMATIC FUNCTION OF THE P GENE PRODUCT IS ESSENTIAL FOR RNA ENCAPSIDATION

As has been discussed above, hepadnaviral reverse transcriptase is not synthesized as a core/polymerase fusion protein and therefore cannot be incorporated into a nascent capsid by

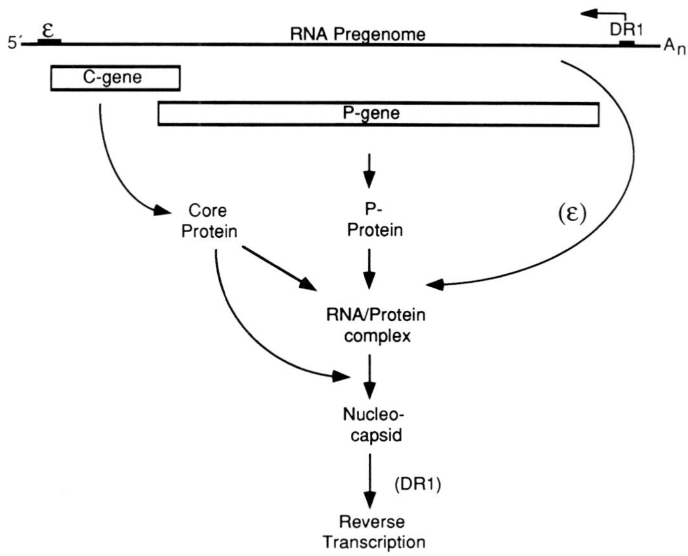

FIGURE 4. Schematic representation of the steps leading to the formation of replication-competent nucleocapsids. ∈, Encapsidation signal; DR1, direct repeat 1. For details, see text.

coassembly of such fusion proteins with core protein molecules. On the other hand, packaging of the RNA pregenome into core particles is a highly specific process in which subgenomic RNAs serving as mRNAs for the preS/S gene products, and also the preC mRNA, which differs from the pregenome only by the presence of about 30 nucleotides at its 5′ end (Figure 1), are excluded.[27-29] The RNA sequences recognized during pregenome encapsidation were only very recently identified at the precore/core gene junction,[28,30] but it remained an open question whether the RNA encapsidation required the core protein alone or whether other viral proteins were needed. As the surface proteins are not part of replicating complexes, and since the X gene is missing in DHBV, a likely candidate is the P protein.

To clarify this point, several nonsense mutations scattered throughout the HBV P gene or the DHBV P gene were transiently expressed in tissue culture cells and encapsidated RNAs were analyzed in nuclease protection reactions or primer extension experiments. The data obtained clearly demonstrated that a P gene product is required for genomic RNA encapsidation since all these mutants produced core particles containing less than 5% of the level of pregenomic RNA as compared to wild type.[30,31] Moreover, packaging of nonviral RNAs fused to the HBV encapsidation signal (∈, Figure 4) also depended on the synthesis of P protein, emphasizing that core polypeptides alone were insufficient.[30] However, RNA encapsidation does not require enzymatic activity since mutations in the highly conserved YMDD amino acid sequence motif that, as mentioned above, is essential for reverse transcriptase activity, had no influence on pregenomic RNA encapsidation.[30,31] Thus, the polymerase/RNase H and the encapsidation functions can be separated genetically.

As described in Section II, trans-complementation studies performed to elucidate the mechanism of P gene expression demonstrated that mutations in *pol* or *core* could be complemented by cotransfection of plasmids directing the expression of unaltered viral proteins. However, the original idea of freely diffusible P and C proteins was soon limited by the finding that cotransfection of a C^+P^- mutant with a wild-type construct (C^+P^+) or a C^-P^+ mutant resulted almost exclusively in the packaging of the P^+ genomes.[30,31] Only about 10%

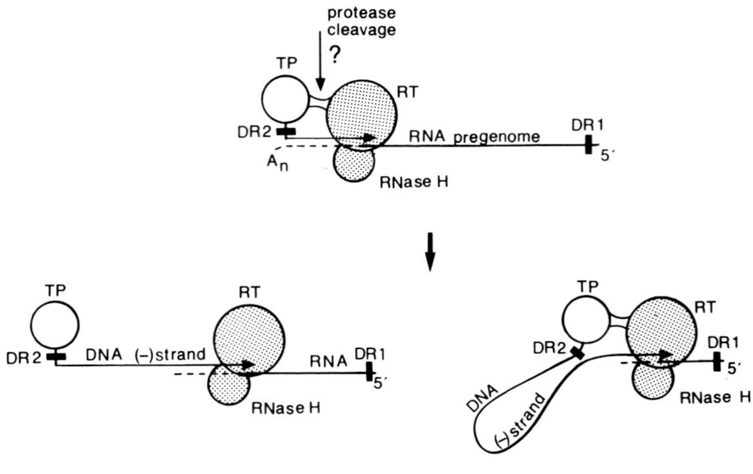

FIGURE 5. Scheme of hepadnaviral reverse transcription with (left) or without (right) protease cleavage of the P gene product. TP, terminal protein; RT, reverse transcriptase; DR, direct repeat sequence.

of the P⁻ genomes were detectable, although both RNAs were present in equal amounts in the cytoplasm. Preferential encapsidation of the pregenomic RNA from which P protein is translated indicated that the P protein acts primarily in cis. In contrast, efficient trans-complementation could be observed when the P gene was expressed either under foreign promoter control or from genomic RNAs not containing an intact encapsidation signal, indicating that diffusible P proteins were able to mediate RNA encapsidation. The mechanism underlying these findings is not understood, but one simple explanation could be that P proteins are normally expressed only in limited amounts and that they have a high affinity for the RNA packaging signal (and/or core protein). This would keep the P gene product preferentially at the RNA from which it is made and only when overexpressed, e.g., from constructs like the ones just mentioned, would high enough quantities be synthesized to allow efficient trans-complementation.

The dependence of genomic RNA encapsidation on P proteins seems to add a further point to the list of differences between hepadnaviruses and retroviruses since, at least in the case of Moloney murine leukemia virus, it could be shown that retroviral Pol mutants produce particles containing normal amounts of 70S RNA.[32] As these mutants have only been characterized phenotypically by the lack of both immunoprecipitable P proteins and reverse transcriptase activity, further studies are being carried out to elucidate the mechanism of retroviral genomic RNA encapsidation.[33]

V. CONCLUSIONS AND UNRESOLVED ISSUES

In Figures 3 through 5 we have summarized the current concepts of biosynthesis and enzymatic activities of the hepadnaviral reverse transcriptase as they have emerged from the experimental data described above. In contrast to retroviruses, hepadnaviral reverse transcriptase is not expressed as a C/Pol fusion protein but instead as an authentic P gene product (Figures 3 and 4). Synthesis of this protein most likely starts by internal initiation within the pregenomic mRNA, which also encodes the core protein and serves as the template for reverse transcription. The P protein thus generated consists of at least two domains that are separated by a spacer region that is largely dispensable for polymerase function. The N-terminal domain encodes the protein that is believed to prime reverse transcription and which remains covalently bound to the 5′ end of the DNA minus strand. The C-terminal domain contains the reverse transcriptase as well as an RNase H activity. According to very recent data this multifunctional protein forms a complex with the RNA pregenome and core protein(s) that is then converted by further addition of core protein subunits into replication-competent nucleocapsids (Figure 4). Whatever the mechanism of these interactions may be, the findings described above indicate that formation of such a complex is a prerequisite for the efficient incorporation of the RNA pregenome, and probably also polymerase, into nucleocapsids. In this way, hepadnaviruses may have developed a mechanism to ensure Pol incorporation into core particles without the necessity of synthesizing core–Pol fusion proteins. It is assumed, although not proven, that reverse transcription takes place only within these particles. Whether elongation of the DNA strand requires cleavage of the P protein is under investigation (Figure 5).

Although much has been learned about the mechanism of hepadnaviral P protein biosynthesis and its functional domains by genetic analysis, there are still many open questions, most of which can only be answered in the appropriate cell-free systems that need to be established. One key problem concerns the nature of the interaction between P protein(s), core protein, and RNA, and the mechanism of incorporation of the RNA/protein complex into nucleocapsids. Furthermore, two major problems concerning initiation of reverse transcription still must be solved: (1) as the encapsidation signal ($\in$, Figure 4), which is assumed to interact (directly or indirectly) with the P protein early in pregenome packaging, resides near the 5′ end of the RNA pregenome, it is difficult to understand how the polymerase initiates reverse transcription at the direct repeat 1 (DR1) located at the 3′ end of the RNA molecule; and (2) what is the mechanism by which terminal protein initiates reverse transcription?

Another enigmatic finding is the extreme template specificity of the polymerase. No evidence could be obtained that hepadnaviral reverse transcriptase can switch to an exogenous template under native conditions. How can an enzyme, which must have a certain mobility in order to synthesize a new DNA strand, be attached so tightly to its template? And last but not least, why has the template-free polymerase still not been unambiguously identified? Is it only the scarcity of this protein or is an unusual processing step or modification responsible for this failure? Obviously, this peculiar protein will keep many people busy for a few more years to come.

REFERENCES

1. **Kaplan, P., Greenman, R., Gerin, J., Purcell, R., and Robinson, W.,** DNA polymerase associated with human hepatitis B antigen, *J. Virol.*, 12, 995, 1973.
2. **Summers, J. and Mason, W.,** Replication of the genome of a hepatitis B-like virus by reverse transcription of an RNA intermediate, *Cell*, 29, 403, 1982.
3. **Ganem, D. and Varmus, H.,** The molecular biology of the hepatitis B viruses, *Annu. Rev. Biochem.*, 56, 651, 1987.
4. **Toh, H., Hayashida, H., and Miyata, T.,** Sequence homology between retroviral reverse transcriptase and putative polymerases of hepatitis-B virus and cauliflower mosaic virus, *Nature (London)*, 305, 827, 1983.
5. **Hull, R. and Covey, S.,** Genome organization and expression of reverse transcribing elements: variations and a theme, *J. Gen. Virol.*, 67, 1751, 1986.
6. **Weiss, R., Teich, N., Varmus, H., and Coffin, J.,** *RNA Tumor Viruses*, CSH Press, New York, 1982.
7. **Jacks, T. and Varmus, H.,** Expression of the Rous sarcoma virus pol gene by ribosomal frame shifting, *Science*, 230, 1237, 1985.

8. **Jacks, T., Townsley, K., Varmus, H., and Majors, J.,** Two efficient ribosomal frameshifting events are required for synthesis of mouse mammary tumor virus gag-related polyproteins, *Proc. Natl. Acad. Sci. U.S.A.*, 84, 4298, 1987.
9. **Jacks, T., Power, M. D., Masiarz, F. R., Luciw, P. A., Barr, P. J., and Varmus, H.,** Characterization of ribosomal frameshifting in HIV-I gag-pol expression, *Nature (London)*, 331, 280, 1988.
10. **Yoshinaka, Y., Katoh, I., Copeland, T. D., and Oroszlan, S.,** Murine leukemia virus protease is encoded by the gag-pol gene and is synthesized through suppression of an amber termination codon, *Proc. Natl. Acad. Sci. U.S.A.*, 82, 1618, 1985.
11. **Yoshinaka, Y., Katoh, I., Copeland, T. D., and Oroszlan, S.,** Translational readthrough of an amber termination codon during synthesis of feline leukemia virus protease, *J. Virol.*, 55, 870, 1985.
12. **Will, H., Salfeld, J., Pfaff, E., Manso, C., Theilmann, L., and Schaller, H.,** Putative reverse transcriptase intermediates of human hepatitis B virus in primary liver carcinomas, *Science*, 231, 594, 1986.
13. **Schlicht, H.-J., Salfeld, J., and Schaller, H.,** The pre-C region of the duck hepatitis B virus is essential for synthesis and secretion of processed core proteins but not for virus formation, *J. Virol.*, 61, 3701, 1987.
14. **Schlicht, H.-J., Radziwill, G., and Schaller, H.,** Synthesis and encapsidation of duck hepatitis B virus reverse transcriptase do not require formation of core-polymerase fusion proteins, *Cell*, 56, 85, 1989.
15. **Chang, L., Pryciak, P., Ganem, D., and Varmus, H.,** Biosynthesis of the reverse transcriptase of hepatitis B viruses involves *de novo* translational initiation not ribosomal frameshifting, *Nature (London)*, 337, 364, 1989.
16. **Jean-Jean, O., Weimer, T., de Recondo, A.-M., Will, H., and Rossignol, J.-M.,** Internal entry of ribosomes and ribosomal scanning involved in hepatitis B virus P gene expression, *J. Virol.*, 63, 5451, 1989.
17. **Roychoudhury, S. and Shih, C.,** *Cis* rescue of a mutated reverse transcriptase gene of human hepatitis B virus by creation of an internal ATG, *J. Virol.*, 64, 1063, 1990.
18. **Larder, B., Purifoy, D., Powell, K., and Darby, G.,** Site-specific mutagenesis of AIDS virus reverse transcriptase, *Nature (London)*, 327, 716, 1987.
19. **Johnson, M., McClure, M., Feng, D., Gray, J., and Doolittle, R.,** Computer analysis of retroviral pol genes: assignment of enzymatic functions to specific sequences and homologies with nonviral enzymes, *Proc. Natl. Acad. Sci. U.S.A.*, 83, 7648, 1986.
20. **Radziwill, G., Tucker, W., and Schaller, H.,** Mutational analysis of the hepatitis B virus P gene product: domain structure and RNase H activity, *J. Virol.*, 64, 613, 1990.
21. **Sprengel, R., Kuhn, C., Will, H., and Schaller, H.,** Comparative sequence analysis of duck and human hepatitis B virus genomes, *J. Mol. Virol.*, 15, 323, 1985.
22. **Bartenschlager, R. and Schaller, H.,** The amino-terminal domain of the hepadnaviral P gene encodes the terminal protein (genome-linked protein) believed to prime reverse transcription, *EMBO J.*, 7, 4185, 1988.
23. **Bosch, V., Bartenschlager, R., Radziwill, G., and Schaller, H.,** The duck hepatitis B virus P gene codes for protein strongly associated with the 5′ end of the viral minus DNA strand, *Virology*, 166, 475, 1988.
24. **Radziwill, G., Zentgraf, H., Schaller, H., and Bosch, V.,** The duck hepatitis B virus DNA polymerase is tightly associated with the viral core structure and unable to switch to an exogenous template, *Virology*, 163, 123, 1988.
25. **Bavand, M. and Laub, O.,** Two proteins with reverse transcriptase activities associated with hepatitis B virus-like particles, *J. Virol.*, 62, 626, 1988.
26. **Bavand, M., Feitelson, M., and Laub, O.,** The hepatitis B virus reverse transcriptase is encoded by the viral pol gene, *J. Virol.*, 63, 1019, 1989.
27. **Enders, G. H., Ganem, D., and Varmus, H. E.,** 5′-Terminal sequences influence the segregation of ground squirrel hepatitis virus RNAs into polyribosomes and viral core particles, *J. Virol.*, 61, 35, 1987.
28. **Junker-Niepmann, M., Bartenschlager, R., and Schaller, H.,** A short *cis*-acting sequence is required for hepatitis B virus pregenome encapsidation and sufficient for packaging of foreign RNA, *EMBO J.*, 9, 3389, 1990.
29. **Nassal, M., Junker-Niepmann, M., and Schaller, H.**, Translational inactivation of RNA function: discrimination against a subset of genomic transcripts during hepatitis B virus nucleocapsid assembly, *Cell*, 63, 1357, 1990.
30. **Bartenschlager, R., Junker-Niepmann, M., and Schaller, H.,** The P-gene product of hepatitis B virus is required as a structural component for genomic RNA encapsidation, *J. Virol.*, 64, 5324, 1990.
31. **Hirsch, R. C., Lavine, J. E., Chang, L.-J., Varmus, H. E., and Ganem, D.,** Polymerase gene-products of hepatitis B virus are required for genomic RNA packaging as well as for reverse transcription, *Nature (London)*, 344, 522, 1990.
32. **Shields, A., Witte, O. N., Rothenberg, E., and Baltimore, D.,** High frequency of aberrant expression of moloney murine leukemia virus in clonal infections, *Cell*, 14, 601, 1978.
33. **Oevtle, S. and Spahr, P. -F.** Role of gag polyprotein precursor in packaging and maturation of Rous sarcoma virus genomic RNA, *J. Virol.*, 64, 5757, 1990.

Chapter 9

THE HEPADNAVIRAL X PROTEIN

Nancy Schek, Markus Fischer, and Heinz Schaller

TABLE OF CONTENTS

I. INTRODUCTION

The smallest open reading frame (ORF) of the human hepatitis B virus (HBV) genome was first identified by Galibert et al.[1] through nucleotide sequencing of the entire viral genome, and later designated by the same group as the ORF-X.[2] The ORF found in the majority of isolates is capable of coding for an "X protein" of 145 to 154 amino acids. Indirect evidence supporting the existence of such a protein has been mounting over the last decade, and includes the conservation of the X ORF in all mammalian hepadnaviruses, the finding of anti-X-specific antisera in HBV-infected patients, and the assignment of a transcriptional trans-activation function to the encoded protein sequence. However, no study examining HBV-infected cells or virions has succeeded in demonstrating an X protein of molecular mass in the range predicted from the X ORF. Presumably, if this protein is indeed a product of hepadnaviral infection, its levels are below the current limit of detection. This situation has led to difficulties in determining the actual role for the X protein in hepadnaviral replication and hepadnaviral-associated disease.

In this chapter, we will review data suggesting that the X protein is indeed produced and that it plays a role in hepadnaviral replication. In the following sections, we will discuss the expression of the X ORF *in vivo* and in cultured cells, and describe results from systems used to characterize the X protein and to study its trans-activation function by the overexpression of the X polypeptide utilizing various heterologous promoters. Finally, we will speculate on the role of X protein in HBV replication, and its possible contributions to carcinogenic transformation of the host cell.

II. PREDICTION OF AN X OPEN READING FRAME-ENCODED GENE PRODUCT

The position and length of the X ORF in the genome of the mammalian prototype hepadnavirus HBV are shown in Figure 1. The position and sequence are well conserved between various HBV subtypes. Lo et al.[3] found 80% homology when comparing the predicted amino acid sequences of X ORFs from 11 HBV DNAs representing the 4 different subtypes.

The genomes of the other two well-studied mammalian hepadnaviruses, woodchuck hepatitis virus (WHV) and ground squirrel hepatitis virus (GSHV), each contain an ORF similar to the HBV X ORF in size and position in the genome.[4,5] Their amino acid sequences are 71% homologous,[4] but the GSHV X ORF amino acid sequence bears only a 31% identity with the HBV X ORF sequence. Strongly conserved homologies were preferentially found at the amino and carboxyl termini of the ORF.[5,6] In addition, internal amino acid sequences of HBV X and WHV X have been predicted to bear a high degree of similarity in their secondary structure, suggesting the evolutionary maintenance of X gene product function.[6]

In seeming contradiction to the above findings, the avian hepadnaviruses duck hepatitis B virus (DHBV) and heron hepatitis B virus lack an ORF corresponding to the mammalian X ORF.[7-9] Although sequences with very limited homology to the HBV X ORF sequences have been detected in regions both within and upstream of the core gene of the avian hepadnaviruses,[10] this could be interpreted to mean that the function of the putative protein product is not absolutely required for hepadnaviral replication. This notion is also supported by a genetic analysis of HBV gene functions demonstrating that mutational inactivation of the X gene only reduced, but did not abolish, virus production from transfected genomes.[11,12] Defective X function could be complemented in trans by cotransfection with complete

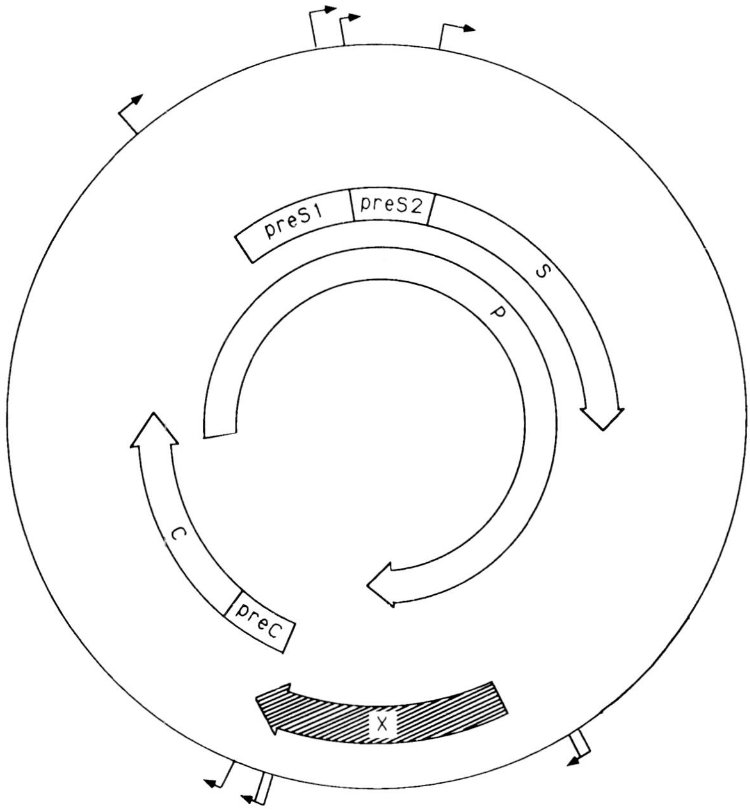

FIGURE 1. Localization of the X ORF in the circular HBV genome. The major arrows represent open reading frames, small arrows attached to the outer circle indicate transcriptional start sites.

genomes of WHV or GSHV, but not of DHBV, again confirming that the avian viral genome lacks an X-like trans-activation function.[13]

What features can be predicted from the primary protein sequence encoded in the X ORF? Computer analyses predict that the X polypeptide is a soluble intracellular protein.[13] No signal sequence, nuclear transport signal, or other such sequence motif is present. However, consensus sites for protein phosphorylation are contained in the X protein sequences of HBV (Figure 2[1]), WHV, and GHSV. The X protein sequence contains nine conserved cysteine residues, which is an unusually high number for a 17-kDa protein, and five of these residues are also present in the sequences of the animal viruses WHV and GSHV (Figure 2). No convincing homologies have been found between the predicted X polypeptide sequence and any known protein sequence.

III. ANTI-X IMMUNE RESPONSE AND THE SEARCH FOR X ANTIGEN PRODUCED DURING HEPADNAVIRAL INFECTION

Following the identification of the X ORF, many groups sought to demonstrate the presence of antibodies specific for the predicted amino acid sequence in HBV-infected patients. Most frequently, screening was done either by enzyme-linked immunosorbent assay

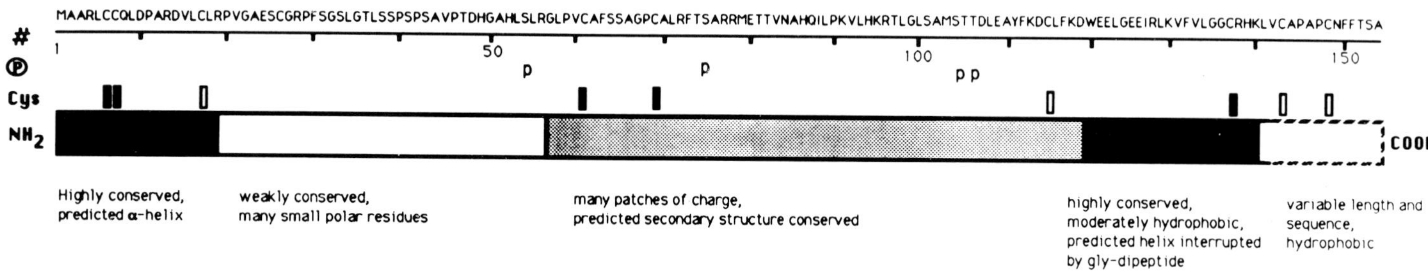

FIGURE 2. Structural features predicted from the primary sequence of the X ORF (partially adapted from Reference 13). The amino acid sequence deduced from HBV-2 DNA (subtype *ayw*)[13a] is given in the single-letter code. Cysteine residues conserved between HBV subtypes *adr*, *ayw*, and *ayr*[3] (vertical open and black bars) and between the HBV sequences and the animal hepadnaviruses WHV and GSHV (black bars) are shown. (HBV subtype *adw* lacks the cysteine at position 6 and contains an additional cysteine at position 78.) Potential protein phosphorylation sites for protein kinase C (amino acids 54 and 75) and casein kinase II (amino acids 104 and 106) are indicated by "p".

(ELISA), using X ORF-specific synthetic peptides,[14-16] or by radioimmunoprecipitation or Western blotting with fusion proteins containing portions of the X ORF.[17-19] In these studies, a fraction of HBV-infected patients ranging from 7 to 54% was found to produce anti-HBx antibodies.[15-22] Antibodies recognizing WHV X and GSHV X sequences have also been demonstrated in WHV-infected woodchucks[16] and GSHV-infected ground squirrels, respectively.[23]

No definitive correlations have been made between the presence of these antibodies and the classical serological markers of HBV infection. For example, anti-HBx has been found in sera from human patients and from experimentally infected chimpanzees that were either positive for HBeAg or positive for anti-HBe. Similarly, anti-HBx antibodies were present in both surface antigen-positive and surface antigen-negative sera.[16-22] Therefore, the initial correlation found in early studies between prevalence of anti-HBx antibodies and chronic HBV-infection[14,15,19,20,22] and, in some cases, hepatocellular carcinoma,[14,15,20] is now being called into question. Finally, from the results of all work to date, it does appear that a certain percentage of HBV-infected patients never mount an immune response to the X antigen. This might reflect the production of only low levels of X protein in HBV-infected cells, an intrinsic weak antigenicity of the protein, and/or a variability in the host immune response.

The serological studies described above argue that the X ORF is expressed in at least some HBV-infected patients; however, the antigen responsible for the anti-HBx response still awaits characterization as the amounts of X protein produced in the HBV-infected liver are probably very low.[20] In two reports, antigen, reacting with anti-X antibodies, was demonstrated in some but not all liver tissue sections from patients with chronic hepatitis and with liver cell carcinoma, appearing predominantly in the cytoplasm.[24,25] Other groups have detected X antigen(s) in HBV-infected liver tissue; however, the apparent molecular mass of the proteins detected (22 kDa[26] or 28 kDa[14]) is not compatible with our present size estimate of the X gene product in the absence of a major size-altering posttranslational modification. Fusion of the X gene sequences with cellular gene sequences upon viral DNA integration cannot be used to explain this discrepancy since similarly sized fusion products cannot in every instance arise by HBV DNA integration at random sites. Thus the only tenable explanation of the data seems to be the induction by HBV infection of cellular proteins that cross react with the particular antiserum used. Recently, Feitelson et al.[27] have presented evidence that X antigen circulates in the serum of more than half of HBV-infected patients tested, with X antigen often appearing in the serum before the appearance of surface antigen, but these results must wait for confirmation by other laboratories. Thus, while all of these studies tentatively support the contention that a protein encoded by the X ORF is expressed during hepadnaviral infection, they have not enlightened us as to the character of the protein or to its role in viral infection.

IV. EXPRESSION OF THE X GENE IN CULTURED CELLS

In the absence of a system allowing hepadnavirus infection of cells in culture, Chang et al.[28] used transient transfection of a human hepatoma cell line with cloned HBV DNA to study viral protein products. Viral replication occurred in these cells, and two proteins could be immunoprecipitated, although in low amounts, with an antiserum specific for X gene products. The 17-kDa product was assumed to be the predicted product of the HBV X ORF, while the 22-kDa product was suggested to be an HBV-encoded fusion product or an X-related cellular protein. Less direct evidence for the production of an X gene product from the HBV genome stems from examination by indirect immunofluorescence of a human hepatoblastoma cell line producing HBV particles[29] and rodent fibroblastoid cells transfected with HBV DNA.[30] Recently, 14- and 17-kDa X antigens were detected in immortalized hepatocytes that became malignantly transformed after transfection with HBV DNA.[31] This latter system offers the opportunity to explore the role that the X protein plays in HBV-associated tumorigenesis.

The mechanism by which the X gene is expressed during viral replication is presently still unclear. The protein could be made either, by analogy to the P protein, by internal translation initiation from one of the major polycistronic mRNAs or, alternatively, it could be translated conventionally from a group of transcripts of about 0.9 kbp[33] that initiate just upstream of the X ORF.[32,33] The production of these transcripts from the intact HBV genome has so far been demonstrated only in transfection experiments with rather unconventional genomic constructs,[33-35] or in the equally unusual environment of the mouse testes.[37] It is, however, produced in significant amounts from subgenomic transcription units containing the X gene and the upstream enhancer element. This suggests that the promoter/enhancer element producing the 0.9-kbp transcript has significant potential activity but is stringently down-regulated (possibly by its own gene product) during normal viral replication.

V. HETEROLOGOUS EXPRESSION OF THE X OPEN READING FRAME

Since the levels of X gene products made in cell culture from HBV regulatory sequences are too low to allow biochemical studies of the X protein, some workers have turned to the use of heterologous promoters to increase expression of the protein. So far, this strategy has mostly yielded data on the subcellular distribution of the X protein. Siddiqui et al.[38] placed the X ORF under the transcriptional control of the adenovirus major late promoter, simian virus 40 (SV40) enhancer, and virus-associated (VA) genes I and II for transient expression in COS cells. A 16-kDa protein reacting with X-specific antisera was found associated with the nucleus in biochemical fractionation. However, on examination of the transfected cells by immunostaining, X antigen was found mainly localized at the nuclear periphery and extending into the cytoplasm with a diffuse, and sometimes punctate, distribution. The authors suggested the association of X protein with the cytoskeleton and nuclear envelope. Levrero et al.[22] obtained comparable results on expression of the X ORF under the control of the adenovirus major late promoter in human embryonic lung 293 cells. These findings in overexpression systems concur with the results of Chisaka et al.,[29] in which staining of X antigen occurred on and around the nuclear membrane of a human hepatoblastoma cell line, and with those of Pugh et al.,[30] in which an HBV DNA-transformed rodent cell line showed cytoplasmic staining for X antigen. Although one other analysis by biochemical fractionation supports a nuclear association of X protein,[31] most of the evidence agrees with that obtained by examination of HBV-infected liver tissue,[24,25] where X antigen is found predominantly in the cytoplasm.

We have reexamined the question of the subcellular distribution of the X protein using a vaccinia virus vector to express the X ORF in human hepatoma HepG2 cells.[39] A major species of 16 kDa reacting with X protein-specific antisera was produced in the recombinant vaccinia virus-infected cells. By immunostaining, X protein appeared distributed throughout the cells, with a tendency to localize at the nuclear periphery and to accumulate in granules as its level increased. This suggests that the pattern of granule formation could be an artifact of overproduction, but more importantly, confirms the results of others that X protein does not accumulate preferentially in nuclei. This finding is curious in view of the assignment of a trans-activation function for X protein (see Section VI). The primary sequence of the X ORF does not contain a known nuclear localization signal, as stated above; however, a protein of 16 to 17 kDa would be expected to diffuse passively into the nucleus.[40] Perhaps only an extremely low number of X protein molecules is required for its intranuclear function, such that no specific accumulation in the nucleus is required. Alternatively, X protein may function in a location external to the nucleus, interacting with cellular factors to upregulate transcription. So far, no published study has clarified the subcellular distribution of X protein under conditions where it is also shown to function as a trans-activator.

The high level of X protein expression obtained using a vaccinia virus recombinant[39] has allowed us to examine the processing of this molecule. In pulse–chase experiments, the X protein was found to turn over with a bimodal half-life of 15 min and 3 h. This short half-life may be a mechanism contributing to the maintenance of an extremely low level of X protein during HBV infection. The X protein was found neither to be myristylated nor palmitylated. However, a fraction of the X protein produced could be metabolically labeled with $^{32}P_i$. The finding that X protein can be phosphorylated in cells of hepatic origin may lead to an understanding of its function as a transcriptional trans-activator.

X ORF-encoded sequences have also been expressed in cell-free systems by using either SP6 transcripts containing the X-coding sequence[20] or mRNA isolated from COS-1 cells transfected with an expression plasmid containing the X ORF.[41] In both cases, a 17-kDa protein was synthesized and recognized by antisera from some HBV-infected patients. Lin and Lo[42] have shown that the 17-kDa X protein synthesized *in vitro* is capable of forming disulfide linkage-dependent dimers. It is not known whether the X protein forms dimers *in vivo*.

VI. TRANSCRIPTIONAL TRANS-ACTIVATION AND CONTROL OF X GENE EXPRESSION

Recently it has been shown that the X gene product is capable of stimulating the expression of target genes coupled to a variety of viral promoters. Activation occurs on the transcriptional level and can affect many different enhancer/promoter combinations, including that of the HBV enhancer/core gene promoter[13] or the S gene promoter,[43] the SV40 enhancer and early promoter,[44] the herpes simplex virus thymidine kinase (HSV-TK) promoter,[45] the long terminal repeats (LTRs) of human T-lymphotropic virus type 1 (HTLV-1),[45] human immunodeficiency virus type 1 (HIV-1),[46] Rous sarcoma virus (RSV),[44] and the β-interferon promoter.[47] Furthermore, as indicated by a recent report,[48] RNA polymerase III (Pol-III) transcription initiation is also markedly stimulated in X gene-expressing cells. Thus, targets for X-mediated activation do not possess obvious common regulatory sequences. Together with further reports describing species-specific[45] and/or cell type-specific[49] contributions to trans-activation, these data suggest that the X gene product interacts (directly or indirectly) with multiple factors involved in transcriptional regulation. Using different approaches, at least three such factors influencing Pol II transcription have been so far identified: NF-κB,[50] AP-1, and AP-2.[51] The mechanism of X action on these transcription factors is still unclear.

During productive viral infection, global transcriptional activation of hepadnaviral gene expression is the most plausible role for the X gene product. A 3- to 10-fold higher level of synthesis of HBsAg, HBcAg, and pregenomic transcripts mediated by X trans-activation has been observed during transient expression of HBV genomes when compared to genomes lacking a functional X gene.[52] Similarly, increased levels of activity have been reported for the corresponding promoters if coupled individually to indicator genes, such as those coding for chloramphenicol acetyltransferase (*cat*) or luciferase (*luc*).[13,53] In this latter experimental setting, strong trans-activation was also observed on the enhancer/X promoter element. Extrapolated to the HBV genome, this suggests that the X gene product could stimulate its own synthesis. If so, this would obviously require a mechanism for down-regulation of X expression after a desired level of X protein has been reached. In agreement with such an autoregulatory scheme, a reduction of HBV gene expression has been observed if high levels of X gene product were supplied in trans.[54] Finally, it should be pointed out that the broad spectrum of X gene activity on transcription suggests the possibility that its expression during viral replication may influence cellular genes that might be involved in regulating HBV transcription.

Several attempts have been made to localize the trans-activating domain in the X protein. The results obtained in cotransfection experiments with deletion mutants show that the carboxyl terminus of the X protein is dispensable for its trans-activating activity.[55] The capability of truncated X proteins to retain their trans-activating properties is noteworthy since, as demonstrated by the sequencing of several cloned HBV integrates, the integration process often leads to the interruption of the X ORF in its 3′ end.[56]

Evidence that a trans-activation domain resides in a cluster of acidic amino acid residues near the X protein carboxyl terminus was recently obtained by Unger and Shaul[57] in a deletion analysis after linking the X protein to the DNA-binding domain of a cellular transcription factor. However, acidic sequences of ubiquitous origin can be used to activate artificially a eukaryotic promoter once they are targeted as fusion proteins to a nearby binding site,[58] and, therefore, these data must presently be interpreted with due caution.

A direct interaction of the X protein with its many targets seems also unlikely in view of the low X protein concentration in the HBV-infected liver. Instead, direct or indirect modulation of factors regulating cellular gene activity appears to be a more realistic alternative to explain the pleiotropic trans-activation properties of the X gene product. In keeping with this notion, very recent data suggest that the X protein may possess protein kinase activity[59] or belong to a family of protease inhibitors.[60]

VII. A ROLE FOR X PROTEIN IN THE DEVELOPMENT OF HEPADNAVIRAL-ASSOCIATED HEPATOCELLULAR CARCINOMA?

The ability of the HBV X protein to function as a trans-activator has led to the suggestion that it might be involved in the development of HBV-associated hepatoma, a concept that has been recently discussed by others.[61,62] However, the long latency period required for development of hepatocellular carcinoma after infection by HBV argues against the likelihood that the X gene acts as an acutely transforming oncogene.

Shirakata et al.[63] have tested the ability of the X gene, placed under the control of the strong SV40 early promoter, to function as a classical oncogene after transfection into mouse NIH 3T3 cells. The resulting clones expressed X mRNA and protein, and their cell saturation density was found to correlate positively with these expression levels. Cells with a high level of X gene expression formed tumors after injection into nude mice, although these tumors had latency periods two to three times longer than that of *ras* oncogene-induced tumors. Transfection of cells with the intact HBV DNA genome containing the X gene has not been found to result in oncogenic transformation. Therefore, it appears that the X gene must be overexpressed in order to have tumorigenic effects.

Aberrant expression patterns of the X gene might occur following HBV infection, as a consequence of viral DNA integration into the cellular genome. X gene sequences are frequently present in the integrated forms of hepadnaviral DNA found in tumor cells.[56,64,65] However, a subset of the integrated HBV sequences examined do not contain a functional X ORF[56] and it is therefore unlikely that continuous X gene activity is essential in every case for HBV-induced tumor formation. Interestingly, the integration of the viral genomes very often occurs between the two direct repeats, DR1 and DR2,[56,66] with the result that the X gene undergoes 3′ terminal truncation and occasionally may also result in fusion of the remaining X gene sequences to cellular coding sequences.[61,67] Since they have not been well characterized, it is unclear whether the antigens detected in hepatoma cell lines[14,29] and tissues[24,25] correspond to the X ORF product or to products of fusions between the X gene and cellular genes.

Hofschneider and co-workers[61,67] have tested the functionality of the X gene product encoded by chromosomally integrated HBV DNA sequences from human hepatoma tissue. On cloning and examination of one such integrate, an X ORF with 28 bp deleted at its 3′ end was found to be in frame with a 3′ flanking cellular ORF of 228 bp. When the cloned integrate was transfected into CC113 liver cells, the fused sequences were found to be cotranscribed from the HBV X promoter and to encode a trans-activating function that was dependent on X gene expression. These experiments show that the X gene product can retain its trans-activating function despite truncation or fusion with cellular-encoded sequences, and suggest that novel forms of X protein with altered properties may be generated on HBV DNA integration in hepatoma cells. Recently, a 3′ truncated preS2/S region of integrated HBV DNA found in a human hepatoma cell line[68] and tissue[69] was also found to encode a transcriptional trans-activator, suggesting that various routes may exist for the generation of trans-acting functions by HBV DNA integration.

Höhne et al.[31] have recently developed an *in vitro* system for the study of HBV-induced transformation. Nonmalignant immortalized hepatocytes expressing the SV40 large tumor antigen were transfected with a plasmid containing HBV DNA inserted downstream of the SV40 enhancer and early promoter. The resulting tumorigenic clones harbored HBV DNA integrates, and most produced X mRNA and protein. The significance of the finding of X gene expression in most of these clones remains to be proven; however, this system should be helpful in defining the role played by the X protein in HBV-associated tumorigenesis.

As originally noted by Miller and Robinson,[70] certain parallels can be drawn between the HBV X gene and the human T-lymphotropic virus type I (HTLV-I) *tax* gene and between the trans-activating functions that they encode. HTLV-I infection is associated with the development of adult T cell leukemia, a disease that also appears after a long latency period. The HTLV-I *tax* gene has recently been shown to induce tumors in Tax-expressing transgenic mice[71] and to cooperate with the *ras* oncogene to induce malignant transformation of cells in culture,[72] and therefore may play a role in HTLV-I-linked oncogenesis. It is conceivable that the X gene product acts in similar ways to stimulate the expression of cellular genes involved in growth control, inducing growth dysregulation and oncogenesis. To address this possibility, it will be important to determine the effects of the X gene product on the expression of cellular genes known to be involved in growth control and cellular transformation.

ACKNOWLEDGMENTS

We thank our many colleagues for providing reprints and preprints, and also Sandra Nazareth and Karin Coutinho for secretarial assistance. The authors' work cited here was supported by the Deutsche Forschungsgemeinschaft.

REFERENCES

1. **Galibert, F., Mandart, E., Fitoussi, F., Tiollais, P., and Charnay, P.,** Nucleotide sequence of the hepatitis B virus genome (subtype *ayw*) cloned in *E. coli*, *Nature (London)*, 281, 646, 1979.
2. **Tiollais, P., Charnay, P., and Vyas, G. N.,** Biology of the hepatitis B virus, *Science*, 213, 406, 1981.
3. **Lo, S. J., Chien, M. L., and Lee, Y. H.,** Characteristics of the X gene of hepatitis B virus, *Virology*, 167, 289, 1988.
4. **Seeger, C., Ganem, D., and Varmus, H. E.,** Nucleotide sequence of an infectious molecularly cloned genome of ground squirrel hepatitis virus, *J. Virol.*, 51, 367, 1984.
5. **Galibert, F., Chen, T. N., and Mandart, E.,** Nucleotide sequence of a cloned woodchuck hepatitis virus genome: comparison with the hepatitis B virus sequence, *J. Virol.*, 41, 51, 1982.
6. **Schaeffer, E. and Sninsky, J. J.,** Predicted secondary structure similarity in the absence of primary amino acid sequence homology: hepatitis B virus open reading frames, *Proc. Natl. Acad. Sci. U.S.A.*, 81, 2902, 1984.
7. **Mandart, E., Kay, A., and Galibert, F.,** Nucleotide sequence of a cloned duck hepatitis B virus genome: comparison with woodchuck and human hepatitis B virus sequences, *J. Virol.*, 49, 782, 1984.
8. **Sprengel, R., Kuhn, C., Will, H., and Schaller, H.,** Comparative sequence analysis of duck and human hepatitis B virus genomes. *J. Med. Virol.*, 15, 323, 1985.
9. **Sprengel, R., Kaleta, E. F., and Will, H.,** Isolation and characterization of a hepatitis B virus endemic in herons, *J. Virol.*, 62, 3832, 1988.
10. **Feitelson, M. A. and Miller, R. H.,** X gene-related sequences in the core gene of duck and heron hepatitis B viruses, *Proc. Natl. Acad. Sci. U.S.A.*, 85, 6162, 1988.
11. **Yaginuma, K., Shirakata, Y., Kobayashi, M., and Koike, K.,** Hepatitis B virus particles are produced in a cell culture system by transient expression of transfected HBV DNA, *Proc. Natl. Acad. Sci. U.S.A.*, 84, 2678, 1987.
12. **Matsubara, K.,** personal communication, 1989.
13. **Colgrove, R., Simon, G., and Ganem, D.,** Transcriptional activation of homologous and heterologous genes by the hepatitis B virus X gene product in cells permissive for viral replication, *J. Virol.*, 63, 4019, 1989.
14. **Moriarty, A. M., Alexander, H., Lerner, R. A., and Thornton, G. B.,** Antibodies to peptides detect new hepatitis B antigen: serological correlation with hepatocellular carcinoma, *Science*, 227, 429, 1985.
15. **Moriarty, A. M., Mitamura, K., and Thornton, G. B.,** Detection of antibody to the hepatitis B viral X determinant in chronic hepatitis B viral infection, in *Viral Hepatitis and Liver Disease*, Zuckerman, A. J., Ed., Alan R. Liss, New York, 1988, 345.
16. **Feitelson, M. A. and Clayton, M. M.,** The X antigen/antibody markers in hepadnavirus infections. II. Antibodies to the X gene product(s); submitted.
17. **Meyers, M. L., Trepo, L. V., Nath, N., and Sninsky, J. J.,** Hepatitis B virus polypeptide X: expression in *Escherichia coli* and identification of specific antibodies in sera from hepatitis B virus-infected humans, *J. Virol.*, 57, 101, 1986.
18. **Kay, A., Mandart, E., Trepo, C., and Galibert, F.,** The HBV HBx gene expressed in *E. coli* is recognised by sera from hepatitis patients, *EMBO J.*, 4, 1287, 1985.
19. **Elfassi, E., Haseltine, W. A., and Dienstag, J. L.,** Detection of hepatitis B virus X product using an open reading frame *Escherichia coli* expression vector, *Proc. Natl. Acad. Sci. U.S.A.*, 83, 2219, 1986.
20. **Pfaff, E., Salfeld, J., Gmelin, K., Schaller, H., and Theilmann, L.,** Synthesis of the X-protein of hepatitis B virus *in vitro* and detection of anti-X antibodies in human sera, *Virology*, 158, 456, 1987.
21. **Hess, J., Stemler, M., Will, H., Schröder, C. H., Kuhn, J., and Braun, R.,** Frequent detection of antibodies to hepatitis B virus X-protein in acute, chronic and resolved infections, *Med. Microbiol. Immunol.*, 177, 195, 1988.
22. **Levrero, M., Jean-Jean, O., Balsano, C., Will, H., and Perricaudet, M.,** Hepatitis B virus (HBV) X gene expression in human cells and anti-HBx antibodies detection in chronic HBV infection, *Virology*, 174, 299, 1990.
23. **Persing, D. H., Varmus, H. E., and Ganem, D.,** Antibodies to pre-S and X determinants arise during natural infection with ground squirrel hepatitis virus, *J. Virol.*, 60, 177, 1986.
24. **Zentgraf, H., Herrmann, G., Klein, R., Schranz, P., Loncarevic, I., Herrman, D., Hübner, K., and Schröder, C. H.,** Mouse monoclonal antibody directed against hepatitis B virus X protein synthesized in *Escherichia coli*: detection of reactive antigen in liver cell carcinoma and chronic hepatitis, *Oncology*, 47, 143, 1990.

25. **Vitvitski, L., Meyers, M. L., Sninsky, J. J., Berthillon, P., Chevalier, P., Sells, M. A., Acs, G., and Trepo, C.**, Expression of the X gene product of hepatitis B virus and WHV in infected livers and transfected 3T3 cells: evidence for cross reactivity and correlation with core/e gene expression, in *Viral Hepatitis and Liver Disease*, Zuckerman, A. J., Ed., Alan R. Liss, New York, 1988, 341.
26. **Pfaff, E., Theilmann, L., Salfeld, J., and Schaller, H.**, X-Sequence related proteins and X-specific antibodies, paper presented at the 1986 meeting on Molecular Biology of Hepatitis B Viruses, Cold Spring Harbor Laboratory, Cold Spring Harbor, New York, 1986 (abstract).
27. **Feitelson, M. A., Clayton, M. M., and Blumberg, B. S.**, The X antigen/antibody markers in hepadnavirus infection. I. Presence and significance of the hepadnavirus X gene product(s) in serum; submitted.
28. **Chang, C., Jeng, K., Hu, C., Lo, S. J., Su, T., Ting, L. P., Chou, C. K., Han, S., Pfaff, E., Salfeld, J., and Schaller, H.**, Production of hepatitis B virus *in vitro* by transient expression of cloned HBV DNA in a hepatoma cell line, *EMBO J.*, 6, 675, 1987.
29. **Chisaka, O., Araki, K., Ochiya, T., Tsurimoto, T., Hiranyawasitte-Attatippaholkun, W., Yanaihara, N., and Matsubara, K.**, Purification of hepatitis B virus gene X product synthesized in *Escherichia coli* and its detection in a human hepatoblastoma cell line producing hepatitis B virus, *Gene*, 60, 183, 1987.
30. **Pugh, J. C., Weber, C., Houston, H., and Murray, K.**, Expression of the X gene of hepatitis B virus, *J. Med. Virol.*, 20, 229, 1986.
31. **Höhne, M., Schaefer, S., Seifer, M., Feitelson, M. A., Paul, D., and Gerlich, W. H.**, Malignant transformation of immortalized transgenic hepatocytes after transfection with hepatitis B virus DNA, *EMBO J.*, 9, 1137, 1990.
32. **Treinin, M. and Laub, O.**, Identification of a promoter element located upstream from the hepatitis B virus X gene, *Mol. Cell. Biol.*, 7, 545, 1987.
33. **Gough, N. M. and Murray, K.**, Expression of the HBV surface, core and e antigen genes by stable rat and mouse cell lines, *J. Mol. Biol.*, 162, 43, 1982.
34. **Saito, I., Oya, Y., and Shimojo, H.**, Novel RNA family structure of hepatitis B virus expressed in human cells, using a helper-free adenovirus vector, *J. Virol.*, 58, 554, 1986.
35. **Asselsbergs, F. A. M., Will, H., Wingfield, P., and Hirschi, M.**, A recombinant Chinese hamster ovary cell line containing a 300-fold amplified tetramer of the hepatitis B genome together with a double selection marker expresses high levels of viral protein, *J. Mol. Biol.*, 189, 401, 1986.
37. **Araki, K., Miyazaki, J. I., Hino, O., Tomita, N., Chisaka, O., Matsubara, K., and Yamamura, K. I.**, Expression and replication of hepatitis B virus genome in transgenic mice, *Proc. Natl. Acad. Sci. U.S.A.*, 86, 207, 1989.
38. **Siddiqui, A., Jameel, S., and Mapoles, J.**, Expression of the hepatitis B virus X gene in mammalian cells, *Proc. Natl. Acad. Sci. U.S.A.*, 84, 2513, 1987.
39. **Schek, N., Kuhn, C., Bartenschlager, R., and Schaller, H.**, Phosphorylation and rapid turnover of hepatitis B virus X protein expressed in HepG2 cells from a recombinant vaccinia virus, *Oncogene*, in press.
40. **Dingwall, C. and Laskey, R. A.**, Protein import into the cell nucleus, *Annu. Rev. Cell. Biol.*, 2, 367, 1986.
41. **Koike, K., Akatsuka, T., and Miyamura, T.**, Characterization of hepatitis B virus X gene: *in vitro* translation of mRNA from COS-1 cells transfected with the X gene, *Virology*, 163, 233, 1988.
42. **Lin, M. H. and Lo, S. J.**, Dimerization of hepatitis B viral X protein synthesized in a cell-free system, *Biochem. Biophys. Res. Commun.*, 164, 14, 1989.
43. **Rossner, M. T., Jackson, R. J., and Murray, K.**, Modulation of expression of the hepatitis B virus surface antigen gene by the viral X gene product; submitted.
44. **Spandau, D. F. and Lee, C. H.**, *Trans*-activation of viral enhancers by the hepatitis B virus X protein, *J. Virol.*, 62, 427, 1988.
45. **Zahm, P., Hofschneider, P. H., and Koshy, R.**, The HBV X-ORF encodes a transactivator, a potential factor in viral hepatocarcinogenesis, *Oncogene*, 3, 169, 1988.
46. **Twu, J. S. and Robinson, W. S.**, Hepatitis B virus X gene can transactivate heterologous viral sequences, *Proc. Natl. Acad. Sci. U.S.A.*, 86, 2046, 1989.
47. **Twu, J. S. and Schloemer, R. H.**, Transcriptional transactivating function of hepatitis B virus, *J. Virol.*, 61, 3448, 1981.
48. **Aufiero, B. and Schneider, R. J.**, The hepatitis B virus X-gene product *trans*-activates both RNA polymerase II and III promoters, *EMBO J.*, 9, 497, 1990.
49. **Seto, E., Zhou, D. X., Peterlin, B. M., and Yen, T. S. B.**, *Trans*-activation by the hepatitis B virus X protein shows cell-type specificity, *Virology*, 173, 764, 1989.

50. **Twu., J. S., Chu, K., and Robinson, W. S.,** Hepatitis B virus X gene activates kappa B like enhancer sequences in the long terminal repeat of human immunodeficiency virus 1, *Proc. Natl. Acad. Sci. U.S.A.*, 86, 5168, 1989.
51. **Seto, E., Mitchell, P. J., and Yen, T. S.,** Transactivation by the hepatitis B virus X protein depends on AP-2 and other transcription factors, *Nature (London)*, 344, 72, 1990.
52. Unpublished results, 1988.
53. **Raney, A. K., Milich, D. R., Easton, A. J., and McLachlan, A.,** Differentiation-specific transcriptional regulation of the hepatitis B virus large surface antigen gene in human hepatoma cell lines, *J. Virol.*, 64, 2360, 1990.
54. Unpublished results, 1989.
55. **Balsano, C., Avantaggiati, M. L., Natoli, G., De Marzio, E., Will, H., and Levrero, M.,** Transactivation of c-myc and viral regulatory sequences by full length and truncated versions of the HBV-X protein, *Oncogene*, submitted.
56. **Nagaya, T., Nakamura, T., Tokino, T., Tsurimoto, T., Imai, M., Mayumi, T., Kamino, K., Yamamura, K., and Matsubara, K.,** The mode of hepatitis B virus DNA integration in chromosomes of human hepatocellular carcinoma, *Genes Dev.*, 1, 773, 1987.
57. **Unger, T. and Shaul, Y.,** The X protein of hepatitis B virus acts as a transcription factor when targeted to its responsive element, *EMBO J.*, 9, 1889, 1990.
58. **Ma, J. and Ptashne, M.,** A new class of yeast transcriptional activators, *Cell*, 51, 113, 1987.
59. **Robinson, W.,** personal communication, 1990.
60. **Koike, T.,** personal communication, 1990.
61. **Koshy, R. and Hofschneider, P. H.,** Transactivation by hepatitis B virus may contribute to hepatocarcinogenesis, *Curr. Top. Microbiol. Immunol.* 144, 265, 1989.
62. **Schroeder, C. H. and Zentgraf, H.,** Hepatitis B virus related hepatocellular carcinoma: chronicity of infection — the opening to different pathways of malignant transformation?, *Biochem. Biophys. Acta*, in press.
63. **Shirakata, Y., Kawada, M., Fujiki, O., Sano, H., Oda, M., Yaginuma, K., Kobayashi, M., and Koike, K.,** The X gene of hepatitis B virus induced growth stimulation and tumorigenic transformation of mouse NIH3T3 cells, *Jpn. J. Cancer Res.*, 80, 617, 1989.
64. **Miyaki, M., Sato, C., Gotanda, T., Matsui, T., Mishiro, S., Imai, M., and Mayumi, M.,** Integration of region X of hepatitis B virus genome in human primary hepatocellular carcinomas propagated in nude mice, *J. Gen. Virol.*, 67, 1449, 1986.
65. **Fuchs, K., Heberger, C., Weimer, T., and Roggendorf, M.,** Characterization of woodchuck hepatitis virus DNA and RNA in the hepatocellular carcinomas of woodchucks, *Hepatology*, 10, 215, 1989.
66. **Yaginuma, K., Kobayashi, H., Kobayashi, M., Morishima, T., Matsuyama, K., and Koike, K.,** Multiple integration sites of hepatitis B virus DNA in hepatocellular carcinoma and chronic active hepatitis tissues from children, *J. Virol.*, 61, 1808, 1987.
67. **Wollersheim, M., Debelka, U., and Hofschneider, P. H.,** A transactivating function encoded in the hepatitis B virus X gene is conserved in the integrated state, *Oncogene*, 3, 545, 1988.
68. **Kekule, A. S., Lauer, U., Meyer, M., Caselmann, W. H., Hofschneider, P. H., and Koshy, R.,** The preS2/S region of integrated hepatitis B virus DNA encodes a transcriptional transactivator, *Nature (London)*, 343, 457, 1990.
69. **Caselmann, W. H., Meyer, M., Kekule, A. S., Lauer, U., Hofschneider, P. H., and Koshy, R.,** A transactivator function is generated by integration of hepatitis B virus preS/S sequences in human hepatocellular carcinoma DNA, *Proc. Natl. Acad. Sci. U.S.A.*, 87, 2970, 1990.
70. **Miller, R. H. and Robinson, W. S.,** Common evolutionary origin of hepatitis B virus and retroviruses, *Proc. Natl. Acad. Sci. U.S.A.*, 83, 2531, 1986.
71. **Nerenberg, M., Hinrichs, S. H., Reynolds, R. K., Khoury, G., and Jay, G.,** The *tat* gene of human T-lymphotropic virus type 1 induces mesenchymal tumors in transgenic mice, *Science*, 237, 1324, 1987.
72. **Pozzatti, R., Vogel, J., and Jay, G.,** The human T-lymphotropic virus type I *tax* gene can cooperate with the *ras* oncogene to induce neoplastic transformation of cells, *Mol. Cell. Biol.*, 10, 413, 1990.

Chapter 10

REGULATION OF HEPADNAVIRUS TRANSCRIPTION

Yosef Shaul

TABLE OF CONTENTS

I. INTRODUCTION

The genome of hepadnaviruses is extremely small and used very economically. There is not a single base pair within the viral DNA that is not included in one of the viral open reading frames, leaving no space devoted exclusively to the regulatory sequences (e.g., promoters and enhancer). This feature has a strong impact on the evolutionary conservation of the regulatory sequences, as underscored by sequence similarity found among the members of the hepadnavirus family. In fact, in the absence of information on the regulatory sequences in the genomes of the other members of hepadnaviruses besides hepatitis B virus (HBV), we must rely on sequence homology in order to define the corresponding elements within the genomes of these viruses. Although the hepadnavirus life cycle is a permuted version of the RNA retroviral life cycle, there is a basic difference between the transcriptional regulation of their genes. Unlike the retroviral genome, which contains a single promoter, the hepadnaviral genome contains multiple promoters, each regulating the synthesis of a distinct transcript. This is perhaps the reason for the minor role that RNA splicing plays in the hepadnaviral life cycle. Four promoters have been found within the genome of HBV and most of them, except the X gene promoter, were also found in the genomes of other members of this family. In this chapter, the promoter that is responsible for the synthesis of the 3.5-kb viral mRNA is referred to as the core promoter; the one that is responsible for the synthesis of the 2.4- to 2.6-kb mRNA as the preS promoter; the promoter that directs the synthesis of the viral 2.1-kb mRNA as the S promoter; and finally the promoter that is positioned next to the X gene and that possibly directs the synthesis of the 0.9-kb mRNA as the X promoter.

The activity of promoters in general is under tight regulation by multiple cis-acting elements found in close vicinity to the transcriptional initiation sites.[1-3] These cis-acting elements are recognized by sequence-specific transcription factors to generate DNA–protein complexes. In HBV, some of the functional cis-acting elements of the different promoters have been characterized, and in a few cases the nature of the binding proteins (the transcription factors) have been defined. Accumulating data clearly suggest that viral transcription is liver (and perhaps kidney) specific. Apparently some of the functional cis-acting elements and their cognate binding proteins are responsible for this tissue specificity, but so far little progress has been made to delineate these elements.

The HBV genome and perhaps the genomes of other members of this family contains an enhancer that seems to be essential for optimal activity of the viral promoters. We do not know yet how the enhancer works but it seems that we are starting to gain clues toward the understanding of the composition and mode of action of this transcriptional regulator.

II. THE CORE PROMOTER

In vitro studies using HeLa cell extracts revealed a transcription initiation site at position –130 (the first nucleotide of the precore open reading frame (ORF) ATG is taken as +1).[4] However, all the subsequent *in vivo* studies could find no such transcription initiation site; instead, three initiation sites in the region +5 to –31 were identified (Figure 1). These sites were identified in cultured cells transfected with fragments of HBV DNA,[5,6] in liver cells infected with HBV,[7] in stable and transient expression systems for HBV production in cultured cells,[8,9] and in transgenic mice that express and replicate the HBV genome.[10] In addition, initiation sites that were positioned at –64 ± 3 and at +90 nucleotides were also reported.[5,9] Progressive deletion mutants were employed in order to delineate the minimal promoter region (Figure 1). In one study this region was mapped between –115 and +52.[11] Removal of the region –115 to –89 resulted in a sharp reduction in promoter activity, suggesting this region

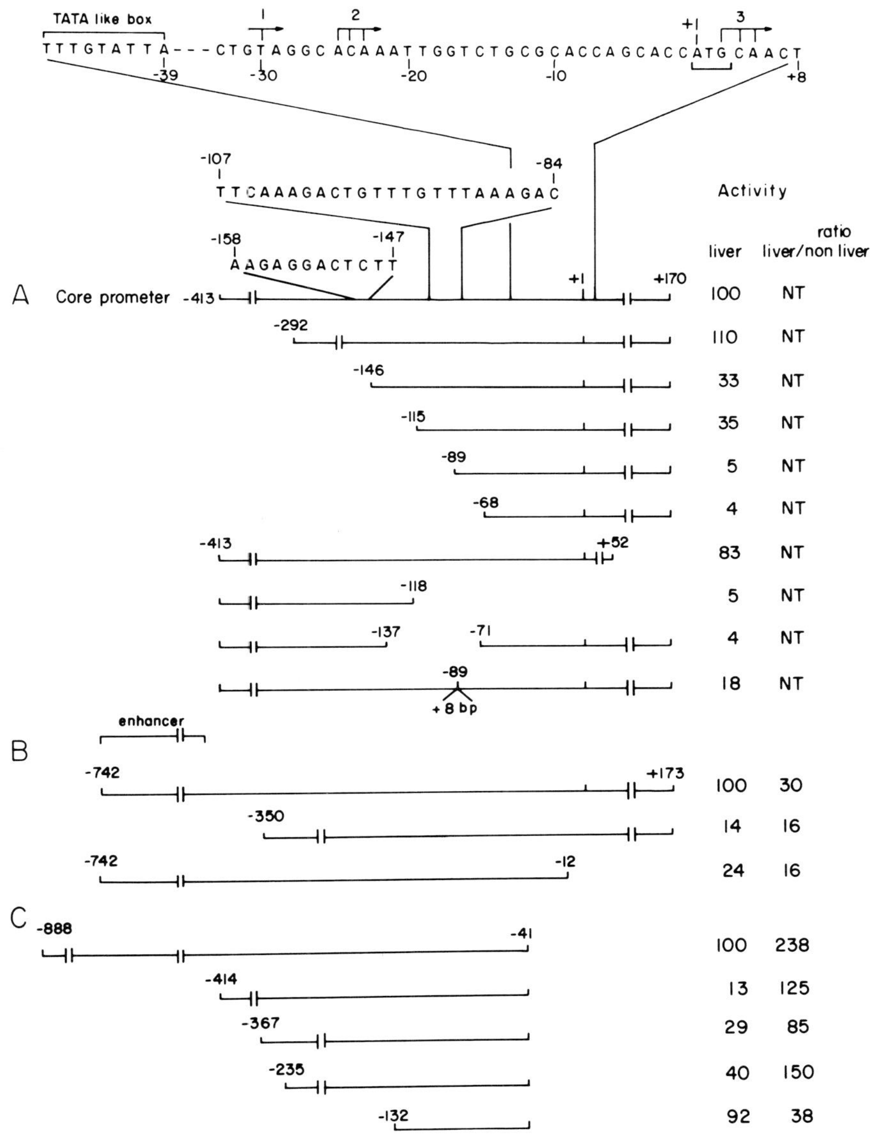

FIGURE 1. The HBV core promoter region. The figure summarizes data from Yaginuma et al.[11] (A), Roossinck et al.[5] (B), and Yee[12] (C). The A residue of the precore ATG codon is represented as +1. The three groups of the transcription initiation sites are shown by small arrows and numbered from 1 to 3. Sequences in the region of the promoter with potentially important promoter elements are indicated. The sequence of the regions from –107 to –84 and from –158 to –147 that bind cellular proteins are shown (see text). The lines represent the promoter region cloned in the chloramphenicol acetyltransferase (CAT) reporter plasmid for functional analysis. The numbers at the right indicate the rate of CAT activity relative to that of the largest DNA fragment used in each study, and the ratio of the activity obtained in liver cell lines vs. nonliver cells. NT, not tested.

was an essential promoter component. Although a comparative study has not been performed it seems that deletion of the promoter from the 3′ end up to –41 is not detrimental for promoter activity.[12] Thus, the minimal promoter fragment is mapped within the –115 to –41 region. Indeed, a –111 to –11 fragment has been shown to contain an efficient promoter activity.[6] Within this fragment a diverged TATA box is found within the –39 to –47 region (Figure 2),

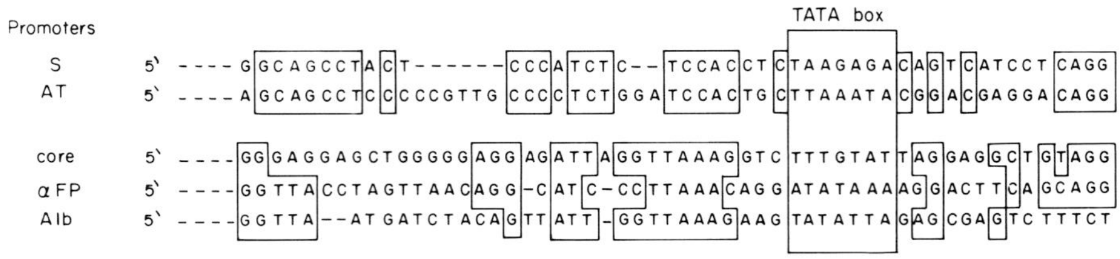

FIGURE 2. The sequence of the HBV core and the S promoters are similar to those of other liver-specific genes. The sequences of the promoters of the following genes were compared: the HBV S (S), the α-1 antitrypsin (AT),[14] the HBV core, the mouse α-fetoprotein (αFP),[17] and the albumin (Alb) promoter.[17] The homologous sequences between the S and AT promoters were boxed separately from those of the core, αFP, and Alb, except for the TATA box region.

which is highly conserved in the genomes of other members of the hepadnavirus family.[11] Interestingly, the sequence of the core promoter at this region is also very similar to that of two liver-specific genes; albumin and α-fetoprotein (Figure 2). A detailed mutagenesis study of this region is required in order to determine its functional significance.

At least two upstream regions are required for the activity of the core promoter. A proximal one is positioned between –115 and –89 and a distal one between –292 and –146. Removal of these regions resulted in a seven- and threefold reduction in the core promoter activity, respectively.[11] However, since deletion of the proximal element alone (deletion of the region from –137 to –71, Figure 1) abolished the promoter activity, it seems that the distal positive element acts cooperatively with the proximal element. The ability of the proximal element to stimulate transcription was confirmed by others. In one study a DNA fragment from –132 to –41 has been shown to activate the thymidine kinase (TK) promoter of herpes simplex virus (HSV) in a liver environment in an enhancer-like manner.[12] However, in this study two additional observations have been made that are in disagreement with the structure of the core promoter that has been described so far. One suggests that in HepG2 cells the distal element does not act as a positive but rather as a strong negative element, so that removal of the –414 to –132 region resulted in about a sevenfold induction in promoter activity (Figure 1). The presence of such a negative element was also proposed by others.[6] The second suggests that in the Huh6 hepatoma cell line the distal element has no regulatory role, and all the promoter activity resides within the –132 to –41 region.[12] Thus, it seems that the role of the core promoter distal element is different among the different hepatoma cell lines.

The binding of two cellular nuclear factors (possible transcription factors) to the core promoter has been documented. The first protein binds the proximal element at the –107 to –84 site,[11] and the second binds the distal element at the –158 to –147 site.[13] The direct role of these proteins in the activation of the core promoter was not documented; however, insertion of an unrelated eight-nucleotide pair sequence at the proximal factor-binding site resulted in a sixfold reduction in core promoter activity, strongly suggesting that this site, and possibly the corresponding binding factor, is required for the promoter activity. Exactly at the site of the eight-nucleotide pair insertion, two copies of the sequence motif 5′-TAGGTTAAAGG-3′ are found that are very similar to that found in the albumin gene promoter, adjacent to the TATA box. Interestingly, however, the core promoter contains a third copy of this sequence at the homologous position next to the TATA-like box (Figure 2). A similar sequence is also found in the promoter of the α-1 antitrypsin (α-1AT) gene. Specific mutation of this site in the α-1AT gene promoter resulted in a sharp reduction in promoter activity, suggesting that this sequence motif is a recognition binding site of an important

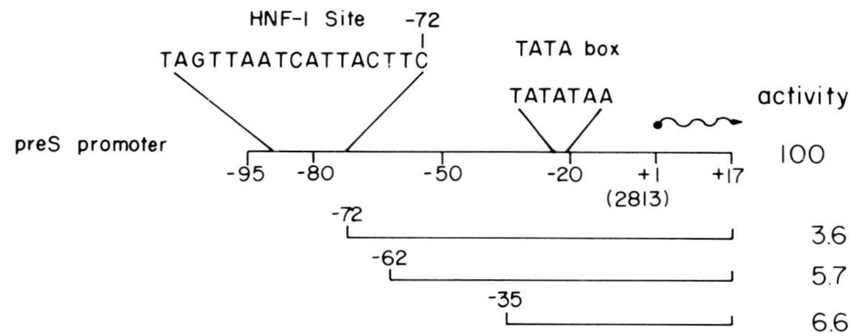

FIGURE 3. The preS promoter. The preS promoter region with some important sequences (see Section III) is shown. The relative activity data were taken from Chang et al.[38] The activity was determined by using the CAT reporter plasmid. The activity of the fragment –95 to +17 was arbitrarily taken as 100%. The relative promoter activities of progressive 5′ end-deletion mutants are indicated. HNF-1, hepatocyte nuclear factor 1.

transcription factor.[14] In the albumin gene promoter this site binds a liver-enriched factor that is either C/EBP or a closely related factor.[15,16] Thus, it is very likely that the proximal element of the core promoter binds this factor at one or more sites.

A. REGULATION OF THE CORE PROMOTER BY THE ENHANCER

The HBV enhancer is positioned at –737 to –583 relative to the core promoter.[18,19] Several groups have reported that the core promoter is regulated by the enhancer.[5,6,12] In HepG2[5,12] but not Huh6[12] cells, the core promoter is about seven- to eightfold less active in the absence of the enhancer (Figure 1). In these studies, however, it is not clear whether this is a direct effect. Removal of the enhancer also inactivates the adjacent X promoter and reduces the production of the X protein, the gene for which is located between the enhancer and the core promoter. Therefore, the low activity of the core promoter may be attributed to the absence of the trans-activator X protein rather than the enhancer.

B. TISSUE-SPECIFIC EXPRESSION OF THE CORE PROMOTER

Although some activity of the core promoter was observed in nonliver cell lines,[20,21] the activity of this promoter seems to be liver specific.[5,6,12] This observation was recently confirmed by the specific expression of the HBV pregenomic transcript in liver and to a lower extent in kidney of transgenic mice carrying the full HBV genome.[10,22] In Figure 1 the liver-specific expression of the core promoter was calculated as a ratio between the activity obtained in liver and nonliver cells. In two independent studies this ratio is twofold higher in the presence of the enhancer. Consistent with these observations, Honigwachs et al.[6] had shown that the HBV enhancer and the core promoter, each under the influence of simian virus 40 (SV40) regulatory elements, are partially active in heterologous cells, but the combination of the two is active only in liver cells. Thus, the combined action of the core promoter and the enhancer seems to produce increased liver-specific expression relative to each of these elements alone. These observations provide an explanation, on the molecular level, for the hepatotropism of hepadnaviruses. The virus cannot replicate in heterologous cells simply because the core promoter is unable to direct the biosynthesis of the pregenomic RNA.[23] Indeed, in two independent studies, replication of both HBV and woodchuck hepatitis B virus (WHV) was obtained in heterologous cells once the lack of synthesis of the pregenomic RNA was circumvented by replacing the viral core promoter with ubiquitously active heterologous regulatory elements.[24,25]

III. THE PreS PROMOTER

The initiation site of the preS gene transcript was mapped to nucleotide 2810 ± 10 on the HBV *adw* genome by *in vitro* runoff transcription analysis using HeLa cell extracts.[4] It was suggested that the A residues at positions 2813 and 2816 were the most likely start site for the preS transcript. A classic TATA box sequence, TATATAA, is present at position 2785 to 2791 which is about 20 to 30 bp upstream from the initiation site, where a functional TATA box is expected to be found (Figure 3). The preS promoter is efficiently utilized in an *in vitro* cell-free transcription system[4] using HeLa cell extracts, but is weakly active *in vivo*.[28,29] The activity of this promoter is stimulated under the regulation of either the SV40 enhancer[26,27] or the HBV enhancer.[30,32] This promoter is responsible for the production of the 2.4-kb viral mRNA. This mRNA is found only in low abundance in liver cells with either integrated[33] or replicative forms of HBV DNA,[8,34] in infected liver cells of chimpanzees,[23] and in transgenic mice that carry the whole HBV DNA,[10,22] suggesting that preS is a weak promoter. Antonucci and Rutter[30] reached the same conclusion by comparative study of all four viral promoters. The synthesis of the 2.4-kb mRNA is highly liver specific. In transgenic mice this RNA species was detected specifically in liver and only in a few cases in the kidney and heart.[10,22] The liver specificity of this promoter was also addressed by employing the reporter chloramphenicol acetyltransferase (CAT) plasmid.[30,31] Chang and Ting[31] reported that the activity of the preS promoter displays a preference of up to 750-fold for differentiated hepatoma cell lines. The reason why this promoter is very active *in vitro* in nonliver HeLa cell extract is not clear. Recently a liver-specific transcription factor has been identified that binds to a specific sequence found within the promoter regions of several liver-specific genes.[35,36] This factor, called HNF-1 (for hepatocyte nuclear factor 1), was purified, and the purified protein binds to the HBV DNA next to the TATA box of the preS promoter[37] (Figure 3). The binding of HNF-1 has been shown to play a role in expression from the preS promoter.[38] A fragment of 112 bp is functional as a liver-specific promoter. This fragment contains the TATA box and the HNF-1-binding site that is positioned about 50 bp upstream of the TATA box. However, a shorter fragment of 89 bp that does not contain the HNF-1-binding site is about 20-fold less active, suggesting that the HNF-1-binding site is required for the activity of the preS promoter. The structures of preS promoters in the genomes of the other members of the hepadnavirus family are similar.[38] However, their function has not been experimentally demonstrated.

The activity of the preS promoter is regulated by the HBV enhancer.[30,31] A fragment of 478 bp that contains the sequence of the preS promoter from +35 to –443 shows a low level of promoter activity. This activity is induced 100-fold when the HBV fragment that contains the enhancer was also inserted into the plasmid.[30] Similar results were obtained using a shorter HBV DNA fragment, with the preS promoter from –377 to +15.[31] In that case, in the absence of the enhancer, the activity of this promoter was 50- to 130-fold higher in differentiated liver cells (HepG2 and HuH-7) than in poorly differentiated liver cells (HA22T/VGH) and nonliver cells (HeLa). In differentiated cells this activity was further induced up to 44-fold in the presence of the enhancer. Thus, the activity of preS promoter in differentiated hepatoma cell lines is regulated by a combinatorial action of the promoter and the viral enhancer.[31,38]

Recently it was proposed that the activity of the preS promoter is down regulated by DNA sequences containing the downstream S promoter.[32] It was suggested that an active S promoter is responsible for this down regulation; however, this possibility has not been tested experimentally. If indeed the activity of the S promoter negatively affects that of the preS promoter, then the overall activity of the preS promoter is determined by a combinatorial action of the preS and the S promoters and the enhancer.

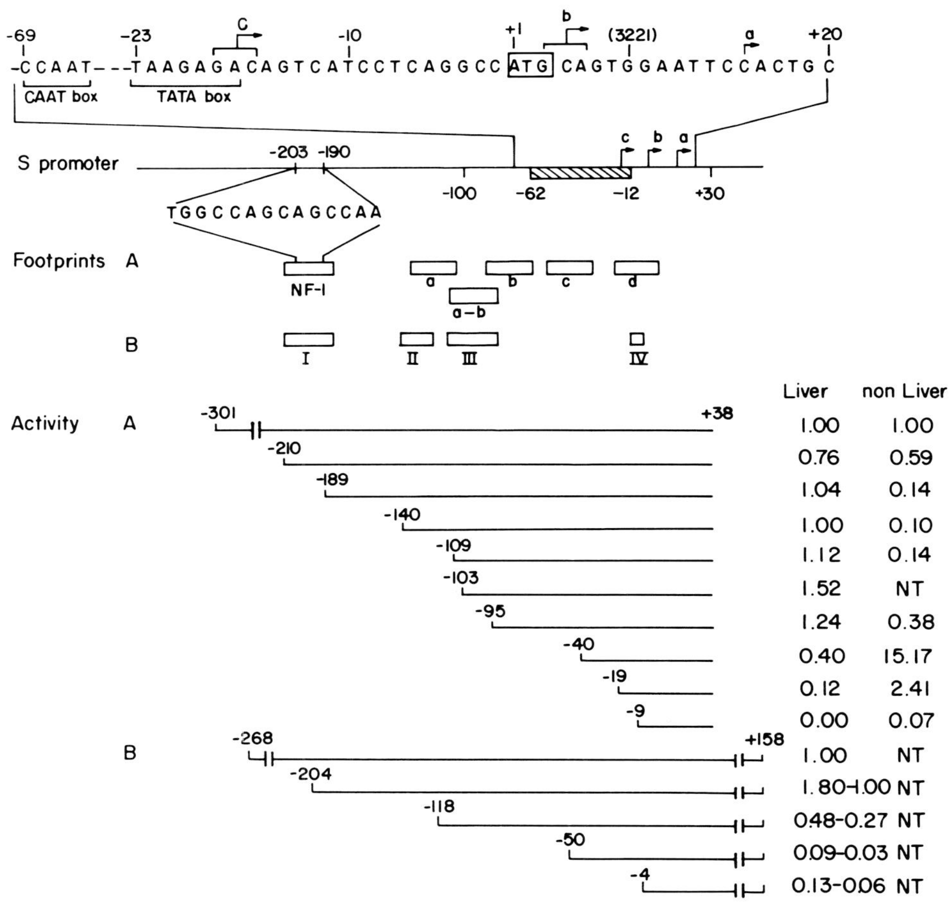

FIGURE 4. The structure and activity of the S promoter. The S promoter region with some functionally important sequences (see Section IV) is shown. The residue of the ATG of the preS2 open reading frame was taken as position +1. The three well-characterized transcription start sites are shown by arrows. The region with sequence similar to the SV40 late promoter is shown by a shaded box. The DNase I-footprinted regions are shown by open boxes at their corresponding positions. Footprints in (A) are from De-Medina and Shaul;[39] those in (B) are from Raney et al.[42] In the former studies the DNA of subtype *adw* was used and in the latter studies the *ayw* subtype was analyzed. At the right-hand side the relative activity of the DNA fragments (shown by lines) in CAT reporter plasmid, in both liver and nonliver cell lines, is indicated. The data summarized here are from De-Medina et al.[29] (A) and Raney et al.[42] (B). In each study the activity of the longest fragment is arbitrarily taken as 1.00, regardless of the cell line used.

IV. THE S PROMOTER

A second promoter that regulates the synthesis of the envelope polypeptides is the S promoter. Three major initiation sites for the S gene transcripts were identified and mapped, designated a,b, and c (Figure 4).[39] In order to allow a convenient comparison among the sequences of different HBV subtypes, the nucleotide A of the preS2 ATG was taken as +1. Using this numbering, the initiation sites a, b, and c were mapped to the following nucleotides: +14 to +16, +3 to +5, and −18 to −16, respectively.[8,39a] *In vitro* and *in vivo*[27,28,42] studies revealed that transcription from these initiation sites is controlled by an HBV DNA fragment that is distinct from the preS promoter. This promoter has no typical TATA box. Based on sequence similarity, it was proposed that the S promoter resembles the SV40 late promoter,[40] which also lacks a TATA box. However, the sequence of the S promoter is also very similar to that of liver-specific α-1 antitrypsin gene (Figure 2). Based on this similarity, the sequence from –23 to –17 (TAAGAGA) of the S promoter seems to be the TATA box homolog. This diverged TATA box is found in the genome of *adw* and *ady* but not in the other subtypes. Analysis of deletion mutants of the S promoter in *adw* revealed that the region –40 to –9, which contains this box, is crucial for the promoter activity.[27] This is not the case, however, in subtype *ayw,* which lacks this box.[42] These data suggest that the TATA box-like element of the S promoter is functionally active.

Interestingly, this sequence motif also bears a strong similarity with the adenovirus EIIaE1 TATA box-like region.[27,43] The similarity between the two promoters is striking also in the sense that in both cases the TATA box region is the initiation site of a second transcript (initiation site c at the S promoter, Figure 4). Recently, a 17-bp element, termed initiator, was identified by Smale and Baltimore[44] that is sufficient for accurate basal transcription *in vivo* and *in vitro.* This motif contains within itself a transcription start site similar to the S promoter. It is therefore possible, although not proven, that an initiator element is found between –9 to –40 and that this element is responsible for the promoter activity of the S promoter. A cellular factor that specifically interacts with the S promoter at the region +6 to –21 was seen (footprint d, Figure 4). Although the nature of this factor was not investigated, it is possible that this factor is a related factor to TFIID, a protein described by Sawadogo and Roeder,[45] which binds to the TATA box in front of many genes and is required for transcription *in vitro.*

Apart from the TATA box, the CCAAT box is also a common DNA sequence that is usually found near the transcription initiation site of many genes. This element is positioned about 50 bp upstream of the TATA box in a variety of promoters, and associated with transcriptional activation of several genes.[46,47] The S promoter contains a CCAAT box at –65 to –69 (Figure 4). The sequence of this box is conserved among all the known subtypes of HBV. In two independent studies[27,42] in hepatoma cell lines, deletion mutants of the S promoter that lack the CCAAT box are poorly active, if at all (Figure 4). In subtype *adw* deletion of the region between –95 and –40 resulted in a threefold reduction in activity[27] and in subtype *ayw* deletion of the –118 to –50 region abolished the activity of the S promoter.[42] Unexpectedly, however, the deletion mutants of *adw* showed a strong activity in nonliver cells, which may suggest that the CCAAT box region played the role of a negative regulator in heterologous cells.[27] This possibility is in accordance with the notion that the CCAAT box and its adjacent nucleotides may also have a more complicated function, such as modulating transcript levels during differentiation or induction.[49]

In most of the analyzed cases, one or more transcription factors have been reported to bind the CCAAT box.[46,48,49] A CCAAT-binding transcription factor (CTF) has been purified from HeLa cell nuclear extracts[48] and found to be indistinguishable from nuclear factor 1 (NF-1).[48] However, CTF is not an exclusive CCAAT-binding protein (CBP) and other CBPs have also been identified.[46,49] Despite the important role that the CCAAT box seems to play in the S promoter, no binding of CBP has been reported. In this region two footprints have been identified (footprints b and c, Figure 4) but the CCAAT box is not located within these footprints. Thus, the possible role of the CCAAT box in the activation of the S promoter should be investigated further in order to reach definitive conclusions.

A distal element that regulates the S promoter is found at region –203 to –190 (Figure 4). This element binds NF-1 with a high affinity.[50] The importance of this element resides in the observation that deletion mutants of the S promoter lacking a portion of or the full NF-1-binding site are significantly less active in HeLa cells.[50] In *ayw*, removal of the –204 to –118 region that contains the NF-1-binding site resulted in a three- to fivefold reduction in promoter activity in liver cells[42] (Figure 4).

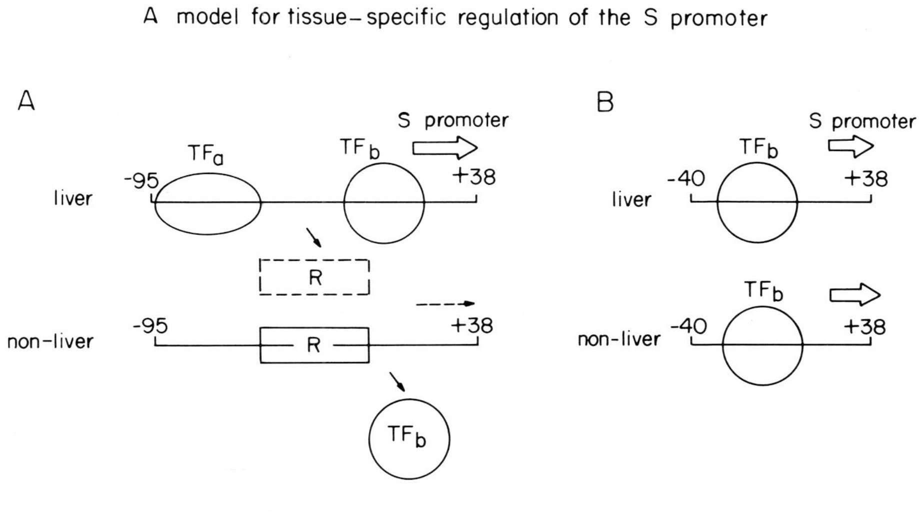

FIGURE 5. A model for the cell type-dependent activity of the S promoter. This model is based on data obtained by De-Medina et al.[29] In this model three factors are involved: TF_a, TF_b, and R. TF_a is a liver-specific transcription factor. TF_b is a ubiquitous transcription factor and R is a ubiquitous repressor. (A). The minimal promoter region is shown. In this model the repressor (R) binds the promoter only in nonliver cell lines. On binding of R, the binding of TF_b is eliminated and the promoter is inactive. In liver cells, however, TF_a and TF_b displace R and the promoter is now fully active. (B). A truncated promoter is active also in nonliver cells due to the fact that the binding site of R is deleted and as a result R cannot interfere with the binding of TF_b. Since a truncated promoter is mostly active in nonliver cells,[29] it is possible that TF_b is found in these cells in higher amounts.

A. CELL TYPE-DEPENDENT EXPRESSION OF THE S PROMOTER

The S promoter is one of the strongest HBV promoters and shows different levels of activity in a variety of cell lines from different tissues and species. However, transgenic mice that contain in their chromosomes the complete HBV genome, except for the core gene coding region, express the 2.1-kb RNA, the synthesis of which is regulated by the S promoter, specifically in liver.[51] Similar results were obtained from the analysis of transgenic mice that carry the full HBV genome.[10] In these experiments the injected DNA contains, in addition to the S promoter, other liver-specific regulatory elements, such as the enhancer, the core, and the preS promoters. Therefore, from this study it is not clear that the S promoter itself is tissue specific. A different set of transgenic mice was generated by Burke et al.[52] In this case liver-specific production of the 2.1-kb transcript was obtained in the absence of the preS and the S promoter. It was suggested that liver specificity is likely to be determined by sequences 3′ to the S mRNA start sites. These could again be the enhancer, the core promoter, or a yet unidentified liver-specific element. Data that suggest that the S promoter itself is mostly active in liver cell lines were obtained by employing the CAT reporter plasmid.[27,31,41] Progressive deletion mutants of the S promoter were generated in order to delineate the region that is responsible for the preferred expression of this promoter in liver-derived cell lines[27] (Figure 4). It was found that in nonliver cells the S promoter is under repression and that the promoter region between –95 and –40 exerts this negative effect.[27] Deletion of this region resulted in a 40- to 50-fold induction in the activity of the S promoter in nonliver cell lines (Figure 4). A model that explains the preferential expression of the S promoter in liver cells in depicted in Figure 5. This model assumes the binding of at least two positive transcription factors (TF_a and TF_b), and a repressor (R). TF_a is liver specific and binds at the region –95 to –40. TF_b is a ubiquitous factor that binds at the region –40 to +38. The repressor (R), which is found either in all the cells or only in heterologous cells, binds somewhere between TF_a and TF_b, at the region that contains the CCAAT box. On binding of TF_a and TF_b, a situation found only in liver cells, R is displaced and the promoter is active. In heterologous cells in the absence of TF_a, R displaces TF_b, and the promoter is poorly active. However, R cannot bind a promoter truncated up to –40. In this case, TF_b binds efficiently and activates the promoter in a cell type-independent manner.

B. REGULATION OF THE S PROMOTER BY THE ENHANCER

The effect of the enhancer on the activity of the S promoter was studied in two ways. In one, both the enhancer and the S promoter were cloned next to a reporter gene. These studies showed that the enhancer stimulated the activity of the S promoter by 20- to 60-fold in various hepatoma cell lines.[30,31,41,53] Some of these studies were complicated, however, by the fact that the HBV DNA fragment chosen to contain the enhancer sequence included also the adjacent X promoter. In this case it is not clear whether the transcription of the CAT gene is derived from the S or by the juxtapositioned X promoter.[30,53] In addition, at least in one case the fragment chosen to contain the sequence of the S promoter was from –22 to –269, which does not contain all the sequence of the minimal promoter region. In contrast, other studies showed that the effect of the enhancer is weak in hepatoma cell lines[41,54] and is much stronger in heterologous cell lines in which the S promoter alone is poorly active.[41]

In the second set of experiments, the effect of the enhancer on the S promoter was determined by quantitative analysis of the S mRNA and HBsAg synthesis in the presence and the absence of the enhancer.[41,42,54] These studies showed that the enhancer induces the activity of the S promoter only by two- to fourfold in hepatoma cell lines,[42,54] and again a more dramatic effect was obtained in a Sk-Hep1 cells, a nondifferentiated liver cell line in which the activity of the S promoter is very low.[41] By using an *in vitro* nuclear run-on assay, this effect was determined to be at the level of transcription.[54] This observation raises the interesting possibility that the enhancer acts to accommodate full activity of the S promoter in the absence of some of the promoter transcription factors. Thus, the role of the enhancer is to ensure an active S promoter under a wide range of physiological conditions.

V. THE X PROMOTER

Two groups have identified a promoter sequence upstream of the X ORF that probably directs the transcription of a 0.7- to 0.9-kb X gene RNA.[55,56] A viral transcript of this size was found in nonliver cell lines, such as the rat fibroblast Rat 2 cells[20] and the testes of transgenic mice that carry the whole HBV genome.[10] The 5′ ends of the X gene transcripts were mapped by primer extension analysis[55] and S1 nuclease mapping.[56] It was found that, like the S and the core promoters, the X promoter initiates transcription from multiple start sites. However, exact mapping of these heterogeneous 5′ ends has not been performed, but approximate mapping localized them between nucleotides 1255 and 1340 on the HBV genome or between –36 and –121 (the A residue of the X gene ATG considered as +1). Since the enhancer is positioned between –294 and –141, it was proposed that the X promoter boundaries are extended into the enhancer region. However, a more detailed functional analysis of the X promoter is required in order to define its sequence boundaries. Deletion of the region preceding –141 (*Sph*I site) abolished the X promoter activity,[55] suggesting that sequences upstream of –141 are important for the activity of this promoter. This region contains the enhancer sequence and therefore it was suggested that the enhancer regulates the X promoter. However, it is also possible that elements regulating the X promoter are distinct from the enhancer but are embedded within the enhancer sequence. To demonstrate that the enhancer per se regulates the X promoter, the individual essential enhancer elements were mutated by

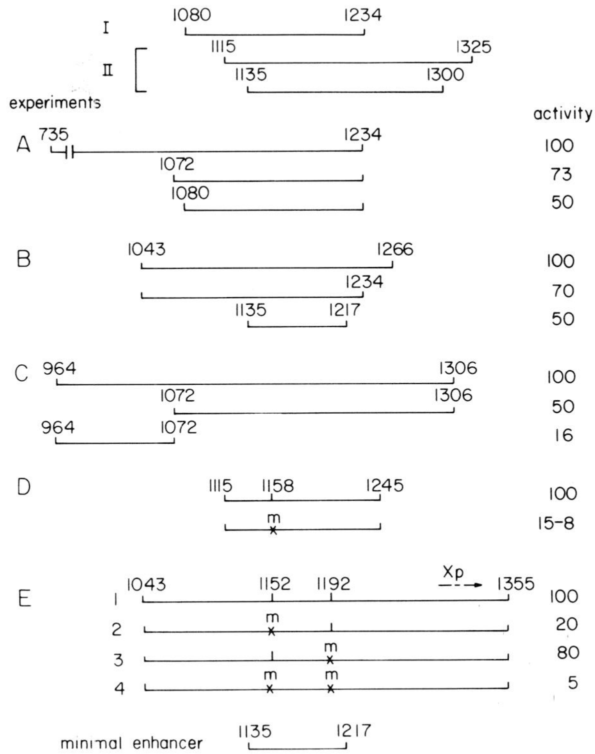

FIGURE 6. The HBV enhancer. At the top, the DNA fragments that exhibit enhancer activity as defined by Shaul et al.[18] (I) and Tognoni et al.[19] (II) are shown. This figure summarizes the data published by Shaul et al.[18] (experiment A), Ben-Levy et al.[57] (experiment B), Patel et al.[58] (experiment C), Ostapchuk et al.[59] (experiment D), and unpublished data from the author's laboratory (experiment E). In each case the enhancer activity of the largest fragment was taken as 100% activity. M, site-directed mutations.

site-directed mutagenesis (Figure 6E). It was shown that each of these enhancer elements is required for the activity of the X promoter and mutations that inactivated the enhancer concomitantly reduced the activity of the X promoter.[56a] Thus, the activity of the X promoter is tightly regulated by the enhancer.

VI. THE HBV ENHANCER

Two different experimental approaches were taken in order to define and to localize the sequence of the HBV enhancer. In one case the bacterial CAT gene was used as a reporter gene. Fragments of HBV DNA were cloned at the 5′ end of the CAT gene, the constructed plasmids were transferred into cultured cells, and enhancer and promoter activities were measured by determination of the CAT activity in the extracts of the transfected cells. These studies revealed that the HBV genome contains an enhancer that, like other known enhancers, acts in an orientation-independent manner and activates heterologous promoters.[18] The HBV enhancer was localized from nucleotides 735 to 1235 on the HBV genome. However, a shorter fragment of 154 bp (from 1080 to 1234, using the numbering system that considers the *Eco*RI site as +1) was shown to contain most of the enhancer activity.[18] The second experimental approach that was employed was called expression selection.[19] Expression selection is based on a vector containing an enhancerless SV40 T antigen gene. Random fragments of HBV

DNA were cloned in this vector and the fragment with the enhancer activity was selected by virtue of its ability to restore T antigen expression and subsequent SV40 replication. Two overlapping fragments of 166 and 214 bp were rescued. The sequence shared by these fragments contains the HBV enhancer and is mapped between nucleotides 1135 and 1300 (using the same numbering system as above; note that in the original paper the numbering system is different).[19] Interestingly, this sequence contains most of the enhancer sequence defined by the first experimental approach, suggesting that the HBV genome contains only one enhancer and that this enhancer is located within the overlapping region determined by the two different mapping procedures at nucleotides 1135 to 1235 (Figure 6).

A. THE ENHANCER BOUNDARIES

Results obtained from different investigators showed that stronger enhancer activity is obtained by larger DNA fragments, suggesting that the sequence adjacent to the enhancer plays some regulatory role.[18,57,58] However, it is clear that the core of the enhancer, in the absence of which no enhancer activity is detected, resides in a short DNA sequence. Data from different studies are summarized in Figure 6. In most of the experiments activation of the SV40 early promoter by HBV DNA fragments was determined in order to monitor enhancer activity (experiments A, C, and D, Figure 6). Similar data were obtained by employing the globin promoter and analyzing RNA levels (experiment B, Figure 6). All the fragments used contain the minimal sequence that resides between nucleotides 1135 and 1217. The 5′ border of the minimal enhancer is not much beyond nucleotide 1135, since point mutations at nucleotides 1158 to 1160 (experiment D, Figure 6) or at 1152 to 1155 (experiment E, Figure 6) sharply reduce the enhancer activity.[59] However, the influence of the sequences between nucleotides 1135 and 1152 has not been determined. The definition of the 3′ border of the enhancer seems to be much more complex due to the X promoter sequence that is located in this region. Tognoni et al.[19] suggested that the sequence from nucleotide 1234 to 1300 is important for enhancer activity. Others showed, however, that an HBV fragment missing this sequence exhibits most (50 to 70%) of the enhancer function.[18,57,59] Perhaps a better definition of the enhancer minimal core sequence comes from experiments in which a fragment of DNA from nucleotide 1043 to 1355, containing in addition to the enhancer the X promoter, was mutated at specific sequences (experiment E, Figure 6). Mutations at two sites, nucleotide 1152 and nucleotide 1192, resulted in a sharp reduction (20 fold) in the enhancer activity, localizing the 3′ border of the enhancer at or very close to nucleotide 1192. Altogether these results suggest that the minimal HBV DNA fragment containing the core enhancer sequence and capable of stimulating the activity of heterologous promoters is located between nucleotides 1135 and 1217.

B. TRANSCRIPTION FACTORS AND THE HBV ENHANCER

The mechanism by which enhancers stimulate transcription is not understood; however, it is widely documented that specific enhancer-binding proteins play a key role in this process.[1,3,60] The HBV enhancer, similar to other well-characterized viral enhancers, binds multiple proteins.[57,58,61] The sequences that bind nuclear proteins have been characterized in detail by employing the DNAse I protection assay (footprinting). A number of protected sequences were mapped outside the minimal enhancer region by Shaul and Ben-Levy.[61] These sequences are designated UE3, UE2, UE1, and NF-1c at the 5′ end and NF-1a at the 3′ end of the enhancer (Figure 7). Similar data were obtained by Patel et al.,[58] who designated the UE2, UE1, NF-1c, and NF-1a homolog sites VIII, VI, IV, and I, respectively. In addition, two new protected sites were seen, one between the UE1 and UE2 sites and designated VII, and the second designated V.

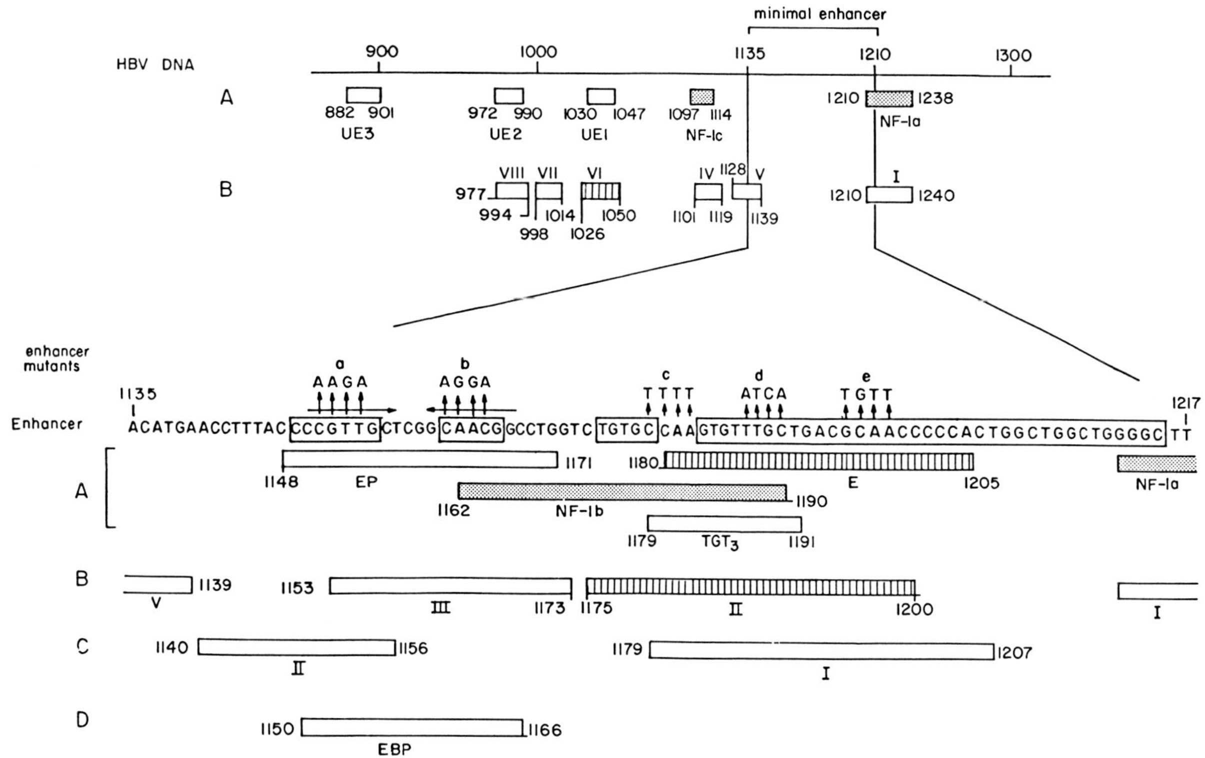

FIGURE 7. Footprint map of the HBV enhancer. Data are summarized from Shaul and Ben-Levy[61] and Ben-Levy et al.[57] (A), Patel et al.[58] (B), Chou et al.[62] (C), and Karpen et al.[14] (D). The footprinted regions are shown by boxes. Dotted boxes represent the binding sites of NF-I and the striped boxes the binding site of C/EBP. In the middle the sequence of the minimal enhancer is shown. The boxed sequences represent the regions conserved in the genome of most of the members of the hepadnavirus family except DHBV. Also shown are the sequences of several enhancer mutants, marked from a to e, used to define the function of the protein-binding sites in enhancer activity.[68]

The footprint pattern of the minimal enhancer sequence (nucleotides 1135 to 1217) is very complex. Two major protected regions contained within the minimal enhancer sequence were seen, one mapped at nucleotides 1148 to 1171 and the second at nucleotides 1180 to 1205 and designated EP and E, respectively.[57] Similar data were obtained by others but with slight differences in determination of the exact protected nucleotides. Footprint EP overlaps with and is perhaps identical to footprint III of Patel et al.,[58] enhancer-binding protein (EBP) site of Karpen et al.,[13] and EF-C-binding site of Ostapchuk et al.[59] However, the footprint II observed at this region by Chou et al.[62] is not similar to the EP and may represent the binding of a second protein whose binding partially overlaps that of the EP site. The second important protected region within the minimal enhancer element is the E site, which is very similar to footprint II of Patel et al.[58] and footprint I of Chou et al.[62] Two additional footprints were detected in that region and designated NF-1b and TGT_3 (A, Figure 7). These latter footprints were detected only after using either purified proteins or specific mutants of the enhancer as probes.[57]

Despite the thorough characterization of enhancer sequences that bind proteins, little is known about the origin of these proteins. It was reported that the proteins that bind to the footprints UE1 and E are heat-stable.[61] Recently a heat stable transcription factor was purified from rat liver that binds to the UE1 and E sites.[63,64] This protein, named C/EBP, is found only in a limited number of tissues and is very abundant in the liver.[65] Thus, it is very likely that the protein that binds to UE1 and E is C/EBP. Indeed, when bacterially expressed C/EBP was used for footprinting analysis a footprint was obtained that is indistinguishable from the E footprint. Interestingly, when a larger amount of this protein was used, several additional sites within the minimal enhancer sequence were obtained, the biological significance of which is not known.[66] Recently it was also reported that the transcription factor AP-1 and related proteins (the Jun proteins) bind to the E footprints.[57,67]

A second transcription factor that binds the HBV enhancer was identified to be NF-1 (the same factor that binds the distal element of the S promoter; see Section IV). Purified NF-1 was shown to bind to three sites at the enhancer region. These sites are NF-1a, NF-1b, and NF-1c (Figure 7). However, since multiple proteins have been identified that recognize and bind to the NF-1 consensus motif[49] it is very likely that these sites bind some other proteins as well.

A third transcription factor binds the EP footprint. An inverted repeat is located within the EP footprint that is similar to that found within the polyoma virus enhancer. The nuclear factor that binds the EP footprint was termed EP protein[57] and the one that binds a similar sequence within the polyoma enhancer was named EF-C.[57,59] It is very likely that EP and EF-C are the same protein; however, purification of these proteins is required to confirm this possibility.

Some, but not all, of the footprinted sequences are functionally important. The assumption is that the enhancer-binding proteins are responsible for the activity of the enhancer. One way to confirm this point is to determine the activity of modified enhancers in which the sequences of the footprinted regions were mutated so that they lose the capacity to bind the corresponding proteins. Using this approach the role of the EP (EF-C) footprint was extensively analyzed.[59] Mutations that changed the length of the spacer region between the arms of the inverted repeat were introduced and analyzed. Introduction of 1 or 2 bp between the repeats did not affect EP (EF-C) binding, but deletion of 1 bp or insertion of 3 to 9 bp reduced binding as well as enhancer activity dramatically. These results suggest that EP (EF-C) binding is stabilized by dimerization, that each unit binds one of the repeats, and that this footprint site is a functional component of the enhancer.[59] Dikstein et al.[68] have reached similar conclusions by mutating each of the inverted repeats (Figure 7, mutants a and b). These enhancer mutants do not bind the EP (EF-C) protein and show very low enhancer activity. Mutation of the E footprint (mutants d and e) resulted in a three- to sixfold reduction in several analyzed cell lines, suggesting that this region is essential for the enhancer activity. The third set of enhancer mutants carries mutations at the NF-1 related motifs. Mutants of the NF-1c and NF-1b sites display almost a wild-type activity. In contrast, mutations at the NF-1a site reduced the activity to 40 to 60%.[68] The observation that mutations in several distinct regions resulted in a 3- to 20-fold reduction in enhancer activity suggests that the HBV enhancer is composed of multiple functional elements that act synergistically to stimulate transcription. This conclusion was further supported by the abolished activity of an enhancer mutant that contains two point mutations, one at the E and the second at the EP footprints.[68] Thus, these footprints represent essential enhancer elements. Based on sequence similarity it appears that a similar structure of enhancer is found within the genome of the other members of hepadnavirus family.[57]

C. THE ROLE OF THE ENHANCER ELEMENTS E AND EP

The minimal enhancer contains two functional elements, the EP and the E element. The function of each of these elements was analyzed separately by preparing synthetic enhancers comprising multimers of either the E- or the EP-binding sites. It was found that multimers (four to six copies) of the E but not of the EP element act as a strong enhancer,[67,68] suggesting that these elements have distinct functions. The activity of the E element is regulated by several transcription factors, including the C/EBP and Jun proteins. Furthermore, the viral trans-activators, X of HBV and Tax of HTLV-1, induce the enhancer activity of the E element.[69] The activity of the E element is also regulated by extracellular stimuli. Both TPA and cAMP, each known to activate a distinct signal transduction pathway, stimulate the activity of the E element.[57,67] E, therefore, is the major HBV enhancer cis-acting element through which multiple cellular and viral transcriptional regulators act. As such, this element plays the role of the enhancer modulator. The capacity of the enhancer to respond to various signals has a strong implication for the life cycle of the virus, which depends on the proliferative potential of its host. The role of the enhancer modulator may be to ensure an optimal activity of the enhancer under different physiological conditions.

In contrast to E, the role of the EP element is obscure. EP, despite being an essential enhancer element, has no intrinsic enhancer activity and cannot stimulate transcription on its own. The data suggest that the interaction of the EP element with a juxtapositioned second enhancer element is essential for its activity. We have proposed that the EP-binding protein acts to organize different enhancer elements into a functional unit and acts as an enhancer coordinator. Conceivably, the role of an enhancer coordinator is to generate protein–DNA secondary complexes that seem to be required for efficient activation of transcription.[68]

D. HBV ENHANCER AND LIVER TROPISM

The hepatotropism of hepadnaviruses extends beyond the ability of the virus to attach to and penetrate its host cell. As discussed above, the complex enhancer-core promoter is preferentially active in liver cells.[6] It was previously proposed that this liver tropism is attributed at least in part to the enhancer. This was determined by the ability of the enhancer to confer tissue-specific expression upon a heterologous, otherwise ubiquitous promoter.[6,18] This observation was confirmed by some[58,70,71] but not by others.[72,73] This controversy could be resolved in part by the information we gained on the nature of the transcription factors that interact with the enhancer in general and the E element in particular. This element binds with very high affinity the transcription factor C/EBP, which is a tissue-restricted factor that is found mostly in tissues where metabolism of lipids is an important part of their normal physiology, such as liver and fat cells.[65] Considering the important role that the E element plays in enhancer function, this interaction is most probably of functional significance and determines the preferred activity of the enhancer in liver cells. Unexpectedly, this element also binds with low affinity the ubiquitous Jun proteins,[67] which are strong transcriptional activators. Furthermore, a plasmid that directs the synthesis of v-Jun activates the HBV enhancer. Thus, in a nonliver environment (i.e., in the absence of the C/EBP), the partial activity of the enhancer is, most likely, attributed to the Jun proteins. The EP-binding site, the second important enhancer element, binds a ubiquitous factor[59] and therefore seems not to play an important role in determination of the liver tropism of the enhancer.

E. REGULATION OF THE HBV ENHANCER BY HORMONES

Hormones are known to induce transcription of cellular and viral genes. One of the best described hormone-dependent viral systems is the mouse mammary tumor virus (MMTV), in which a hormone-receptor complex binds to a glucocorticoid-responsive element (GRE), which can then up-regulate expression of both homologous and heterologous genes. The sex-dependent frequency of the HBV chronic carrier state raised the possibility of the involvement of hormonal regulation in viral expression. It was found that glucocorticoids stimulate HBV enhancer activity through a GRE that is located at nucleotide position 341 to 370 in the ORF of the S gene.[71,74] This element lacks autonomous enhancer activity and acts only in the presence of the HBV enhancer.[74] The HBV GRE contains the sequence that is shared by nearly all of the characterized GREs and binds purified glucocorticoid receptor.[71,74]

F. ENHANCER AND RNA STABILITY

The hepadnavirus enhancer is unusual in that it is located in a region of the viral genome that is not only transcribed but also translated as part of the reverse transcriptase polypeptide. This unusual position of the enhancer raised the possibility that the enhancer may have some other roles such as regulating the stability of transcripts within which the enhancer sequence is found. Vannice and Levinson[73] tested this possibility by inserting the enhancer in a transcribed region of chimeric reporter genes. They found that the HBV enhancer dramatically increased expression levels of reporter genes controlled by the SV40 promoter-enhancer only when the enhancer was located in the transcribed region but not outside of it. These data raised the possibility, which has not been experimentally tested, that the enhancer sequence may be involved in increasing the stability of the HBV transcripts. This enhancer property may explain the difficulty of detecting the 0.9-kb viral mRNA. Unlike the other viral transcripts this transcript lacks the enhancer sequence and therefore may be unstable.

VII. FUTURE DIRECTIONS

Data accumulated during the last decade on regulation of gene expression of the hepadnaviruses suggest that these viruses rely mostly on transcriptional regulation to obtain differential gene expression. Thus, understanding of transcriptional regulation is one of the key steps in exploring the virus life cycle and its interactions with its host. We have good knowledge of the transcription initiation sites but our understanding of the promoters that control the machinery responsible for the initiation of transcription is very limited. Most of the viral promoters are unusual among RNA polymerase II promoters in that the TATA box homology is a poor fit to the consensus TATA box sequence. Does it suggest that the virus uses a novel type of TATA box-binding factor (TFIID)? Now that the TFIID gene of yeast has been cloned, we are in a position to explore this important question. Also, several transcription factor genes have been cloned, some of which may be involved in the regulation of the genes in this viral family. How and when they act to regulate the expression of the viral genes are important questions that may now be addressed. How does the enhancer work? Is the enhancer important at all for the viral life cycle? Does the enhancer stimulate the activity of all the viral promoters simultaneously or in a developmentally regulated manner? When the molecular mechanisms underlying transcription are better understood, we will be in a better position to answer these many interesting questions raised by the study of hepadnaviruses.

REFERENCES

1. **Dynan, W. S. and Tjian, R.**, Control of eukaryotic messenger RNA synthesis by sequence-specific DNA-binding proteins, *Nature (London)*, 316, 774, 1985.
2. **Ptashne, M.**, How eukaryotic transcriptional activators work, *Nature (London)*, 335, 683, 1988.
3. **Mitchell, P. J. and Tjian, R.**, Transcriptional regulation in mammalian cells by sequence-specific DNA binding proteins, *Science*, 245, 371, 1989.
4. **Rall, L. B., Standring, D. N., Laub, O., and Rutter, W. J.**, Transcription of hepatitis B virus by RNA polymerase II, *Mol. Cell. Biol.*, 3, 1766, 1983.
5. **Roossinck, M. J., Jameel, S., Loukin, S. H., and Siddiqui, A.**, Expression of hepatitis B viral core region in mammalian cells, *Mol. Cell. Biol.*, 6, 1393, 1986.
6. **Honigwachs, J., Faktor, O., Dikstein, R., Shaul, Y., and Laub, O.**, Liver-specific expression of hepatitis B virus is determined by the combined action of the core gene promoter and the enhancer, *J. Virol.*, 63, 919, 1989.
7. **Will, H., Reiser, W., Weimer, T., Pfaff, E., Büscher, M., Sprengel, R., Cattaneo, R., and Schaller, H.**, Replication strategy of human hepatitis B virus, *J. Virol.*, 61, 904, 1987.
8. **Yaginuma, K., Shirakata, Y., Kobayashi, M., and Koike, K.**, Hepatitis B virus (HBV) particles are produced in a cell culture system by transient expression of transfected HBV DNA, *Proc. Natl. Acad. Sci. U.S.A.*, 84, 2678, 1987.
9. **Sells, M. A., Zelent, A. Z., Shvartsman, M., and Acs, G.**, Replicative intermediates of hepatitis B virus in HepG2 cells that produce infectious virions, *J. Virol.*, 62, 2836, 1988.
10. **Araki, K., Miyazaki, J. I., Hino, O., Tonita, N., Chisaka, O., Matsubara, K., and Yamamura, K. I.**, Expression and replication of hepatitis B virus genome in transgenic mice, *Proc. Natl. Acad. Sci. U.S.A.*, 86, 207, 1989.
11. **Yaginuma, K. and Koike, K.**, Identification of a promoter region for 3.6 kilobase mRNA of hepatitis B virus and specific cellular binding protein, *J. Virol.*, 63, 2914, 1989.
12. **Yee, J. K.**, A liver-specific enhancer in the core promoter region of human hepatitis B virus, *Science*, 246, 658, 1989.
13. **Karpen, S., Banergee, R., Zelent, A., Price, P., and Acs, G.**, Identification of protein-binding sites in the hepatitis B virus enhancer and core promoter domains, *Mol. Cell. Biol.*, 8, 5159, 1988.
14. **Monaci, P., Nicosia, A., and Cortese, R.**, Two different liver-specific factors stimulate *in vitro* transcription from the human α-1-antitrypsin promoter, *EMBO J.*, 7, 2075, 1988.
15. **Lichtsteiner, S., Wuarin, J., and Schibler, U.**, The interplay of DNA-binding proteins on the promoter of the mouse albumin gene, *Cell*, 51, 963, 1987.
16. **Friedman, A. D., Landschulz, W. H., and McKnight, S.L.**, CCAAT/enhancer binding protein activates the promoter of the serum albumin gene in cultured hepatoma cells, *Genes Dev.*, 3, 1314, 1989.
17. **Scott, R. W. and Tilghman, S. M.**, Transient expression of a mouse α-fetoprotein minigene: deletion analysis of promoter function, *Mol. Cell. Biol.*, 3, 1395, 1983.
18. **Shaul, Y., Rutter, W. J., and Laub, O.**, A human hepatitis B viral enhancer element, *EMBO J.*, 4, 427, 1985.
19. **Tognoni, A., Cattaneo, R., Serfling, E., and Schaffner, W.**, A novel expression selection approach allows precise mapping of the hepatitis B virus enhancer, *Nucleic Acids Res.*, 13, 7457, 1985.
20. **Gouth, N. M.**, Core and E antigen synthesis in rodent cells transformed with hepatitis B virus DNA is associated with greater than genome length viral messenger RNAs, *J. Mol. Biol.*, 165, 683, 1983.
21. **Zelent, A. Z., Sells, M. A., Price, P. M., Mehamad, A., Acs, G., and Christman, J. K.**, Murine cells carrying integrated tandem genomes of hepatitis B virus DNA transcribe RNAs from endogenous promoters on both viral strands and express middle and major viral envelope proteins, *J. Virol.*, 61, 1108, 1987.
22. **Farza, H., Hadchouel, M., Scotto, J., Tiollais, P., Babinet, C., and Pourcel, C.**, Replication and gene expression of hepatitis B virus in a transgenic mouse that contains the complete viral genome, *J. Virol.*, 62, 4144, 1988.
23. **Ganem, D. and Varmus, E.**, The molecular biology of the hepatitis B viruses, *Annu. Rev. Biochem.*, 56, 651, 1987.
24. **Junker, M., Gall, P., and Schaller, H.**, Expression and replication of the hepatitis B virus genome under foreign promoter control, *Nucleic Acids Res.*, 15, 10117, 1987.
25. **Seeger, C., Baldwin, B., and Tennant, B. C.**, Expression of infectious woodchuck hepatitis virus in murine and avian fibroblasts, *J. Virol.*, 63, 4665, 1989.
26. **Laub, O., Rall, L. B., Truett, M., Shaul, Y., Standring, D. N., Valenzuela, P., and Rutter, W. J.**, Synthesis of hepatitis B surface antigen in mammalian cells: expression of the entire gene and the coding region, *J. Virol.*, 48, 271, 1983.
27. **De-Medina, T., Faktor, O., and Shaul, Y.**, The S promoter of hepatitis B virus is regulated by positive and negative elements, *Mol. Cell. Biol.*, 8, 2449, 1988.

28. **Siddiqui, A., Shahid, J., and Mapoles, J.,** Transcriptional control elements of hepatitis B surface antigen gene, *Proc. Natl. Acad. Sci. U.S.A.*, 83, 566, 1986.
29. **Malpiece, Y., Michel, M. L., Carloni, G., Revel, M., Tiollais, P., and Weissenbach, J.,** The gene S promoter of hepatitis B virus confers constitutive gene expression, *Nucleic Acids Res.*, 11, 4645, 1983.
30. **Antonucci, T. K. and Rutter, W. J.,** Hepatitis B virus (HBV) promoters are regulated by the HBV enhancer in a tissue-specific manner, *J. Virol.*, 63, 579, 1989.
31. **Chang, H.-K., and Ting, L.-P.,** The surface gene promoter of the human hepatitis B virus displays a preference for differentiated hepatocytes, *Virology*, 170, 176, 1989.
32. **Bulla, G. A. and Siddiqui, A.,** Negative regulation of the hepatitis B virus pre-S promoter by internal DNA sequences, *Virology*, 170, 251, 1989.
33. **Ou, J. and Rutter, W. J.,** Hybrid hepatitis B virus-host transcripts in a human hepatoma cell, *Proc. Natl. Acad. Sci. U.S.A.*, 82, 83, 1985.
34. **Sureau, C., Romet-Lemonne, J.-L., Mullins, J. I., and Essex, M.,** Production of hepatitis B virus by a differentiated human hepatoma cell line after transfection with cloned circular HBV DNA, *Cell*, 47, 37, 1986.
35. **Courtois, G., Morgan, J. G., Campbell, L. A., Fourel, G., and Crabtree, G. R.,** Interactions of a liver-specific nuclear factor with the fibrinogen and α-1-antitrypsin promoter, *Science*, 238, 688, 1987.
36. **Hardon, E., Farin, M., Paonessa, G., and Cortese, R.,** Two distinct factors interact with the promoter regions of several liver-specific genes, *EMBO J.*, 7, 1711, 1988.
37. **Courtois, G., Baumhueter, S., and Crabtree, G. R.,** Purified hepatocyte nuclear factor 1 interacts with a family of hepatocyte-specific promoters, *Proc. Natl. Acad. Sci. U.S.A.*, 85, 7937, 1988.
38. **Chang, H.-K., Wang, B.-Y., Yuh, C.-H., WSei, C.-L., and Ting, L.-P.,** A liver-specific nuclear factor interacts with the promoter region of the large surface protein gene of human hepatitis B virus, *Mol. Cell. Biol.*, 9, 5189, 1989.
39. **De-Medina, T. and Shaul, Y.,** unpublished observations, 1990.
39a. **Standring, D. N., Rutter, W. J., Varmus, H. E., and Ganem, D.,** Transcription of the hepatitis B surface antigen gene in cultured murine cells initiates within the presurface region, *J. Virol.*, 50, 563, 1984.
40. **Cattaneo, R., Will, H., Hernandez, N., and Schaller, H.,** Signals regulating hepatitis B surface antigen transcription, *Nature (London)*, 305, 336, 1983.
41. **Faktor, O., De-Medina, T., and Shaul, Y.,** Regulation of hepatitis B virus S gene promoter in transfected cell lines, *Virology*, 162, 362, 1988.
42. **Raney, A. K., Milich, D. R., and McLachlan, A.,** Characterization of hepatitis B virus major surface antigen gene transcriptional regulatory elements in differentiated hepatoma cell lines, *J. Virol.*, 63, 3919, 1989.
43. **Zaychowski, D. A., Bouef, A. H., and Kedinger, C.,** The adenovirus-2 early EIIa transcription unit possesses two overlapping promoters with different sequence requirements for EIa dependent stimulation, *EMBO J.*, 4, 1293, 1985.
44. **Smale, S. T. and Baltimore, D.,** The "initiator" as a transcription control element, *Cell*, 57, 103, 1989.
45. **Sawadogo, M. and Roeder, R. G.,** Interaction of a gene specific transcription factor with the adenovirus major late promoter upstream of the TATA box region, *Cell*, 43, 165.
46. **McKnight, S. and Tjian, R.,** Transcriptional selectivity of viral genes in mammalian cells, *Cell*, 46, 795, 1986.
47. **Myers, R. M., Tilly, K., and Maniatis, T.,** Fine structure genetic analysis of a b-globin promoter, *Science*, 232, 613, 1986.
48. **Jones, K. A., Kadonaga, J. T., Rosenfeld, P. J., Kelly, T. J., and Tjian, R.,** A cellular DNA-binding protein that activates eukaryotic transcription and DNA replication, *Cell*, 48, 79, 1987.
49. **Dorn, A., Bollekens, J., Staub, A., Benopist, C., and Mathis, D.,** A multiplicity of CCAAT box-binding proteins, *Cell*, 50, 863, 1987.
50. **Shaul, Y., Ben-Levy, R., and De-Medina, T.,** High affinity binding site for nuclear factor 1 next to the hepatitis B virus S gene promoter, *EMBO J.*, 5, 1967, 1986.
51. **Babinet, C., Farza, H., Morello, D., Hadchouel, M., and Pourcel, C.,** Specific expression of hepatitis B surface antigen (HBsAg) in transgenic mice, *Science*, 230, 1160, 1985.
52. **Burk, R. D., Deloia, J. A., Eawady, M. K., and Gearhart, J. D.,** Tissue preferential expression of the hepatitis B virus (HBV) surface antigen gene in two lines of HBV transgenic mice, *J. Virol.*, 62, 649, 1988.
53. **Chang, H.-K., Chou, C.-K, Chang, C., Su, T.-S., Hu, C.-P., Yoshida, M., and Ting, L.-P.,** The enhancer sequence of human hepatitis B virus can enhance the activity of its surface gene promoter, *Nucleic Acids Res.*, 15, 2261, 1987.
54. **Bulla, G. A. and Siddiqui, A.,** The hepatitis B virus enhancer modulates transcription of the hepatitis B virus surface antigen gene from an internal location, *J. Virol.*, 62, 1437, 1988.
55. **Treinin, M. and Laub, O.,** Identification of a promoter element located upstream from the hepatitis B virus X gene, *Mol. Cell. Biol.*, 7, 545, 1987.
56. **Siddiqui, A., Jameel, S., and Mapoles, J.,** Expression of the hepatitis B virus X gene in mammalian cells, *Proc. Natl. Acad. Sci. U.S.A.*, 84, 2513, 1987.

56a. **Shaul, Y.**, unpublished observation, 1990.
57. **Ben-Levy, R., Faktor, O., Berger, I., and Shaul, Y.**, Cellular factors that interact with the hepatitis B virus enhancer, *Mol. Cell. Biol.*, 9, 1804, 1989.
58. **Patel, N. U., Jameel, S., Isom, H., and Siddiqui, A.**, Interactions between nuclear factors and the hepatitis B virus enhancer, *J. Virol.*, 63, 5293, 1989.
59. **Ostapchuk, P., Scheirle, G., and Hearing, P.**, Binding of nuclear factor EFC to a functional domain of the hepatitis B virus enhancer region, *Mol. Cell. Biol.*, 9, 2787, 1989.
60. **Dynan, W. S.**, Modulating in promoters and enhancers, *Cell*, 58, 1, 1989.
61. **Shaul, Y. and Ben-Levy, R.**, Multiple nuclear proteins in liver cells are bound to hepatitis B virus enhancer element and its upstream sequences, *EMBO J.*, 6, 1913, 1987.
62. **Chou, H.-W., Harrel, D., Forough, R., and Watabe, K.**, Binding of tissue-specific factors to the enhancer sequence of hepatitis B virus, *FEBS Lett.*, 229, 349, 1988.
63. **Johnson, P. F., Landschulz, W. H., Graves, B. J., and McKnight, S. L.**, Identification of a rat liver nuclear protein that binds to the enhancer core element of three animal viruses, *Genes Dev.*, 1, 133, 1987.
64. **Landschulz, W. H., Johnson, P. F., Adashi, E. Y., Graves, B. J., and McKnight, S. L.**, Isolation of a recombinant copy of the gene encoding C/EBP, *Genes Dev.*, 2, 786, 1988.
65. **Birkenmeir, E. H., Gwynn, B., Howard, S., Jerry, J., Gordlon, J. I., Landschulz, W. H., and McKnight, S. L.**, Tissue-specific expression, developmental regulation, and genetic mapping of the gene encoding CCAAT/enhancer binding protein, *Genes Dev.*, 3, 1146, 1989.
66. **Dikstein, R. and Shaul Y.**, Hierarchic and cooperative binding of the rat liver nuclear protein C/EBP at the hepatitis B virus enhancer, *Mol. Cell. Biol.*, 10, 4427, 1990.
67. **Faktor, O., Budlovsky, S., Ben-Levy, R., and Shaul, Y.**, Single element within the hepatitis B virus enhancer binds multiple proteins and responds to multiple stimuli, *J. Virol.*, 64, 1861, 1990.
68. **Dikstein, R., Faktor, O., Ben-Levy, R., and Shaul, Y.**, Functional organization of the hepatitis B virus enhancer, *Mol. Cell. Biol.*, 10, 3683, 1990.
69. **Faktor, O. and Shaul, Y.**, The identification of hepatitis B virus X gene responsive elements reveals functional similarity of X and HTLV-I box, *Oncogene*, 5, 867, 1990.
70. **Jameel, S. and Siddiqui, A.**, The human hepatitis B virus enhancer requires trans-acting cellular factor(s) for activity, *Mol. Cell. Biol.*, 6, 1710, 1986.
71. **Tur-Kaspa, R., Burk, R. D., Shaul, Y., and Shafritz, D. A.**, Hepatitis B virus DNA contains a glucocorticoid-responsive element, *Proc. Natl. Acad. Sci. U.S.A.*, 83, 1627, 1986.
72. **Elfassi, E.**, Broad specificity of the hepatitis B enhancer function, *Virology*, 160, 259, 1987.
73. **Vannice, J. L. and Levinson, A. D.**, Properties of the human hepatitis B virus enhancer: position effects and cell-type nonspecificity, *J. Virol.*, 62, 1305, 1988.
74. **Tur-Kaspa, R., Shaul, Y., Moore, D. D., Burk, R. D., Ikret, S., Poellinger, L., and Shafritz, D. A.**, The glucocorticoid receptor recognizes a specific nucleotide sequence in hepatitis B virus DNA causing increased activity of the HBV enhancer, *Virology*, 167, 630, 1988.

Chapter 11

HEPADNAVIRUS REPLICATION

Christoph Seeger

TABLE OF CONTENTS

I. INTRODUCTION

Hepadnaviruses have been isolated from humans, woodchucks, ground squirrels, pekin ducks, and herons.[1,2] Hepadnaviruses are species specific and replicate primarily in the liver of the infected animal. Depending on the host, secondary sites for virus replication can include the pancreas, kidney, spleen, and peripheral blood lymphocytes.[3-7] Hepadnaviruses are transmitted horizontally by contaminated blood, and vertically by congenital infection of progeny. Infection can be self-limited or chronic. The former is characterized by the production of virus-neutralizing antibodies, the latter by the continuous shedding of infectious virus into the blood.[1,8]

The discovery of a DNA polymerase activity in virions purified from human hepatitis B virus (HBV)-infected individuals occurred 3 years after the characterization of a reverse transcriptase activity in murine and avian retroviruses.[9-11] However, it took almost another decade until the concept of hepadnavirus replication via reverse transcription was unraveled.[12] The DNA genome of hepadnaviruses helped to disguise the mechanism for genome replication. The deficiency of the viral DNA polymerase to accept exogenous RNA or DNA templates rendered the polymerase inaccessible to a biochemical analysis. Finally, the failure to grow HBV in cultured cells and to infect animals other than the chimpanzee severely limited the source of infected material necessary for a molecular analysis.

The discovery of the HBV-like viruses, woodchuck hepatitis virus (WHV), ground squirrel hepatitis virus (GSHV), and duck hepatitis B virus (DHBV), energized a rapid investigation of the hepadnavirus replication cycle.[13-15] The mystery of hepadnavirus replication was uncovered with the DHBV system. It allowed the isolation of replication complexes, from infected livers, that were competent for the synthesis of viral DNA from an RNA template.[12,16] The recent development of tissue culture systems for the propagation of hepadnaviruses marked a second crucial step for hepadnavirus research.[17-21] It is now possible to perform detailed genetic analyses of the functions encoded by the different viral genomes.

Employing reverse transcription as a mechanism for genome replication, hepadnaviruses are members of the growing group of retroid elements. Together with caulimoviruses (Chapter 14, this volume), they are the only DNA viruses of this superfamily characterized so far. Like caulimoviruses, but in contrast to retroviruses, hepadnavirus replication does not require the integration of viral DNA into the host genome.[2] Finally (and as will be discussed in detail in this chapter), reverse transcription of hepadnavirus RNA passes through several steps that are unprecedented among retroid elements. The most prominent among them are the covalent coupling between protein and DNA and the translocation of an RNA oligomer essential for the priming of plus-strand DNA synthesis.[22-24]

II. VIRION DNA

Sera of HBV-infected individuals contain three distinct types of lipid containing virus-specific particles: spheres of apparent icosahedral symmetry, 42 and 22 nm in diameter, and rodlike structures, several hundred nanometers in length.[25] All forms react with antibodies directed against the viral surface antigen, but only the 42-nm particle, termed the Dane particle, bears an inner core body that harbors the viral DNA genome. The Dane particle represents the infectious hepadnavirus. It contains a virus-encoded DNA polymerase activity that can extend the incomplete DNA strand in virions in the presence of dNTPs and magnesium.[11]

The hepadnavirus genome is a relaxed circular, partially double-stranded DNA molecule, approximately 3.2 kbp in length.[26,27] It is composed of a complete minus-strand DNA base paired with an incomplete plus-strand DNA of variable length.[27] The 5′ ends of minus- and plus-strand DNA are separated by approximately 220 nucleotides in the mammalian viruses and 50 nucleotides in the avian viruses.[27-30] They secure the cohesive conformation of the hepadnavirus genome.

Incubation of virions with pronase prior to extraction with phenol significantly increased the efficiency with which HBV DNA was purified from concentrated virus particles, suggesting that protein was tightly associated with viral DNA.[11,22] This notion was substantiated by demonstrating that the DNA–protein complex was resistant to heat treatment in the presence of detergent as well as to treatment with strong alkali. The failure of polynucleotide kinase to phosphorylate the minus-strand DNA following incubation with phosphatase indicated that the protein was attached to the 5′ end of minus-strand DNA.[22] Ensuing the findings reported for HBV, protein was identified at the 5′ end of the minus-strand DNA of DHBV and GSHV.[29,31,32]

III. REPLICATION VIA REVERSE TRANSCRIPTION

The pivotal results dismantling the mechanism for the synthesis of virion DNA isolated from Dane particles emerged from the analysis of virus-specific DNA species present in infected livers. Southern blot analysis of liver DNA from HBV-infected individuals revealed covalently closed, circular (ccc) DNA species that could be converted into relaxed circular and 3.2-kbp long linear DNA species following treatment with S1 nuclease.[33] In addition to cccDNA, shorter DNA species were observed that were heterogeneous in size.[34] Mason et al. identified these heterogeneous DNA forms in DNA extracted from livers of DHBV-infected ducks and demonstrated that they consisted predominantly of complete and nascent single-stranded minus-strand DNA (ssDNA) species.[16] The 5′ ends of the nascent minus-strand DNA are covalently linked to protein.[32] Just as in the DHBV system, cccDNA and heterogeneous DNA forms linked to protein accumulate in GSHV- and HBV-infected livers, implying that avian and mammalian hepadnaviruses use similar strategies for the replication of their genomes.[31,35,36]

The unequal accumulation of viral minus- and plus-strand DNA species was incompatible with a model for semiconservative replication of double-stranded DNA, as exemplified by the papovavirus or adenovirus groups. It suggested that the synthesis of minus-strand DNA was independent from the synthesis of plus-strand DNA. Summers and Mason[12] uncovered the mechanism responsible for these observations by demonstrating that the template for minus-strand DNA synthesis is RNA and not DNA and that hepadnaviruses thus employ reverse transcriptase for genome multiplication. Using replication complexes present in the cytoplasm of infected duck livers, the authors substantiated their claim by demonstrating that (1) plus-strand DNA but not minus-strand DNA synthesis was inhibited by actinomycin D, implicating RNA as the template for minus-strand synthesis; (2) nascent minus-strand DNA was base paired with RNA; and (3) the length of the RNA template decreased with increasing length of minus strands, inferring that RNA is degraded following DNA synthesis.

Based on the structural analyses of virion DNA purified from serum and on replicative intermediates present in infected livers, the following model for hepadnavirus replication was proposed. On infection, relaxed circular virion DNA is converted into cccDNA. cccDNA directs the synthesis of viral mRNAs and the synthesis of the template for reverse transcription, termed pregenomic RNA. Pregenomic RNA is packaged into core particles and subsequently copied into minus-strand DNA, followed by plus-strand DNA synthesis. The final product of this reaction is relaxed circular virion DNA (Figure 1).

To substantiate individual steps of this model and to formulate a more detailed model for

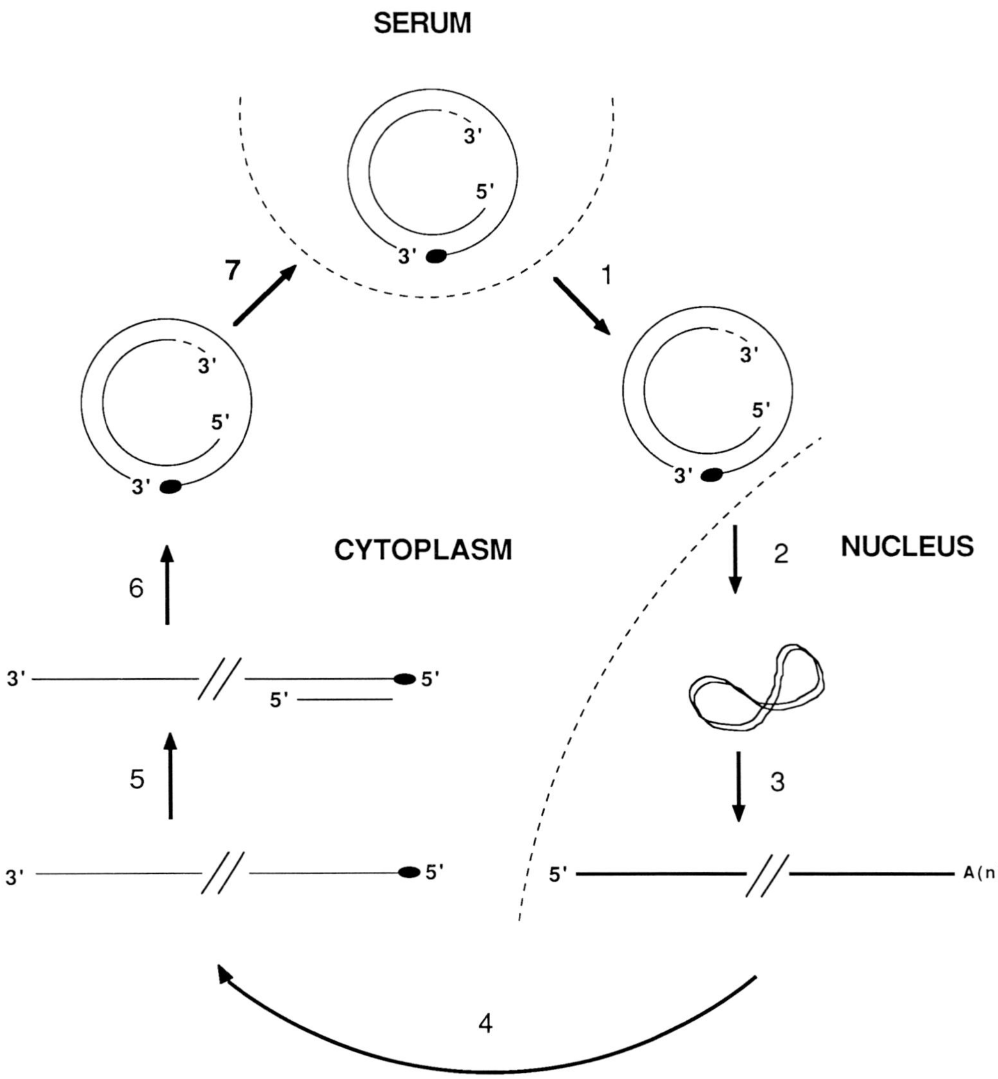

FIGURE 1. The replication cyle of hepadnavirus genomes. After infection (1) virion DNA is converted into covalently closed circular (ccc) DNA (2). cccDNA is transcribed into pregenomic RNA in the nucleus of the infected cell (3). Pregenomic RNA is packaged into cytoplasmic core particles and reverse transcribed into minus-strand DNA (4), which subsequently serves as the template for plus-strand DNA synthesis (5 and 6). Virion DNAs containing core particles assemble with surface antigen and exit the cell (7).

the mechanism by which hepadnaviruses synthesize DNA from pregenomic RNA, it was essential to determine the positions of the exact 5′ and 3′ ends of pregenomic RNA and of virion DNA. This undertaking was greatly facilitated by the availability of the nucleotide sequences of HBV and the animal models WHV, GSHV, and DHBV.[37-41] Landmarks for replication on the 3- to 3.3-kbp long hepadnavirus genomes are direct repeats, termed DR1 and DR2. They are 10 to 12 nucleotides in length and separated by approximately 220 nucleotides in the mammalian hepadnaviruses and by 50 nucleotides in the avian hepadnaviruses. As will be discussed in detail below, DR1 and DR2 are the sites for initiation of minus- and plus-strand DNA synthesis.

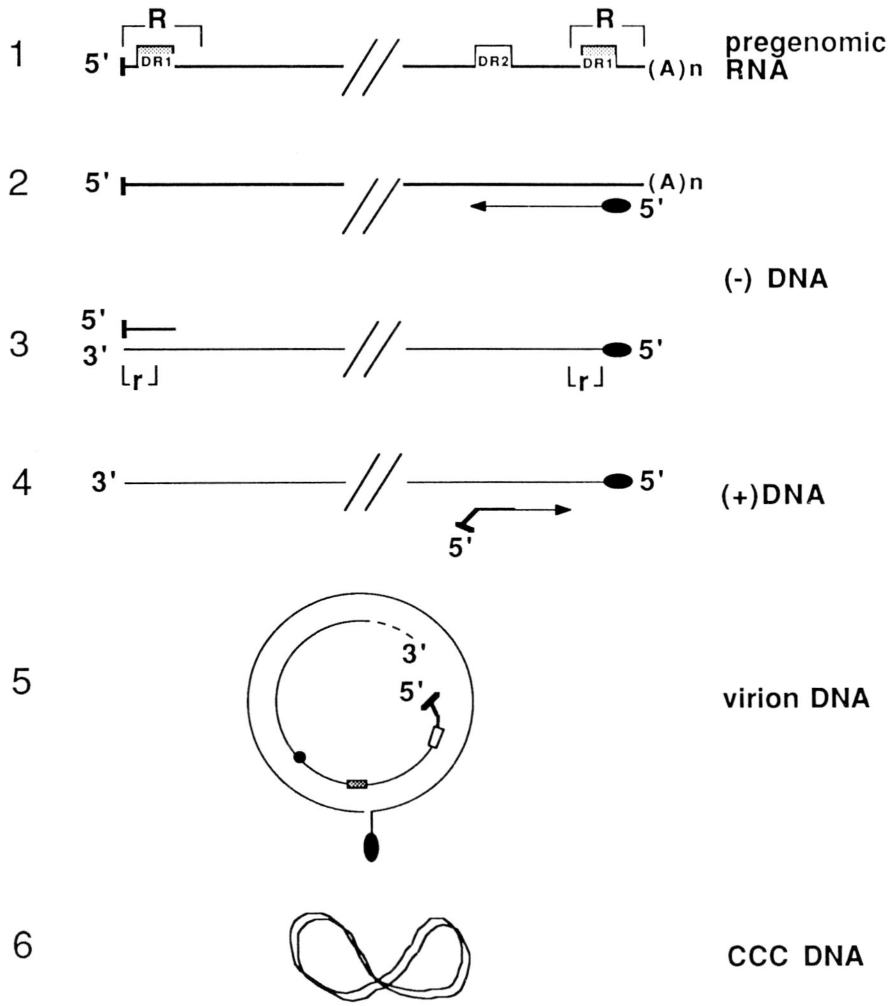

FIGURE 2. Model for the synthesis of hepadnaviral DNA from pregenomic RNA. (1) Pregenomic RNA with the terminal redundancy (R), cap structure (vertical bar), polyadenylic acid tract (A_n), and the direct repeats, DR1 and DR2. (2) Initiation of minus-strand DNA synthesis at DR1, close to the 3′ end of pregenomic RNA. The filled oblong symbol depicts the terminal protein at the 5′ end of minus-strand DNA. (3) Completion of minus-strand DNA synthesis and the formation of the 3′ end of the RNA primer for plus-strand DNA synthesis. The terminal redundancy on minus-strand DNA is indicated (r). (4) Priming and extension of plus-strand DNA. The RNA primer is translocated from the 3′ end of minus-strand DNA to the initiation site for plus-strand DNA at DR2. (5) Continuation of plus-strand DNA synthesis following the transfer of nascent plus-strand DNA from the 5′ end of minus-strand DNA to the 3′ end. The positions of DR1 (hatched box), DR2 (open box), and the polyadenylation signal (filled circle) are indicated. (6) Conversion of the RC genome to cccDNA, the template for the synthesis of pregenomic RNA.

IV. SYNTHESIS OF THE MAJOR REPLICATIVE RNA AND DNA INTERMEDIATES

A. THE RNA PREGENOME

The analysis of viral RNA present in the liver of infected animals revealed a 3.5-kb long RNA species and one or two shorter transcripts, 2.1 to 2.5-kb in length[42-44] (Chapter 5, this

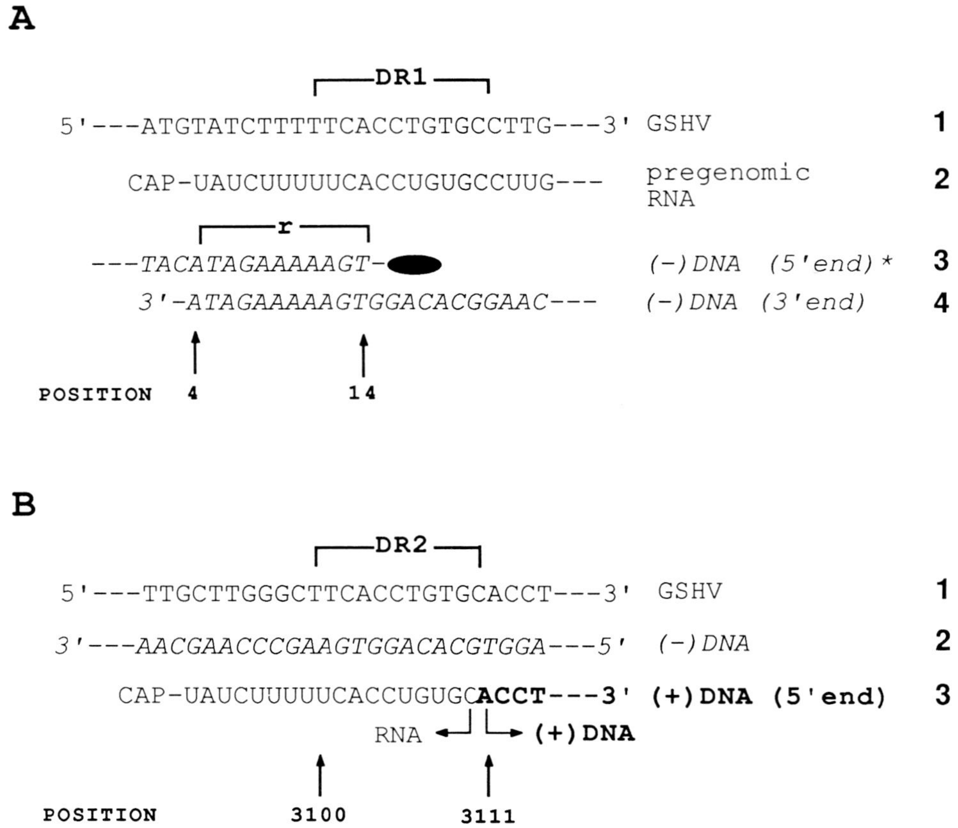

FIGURE 3. The positions of the 5′ and 3′ ends of the replicative intermediates at DR1 (A) and DR2 (B). (A) 1: Nucleotide sequence of GSHV from position 1 to 25, including DR1. 2: The 5′ end of pregenomic RNA at position 4. 3: The 5′ end of minus-strand DNA at position 14 within DR1 close to the 3′ end of pregenomic RNA (*). The filled, oblong symbol indicates the terminal protein at the 5′ end of minus-strand DNA. 4: The 3′ end of minus-strand DNA coincides with the 5′ end of pregenomic RNA at position 4. (B) 1 and 2: Nucleotide sequence of plus (1) and minus (2) strand DNA of GSHV from position 3090 to 3114, including DR2. 3: The 5′ end of plus-strand DNA with the RNA primer derived from the 5′ end of pregenomic RNA (A, line 2). The numbering of the positions is according to the data described in References 24, 37, and42.

volume). The 3.5-kb long transcript, synthesized from cccDNA (see Section IV.D), has heterogeneous 5′ ends. The shortest species of these transcripts initiates six to seven nucleotides 5′ to DR1. It is the template for reverse transcription of minus-strand DNA[24,42] (Figure 2, line 1 and Figure 3A, line 2) and is termed pregenomic RNA. The 3′ end of this transcript is located approximately 100 to 200 nucleotides 3′ to DR1.[42,44] As a consequence of the position of its 5′ and 3′ ends, pregenomic RNA has terminally redundant ends and bears two copies of DR1, one at its 5′ end and the other close to its 3′ end (Figure 2, line 1).

B. MINUS-STRAND DNA

As a result of the reverse transcription reaction, the 5′ end of minus-strand DNA is covalently linked to the terminal protein. Unlike other known retroid elements, which use a cellular transfer RNA as a primer for minus-strand DNA synthesis, hepadnaviruses may use this protein as a primer for DNA synthesis. The evidence favoring this assumption is primarily based on the observation that nascent minus-strand DNA species, as short as 30 nucleotides in length, are covalently linked to the terminal protein.[32]

The linkage between the terminal protein and DNA has not yet been fully characterized. Based on the resistance of the protein–DNA bond to treatment with strong alkali, it was proposed that the DNA was linked to the protein via tyrosine.[45] Replicative intermediates of DHBV can be immunoprecipitated with antibodies directed against the polymerase gene product, suggesting that the terminal protein of DHBV is of viral origin (Chapter 8, this volume).[45]

The 5′ end of deproteinized minus-strand DNA isolated from virions maps to the third or fourth nucleotide of DR1.[24,30,46,47] The 3′ end of minus-strand DNA coincides with the 5′ end of the shortest of the three 3.5-kb RNAs. The constellation of the 5′ and 3′ ends creates a 9- to 11-nucleotide long terminal redundancy (r) on minus-strand DNA[24,46,47] (Figure 3A, lines 3 and 4).

Since two copies of DR1 are present on pregenomic RNA, the initiation site for minus-strand DNA cannot be inferred solely from the analysis of its 5′ end. The expression of a cDNA equivalent of WHV pregenomic RNA with the help of a foreign promoter permitted the genetic manipulation of the two copies of DR1 independently from each other (see Section V). With this approach, the DR1 close to the 3′ end of pregenomic RNA was identified as the origin for minus-strand DNA synthesis.[48,48a] Thus, minus-strand DNA synthesis can proceed continuously beginning at the 3′ end of the RNA template and terminating at its 5′ end. Concomitant with minus-strand DNA synthesis, the RNA template is degraded, presumably by an RNase H-like activity associated with the viral reverse transcriptase[12] (Figure 2, lines 2 and 3).

C. PLUS-STRAND DNA

The preponderance of minus-strand DNA over plus-strand DNA in virus-producing cells reflects the strategy for viral DNA synthesis. Since minus-strand DNA is the template for plus-strand DNA synthesis, the priming reactions for the synthesis of the two DNA strands are temporally separated from each other and occur by different mechanisms. Plus-strand DNA synthesis is primed by a 17- to 18-nucleotide long RNA oligomer, which remains covalently attached to plus-strand DNA isolated from virions.[23,24,47] The 5′ end of the RNA primer is capped, explaining the failure to phosphorylate the 5′ ends of plus-strand DNA with polynucleotide kinase.[22]

The 5′ end of RNase-treated plus-strand DNA maps to the first nucleotide 3′ to DR2. The 3′ portion of the RNA oligomer is identical with DR2 and abuts the 5′ end of plus-strand DNA. Unexpectedly, the 5′ end of the RNA primer corresponds to the 5′ end of pregenomic RNA (Figure 3B, line 3). Thus, the RNA primer originates from the 5′ end of pregenomic RNA, including DR1. To prime plus-strand DNA synthesis it must be translocated to DR2 on minus-strand DNA. Annealing of the primer is facilitated by the homology between DR1 and DR2 (Figure 2, lines 3 and 4). The 3′ end of the RNA primer is most likely determined by an RNase H-like activity, prior to its transfer to DR2.[49]

The validity of the model for the priming of plus-strand DNA synthesis was substantiated with genetically engineered mutants of GSHV and WHV. GSHV DNA with a single base change in DR1 led to the expression of virion DNA with a corresponding base change in the RNA primer.[24] Expression of pregenomic RNA of WHV with a one-base deletion at the 5′ end led to the expression of virion DNA with an RNA primer that was one nucleotide shorter.[49] Thus, the position of the 5′ end of of pregenomic RNA dictates the position of the 5′ end of the RNA primer attached to plus-strand DNA.

As a consequence of the mechanism for plus-strand DNA synthesis, the 5′ portion of pregenomic RNA has a dual function; first, it must serve as a template for the synthesis of the 3′ end of minus-strand DNA, and then, second, as a primer for the synthesis of plus-strand DNA. This model predicts that plus-strand DNA synthesis can initiate at DR2 only after completion of minus-strand DNA synthesis. Alternatively, two RNA templates would be required for the synthesis of the complete viral genome (Figure 2, lines 3 and 4).

Plus-strand DNA synthesis proceeds to the 5′ end of minus-strand DNA. To continue DNA synthesis, the 3′ ends of nascent plus strands must disengage from the 5′ terminal portion of minus-strand DNA and reanneal to its 3′ end. As in retroviruses, this transfer is likely to be facilitated by the short terminal redundancy (r) on minus-strand DNA (Figure 2, line 4). The protein at the 5′ end of minus-strand DNA may provide the scaffold necessary to keep the two ends of minus-strand DNA in close proximity. However, the determinants that play a role in this event await experimental characterization. The continuation of plus-strand DNA synthesis beyond the 5′ end of its template accounts for the stabilization of the cohesive structure of virion DNA (Figure 2, line 5). For reasons that are not yet understood, DNA synthesis in the mammalian hepadnaviruses terminates 30 to 50% short of its completion, whereas in DHBV plus-strand DNA synthesis appears to be almost complete, terminating at or a few nucleotides 5′ of DR2.[26,27,29,46]

The analysis of the structure of the major replicative intermediates has identified the region including DR1 as a central site for DNA replication (Figure 3). It bears the initiation sites for the synthesis of pregenomic RNA as well as the initiation and termination sites for minus-strand DNA. It encodes the sequence for the RNA primer for plus-strand DNA synthesis and finally serves as a coding region for the precore gene product (Chapter 7, this volume). To execute these functions, the DR1 region may also provide signals for several biochemical reactions. Among them are the signals for (1) initiation of minus-strand DNA synthesis, (2) the formation of the 3′ end of the RNA primer, (3) the transfer of the RNA primer to DR2, and (4) the template switch, necessary for the continuation of plus-strand DNA synthesis beyond the 5′ end of the minus strand. Additionally, this region may provide signals important for the packaging of pregenomic RNA.[50] As will be described in Section V, the possibility of performing a genetic analysis of the hepadnavirus genome in cultured cells will provide a vital tool for a dissection of this region into its functional components.

D. cccDNA

cccDNA accumulates in the nucleus of infected cells.[51] It originates from virion DNA and serves as a template for the synthesis of pregenomic RNA and viral mRNAs. This hypothesis is substantiated by the following observations. First, cccDNA is the only identified viral replicative intermediate that can serve as a template for the transcription of pregenomic RNA. Cloned, recircularized hepadnavirus genomes, identical in their structure with cccDNA, are infectious when inoculated directly into the livers of susceptible animals.[52-54] Consequently, cloned cccDNA equivalents can direct the synthesis of pregenomic RNA and initiate a complete replication cycle in the absence of any other viral components. Second, cccDNA is the first viral replicative intermediate that appears in the liver of DHBV-infected ducklings. cccDNA is detectable prior to the synthesis of single-stranded minus-strand DNA species (ssDNA), the initial products of the reverse transcription reaction.[5,55]

It remains to be shown how relaxed circular virion DNA is converted into cccDNA. This step requires the completion of plus-strand DNA synthesis, the removal of the terminal protein and the RNA primer from the 5′ ends of minus- and plus-strand DNA, followed by the ligation of the 5′ and 3′ ends of the two DNA molecules. These reactions may be catalyzed either by cellular DNA-modifying enzymes or by viral components or both. The terminal protein may provide functions required for accurately joining the ends of minus-strand DNA, thus assuring the correct formation of at least one of the two DNA strands.

How is cccDNA maintained in the nucleus of the infected cell? While retroviruses perpetuate their genomes by integration of the provirus into the chromosome of the infected cell, hepadnaviruses must use an alternative strategy for the conservation of their episomal genome in the infected cell. The copy number of cccDNA in the infected liver is in the order of 5 to 30 copies per host genome equivalent.[55a] Accumulation of cccDNA occurs in cultured primary duck hepatocytes within 5 days after infection with DHBV[56] (see Section V). At least two mechanisms could account for these observations. cccDNA could be maintained by semiconservative replication or by *de novo* synthesis of virion DNA. The second strategy would require either the superinfection of infected cells or, alternatively, the intracellular transport of newly synthesized DNA into the nucleus of the virus-producing cell.

Tuttleman et al. elegantly demonstrated that the pool of cccDNA in cultured hepatocytes is synthesized from RNA and not from DNA.[56] Density labeling of cccDNA through the addition of the nucleotide analog 5-bromodeoxyuridine (BUdR) to the culture medium revealed that BUdR was present only in plus strands of heavy/light (H/L) cccDNA. This result excluded the possibility that semiconservative replication accounted for the accumulation of cccDNA. cccDNA synthesis occurred prior to the release of virus into the culture medium and was not blocked by the addition of suramin, an inhibitor of DHBV infection.[56,57] Thus, cccDNA accumulated in cultured hepatocytes predominantly through an intracellular pathway and not through the reinfection of hepatocytes with newly released DHBV.

In summary, replicative complexes in the cytoplasm of infected hepatocytes can follow two different routes during maturation, one leading to the secretion of the virus particle, the other to amplification of cccDNA in the infected cell. This strategy may allow the virus to maintain a constant pool of cccDNA in the infected cell. The cellular or viral components that determine the fate of the replication complex remain to be determined.

V. REPLICATION IN TISSUE CULTURE CELLS

The ability to transfect tissue culture cells with cloned DNA was exploited early on as a means of expressing the HBV genome in cell culture. While surface antigen expression was observed in transfected cells, production of core antigen, replicative intermediates, and virus was not detectable in cells transfected with cloned HBV DNA.[58-61]

Tuttleman et al. demonstrated that primary duck hepatocytes are competent for all steps of the DHBV replication cycle.[17] Hepatocytes isolated from congenitally infected ducks continued DHBV production for several weeks. Hepatocytes isolated from uninfected ducks are susceptible to DHBV infection *in vitro*. Primary hepatocytes obtained from woodchuck and human livers also support the replication of WHV and HBV, respectively.[62-64]

Infection of primary hepatocyte cultures with DHBV *in vitro* leads to the synthesis of intracellular, replicative intermediates as observed following the infection of duck embryos with DHBV.[5,17,55] They include cccDNA, ssDNA, and relaxed circular (rc) DNA. DHBV secreted into the supernatant of infected hepatocytes is infectious in the animal and can be propagated *in vitro* by infecting fresh hepatocyte cultures.[17]

Interestingly, primary duck hepatocytes can be infected only during the first 4 d of culture, indicating that early steps of virus replication depend on hepatocyte-specific factors, which are lost rapidly during the culture of these cells. Identification of these factors may provide clues about hepatocyte-specific functions required for virus infection, one of them possibly being the production of the viral receptor.

Although the culture of primary hepatocytes supports all steps of the hepadnavirus replication cycle, attempts to introduce cloned DNA into these cells have so far not been successful. Therefore, the primary hepatocyte system is not accessible to a genetic analysis of the viral replication cycle. As discussed below, the combination of this system with the method of producing hepadnavirus from cloned DNA in hepatoma cell lines provides the necessary bypass to this obstacle.

Efforts to produce HBV in cultured cells with cloned DNA led to the identification of three differentiated human hepatoma cell lines (HepG2,[18,19,65] Huh6[20,66] and Huh7[21,67,68]), and one rat hepatoma cell line (Q7[69]) capable of virus production. Using these cell lines, several laboratories reported the isolation of cell clones that constitutively express HBV from transfected, integrated DNA copies.[18-20] Alternatively, HBV and other hepadnaviruses can be transiently expressed in hepatoma cell lines transfected with cloned viral DNA.[21,49,67,70,71] Since virus from cultured cells is infectious in the animal, the system is competent for all steps in virus replication subsequent to cccDNA synthesis.[70-73]

Pugh et al. demonstrated that DHBV produced in transfected hepatoma cells can be used to infect primary duck hepatocytes.[74] The combination of the two tissue culture systems provides a powerful approach for the genetic analysis of functions encoded by the hepadnavirus genome that play a role during early steps in virus replication.

The limitation of hepadnavirus production to a few selected hepatoma cell lines appears to be the result of the dependency of the pregenomic RNA promoter on specific transcription factors. The expression of pregenomic RNA under the control of foreign promoters leads to virus production in tissue culture cells that otherwise do not support virus production. The fusion of HBV sequences with the metallothionein promoter allows the synthesis of DNA polymerase containing virus-like particles in transfected HeLa cells.[75] Similarly, expression of WHV with the cytomegalovirus immediate early promoter permits the production of WHV in mouse and avian fibroblast cells.[76] WHV expressed in fibroblasts is infectious in woodchucks.[76] Thus, unlike the synthesis of pregenomic RNA, posttranscriptional events of the hepadnavirus replication cycle are not dependent on hepatocyte-specific functions.

VI. SUMMARY AND PERSPECTIVES

Our understanding of the hepadnavirus replication cycle has dramatically improved during the last decade. Whereas major research efforts were dedicated to the process of viral DNA synthesis from pregenomic RNA, insufficient attention has been devoted to the analysis of early steps in virus replication. They include the adsorption of the virus to the target cell, transport across the cytoplasmic membrane, uncoating, delivery of the genome into the nucleus, and conversion of the relaxed circular genome into cccDNA. Some of these events bear one of the major characteristics of hepadnavirus infection, tissue and species specificity. To fill this gap, it will be necessary to identify the viral receptor and to determine the mechanism of virus uptake, either by membrane fusion or receptor-mediated endocytosis. It will be critical to identify the enzymatic components engaged in the synthesis of cccDNA from virion DNA. Is this event mediated by viral functions provided by the polymerase gene product, cellular enzymes, or both? The combination of the two available tissue culture systems should provide the necessary tools for the design of experiments that will grant meaningful answers to these questions.

Although a detailed model for the replication strategy of hepadnaviruses has emerged from the careful analyses of the structure of the viral replicative intermediates, little is known about the signals on the hepadnaviral genome that orchestrate synthesis of the viral genome from an RNA template. Among them are the signals important for packaging of pregenomic RNA and the viral reverse transcriptase and for initiation of minus-strand DNA synthesis. Furthermore, it remains to be demonstrated whether the terminal protein is the primer for minus-strand DNA synthesis or whether it is attached to minus-strand DNA shortly after the commencement of DNA synthesis.

The molecular mechanism underlying the translocation of the RNA primer from the 5′ end of pregenomic RNA to the primer binding site at DR2 on minus-strand DNA is not yet understood. Although 5′ ends of RNA polymerase transcripts can be used as primers for DNA synthesis,[77] the translocation of an RNA primer for DNA synthesis is so far without a precedent. Why do hepadnaviruses employ this seemingly complicated mechanism for the priming of plus-strand DNA synthesis and why do they not rely on polypurine tracts for the formation of primers as do other retroid elements?

As with the early steps of the replication cycle, little is known about the final steps of virus replication. How are the replication complexes transported to the cytoplasmic membrane and where do they assemble with the envelope proteins to form mature virus particles? Finally, it will be important to identify the signals that determine the fate of the core particles, either to mature into virions that exit the cell or to recycle to the nucleus and thus provide a means for the persistence of virus infection.

ACKNOWLEDGMENTS

I thank M. A. Delaney, Laurel Southard, and W. S. Mason for critical reading of the manuscript and for helpful suggestions. The author is supported by Public Health Service Grant AI-24972 from the National Institutes of Health.

REFERENCES

1. **Ganem, D. and Varmus, H.E.**, The molecular biology of the hepatitis B viruses, *Annu. Rev. Biochem.*, 56, 651, 1987.
2. **Mason, W.S., Taylor, J.M., and Hull, R.**, Retroid virus genome replication, *Adv. Vir. Res.*, 32, 35, 1987.
3. **Korba, B.E., Wells, F., Tennant, B.C., Cote, P.J., and Gerin, J.L.**, Lymphoid cells in the spleens of woodchuck hepatitis virus-infected woodchucks are a site of active viral replication, *J. Virol.*, 61, 1318, 1987.
4. **Tagawa, M., Omata, M., Yokosuka, O., Uchiumi, K., Imazeki, F., and Okuda, K.**, Early events in duck hepatitis B virus infection: sequential appearance of viral deoxyribonucleic acid in the liver, pancreas, kidney, and spleen, *Gastroenterology*, 89, 1224, 1985.
5. **Mason, W.S., Halpern, M.S., England, J.M., Seal, G., Egan, J., Coates, L., Aldrich, C., and Summers, J.**, Experimental transmission of duck hepatitis B virus, *Virology*, 131, 375, 1983.
6. **Ogston, C.W., Schechter, E.M., Humes, C.A., and Pranikoff, M.B.**, Extrahepatic replication of woodchuck hepatitis virus in chronic infection, *Virology*, 169, 9, 1989.
7. **Halpern, M.S., Egan, J., McMahon, S.B., and Ewert, D.L.**, Duck hepatitis B virus is tropic for exocrine cells of the pancreas, *Virology*, 146, 157, 1985.
8. **Mason, W.S. and Taylor, J.M.**, Experimental systems for the study of hepadnavirus and hepatitis delta virus infections, *Hepatology*, 9, 635, 1989.
9. **Baltimore, D.**, RNA-dependent DNA polymerase in virions of RNA tumour viruses, *Nature (London)*, 226, 1209, 1970.
10. **Temin, H. and Mizutani, S.**, RNA-dependent DNA polymerases in virions of Rous sarcoma virus, *Nature (London)*, 226, 1211, 1970.
11. **Kaplan, P.M., Greenman, R.L., Gerin, J.L., Purcell, R.H., and Robinson, W.S.**, DNA polymerase associated with human hepatitis B antigen, *J. Virol.*, 12, 995, 1973.
12. **Summers, J. and Mason, W.S.**, Replication of the genome of a hepatitis B-like virus by reverse transcription of an RNA intermediate, *Cell*, 29, 403, 1982.
13. **Summers, J., Smolec, J.M., and Snyder, R.**, A virus similar to human hepatitis B virus associated with hepatitis and hepatoma in woodchucks, *Proc. Natl. Acad. Sci. U.S.A.*, 75, 4533, 1978.

14. **Marion, P.L., Oshiro, L.S., Regnery, D.C., Scullard, G.H., and Robinson, W.S.**, A virus in Beechey ground squirrels that is related to hepatitis B virus of humans, *Proc. Natl. Acad. Sci. U.S.A.*, 77, 2941, 1980.
15. **Mason, W.S., Seal, G., and Summers, J.**, Virus of Pekin ducks with structural and biological relatedness to human hepatitis B virus, *J. Virol.*, 36, 829, 1980.
16. **Mason, W.S., Aldrich, C., Summers, J., and Taylor, J.M.**, Asymmetric replication of duck hepatitis B virus DNA in liver cells: free minus-strand DNA, *Proc. Natl. Acad. Sci. U.S.A.*, 79, 3997, 1982.
17. **Tuttleman, J.S., Pugh, J.C., and Summers, J.W.**, *In vitro* experimental infection of primary duck hepatocyte cultures with duck hepatitis B virus, *J. Virol.*, 58, 17, 1986.
18. **Sureau, C., Romet-Lemonne, J., Mullins, J.I., and Essex, M.**, Production of hepatitis B virus by a differentiated human hepatoma cell line after transfection with cloned circular HBV DNA, *Cell*, 47, 37, 1987.
19. **Sells, M.A., Chen, M., and Acs, G.**, Production of hepatitis B virus particles in hepG2 cells transfected with cloned hepatitis B virus DNA, *Proc. Natl. Acad. Sci. U.S.A.*, 84, 1005, 1987.
20. **Tsurimoto, T., Fujiyama, A., and Matsubara, K.**, Stable expression and replication of hepatitis B virus genome in an integrated state in a human hepatoma cell line transfected with the cloned viral DNA, *Proc. Natl. Acad. Sci. U.S.A.*, 84, 444, 1987.
21. **Yaginuma, K., Shirakata, Y., Kobayashi, M., and Koike, K.**, Hepatitis B virus (HBV) particles are produced in a cell culture system by transient expression of transfected HBV DNA, *Proc. Natl. Acad. Sci. U.S.A.*, 84, 2678, 1987.
22. **Gerlich, W.H. and Robinson, W.S.**, Hepatitis B virus contains protein attached to the 5′ terminus of its complete DNA strand, *Cell*, 21, 801, 1980.
23. **Lien, J., Aldrich, C.E., and Mason, W.S.**, Evidence that a capped oligoribonucleotide is the primer for duck hepatitis B virus plus strand DNA synthesis, *J. Virol.*, 57, 229, 1986.
24. **Seeger, C., Ganem, D., and Varmus, H.E.**, Biochemical and genetic evidence for the hepatitis B virus replication strategy, *Science*, 232, 477, 1986.
25. **Dane, D.S., Cameron, C.H., and Briggs, M.** Virus-like particles in serum of patients with Australia-antigen-associated hepatitis, *Lancet* , i, 695, 1970.
26. **Robinson, W.S., Clayton, D.A., and Greenman, R.L.**, DNA of a human hepatitis B virus candidate, *J. Virol.*, 14, 384, 1974.
27. **Summers, J., O'Connell, A., and Millman, I.**, Genome of hepatitis B virus: restriction enzyme cleavage and structure of DNA extracted from Dane particles, *Proc. Natl. Acad. Sci. U.S.A.*, 72, 4597, 1975.
28. **Sattler, F. and Robinson, W.S.**, Hepatitis B DNA has cohesive ends, *J. Virol.*, 32, 226, 1979.
29. **Ganem, D., Greenbaum, L., and Varmus, H.E.**, Virion DNA of ground squirrel hepatitis virus: structural characterization and molecular cloning. *J. Virol.*, 44, 374, 1982.
30. **Molnar-Kimber, K.L., Summers, J.W., and Mason, W.S.**, Mapping of the cohesive overlap of duck hepatitis B virus DNA and of the site of initiation of reverse transcription, *J. Virol.*, 51, 181, 1984.
31. **Weiser, B., Ganem, D., Seeger, C., and Varmus, H.E.**, Closed circular viral DNA and asymmetric heterogeneous forms in livers from animals infected with ground squirrel hepatitis virus, *J. Virol.*, 48, 1, 1983.
32. **Molnar-Kimber, K.L., Summers, J.W., Taylor, J.M., and Mason, W.S.**, Protein covalently bound to minus strand DNA intermediates of duck hepatitis B virus, *J. Virol.*, 45, 165, 1983.
33. **Ruiz-Opazo, N., Chakraborty, P.R., and Shafritz, D.A.** Evidence for supercoiled hepatitis B virus DNA in chimpanzee liver and serum Dane particles: possible implications in persistent HBV infection, *Cell*, 29, 129, 1982.
34. **Brechot, C., Scotto, J., Charnay, P., Hadchouel, M., Degos, F., Trepo, C., and Tiollais, P.**, Detection of hepatitis B virus DNA in liver and serum: a direct appraisal of the chronic carrier state, *Lancet*, i, 765, 1981.
35. **Miller, R.H., Marion, P.L., and Robinson, W.S.**, Hepatitis B viral DNA-RNA hybrid molecules in particles from infected liver are converted to viral DNA molecules during an endogenous DNA polymerase reaction, *Virology*, 139, 64, 1984.
36. **Blum, H.E., Haase, A.T., Harris, J.D., Walker, D., and Vyas, G.N.**, Asymmetric replication of hepatitis B virus DNA in human liver: demonstration of cytoplasmic minus strand DNA by blot analysis and *in situ* hybridization, *Virology*, 139, 87, 1984.
37. **Seeger, C., Ganem, D., and Varmus, H.E.**, The nucleotide sequence of an infectious molecularly cloned genome of ground squirrel hepatitis virus, *J. Virol.*, 51, 367, 1984.
38. **Galibert, F., Chen, T.N., and Mandart, E.**, Nucleotide sequence of a cloned woodchuck hepatitis virus genome: comparison with the hepatitis B virus sequence, *J. Virol.*, 41, 51, 1982.
39. **Mandart, E., Kay, A., and Galibert, F.** Nucleotide sequence of a cloned duck hepatitis B virus genome: comparison with woodchuck and human hepatitis B virus sequences, *J. Virol.*, 49, 782, 1984.
40. **Valenzuela, P., Quiroga, M., Zaldivar, J., Gray, P., and Rutter, W.J.**, The nucleotide sequence of the hepatitis B viral genome and the identification of the major viral genes, in *Animal Virus Genetics*, Fields, B., Jaenisch, R., and Fox, C. F., Eds., Academic Press, New York, 1980, 70.

41. **Galibert, F., Mandart, E., Fitoussi, F., and Charnay, P.**, Nucleotide sequence of the hepatitis B virus genome (subtype ayw) cloned in *E. coli, Nature (London)*, 281, 646, 1979.
42. **Enders, G.H., Ganem, D., and Varmus, H.E.**, Mapping the major transcripts of ground squirrel hepatitis virus: the presumptive template for reverse transcriptase is terminally redundant, *Cell*, 42, 297, 1985.
43. **Moroy, T., Etiemble, J., Trepo, C., Tiollais, P., and Buendia, M.A.**, Transcription of woodchuck hepatitis virus in the chronically infected liver, *EMBO J.*, 4, 1507, 1985.
44. **Buscher, M., Reiser, W., Will, H., and Schaller, H.**, Transcripts and the putative RNA pregenome of duck hepatitis B virus: implications for reverse transcription, *Cell*, 40, 717, 1985.
45. **Bartenschlager, R. and Schaller, H.**, The amino-terminal domain of the hepadnaviral P-gene encodes the terminal protein (genome-linked protein) believed to prime reverse transcription, *EMBO J.*, 7, 4185, 1988.
46. **Lien, J., Petcu, D.J., Aldrich, C.E., and Mason, W.S.**, Initiation and termination of duck hepatitis B virus DNA synthesis during virus maturation, *J. Virol.*, 61, 3832, 1987.
47. **Will, H., Reiser, W., Weimer, T., Pfaff, E., Buscher, M., Sprengel, R., Cattaneo, R., and Schaller, H.**, Replication strategy of human hepatitis B virus, *J. Virol.*, 61, 904, 1987.
48. **Seeger, C. and Maragos, J.**, Identification and characterization of the woodchuck hepatitis virus origin of DNA replication, *J. Virol.*, 64, 1990.
48a. **Condreay, L.D., Mason, W.S., and Summers, J.**, personal communication, 1990.
49. **Seeger, C. and Maragos, J.**, Molecular analysis of the function of direct repeats and a polypurine tract for plus-strand DNA priming in woodchuck hepatitis virus, *J. Virol.*, 63, 1907, 1989.
50. **Enders, G.H., Ganem, D., and Varmus, H.E.**, 5′-Terminal sequences influence the segregation of ground squirrel hepatitis virus RNA into polyribosomes and viral core particles, *J. Virol.*, 61, 35, 1987.
51. **Miller, R.H. and Robinson, W.S.**, Hepatitis B virus DNA in nuclear and cytoplasmic fractions of infected human liver, *Virology*, 137, 390, 1984.
52. **Will, H., Cattaneo, R., Koch, H., Darai, G., Schaller, H., Schellekens, H., van Eerd, P., and Deinhardt, F.**, Cloned HBV DNA causes hepatitis in chimpanzees, *Nature (London)*, 299, 740, 1982.
53. **Seeger, C., Ganem, D., and Varmus, H.E.**, The cloned genome of ground squirrel hepatitis virus is infectious in the animal, *Proc. Natl. Acad. Sci. U.S.A.*, 81, 5849, 1984.
54. **Sprengel, R., Kuhn, C., Manso, K., and Will, H.**, Cloned duck hepatitis B virus DNA is infectious in Pekin ducks, *J. Virol.*, 52, 932, 1984.
55. **Tagawa, M., Omata, M., and Okuda, K.**, Appearance of viral RNA transcripts in the early stage of duck hepatitis B virus infection, *Virology*, 152, 477, 1986.
55a. **Petcu, D.J. and Mason, W.S.**, personal communication, 1990.
56. **Tuttleman, J.S., Pourcel, C., and Summers, J.**, Formation of the pool of covalently closed circular viral DNA in hepadnavirus-infected cells, *Cell*, 47, 451, 1986.
57. **Wu, T.-T., Coates, L., Aldrich, C.E., Summers, J., and Mason, W.S.**, In hepatocytes infected with duck hepatitis B virus, the template for viral RNA synthesis is amplified by an intracellular pathway, *J. Virol.*, 1990.
58. **Dubois, M.-F., Pourcel, C., Rousset, S., Chany, C., and Tiollais, P.**, Excretion of hepatitis B surface antigen particles from mouse cells transformed with cloned viral DNA, *Proc. Natl. Acad. Sci. U.S.A.*, 77, 4549, 1980.
59. **Zelent, A.Z., Sells, M.A., Price, P.M., Mohamad, A., and Acs, G.**, Murine cells carrying integrated tandem genomes of hepatitis B virus DNA transcribe RNAs from endogenous promoters on both viral strands and express middle and major viral envelope proteins, *J. Virol.*, 61, 1108, 1987.
60. **Nozaki, C., Miyanohara, A., Fujiyama, A., Hamada, F., Ohtomo, N., and Matsubara, K.**, Two mammalian cell systems for propagation of the hepatitis B virus genome in extrachromosomal and chromosomally integrated states: production of the surface and e antigens, *Gene*, 38, 39, 1985.
61. **Gough, N.M**, Core and e antigen synthesis in rodent cells transformed with hepatitis B virus DNA is associated with greater than genome length viral messenger RNAs, *J. Mol. Biol.*, 165, 683, 1983.
62. **Aldrich, C.E., Coates, L., Wu, T.-T., Newbold, J., Tennant, B.C., Summers, J., Seeger, C., and Mason, W.S.**, *In vitro* infection of woodchuck hepatocytes with woodchuck hepatitis virus and ground squirrel hepatitis virus, *Virology*, 172, 247, 1989.
63. **Gripon, P., Diot, C., Theze, N., Fourel, I., Loreal, O., Brechot, C., and Guguen-Guillouzo, C.**, Hepatitis B virus infection of adult human hepatocytes cultured in the presence of dimethyl sulfoxide, *J. Virol.*, 62, 4136, 1988.
64. **Ochiya, T., Tsurimoto, T., Ueda, K., Okubo, K., Shiozawa, M., and Matsubara, K.**, An *in vitro* system for infection with hepatitis B virus that uses primary human fetal hepatocytes, *Proc. Natl. Acad. Sci. U.S.A.*, 86, 1875, 1989.
65. **Aden, D.P., Fogel, A., Plotkin, S., Damjanov, I., and Knowles, B.**, Controlled synthesis of HBsAg in a differentiated human liver carcinoma-derived cell line, *Nature (London)*, 282, 615, 1979.
66. **Doi, I.**, Establishment of a cell line and its clonal sublines from a patient with hepatoblastoma, *Gann*, 67, 1, 1976.

67. **Chang, C.M., Jeng, K.S., Hu, C.P., Lo, S.J., Su, T.S., Ting, L.P., Chou, C.K., Han, S.H., Pfaff, E., Salfeld, J. et al.**, Production of hepatitis B virus *in vitro* by transient expression of cloned HBV DNA in a hepatoma cell line, *EMBO J.*, 6, 675, 1987.
68. **Nakabayashi, H., Taketa, K., Miyano, K., Yamane, T., and Sato, J.**, Growth of human hepatoma cell lines with differentiated functions in chemically defined medium, *Cancer Res.*, 42, 3858, 1982.
69. **Shih, C., Li, L.-S., Roychoudhury, and Ho, M.-H.**, *In vitro* propagation of human hepatitis B virus in a rat hepatoma cell line, *Proc. Natl. Acad. Sci. U.S.A.*, 86, 6323, 1989.
70. **Hirsch, R., Colgrove, R., and Ganem, D.**, Replication of duck hepatitis B virus in two differentiated human hepatoma cell lines after transfection with cloned viral DNA, *Virology*, 167, 136, 1988.
71. **Galle, P.R., Schlicht, H.J., Fischer, M., and Schaller, H.**, Production of infectious duck hepatitis B virus in a human hepatoma cell line, *J. Virol.*, 62, 1736, 1988.
72. **Acs, G., Sells, M.A., Purcell, R.H., Price, P., Engle, R., Shapiro, M., and Popper, H.**, Hepatitis B virus produced by transfected Hep G2 cells causes hepatitis in chimpanzees, *Proc. Natl. Acad. Sci. U.S.A.*, 84, 4641, 1987.
73. **Sureau, C., Eichberg, J.W., Hubbard, G.B., Romet-Lemonne, J.L., and Essex, M.**, A molecularly cloned hepatitis B virus produced *in vitro* is infectious in a chimpanzee, *J. Virol.*, 62, 3064, 1988.
74. **Pugh, J.C., Yaginuma, K., Koike, K., and Summers, J.**, Duck hepatitis B virus (DHBV) particles produced by transient expression of DHBV DNA in a human hepatoma cell line are infectious *in vitro*, *J. Virol.*, 62, 3513, 1988.
75. **Junker, M., Galle, P., and Schaller, H.**, Expression and replication of the hepatitis B virus genome under foreign promoter control, *Nucleic Acids Res.*, 15, 10117, 1987.
76. **Seeger, C., Baldwin, B., and Tennant, B.C.**, Expression of infectious woodchuck hepatitis virus in murine and avian fibroblasts, *J. Virol.*, 63, 1989.
77. **DePamphilis, M.L.**, Transcriptional elements as components of eukaryotic origins of DNA replication, *Cell*, 52, 635, 1988.

Chapter 12

EVOLUTIONARY RELATIONSHIP BETWEEN HEPADNAVIRUSES AND RETROVIRUSES

Roger H. Miller

TABLE OF CONTENTS

I. INTRODUCTION: EVIDENCE OF GENETIC RELATEDNESS BETWEEN HEPADNAVIRUSES AND RETROVIRUSES

Hepatitis B virus (HBV) of humans is the prototype member of the hepadnavirus family, which includes viruses that infect avian (duck and heron) and mammalian (ground squirrel, tree squirrel, and woodchuck) hosts.[1] Prior to 1982 little was known about HBV replication due to the limited host range of the virus and its inability to be grown in tissue culture. An important breakthrough was reached when Summers and Mason,[2] studying duck hepatitis virus-infected liver, demonstrated that hepadnaviruses replicate by reverse transcription of RNA. This was an unexpected finding since hepadnaviruses possess DNA genomes in mature serum virions and were classified as double-stranded DNA viruses. Evidence suggests that hepadnaviruses are related to retroviruses because both virus families use a rare method of replication by reverse transcription of RNA,[2,3] possess homology in important nucleotide and amino acid sequences,[4-12] and share unique features of genome organization.[10,13]

II. REPLICATION BY REVERSE TRANSCRIPTION

Although there are striking similarities in the replication strategies of hepadnaviruses and retroviruses, which both involve reverse transcription, the two mechanisms are by no means identical. Since a more detailed comparison of the replication cycle of these viruses is presented elsewhere[14] (Chapter 11, this volume), only some general aspects of the cycle will be discussed here. HBV genome replication begins when the partially double-stranded DNA genome is transported to the cell nucleus, where it is converted to fully double-stranded covalently closed (supercoiled) DNA (Figure 1). The supercoiled DNA intermediate is the template for transcription by the host RNA polymerase II, which synthesizes the virus pregenomic RNA. Pregenomic RNA has a linear topology and, unlike the other viral replicative intermediates, possesses terminal direct repeats somewhat analogous to the long terminal repeats of retroviruses. Next, pregenomic RNA is packaged into nucleocapsid particles and the reverse transcriptase activity of the virus polymerase synthesizes the first strand of DNA using the pregenomic RNA as a template. Unlike retrovirus reverse transcription, this reaction is primed by a protein that becomes covalently joined to the 5′ end of the first (i.e., minus)-strand DNA. Next, the second DNA strand (of the plus polarity) is synthesized, using the first strand of DNA as a template, by the DNA-dependent DNA polymerase activity of the virus polymerase. Finally, mature virions are produced by the addition of the virus envelope and are exported from the cell. Thus, although HBV packages RNA genomes into nucleocapsid particles, serum virions contain only DNA genomes.

The replication mechanism of retroviruses differs from that of HBV in several important areas. First, retroviruses generally utilize integration of a DNA copy of their genome into host cell DNA as a defined step in their replication cycle. Second, first-strand DNA synthesis is primed using tRNA and not a protein prior to replication using reverse transcriptase. Third, retrovirus long terminal repeats are found in the DNA replicative intermediate. Fourth, RNA genomes are found in mature serum virions. Thus, although HBV and retroviruses replicate by reverse transcription there are significant differences in the pathways.

III. GENOME ORGANIZATION

In the HBV genome, six long open reading frames (ORFs) have been found. Four of these ORFs represent the gene sequences that encode the virus core (nucleocapsid), surface (envelope), and polymerase proteins, and a protein termed X that appears to be involved in trans-

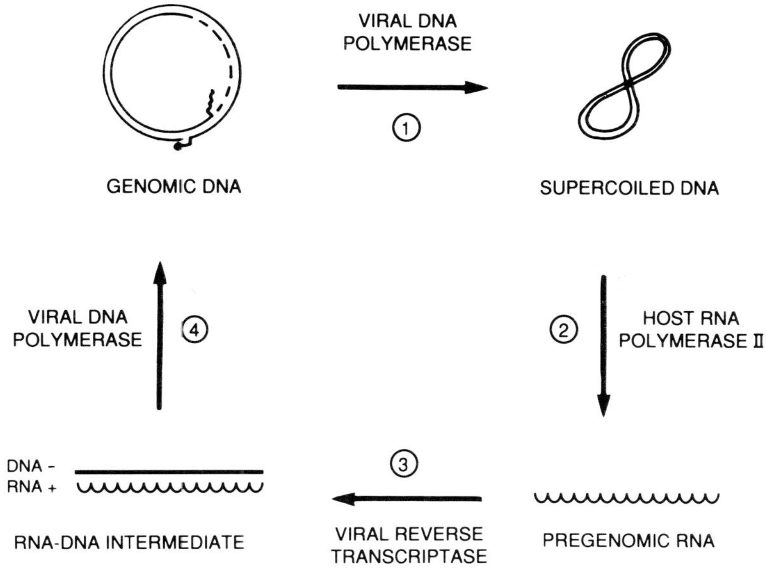

FIGURE 1. Replication cycle of hepadnaviruses. Step 1: Virus nucleic acid enters cell nucleus and is converted to double-stranded, covalently closed circular DNA (cccDNA). Step 2: cccDNA is the template for transcription by the host RNA polymerase II, which synthesizes the virus mRNA and pregenomic RNA. Step 3: The reverse transcriptase activity of the virus polymerase synthesizes the first DNA strand (of negative polarity) using the plus-strand pregenomic RNA as a template. Step 4: The DNA-dependent DNA polymerase activity of the virus polymerase synthesizes the second DNA strand (of plus polarity) using the first strand of DNA as a template.

activation of virus gene expression. The gene order of the linear HBV pregenomic RNA is nucleocapsid, polymerase/envelope, and trans-activation protein gene. Similarly, the organization of the linear retrovirus RNA genome is nucleocapsid, polymerase, envelope, and, in some retroviruses, a trans-activation protein gene. Thus, the nucleocapsid gene in both virus families is positioned at the extreme 5′ end of the linear RNA genome and overlaps with the polymerase gene sequence, which composes the central region of the genome. Also, the gene encoding the trans-activation protein(s) is located at the 3′ end of the genome. Therefore, with the exception of the location of the envelope gene, the basic organization of the genomes of both virus families is similar.

A. NUCLEOCAPSID AND POLYMERASE GENES

Sequence analysis indicates that the nucleocapsid and polymerase genes for HBV, cauliflower mosaic virus (CaMV) of plants, and most retroviruses are located in overlapping translation frames. In some retroviruses the genes are read in the same translation frame but are separated by a termination codon. In retroviruses the products of these genes are expressed as a fusion polyprotein by a frameshifting or termination-suppression mechanism. Individual proteins are released by proteolytic processing of the polyprotein by a virus-encoded protease, the gene sequence of which is located between the nucleocapsid and polymerase genes. Protease translation frames can be part of the nucleocapsid, the polymerase, or a third translation frame. The mechanism for the expression of the HBV and CaMV polymerase proteins is still in question. An early report provided evidence for the production of HBV nucleocapsid and polymerase fusion proteins analogous to those of retroviruses.[15] However, more recent experiments suggest that the synthesis of nucleocapsid and polymerase fusion proteins may not be essential for hepadnavirus replication.[16,17] Also, recent experiments with

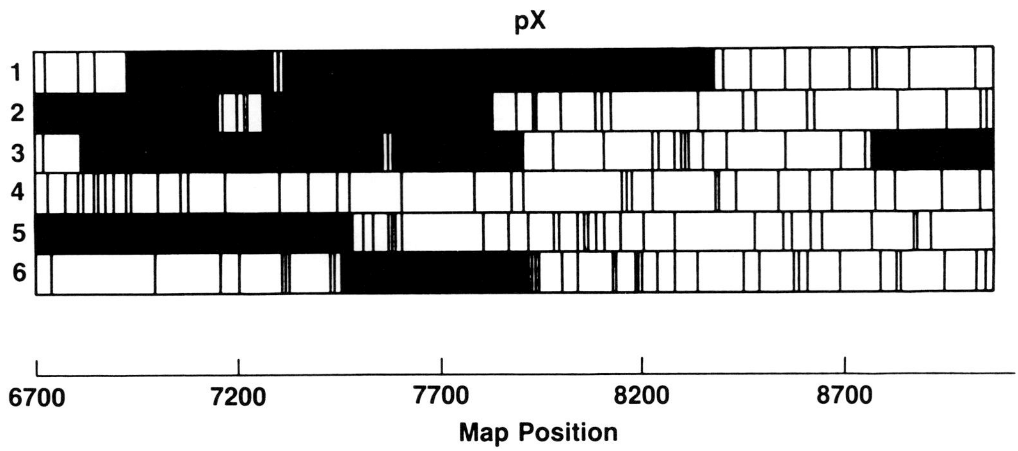

FIGURE 2. Open reading frames of the 3′ terminus of the human T cell leukemia virus (type I). All six translation frames are shown for the region of the genome of HTLV I[25,26] that encodes the trans-activation protein (pX). Termination codons are depicted as vertical lines and open reading frames longer than 100 amino acids are shown as black boxes.

CaMV suggest that the polymerase protein of this retrovirus-like plant virus is expressed by ribosomal reinitiation of translation.[18] Thus, the majority of the current data support the hypothesis that the HBV polymerase protein is expressed by reinitiation of translation after synthesis of the virus core protein using the greater than genome length mRNA as a template. Therefore, HBV and CaMV may have evolved a similar mechanism for the expression of their replicases.

B. TERMINAL REPEATS

The RNA genomes of retroviruses found in mature virions possess short terminal repeats, termed R. The DNA intermediates, found extrachromosomally or as proviruses, possess long terminal repeats (LTRs) consisting of unique 3′ (U3), R, and unique 5′ (U5) elements. The LTRs do not possess significant ORFs and, thus, are noncoding regions of the virus genome. Instead, they contain the control elements used in virus replication and gene expression (i.e., the promoter, enhancer, etc.). Thus, retrovirus LTRs are organized to house the cis-acting regulatory elements of the virus genome.

In contrast, the HBV terminal repeats are protein-encoding regions. Unlike the retrovirus LTRs, the terminal repeats of HBV are present in the intracellular RNA genomes while the DNA intermediates, found in virions, contain no such repeats. Although the HBV terminal repeats do house several cis-acting elements, they do not contain the same profusion of regulatory elements as do the retrovirus LTRs. Thus, the HBV terminal repeats have a different organization and function than the retrovirus LTRs.

C. 3′ END OF THE GENOME

Analysis indicates that the 3′ ends of the pregenomic RNA of HBV and the genomic RNAs of certain retroviruses have an unusual nucleotide sequence characterized by the presence of relatively few termination codons in the six possible translation frames of the double-stranded DNA replicative intermediates synthesized from these regions of the RNA templates (i.e., three translation frames each for the minus and plus DNA strands). Thus, the 3′ ends of these viruses possess multiple ORFs of sizes greater than would be expected to occur by chance. All hepadnavirus isolates examined contain an ORF, designated ORF5, at the 3′ terminus of the

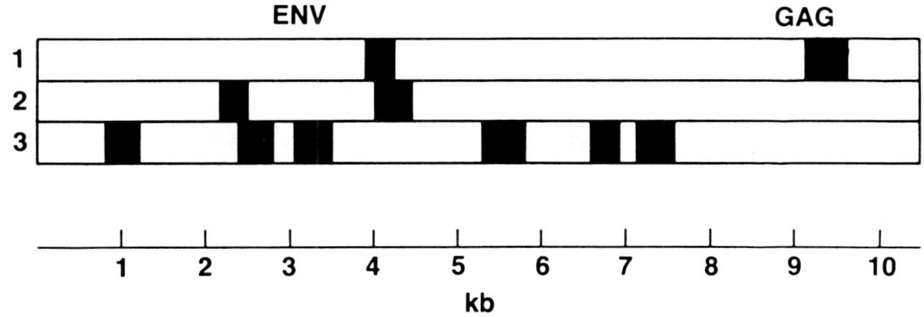

FIGURE 3. Minus DNA strand open reading frames of the simian immunodeficiency virus (SIV). Open reading frames longer than 100 amino acids are shown as black boxes for the genome of SIV[36] translated from the minus DNA strand. The relative positions of the nucleocapsid (GAG) and envelope (ENV) genes are shown.

genome.[19,20] ORF5 ranges from 70 to over 100 codons in length and overlaps with both the polymerase and X gene ORFs in the mammalian hepadnaviruses. Thus, all three possible translation frames at the 3′ end of the message-sense strand of the mammalian hepadnavirus genome possess ORFs. Since ORF5 does not have an in-phase AUG initiation codon, it may be translated by unusual means. One possibility is that a codon other than the standard AUG serves as the initiation codon. This is possible since the codon ACG, known to initiate protein synthesis for expression of several proteins,[21-24] is present near the beginning of all hepadnavirus ORF5 sequences. It is also possible that ORF5 is expressed by a frameshift mechanism during translation of the polymerase protein, and/or the X protein in the case of the mammalian hepadnaviruses, or that virus mRNA is spliced and an initiation codon is fused in phase to the 5′ end of ORF5.

Hepadnaviruses also possess an ORF, and in some isolates multiple ORFs, on the strand of opposite polarity to the known virus genes.[13,20] Termed ORF6, the ORF is more than 200 amino acid residues in length in the HBV genome. An AUG initiation codon is present in phase near the beginning of the ORF in most isolates. Although it is possible that the ORF is translated into protein, it may represent the vestige of a former ORF present in the ancestor of the hepadnaviruses.

The trans-activation, or *tat*, gene region of the retrovirus genome is also located at the 3′ end of the linear genome. Analysis indicates an organization similar to that described for hepadnaviruses. Specifically, the *tat* gene region of the human T cell leukemia virus type I (HTLV I)[25,26] contains 9 ORFs of ≥ 100 codons in length (Figure 2). Seven of the ORFs are located on the RNA strand with plus (message sense) polarity and two of the overlapping ORFs are known to encode proteins. Also, two ORFs are located on the strand with opposite polarity to that of the known genes. This is an unusual example of gene organization in two ways: (1) it is rare to have overlapping ORFs greater than 100 codons in length,[27] and (2) it is rare to find an ORF greater than 100 amino acids in length on the strand of opposite polarity to that of known genes.[28,29] However, most, if not all, retroviruses that encode Tat proteins possess these novel organizational features.

Analysis of the regions of the genomes of other retroviruses that encode proteins involved in trans-activation indicates the presence of multiple overlapping ORFs on the plus-DNA strand as well as one or more ORFs located on the minus DNA strand (i.e., the strand that is transcribed into the RNAs that encode the other virus proteins). Minus-strand DNA ORFs are found on the genomes of human T cell leukemia virus type 2,[30] bovine leukemia virus,[31] simian immunodeficiency virus (SIV),[32] human immunodeficiency virus types 1 (HIV-1) and 2 (HIV-2),[13] visna virus,[33] and equine infectious anemia virus.[34,35] These ORFs are heterogeneous in size and usually do not possess initiation codons within the 5′ region of the ORF. Since there is no evidence that retroviruses produce an antisense mRNA species during the course of replication or gene expression it is unlikely that the ORFs represent functional gene sequences. The one possible exception may be the ORF found on the HIV-1 genome. This ORF is approximately 200 amino acids in length and, of all minus DNA strand ORFs analyzed, is most likely to represent a genuine gene sequence.[13] This is the case because the region 5′ to the ORF contains conserved elements commonly found in eukaryotic promoter elements and the region 3′ to the ORF contains the signal sequences necessary for poly(A) addition to mRNA. Also, the ORF possesses the codon periodicity found in protein-encoding sequences, and it is much longer than would be expected to occur by chance. Thus, the HIV-1 ORF may encode a protein now, or did encode a protein from the minus-strand ORF relatively recently in its evolution. Overall, hepadnaviruses and retroviruses that encode *tat* proteins appear to have evolved from an ancestral virus that possessed ORFs on both DNA strands in the 3′ terminus of the linear genome.

Analysis suggests that the minus-strand ORF of the ancestral virus of the retroviruses was once much longer, and may have extended over all, or nearly all, of the length of the genome. Evidence for this comes from the analysis of the antisense ORFs of several retroviruses, including HIV-1 and SIV. The antisense strand of the genome of one isolate of SIV[36] contains 10 ORFs greater than 100 amino acids in length (Figure 3). One of the three possible translation frames contains six ORFs spanning eight of the 10.5-kb length of the virus genome. Thus, it is probable that at least one antisense translation frame was utilized by the ancestor of the retroviruses to encode a protein. However, it appears the ORF is shrinking in size as retroviruses evolve, because the protein product of this ORF may not be essential for virus replication and the gene has lost its original function and importance.

IV. GENOME SEQUENCE SIMILARITY

Soon after the discovery by Summers and Mason of hepadnavirus replication by reverse transcription, Toh and co-workers[4] reported amino acid sequence similarity between the polymerase of HBV and the reverse transcriptase protein of retroviruses. Subsequently, similarity between hepadnavirus and retrovirus nucleotide and protein sequences were reported in other genome regions.[5-12] Specifically, the U5 domain of the retrovirus LTR was found to share similarity with sequences within the core gene of hepadnaviruses.[7] Further, the predicted amino acid sequences of the major retrovirus nucleocapsid protein ($P30^{gag}$) and additional regions of the polymerase protein were shown to be similar to hepadnavirus core[7,9] and polymerase protein sequences,[5,6,10-12] respectively. Overall, hepadnavirus and retrovirus nucleotide and amino acid sequences share similarities that may represent true homology[37] between the sequences of the members of these two virus families.

A. NUCLEOTIDE SIMILARITY

The most highly conserved nucleotide sequence of the hepadnavirus genome is located in the precore gene and the 5′ end of the core gene sequence.[7] Within a 62-nucleotide domain, 54 (87%) of the nucleotides are identical at each position among 16 mammalian hepadnavirus genomes analyzed. Within this domain is a region of 25 consecutive nucleotides that are identical. There is no other region that is conserved to this degree among hepadnaviruses. This suggests that this region has an important biological function. This function could be at the nucleotide (i.e., a cis-acting function) or the amino acid level (i.e., a trans-acting function). Since experiments by two groups[38,39] demonstrate that production of the precore protein is not essential for virus replication, it is probable that the sequence is conserved for a crucial cis-acting function.

Analysis indicates that this conserved region of the genome is homologous to the U5 domain of the LTR of type C retroviruses.[7] The homology encompasses nearly the entire central region of the Moloney murine leukemia virus (MuLV) U5 sequence, but there is no homology at the extreme 5′ and 3′ ends of the sequence. The most distinctive features in this region of homology are the presence of alternating purine and pyrimidine tracts, and a highly conserved region of 16 nucleotides, including a heptanucleotide sequence CCTTGGG that is nearly identical (i.e., 97% homologous) among 27 hepadnavirus and retrovirus isolates.

U5 is recognized as the most conserved, but least understood, element of the retroviral LTR. It has been hypothesized that U5 plays a role in retrovirus replication because of its location adjacent to the primer-binding site, which is the initiation point for the synthesis of the first strand of DNA by reverse transcription. Mutational analysis of the MuLV U5 region by Murphy and Goff[40] has provided an intriguing view of this signal sequence. Deletions in the 5′ domain of U5 severely reduce packaging of virus RNA into nucleocapsids, while mutations in the 3′ domain of U5 effect reverse transcription. These results demonstrate that U5 plays dual roles in the retrovirus life cycle. Interestingly, the greatest similarity between the U5-like sequence of HBV and the U5 sequence of retroviruses corresponds to the 5′ domain essential for packaging the viral RNA into nucleocapsids. Mutational analysis of the HBV genome demonstrates that an 85-nucleotide domain surrounding the U5-like region is both necessary and sufficient for RNA encapsidation.[41] Thus, since hepadnaviruses use a protein primer for initiation of first-strand DNA synthesis, and not a tRNA primer as do retroviruses, they may have maintained only the essential packaging signal sequence necessary for assembly of virus core particles.

B. AMINO ACID SIMILARITY

1. Core Protein

Comparison of hepadnavirus and retrovirus nucleocapsid protein sequences reveals that the proteins share amino acid similarity. The carboxyl terminus of the hepadnavirus core protein is similar to the carboxyl terminus of the P30*gag* protein of MuLV.[7] Within a 98-amino acid domain the proteins share 28 identical and 12 similar[42] amino acids, representing a similarity of 41%.

In MuLV the translation product of the nucleoscapsid, or *gag*, gene is a polyprotein of 539 amino acids that is cleaved proteolytically into 4 smaller proteins of 15, 12, 30, and 10 kDa by a virally encoded protease. Previous studies have shown that the P30 nucleocapsid protein is the essential *gag* protein. Analogous to the hepadnavirus core protein, the carboxyl terminus of the retroviral P30 protein is rich in basic amino acids and is involved in the essential function of nucleic acid binding for nucleocapsid formation.[7]

2. Protease-Like Sequences

Two highly conserved domains in hepadnavirus proteins share sequence similarity with retrovirus protease sequences.[9,10] One domain is located between the core and polymerase gene sequences, while the other domain is located at the amino terminus of the core gene. The amino acid sequence of the first domain is approximately 37% similar to the aspartic acid protease sequence of MuLV. The second hepadnavirus protease-like sequence, located at the amino terminus of the HBV core protein, shares even more similarity to retrovirus protease sequences. Approximately 45% of the amino acids in this region are similar to the HIV-1

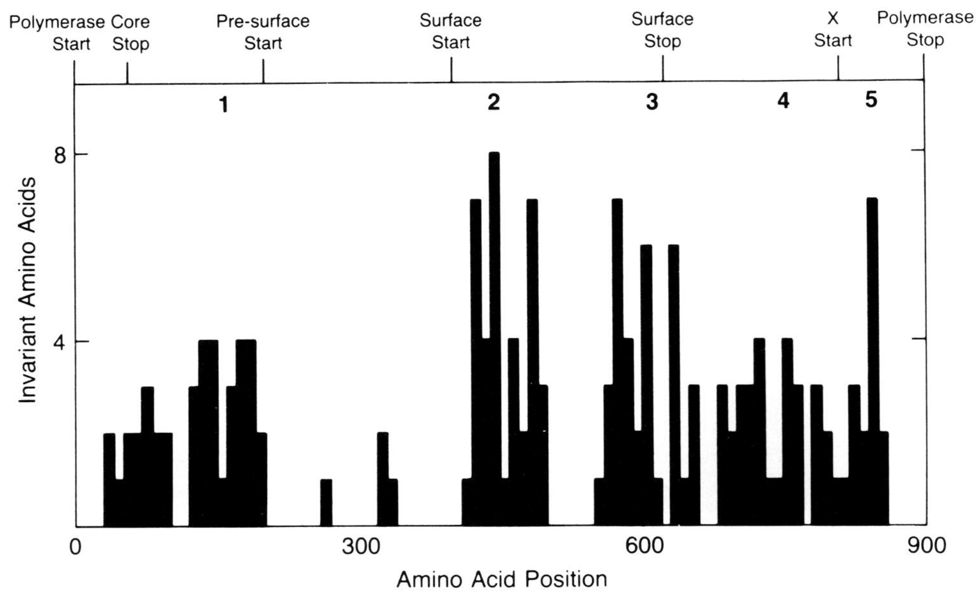

FIGURE 4. Histogram of the conserved regions of the hepadnavirus polymerase protein sequence. The predicted amino acids of the polymerase protein of 14 hepadnaviruses were aligned and the number of identical amino acids in every block of 10 contiguous residues plotted as a function of their position on the molecule.[10] The positions of the initiation and termination codons of the overlapping virus genes are shown. Numbers 1 to 5 refer to domains of high conservation that are discussed in the text.

protease sequence. The domain of the HIV protease that contains the amino acids responsible for catalytic activity is very similar to those in the corresponding region of the HBV protease-like sequence.[9] Unlike the proteases of retroviruses, the hepadnavirus protease-like sequences do not appear to be involved in the cleavage of nucleocapsid–polymerase fusion polyproteins.[43]

CaMV, a retrovirus-like plant virus,[4,44] also possesses a protease-like domain.[45] Apparently, like hepadnaviruses, this protease-like gene is not utilized for the cleavage of polymerase-associated fusion polyproteins. It appears that the CaMV polymerase is expressed by ribosomal reinitiation.[18] Thus, since neither hepadnaviruses nor CaMV utilize a protease to cleave polymerase-associated polyproteins, it remains to be determined whether these protease-like sequences possess proteolytic activity that is utilized in other stages of virus replication, or whether they are analogous to the pseudoprotease sequences discovered in other viruses.

It is possible that one, or both, of the hepadnavirus protease-like sequences are similar to the pseudoprotease sequences identified in retrovirus[46] and vaccinia virus[47] genomes. Computer analysis of lentivirus sequences demonstrated that a domain between the RNase H and integrase genes of the visna and equine infectious anemia virus (EIAV) genomes shares considerable similarity with the protease sequence of the simian retrovirus type I (SRV-I)[46] and not with the protease sequences of more closely related retroviruses. An analogous pseudoprotease sequence was also identified within the *gag* gene of the hamster intracisternal particle (IAP). In addition, Slabaugh and Roseman[47] found that the *Hin*dIII F fragment of the vaccinia virus genome contains a protease-like sequence that shares 31 to 34% amino acid sequence similarity with the above retroviral protease sequences. The function of the pseudoprotease sequences is uncertain. There are no data available accessing the catalytic activity of the protein products of these gene sequences.

The hepadnavirus protease-like sequence located between the core and polymerase gene sequence may represent the original protease sequence inherited from a retrovirus-like ancestor. This protease-like sequence occupies the same genome location as the protease genes of retroviruses, and like other nucleotide and amino acid sequences shares similarity with MuLV sequences. It is clear that the amino acids located at the catalytic active site are poorly conserved since some, but not all, hepadnavirus sequences possess the aspartic acid residue known to reside at the active site of the retrovirus proteases. Thus, it is possible that this protease-like sequence once encoded a protease but that the protein product of this gene has gradually lost its original enzymatic function.

The protease-like sequence located at the amino terminus of the core gene sequence is analogous to the pseudoprotease genes of lentiviruses and vaccinia virus. Like the pseudoproteases, the hepadnavirus protease-like sequence is in an unusual genome position relative to the protease gene sequences of retroviruses. Also, in contrast to other hepadnavirus similarities, this amino acid sequence is more closely related to the protease sequence of HIV-1, and not that of MuLV. It is possible that this hepadnavirus protease-like sequence possesses catalytic activity because the amino acid residues necessary for protease function are well conserved. Although there is experimental evidence indicating that this sequence is not involved in the generation of hepatitis e antigen,[43] it is possible that hepadnaviruses utilize a virally encoded protease in some phase of the virus replication cycle.

3. Polymerase

The longest ORF in the HBV genome encodes the virus polymerase protein.[48-52] The polymerase gene sequence spans approximately three quarters of the virus genome and overlaps all of the other virus genes. When the amino acid sequences of the polymerase proteins of the avian and mammalian hepadnaviruses are aligned, and the invariant residues are plotted in the form of a histogram as a function of their location, five regions within the protein exhibit high conservation of amino acid sequence[10] (Figure 4). Thus, these regions may represent biologically important domains of the protein molecule because they were conserved during hepadnavirus evolution.

Region 1 comprises amino acids 50 to 190 and is located at the amino terminus of the predicted polymerase protein. Although this region is only moderately conserved, there is evidence that this domain plays an important role in the virus replication cycle. In addition to containing one of the protease-like sequences, there is experimental evidence that this region of the hepadnavirus polymerase protein becomes covalently attached to the minus strand of virus DNA.[53] Thus, this domain may possess sequences responsible for priming first-strand DNA synthesis during virus replication by reverse transcription. In support of this hypothesis, a second group has reported similarity between the amino acid sequences of this region and the genome-linked protein of picornaviruses.[12] Thus, it is likely that region 1 is conserved for a function in virus replication.

Polymerase region 2 consists of amino acid residues 420 to 500. This region is the most conserved domain of the hepadnavirus polymerase sequence. Approximately 45% of the amino acids are identical among the sequences examined. Region 2 exhibits considerable similarity to the amino terminus of the reverse transcriptase domain of the polymerase protein sequence of retroviruses where the gene order has been determined to be N-reverse transcriptase-ribonuclease H-endonuclease-C.[54-56] The hepadnavirus sequence is 45% similar to the MuLV polymerase protein.[10] Within an 82-residue domain there are 22 amino acids identical and 15 residues similar between the hepadnavirus and retrovirus polymerase proteins.

Regions 2 and 3 are separated by a segment of about 50 amino acids in the mammalian viruses, but are adjacent in the avian hepadnaviruses. The significance of this observation will be discussed in the next section.

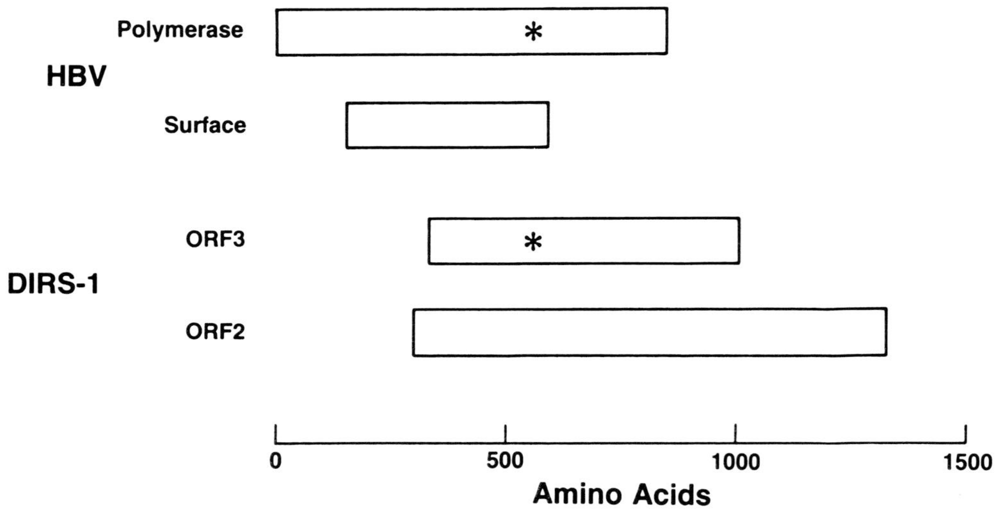

FIGURE 5. Comparison of the long overlapping open reading frames of hepatitis B virus and DIRS-1. The overlapping open reading frames of hepatitis B virus (HBV) and DIRS-1 are shown as boxes. The position of the invariant Asp–Asp moiety of retrovirus reverse transcriptases[4] is depicted as an asterisk (*) in the appropriate open reading frame (ORF).

The next conserved region, region 3, consists of amino acids 560 to 660, where 35% of the amino acid residues are identical at each position. The similarity between region 3 of the hepadnavirus polymerase and the carboxyl terminus of the reverse transcriptase domain of the retrovirus polymerase was first documented by Toh et al.[4] They found that hepadnaviruses and MuLV polymerase protein sequences shared 40% sequence similarity. Within a 94-residue region, 21 of the amino acids are identical and 17 residues similar. It is clear that the central domain of the hepadnavirus polymerase (regions 2 and 3) is most conserved, and probably contains the amino acids located at the catalytic active site of the virus replicase.

Regions 4 and 5 are located at the carboxyl terminus of the hepadnavirus polymerase. These regions, encompassing amino acids 675 to 775 and 810 to 850, respectively, also share similarity to retrovirus polymerase sequences. Region 4 is reported to share similarity with both the reverse transcriptase[10] and the ribonuclease H (RNase H) domains[11,12] of the MuLV polymerase sequence. The homology of the 5′ half of region 4 is clearly most similar to the reverse transcriptase domain where, over a 22-amino acid domain, 54% of the residues are homologous (7 identical and 5 similar amino acids).[10] Region 5 has also been shown to share similarity with the RNase H domain of the retrovirus polymerase protein.[11,12] However, this similarity is very limited, which is illustrated by the fact that the two published alignments differ.[11,12] Overall, amino acid sequence analysis predicts that the hepadnavirus and retrovirus polymerase proteins are organized in a similar fashion with the gene order of reverse transcriptase-RNase H found in both virus families.

It is significant that there has been no report of similarity between the hepadnavirus polymerase protein and the well-conserved integrase domain of the retrovirus polymerase protein. This finding implies that hepadnaviruses have not maintained the integrase gene sequence during the course of their evolution. This hypothesis agrees with existing experimental evidence indicating that the hepadnavirus replication cycle does not include a step requiring integration into host cell DNA.

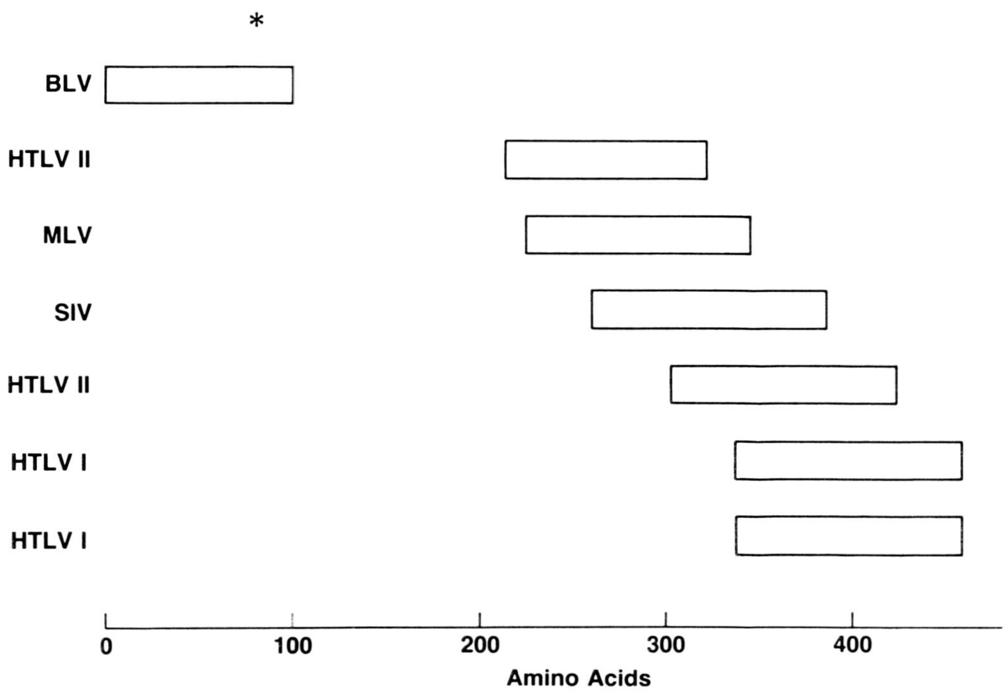

FIGURE 6. Overlapping open reading frames within the polymerase gene sequence of retroviruses. The open reading frames longer than 100 amino acids in alternative translation frames of the polymerase gene are shown as open boxes. The relative position of the invariant Asp–Asp moiety of the reverse transcriptase of retroviruses[4] is shown as an asterisk (*). Virus sequences depicted are bovine leukemia virus (BLV),[31] human T cell leukemia virus type II (HTLV II),[30] murine leukemia virus (MuLV),[60] simian immunodeficiency virus (SIV),[32] and human immunodeficiency virus type I (HTLV I).[25,26] It should be noted that the two ORFs labeled HTLV II are two independent ORFs found by translating one genome isolate while the two ORFs labeled HTLV I are separate, identical ORFs found by translating two different genome sequences.

V. NOVEL FEATURES

There are several novel features found in the organization of the hepadnavirus genome that are shared with the retroviruses and retrovirus-like genetic elements that lend support to the hypothesis that these viruses shared a common ancestor in their evolution. One such feature is the presence of an overlapping ORF within the polymerase gene sequence. The HBV surface protein gene sequence, which consists of approximately 465 amino acids (including the presurface region), is contained completely within the coding sequences of the polymerase protein in an alternate translation frame. Analysis by Ohno[27] demonstrates that this overlapping reading frame did not arise by chance, but descended from a sequence without termination codon sequences. Therefore, the predecessor of hepadnaviruses possessed this unique gene arrangement.

If hepadnaviruses and retroviruses diverged from a common ancestral virus it would be predicted that the retrovirus polymerase gene should possess an analogous overlapping ORF, or a vestige of a former ORF, in the corresponding region of the polymerase gene sequence. Analysis of the genomes of retrovirus-like genetic elements and retroviruses verifies this prediction. Sequence analysis of the retrovirus-like elements Bs1[57,58] and *Dictyostelium* inverted repeat sequence 1 (DIRS-1)[59] reveals the presence of a gene that shares amino acid sequence similarity with the polymerase protein of retroviruses. Both elements possess an overlapping ORF within the polymerase-like gene sequence. In DIRS-1 the two ORFs overlap by more than 500 residues (Figure 5). Therefore, it is likely that the overlapping gene arrangements of Bs1 and DIRS-1 are analogous to that of the surface and polymerase genes of hepadnaviruses and were inherited from a common ancestral virus.

Although a corresponding overlapping ORF has not yet been found within the polymerase gene of retroviruses, there is evidence for the presence of a vestigial ORF that overlaps the polymerase gene sequence. This is characterized by the presence of overlapping ORFs within the polymerase gene with lengths that are greater than expected to occur by chance. Such ORFs are found in the polymerase gene of MuLV, bovine leukemia virus, human T cell leukemia virus types 1 and 2, and simian immunodeficiency virus (Figure 6). Thus, there is evidence that retroviruses, and retrovirus-like genetic elements, evolved from an ancestral virus that possessed a long ORF that overlapped with the polymerase gene sequence.

There is a second line of evidence that suggests that at least one retrovirus, or its immediate ancestor, once encoded a protein in an overlapping gene within the polymerase gene. The domain of the hepadnavirus polymerase sequence between regions 1 and 2 is utilized as a binding site for cellular nuclear factor I (NF-I). NF-I is involved in the regulation of the surface protein gene. The polymerase gene of MuLV contains a domain of approximately 150 nucleotides that correspond to sequences in the NF-I binding site of hepadnaviruses.[10] Thus, MuLV exhibits both a vestigial overlapping gene ORF and sequences analogous to the NF-I-binding site of the surface protein gene of hepadnaviruses. Thus, it is likely that both retroviruses and hepadnaviruses evolved from an ancestral virus that encoded a protein from an alternate translation frame within its polymerase gene. Since this type of gene arrangement is extremely rare, it is probable that these virus families shared this common ancestor in their evolution.

Interestingly, the amino acid similarities between the avian hepadnavirus and the MuLV polymerase sequences are colinear. As mentioned earlier, mammalian hepadnavirus genome sequences contain additional sequences located between conserved regions 2 and 3 that are not found in the avian viruses. Comparison of the polymerase sequence of the avian hepadnaviruses to that of MuLV reveals that the two sequences are nearly identical in length in the conserved reverse transcriptase domain of the MuLV polymerase.[10] The extra sequences found in the mammalian hepadnavirus genome may represent an insertion of bases that occurred in the ancestor of the mammalian hepadnaviruses. Overall, organization of the polymerase sequences of the avian hepadnaviruses and MuLV are remarkably similar.

VI. MUTATION RATE OF THE HEPADNAVIRUS GENOME

The mutation rate of cellular genes and DNA viruses is approximately 10^{-9} base substitutions per site per year.[61] However, the mutation rate for the RNA viruses exceeds this value by a factor of about 1 million. One of the main reasons for this large difference in mutation rates is the lack of proofreading, or editing, functions associated with the polymerase of the RNA viruses.[62]

An essential factor for calculating the evolutionary divergence of hepadnaviruses and retroviruses is an estimation of the mutation rate of the virus genome. The mutation rate is defined as the number of base substitutions that occur in the genome per site per year of continuous virus replication in the host. Experiments performed in our laboratory using the

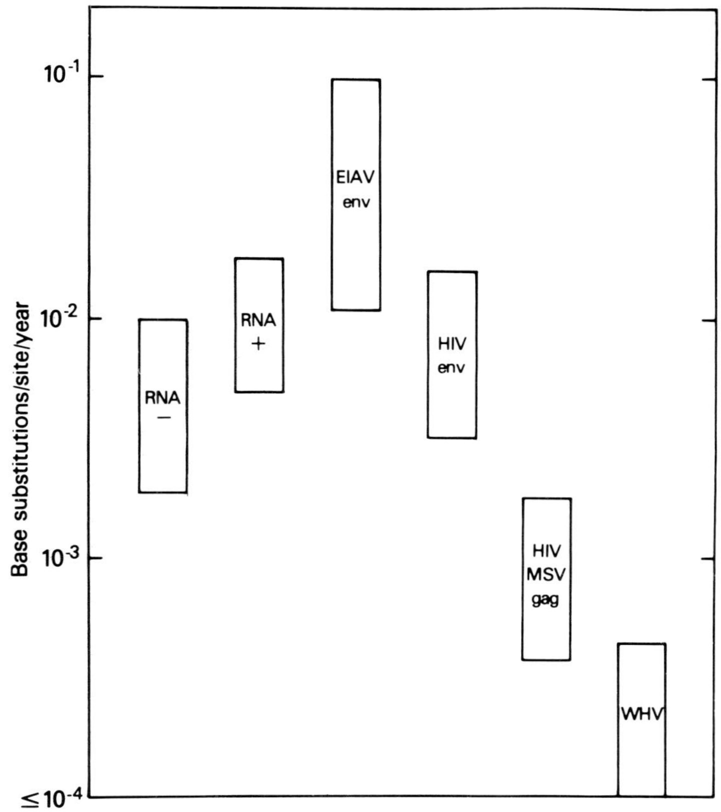

FIGURE 7. Comparison of the mutation rates of viruses that lack polymerase proofreading functions. The range of mutation rates for genome replication in animal or human hosts is shown for the following viruses: equine infectious anemia virus (EIAV),[71] human immunodeficiency virus type 1 (HIV),[70] murine sarcoma virus (MSV),[69] the positive-strand RNA (+RNA) viruses,[64,65] and the negative-strand RNA (–RNA) viruses.[62,66-68]

woodchuck hepatitis virus animal model system indicate that the mutation rate of the hepadnavirus genome is approximately 2×10^{-4} base substitutions per site per year.[63]

Compared to the genomes of other viruses that replicate using a polymerase that lacks proofreading function, the hepadnavirus genome is relatively stable during virus replication. The low mutation rate of the virus genome may be the direct result of a relatively high fidelity of the hepadnavirus polymerase compared to that of retroviruses or other RNA viruses. However, it is also possible that several unique features of the hepadnavirus genome contribute to an underestimation of the mutation rate. This could arise if many of the point mutations that arise at random during virus replication are deleterious or lethal for the virus. This is more likely to be true for hepadnaviruses than other viruses because the hepadnavirus genome is extremely compact in its organization: proteins are encoded from overlapping translation frames and all cis-acting regulatory signal sequences are contained within gene sequences.[20] Therefore, compared to other viruses, hepadnaviruses have unusually high constraints imposed upon the genome sequence and, thus, may be more susceptible to lethal mutations.

The mutation rates of the most intensively studied viruses, the positive- and negative-strand RNA viruses, have been estimated to be between 10^{-2} and 10^{-3} base substitutions per site per year. Nottay et al.[64] reported that the polio virus genome had a mutation rate of 1 to 2% during a 13-month outbreak in 1978. Coupled with results from studies on foot-and-mouth disease virus,[65] the mutation rate of plus-strand RNA viruses has been estimated at 10^{-2} to 10^{-3} base substitutions per site per year. A similar result has been found for negative-strand RNA viruses. Buonagurio et al.[66] sequenced the NS gene of influenza A viruses isolated over 50 years and found a uniform rate of mutation of 1.9×10^{-3} base substitutions per site per year. Consistent with this finding, mutation rates of between 10^{-2} and 10^{-3} have been calculated in other studies on influenza A, as well as for other negative-strand RNA viruses.[62,67,68] Thus, the mutation rate of the genomes of positive- and negative-strand RNA viruses in susceptible hosts appears to be higher than that of the hepadnavirus genome (Figure 7).

A number of excellent studies have been reported regarding the mutation rate of selected genes of retroviruses during replication of the virus in susceptible hosts. The mutation rate of individual retrovirus genes has been estimated as between 10^{-2} and 10^{-4} base substitutions per site per year. Specifically, Gojobori and Yokoyama[69] determined the rate of mutation of several genes of the murine sarcoma virus (MuSV) during propagation in mice. They found that the divergence of the Moloney and Abelson strains of MuSV separated by 14 years of growth in mice was 1.3×10^{-3} and 6.3×10^{-4} base substitutions per site per year in the v-*mos* and *gag* genes, respectively. Also, Hahn et al.,[70] studying the human immunodeficiency virus (HIV), estimated a mutation rate of 1.6×10^{-2} to 3.2×10^{-3} base substitutions per site per year for the *env* gene and 1.8×10^{-3} to 3.7×10^{-4} base substitutions per site per year for the *gag* gene during the course of a natural infection. In a third study, the mutation rate of the *env* gene of three isolates of equine infectious anemia virus (EIAV) was found to be between 1.1×10^{-2} and 1.9×10^{-3} base substitutions per site per year.[71] Taken together, these studies suggest that regions of the retrovirus genome evolve at different rates, and that *gag* gene sequences have the lowest mutation rates of the retrovirus genes examined (Figure 7). At 2.3×10^{-4} base substitutions per site per year, the mutation rate of the hepadnavirus genome is equal to, or slightly lower than, the mutation rate of the most slowly evolving gene (i.e., *gag*) of retroviruses (Figure 7).

Hepadnaviruses and retroviruses utilize at least two polymerases in their replication cycle.[14] One is a host cell-derived DNA-dependent RNA polymerase (RNA polymerase II) that is normally used by the host for transcription of chromosomal genes, producing messenger RNAs used in the synthesis of cellular proteins. The second polymerase used in virus replication is a virally encoded polymerase. The viral polymerase has two catalytic activities: (1) an RNA-dependent DNA polymerase, or reverse transcriptase, activity and (2) a DNA-dependent DNA polymerase activity (Figure 1). It is not known whether the catalytic active sites for these two activities are at identical or different sites on the polymerase protein. Therefore, the mutation rate of hepadnaviruses and retroviruses is a measure of the fidelity of two polymerases with a total of three catalytic activities. Thus, another point of similarity between these virus families is a comparable mutation rate that is reflective of the similarities in their replication mechanism.

VII. CONCLUSIONS

How did the hepadnaviruses evolve from a retrovirus-like ancestor? One likely explanation is that they evolved through a process of deletion from a larger virus. How the specific deletions occurred that resulted in the evolution of the hepadnaviruses is still open to speculation. It is known that specific regions in retrovirus genomes are favored sites for homologous recombination. Intramolecular recombination, followed by deletion, has been known to produce truncated retroviruses and endogenous retrovirus-like elements that possess genome organizations analogous to that of the hepadnaviruses.[7] For example, spleen necrosis virus mutants have been characterized and found to possess deletions between homologous sequences in the 3′ end of U5 and the amino-terminal end of the $P30^{gag}$ gene sequence. Thus, this is an example of how hepadnaviruses may have evolved with a U5-like domain fused to the P30^{gag} nucleocapsid gene. Also, endogenous retrovirus-like elements have been characterized that lack the integrase domain of the polymerase gene. Thus, it is possible that hepadnaviruses evolved from a larger, retrovirus-like ancestor.

During the course of evolution, the hepadnaviruses appear to have retained the essential sequences necessary for virion production and genome replication by reverse transcription of RNA that were present in its ancestor. Such sequences were the putative packaging signal sequence in the U5-like region, the nucleic acid-binding domain of the nucleocapsid protein, a gene (i.e., X) involved in regulation of virus gene expression, the reverse transcriptase and RNase H domains of the polymerase, and the overlapping gene located within the polymerase gene that the hepadnaviruses utilize as an envelope protein. However, hepadnaviruses appear to have lost the functions involved in specific integration of genomic DNA. Further study of the relationship of hepadnaviruses and retroviruses should continue to provide insight into the process of virus evolution.

ACKNOWLEDGMENTS

The author gratefully acknowledges the contributions of R. Purcell, R. Chanock, R. Girones, S. Kaneko, D. Wong, T. Chesnut, J. Woodcock, B. Baldwin, W. Hornbuckle, B. Tennant, P. Cote, and J. Gerin. Computer-assisted nucleic and amino acid analysis was through the BIONET National Computer Resource for Molecular Biology, which is supported by National Institutes of Health Grant U41-01685-05.

REFERENCES

1. **Marion, P. L.,** Use of animal models to study hepatitis B virus, *Prog. Med. Virol.*, 35, 43, 1988.
2. **Summers, J. and Mason, W. S.,** Replication of the genome of a hepatitis B-like virus by reverse transcription of an RNA intermediate, *Cell*, 29, 403, 1982.
3. **Miller, R. H., Marion, P. L., and Robinson, W. S.,** Hepatitis B viral DNA-RNA hybrid molecules in particles from infected liver are converted to viral DNA molecules during an endogenous DNA polymerase reaction, *Virology*, 139, 64, 1984.
4. **Toh, H., Hayashida, H., and Miyata, T.,** Sequence homology between retroviral reverse transcriptase and putative polymerase of hepatitis B virus and cauliflower mosaic virus, *Nature (London)*, 305, 827, 1983.
5. **Patarca, R. and Haseltine, W. A.,** Sequence similarity among retroviruses — Erratum, *Nature (London)*, 309, 728, 1984.
6. **Toh, H., Kikuno, R., Hayashida, H., Miyata, T., Kunimiya, W., Inouye, S., Yuki, S., and Saigo, K.,** Close structural resemblance between putative polymerase of a Drosophila transposable genetic element 17.6 and pol gene product of Moloney murine leukaemia virus, *EMBO J.*, 4, 1267, 1985.
7. **Miller, R. H. and Robinson, W. S.,** Common evolutionary origin of hepatitis B virus and retroviruses, *Proc. Natl. Acad. Sci. U.S.A.*, 83, 2531, 1986.
8. **Robinson, W. S., Miller, R. H., and Marion, P. L.,** Hepadnaviruses and retroviruses share genome homology and features of replication, *Hepatology*, 7, 64S, 1987.
9. **Miller, R. H.,** Proteolytic self-cleavage of hepatitis B virus core protein may generate serum e antigen, *Science*, 236, 722, 1987.
10. **Miller, R. H.,** Close evolutionary relatedness of the hepatitis B virus and murine leukemia virus polymerase gene sequences, *Virology*, 164, 147, 1988.
11. **Schodel, F., Weimer, T., Will, H., and Sprengel, R.,** Amino acid sequence similarity between retroviral and *E. coli* RNase H and hepadnaviral gene products, *AIDS Res. Hum. Retroviruses*, 4, 9, 1988.
12. **Khudyakov, Y. E. and Makhov, A. M.,** Prediction of terminal protein and ribonuclease H domains in the gene P product of hepadnaviruses, *FEBS Lett.*, 243, 115, 1989.

13. **Miller, R. H.,** Human immunodeficiency virus may encode a novel protein on the genomic DNA plus strand, *Science,* 239, 1420, 1988.
14. **Ganem, D. and Varmus, H. E.,** The molecular biology of the hepatitis B viruses, *Annu. Rev. Biochem.,* 56, 651, 1987.
15. **Will, H., Salfeld, J., Pfaff, E., Manso, C., Theilmann, L., and Schaller, H.,** Putative reverse transcriptase intermediates of human hepatitis B virus in primary liver carcinoma, *Science,* 231, 594, 1986.
16. **Chang, L.-J., Pryciak, P., Ganem, D., and Varmus, H. E.,** Biosynthesis of the reverse transcriptase of hepatitis B viruses involves *de novo* translation initiation not ribosomal frameshifting, *Nature (London),* 337, 364, 1989.
17. **Schlicht, H.-J., Radziwill, G., and Schaller, H.,** Synthesis and encapsidation of duck hepatitis B virus reverse transcriptase do not require formation of core-polymerase fusion proteins, *Cell,* 56, 85, 1989.
18. **Penswick, J., Hubler, R., and Hohn, T.,** A viable mutation in cauliflower mosaic virus, a retroviruslike plant virus, separates its capsid protein and polymerase genes, *J. Virol.,* 62, 1460, 1988.
19. **Kaneko, S. and Miller, R. H.,** X-region-specific transcript in mammalian hepatitis B virus-infected liver, *J. Virol.,* 62, 3979, 1988.
20. **Miller, R. H., Kaneko, S., Chung, C. T., Girones, R., and Purcell, R. H.,** Compact organization of the hepatitis B virus genome, *Hepatology,* 9, 322, 1989.
21. **Anderson, C. W. and Buzash-Pollert, E.,** Can ACG serve as an initiation codon for protein synthesis in eucaryotic cells? *Mol. Cell. Biol.,* 5, 3621, 1985.
22. **Peabody, D. S.,** Translation initiation at an ACG triplet in mammalian cells, *J. Biol. Chem.,* 262, 11847, 1987.
23. **Curran, J. and Kolakofsky, D.,** Ribosomal initiation from an ACG codon in the Sendai virus P/C mRNA, *EMBO J.,* 7, 245, 1988.
24. **Curran, J. and Kolakofsky, D.,** Scanning independent ribosomal initiation of the Sendai virus Y proteins *in vitro* and *in vivo, EMBO J.,* 8, 521, 1989.
25. **Seiki, M., Hattori, S., Hirayama, Y., and Yoshida, M.,** Human adult T-cell leukemia virus: complete nucleotide sequence of the provirus genome integrated in leukemia cell DNA, *Proc. Natl. Acad. Sci. U.S.A.,* 80, 3618, 1983.
26. **Malik, K., Even, J., and Karpas, A.,** Molecular cloning and complete nucleotide sequence of an adult T cell leukaemia virus/human T cell leukaemia virus type I (ATLV/HTLV-I) isolate of Caribbean origin: relationship to other members of the ATLV/HTLV-I subgroup, *J. Gen. Virol.,* 69, 1695, 1988.
27. **Ohno, S.,** Segmental homology and internal repetitiousness identified in putative nucleic acid polymerase and human hepatitis B surface antigen of human hepatitis B virus, *Proc. Natl. Acad. Sci. U.S.A.,* 81, 3781, 1984.
28. **Casino, A., Cipollaro, M., Guerrini, A. M., Mastrocinque, G., Spena, A., and Scarlato, V.,** Coding capacity of complementary DNA strands, *Nucleic Acids Res.,* 9, 1499, 1981.
29. **Sharp, P. M.,** Does the "non-coding" strand code?, *Nucleic Acids Res.,* 13, 1389, 1985.
30. **Shimotohno, K., Takahashi, Y., Shimizu, N., Gojobori, T., Golde, D. W., Chen, I., Miwa, M., and Sugimura, T.,** Complete nucleotide sequence of an infectious clone of human T-cell leukemia virus type II: an open reading frame for the protease gene, *Proc. Natl. Acad. Sci. U.S.A.,* 82, 3101, 1985.
31. **Sagata, N., Yasunaga, T., Tsuzuku-Kawamura, J., Ohishi, K., Ogawa, Y., and Ikawa, Y.,** Complete nucleotide sequence of the genome of bovine leukemia virus: its evolutionary relationship to other retroviruses, *Proc. Natl. Acad. Sci. U.S.A.,* 82, 677, 1985.
32. **Fukasawa, M., Miura, T., Hasegawa, A., Morikawa, S., Tsujimoto, H., Miki, K., Kitamura, T., and Hayami, M.,** Sequence of simian immunodeficiency virus from African Green monkey, a new member of HIV/SIV group, *Nature (London),* 333, 457, 1988.
33. **Sonigo, P., Alizon, M., Staskus, K., Klatzmann, D., Cole, S., Danos, O., Retzel, E., Tiollais, P., Haase, A., and Wain-Hobson, S.,** Nucleotide sequence of the visna lentivirus: relationship to the AIDS virus, *Cell,* 42, 369, 1985.
34. **Stephens, R. M., Casey, J. W., and Rice, N. R.,** Equine infectious anemia virus gag and pol genes: relatedness to visna and AIDS virus, *Science,* 231, 589, 1986.
35. **Kawakami, T., Sherman, L., Dahlberg, J., Gazit, A., Yaniv, A., Tronick, S. R., and Aaronson, S. A.,** Nucleotide sequence analysis of equine infectious anemia virus proviral DNA, *Virology,* 158, 300, 1987.
36. **Hirsch, V. M., Olmsted, R. A., Murphey-Corb, M., Purcell, R. H., and Johnson, P. R.,** An African primate lentivirus (SIV_{sm}) closely related to HIV-2, *Nature (London),* 339, 389, 1989.
37. **Reeck, G. R., deHaen, C., Teller, D. C., Doolittle, R. F., Fitch, W. M., Dickerson, R. E., Chambon, P., McLachlan, A. D., Margoliash, E., Juker, T. H., and Zuckerkandl, E.,** "Homology" in proteins and nucleic acids: a terminology muddle and a way out of it, *Cell,* 50, 667, 1987.
38. **Chang, C., Enders, G., Sprengel, R., Peters, N., Varmus, H. E., and Ganem, D.,** Expression of the precore region of an avian hepatitis B virus is not required for viral replication, *J. Virol.,* 61, 3322, 1987.
39. **Schlicht, H., Salfeld, J., and Schaller, H.,** The duck hepatitis B virus pre-C region encodes a signal sequence which is essential for synthesis and secretion of processed core proteins but not for virus formation, *J. Virol.,* 61, 3701, 1987.

40. **Murphy, J. E. and Goff, S. P.**, Construction and analysis of deletion mutations in the U5 region of Moloney murine leukemia virus: effects on RNA packaging and reverse transcription, *J. Virol.*, 63, 319, 1989.
41. **Junker-Niepmann, M., Bartenschlager, R., and Schaller, H.**, A short *cis*-acting sequence is required for hepatitis B virus pregenome encapsidation and sufficient for packaging of foreign RNA, *EMBO J.*, 9, 3389, 1990.
42. **Taylor, W. R.**, The classification of amino acid conservation, *J. Theor. Biol.*, 119, 205, 1986.
43. **Nassal, M., Galle, P. R., and Schaller, H.**, Proteaselike sequence in hepatitis B virus core antigen is not required for e antigen generation and may not be part of an aspartic acid-type protease, *J. Virol.*, 63, 2598, 1989.
44. **Covey, S. N.**, Amino acid sequence homology in gag region of reverse transcribing elements and the coat protein gene of cauliflower mosaic virus, *Nucleic Acids Res.*, 14, 623, 1986.
45. **Toh, H., Ono, M., Saigo, K., and Miyata, T.**, Retroviral protease-like sequence in the yeast transposon Ty 1, *Nature (London)*, 315, 691, 1985.
46. **McClure, M. A., Johnson, M. S., and Doolittle, R. F.**, Relocation of a protease-like gene segment between two retroviruses, *Proc. Natl. Acad. Sci. U.S.A.*, 84, 2693, 1987.
47. **Slabaugh, M. B. and Roseman, N. A.**, Retroviral protease-like gene in the vaccinia virus genome, *Proc. Natl. Acad. Sci. U.S.A.*, 86, 4152, 1989.
48. **Bavand, M. R., and Laub, O.**, Two proteins with reverse transcriptase activities associated with hepatitis B virus-like particles, *J. Virol.*, 62, 626, 1988.
49. **Bavand, M., Feitselson, M., and Laub, O.**, The hepatitis B virus-associated reverse transcriptase is encoded by the viral *pol* gene, *J. Virol.*, 63, 1019, 1989.
50. **Feitelson, M. A., Millman, I., Duncan, G. D., and Blumberg, B. S.**, Presence of antibodies to the polymerase gene product(s) of hepatitis B and woodchuck hepatitis virus in natural and experimental infections, *J. Med. Virol.*, 24, 121, 1988.
51. **Mack, D. H., Block, W., Nath, N., and Sninsky, J. J.**, Hepatitis B virus particles contain a polypeptide encoded by the largest open reading frame: a putative reverse transcriptase, *J. Virol.*, 62, 4786, 1988.
52. **Stemler, M., Hess, J., Braun, R., Will, H., and Schroder, C. H.**, Serological evidence for expression of the polymerase gene of human hepatitis B virus *in vivo*, *J. Gen. Virol.*, 69, 689, 1988.
53. **Bosch, V., Bartenschlager, R., Radziwill, G., and Schaller, H.**, The duck hepatitis B virus P-gene codes for protein strongly associated with the 5′-end of the viral DNA minus strand, *Virology*, 166, 475, 1988.
54. **Johnson, M. S., McClure, M. A., Feng, D.-F., Gray, J., and Doolittle, R. F.**, Computer analysis of retroviral *pol* genes: assignment of enzymatic functions to specific sequences and homologies with nonviral enzymes, *Proc. Natl. Acad. Sci. U.S.A.*, 83, 7648, 1986.
55. **Prasad, V. R. and Goff, S. P.**, Linker insertion mutagenesis of the human immunodeficiency virus reverse transcriptase expressed in bacteria: definition of the minimal polymerase domain, *Proc. Natl. Acad. Sci. U.S.A.*, 86, 3104, 1989.
56. **Repaske, R., Hartley, J. W., Kavlick, M. F., O'Neill, R. R., and Austin, J. B.**, Inhibition of RNase H activity and viral replication by single mutations in the 3′ region of Moloney murine leukemia virus reverse transcriptase, *J. Virol.*, 63, 1460, 1989.
57. **Johns, M. A., Babcock, M. S., Fuerstenberg, S. M., Fuerstenberg, S. I., Freeling, M., and Simpson, R. B.**, An unusually compact retrotransposon in maize, *Plant Mol. Biol.*, 12, 633, 1989.
58. **Jin, Y.-K. and Bennetzen, J. L.**, Structure and coding properties of Bs1, a maize retrovirus-like transposon, *Proc. Natl. Acad. Sci. U.S.A.*, 86, 6235, 1989.
59. **Cappello, J., Handelsman, K., and Lodish, H.**, Sequence of *Dictyostelium* DIRS-1: an apparent retrotransposon with inverted terminal repeats and an internal circle junction sequence, *Cell*, 43, 105, 1985.
60. **Shinnick, T. M., Lerner, R. A., and Sutcliffe, J. G.**, Nucleotide sequence of Moloney murine leukemia virus, *Nature (London)*, 293, 543, 1981.
61. **Britten, R. J.**, Rates of DNA sequence evolution differ between taxonomic groups, *Science*, 231, 1393, 1986.
62. **Holland, J., Spindler, K., Horodyski, F., Grabau, E., Nichol, S., and VandePol, S.**, Rapid evolution of RNA genomes, *Science*, 215, 1577, 1982.
63. **Girones, R. and Miller, R. H.**, Mutation rate of the hepadnavirus genome, *Virology*, 170, 595, 1989.
64. **Nottay, B. K., Kew, O. M., Hatch, M. H., Heyward, J. T., and Obijeski, J. F.**, Molecular variation of type 1 vaccine-related and wild poliovirus during replication in humans, *Virology*, 108, 405, 1981.
65. **Piccone, M. E., Kaplan, G., Giavedoni, L., Domingo, E., and Palma, E. L.**, VP1 of serotype C foot-and-mouth disease viruses: long-term conservation of sequences, *J. Virol.*, 62, 1469, 1988.
66. **Buonagurio, D. A., Nakada, S., Parvin, J. D., Krystal, M., Palese, P., and Fitch, W. M.**, Evolution of human influenza A viruses over 50 years; rapid, uniform rate of change in NS gene, *Science*, 232, 980, 1986.
67. **Hayashida, H., Toh, H., Kikuno, R., and Miyata, T.**, Evolution of influenza virus genes, *Mol. Biol. Evol.*, 2, 289, 1985.
68. **Saitou, N. and Nei, M.**, Polymorphism and evolution of influenza A virus genes, *Mol. Biol. Evol.*, 3, 57, 1986.

69. **Gojobori, T. and Yokoyama, S.,** Rates of evolution of the retroviral oncogene of Molony murine sarcoma virus and of its cellular homologues, *Proc. Natl. Acad. Sci. U.S.A.*, 82, 4198, 1985.
70. **Hahn, B. H., Shaw, G. M., Taylor, M. E., Redfield, R. R., Markham, P. D., Salahuddin, S. Z., Wong-Staal, F., Gallo, R. C., Parks, E. S., and Parks, W. P.,** Genetic variation in HTLV-III/LAV over time in patients with AIDS or at risk for AIDS, *Science*, 232, 1548, 1986.
71. **Payne, S. L., Fang, F.-D., Liu, C.-P., Dhruva, B. R., Rwanbo, P., Issel, C. J., and Montelaro, R. C.,** Antigenic variation and lentivirus persistence in envelope gene sequences during EIAV infection resemble changes for sequential isolates of HIV, *Virology*, 161, 321, 1987.

Chapter 13

CHROMOSOMAL CHANGES ASSOCIATED WITH HEPATITIS B VIRUS DNA INTEGRATION AND HEPATOCARCINOGENESIS

Kenichi Matsubara

TABLE OF CONTENTS

I. INTRODUCTION

Because epidemiologic studies have shown that hepatocellular carcinoma (HCC) is closely associated with chronic hepatitis B virus (HBV) infection,[1-4] much attention has been paid to obtaining direct evidence that HBV is causally related to HCC. Although direct evidence has yet to be obtained, there are three key observations that help establish a link between HBV infection and HCC formation.

The first observation relates to discoveries made in the 1980s that show that HBV DNA is integrated into cellular DNA.[5-12] As described here, the majority of the HCCs related to HBV infection contain HBV DNA integrated into the chromosomal DNA, and integration seems to occur during the early stage of virus multiplication. The question is whether the integration process itself, or the consequences thereof, diverts the cellular regulatory circuit toward hepatocarcinogenesis.

The second observation relates to the discoveries of cis- and trans-acting genetic elements in the HBV genome in studies that used cultured cells for the replication of HBV. The cultured cell lines established in 1982 to 1983 have made possible detailed molecular biological studies of the expression, replication, and regulation of this virus.[13-15] Some of the integrated viral genetic elements are likely to affect one or more of the regulatory processes of the host cells.

The third observation relates to recent advancements in our knowledge of genetic and epigenetic changes that occur in hepatoma cells and in hepatocytes subjected to regenerative conditions (for example, chronic hepatitis or other damage). The appearance of clonal cells (with or without formation of hyperplastic nodules) from which transformed cells often develop reflects these changes. Although molecular biological studies that deal with these changes have only just begun, the concept of multistep changes and cell selection is of crucial importance. In this chapter, recent advances in related research are summarized and the implications for hepatocarcinogenesis are evaluated.

II. HBV DNA INTEGRANTS IN HEPATOCELLULAR CARCINOMA CELLS

Integrated HBV genomes are found in about 80% or more of the HCCs that develop as a consequence of HBV-mediated chronic active hepatitis. In Southern blotting analyses of HCC cellular DNA, cleaved by an enzyme (such as *Hin*dIII) that does not cut the HBV genome DNA sequence, the integrants are readily detectable as discrete bands when probed with HBV DNA. In most cases there are multiple integrants, usually 3 to 4 but sometimes more than 10. A well-known HCC cell line, PLC/PRF/5 (Alexander),[16] carries at least eight integrants;[17] but HCCs that carry an excessively large number of integrants are seldom encountered, which suggests that there may be counterselection for hepatocytes that carry too many integrants, although the infection and replication of this virus per se is not cytocidal. HCCs that carry only one integrant are also not often encountered, which indicates that the founder cell must be subjected to recurring integrations. Even with a sample that displays only one band of hybridization, one cannot rule out the possibility that there might have been other integrants lost by deletion or chromosomal rearrangement (see Section IV). Therefore, selection of an HCC that shows a single band of hybridization and the direct correlation of this integrant to the carcinogenic process must be approached with caution.

Hybridization band profiles differ from HCC to HCC, no common band existing. This indicates that the majority of HCC cells are monoclonal and suggests that integration occurs at various sites along the cellular DNA. In contrast, DNA samples from acute-phase hepatitis livers commonly show a "smear Southern profile" as the integrants are not monoclonal.

Interestingly, many HCC cells do not carry extrachromosomal HBV DNA, a sign that they do not allow replication of the viral DNA. This is in sharp contrast to chronic hepatitis cases, in which extrachromosomal HBV DNAs predominate. Presumably, most HCC cells are genetically, epigenetically nonpermissive for HBV replication and/or the mechanism of which is imperfectly understood. This property, however, must bestow growth advantages to the cell above those of normal hepatocytes that suffer recurring virus replication and attack by immune cells.

III. EARLY ONSET OF HBV DNA INTEGRATION

Several reports have shown that integration can occur in the liver tissues of chronic or acute-phase hepatitis patients.[5,18-25] Interestingly, discrete Southern hybridization band profiles of integrated HBV DNA occasionally are observed in DNA samples from such tissues that do not carry detectable HCCs. These findings demonstrate that there may be clonal expansion of hepatocytes that have this integrated HBV DNA.[26-29] Conceivably, massive death of hepatocytes in combination with their subsequent regeneration may cause the selection or enrichment of cells that have certain growth advantages. In some cases, a clone of such cells covers a fairly large region in the liver.

DNAs from the lesions of HCC and the surrounding adenomatous hyperplasia sometimes show identical Southern hybridization patterns, an indication that both types of cells originate from the same clone.[30] On forming HCC, the Southern hybridization profile remains stable and is not affected by metastasis or propagation in nude mice.[9,31,32]

Okubo et al.[33] examined an *in vitro* culture system of primary human hepatocytes infected with HBV virions, a system that allows replication of the virus.[34] Although the number of cases examined were limited, they were able to detect integrant HBV DNA in the cells a few days after infection. Because the cells were not subjected to a second round of infection, their results demonstrate that integration can occur in primary infected cells. Hepatocytes in affected livers may behave in the same way.

IV. STRUCTURE OF THE INTEGRANTS AND THE SUGGESTED MECHANISM OF INTEGRATION

Since the pioneering work of Ogston et al.[35] a number of analyses have been performed on the structure of HBV integrants in HCCs. Some integrants are complex, whereas others are relatively simple. Figure 1[35a,35b] shows all the published data that are provided with essential parameters and information. A comparison of the structures of the integrant HBV genomes shows common features. (1) Without exception, HBV genomes show deletions. Some of the complex integrants seem to have been assembled from multiple viral genomes, but each unit in the complex is of less than genome size. The lack of integrity of the integrant viral genomes indicates that they cannot serve as templates for productive replication of HBV. (2) A considerable fraction of the integrated HBV genomes terminate within or at the cohesive end region that lies between the direct repeat 1 and 2 (DR1 and DR2) sequences of the HBV genome. These sequences represent sites for the initiation and termination of replication (Chapter 11, this volume). Not a single integrant HBV genome has clearly been proven to carry the intact cohesive end region. Thus, the sequence within the cohesive end region is preferred for the recombination between viral and cellular DNA. (3) The integrant HBV genomes are classifiable in two groups: simple-type integrants that consist of a linear, contiguous region of HBV DNA originating from one genome, and complex-type integrants that consist of regions of viral DNA originating from multiple genomes. In the latter, the

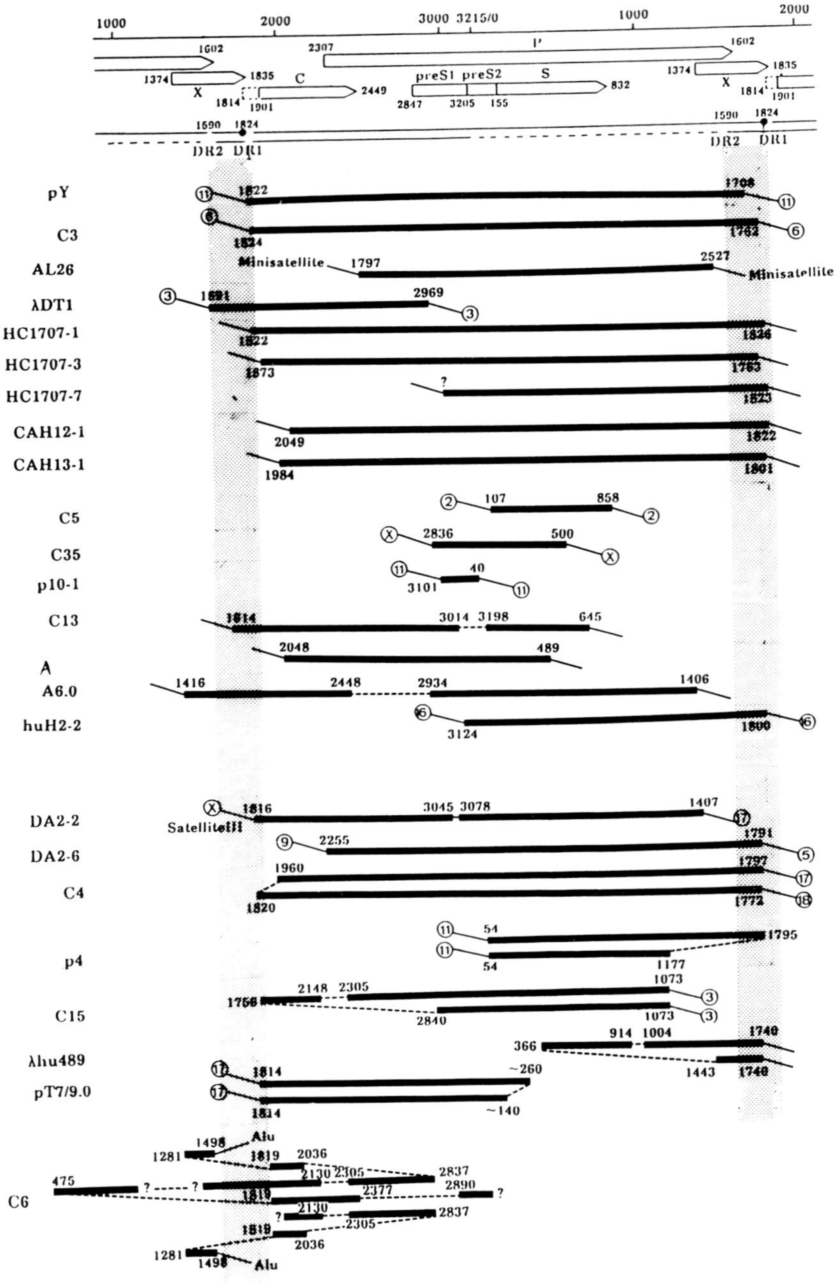

FIGURE 1. Structure of the integrated HBV DNA sequences. *Left:* The entire HBV genome as a tandemly arranged structure. The genome is indicated in base pairs numbered from the hypothetical *Eco*RI site of subtype *adr*.[35a,35b] Open reading frames and their direction of transcription are shown by open arrows. The long thin horizontal lines show the minus strand with the DNA-linked protein (•), and the plus strand. Direct repeat 1 (DR1) and DR2 are the 11-bp direct repeats where viral DNA synthesis is initiated. The cohesive end region is shaded. *Right:* HBV genome regions that are observed in the integrated viral DNA are shown by solid bars. The viral DNA that has one continuous portion(s) of a circular HBV genome is represented by a single solid bar aligned at the same level. Deletions in the genome are shown by dashed lines. The complex-type virus DNAs that consist of two or more HBV genomes whose regions overlap (at least in part) are represented by multiple solid bars aligned at different levels, each representing different units of the virus genome, and are connected by tilted dashed lines. Solid, thin tilted lines represent flanking human cellular DNA, and the encircled numerals show the human chromosome number. L, The cloning vector is joined directly to the virus DNA; Alu, SatelliteIII, or Minisatellites, cell DNA at the junction is homologous to the *Alu* -repeat, satellite III, or minisatellite sequence, respectively. (Modified from Matsubara and Tokino,[61] in which all the references to each clone are cited.)

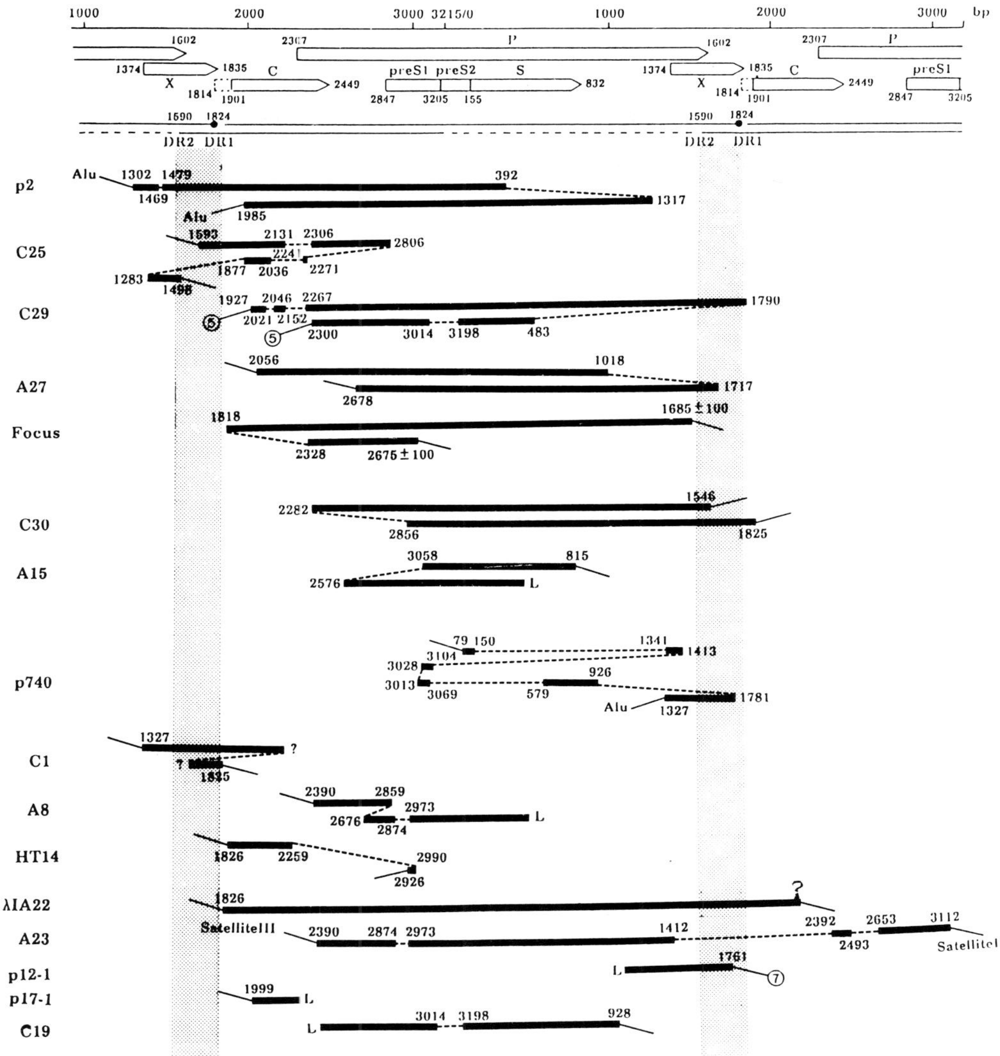

FIGURE 1 (continued).

multiple genome does not necessarily originate from independent genomes; sister genomes made by replication could serve as well. (4) These complex-type integrants carry at least one virus–virus junction. The viral DNA sequence that appears at this junction often lies within or at the cohesive end region. It appears that a mechanism similar to that for the formation of a virus–cell junction may operate in the formation of the virus–virus junction. These structural features must reflect the substrates used for integration, the process of integration, and the complex reactions that take place after the formation of the integrants.

Nakamura et al.[36] noted that some of the simple-type integrants constitute a unique category in that the integrity of the genome is broken within the cohesive end region (when DR1 is in the integrant, DR2 is missing, and vice versa) and that at the site of integration the cell DNA has invariably undergone a microdeletion of about 10 bp. Integrants with identical properties constituted five cases. As judged by their structural simplicity, these integrants most probably represent the primary products of integration, not the secondary rearranged products. From the results of a comparison of these integrants with the known structures of the replication intermediates of HBV genomes, it can be concluded that these intermediates may serve as the substrates for integration.

The virus and cell DNA sequences at the junctions do not show overall homologies, except for sporadic cases in which a short region of homology between the integrating HBV DNA and flanking cellular DNA sequences occurs.[17,37-39] This strongly suggests that the recombination that takes place in the integration process is illegitimate.[40]

In contrast to the idea that the replication intermediates are the substrates for integration, Hino et al.[41] proposed that the circular DNA that appears in infected cells is cleaved at the DR sequence by the cellular topoisomerase I enzyme and then serves as the integration substrate. But in this case the simple topoisomerase I-mediated strand exchanges demonstrated with prokaryotic systems[42] do not take place. Because of the paucity of our knowledge about the actual mechanism of illegitimate recombination, it is premature to speculate which of these proposed mechanisms better accounts for the integration. The integration of other DNA viruses also is associated with the deletion of cellular DNA,[43] an indication that deletion formation is an obligatory reaction associated with illegitimate recombination in eukaryotes. The fact that both viral and cellular DNAs undergo deletions of nonuniform sizes implies that no unique enzyme is at work, unlike in the case of the endonuclease that acts in the integration of retroviral genomes.

An important point is that the integration of HBV DNA is always mutagenic irrespective of how the integrants are made. First, the integration involves the insertion of foreign DNA into the cellular DNA. Second, the integration process invariably creates a deletion in the cellular DNA. Third, as shown later, some integrations appear to mediate further chromosomal rearrangements.

V. THE CHROMOSOMAL SITE AND TARGET SEQUENCE FOR INTEGRATION

Whether HBV integration takes place at a special chromosomal site(s), such as one near an oncogene or an oncosuppressor gene, is a key issue for hepatocarcinogenesis. Whether during the integration process the HBV genome selects a unique cellular DNA sequence, (i.e., a "target for landing") is another issue of interest. To avoid confusion, the terms "chromosomal site" and "target sequence" will be used.

Early expectations that there might be a common target sequence led investigators to examine the junction and flanking cellular DNA sequences of different integrants. As the existence of a common target sequence has been disproved, arguments have been made that HBV DNA integration does not select a unique target sequence. Although this conclusion may be correct, it was based on the use of erroneous data; some cell–virus junctions are produced by secondary rearrangements that occur after integration. The real target sequence may be deduced from the sequence in the microdeletion from the cellular DNA and its flanking sequence (see Section IV). The data available now show clearly that the target sequences defined as such are not unique.

Taking into account the above limitations, the flanking cellular DNA sequences of the various integrants still provide auxiliary information. Of the 39 integrants analyzed so far (Figure 1), 2 are integrants within the satellite III sequence.[44,45] Ogata[46] has found another

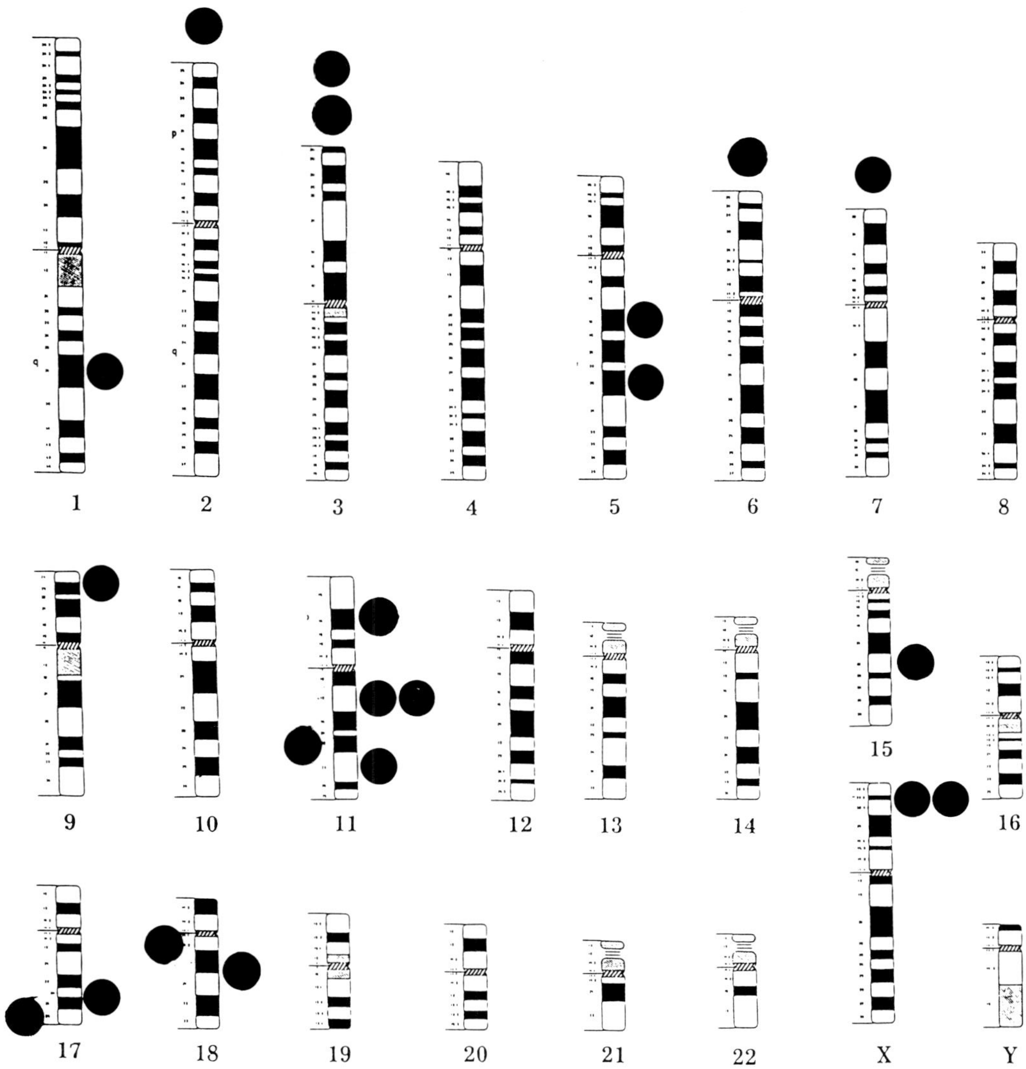

FIGURE 2. The chromosomal sites at which HBV DNA is integrated into the genome of the hepatocellular carcinomas (•). Those integrants whose sites have been assigned only by chromosome number are represented above the chromosome. (Modified from Matsubara and Tokino.[61])

instance in which the HBV DNA is flanked by α-satellite DNA. Some integrants are flanked by cellular *Alu* family sequences (Figure 1). Thus, a considerable fraction of the integrants are found within cellular repeating sequences. Berger and Shaul[40] found two cases (AL14 and AL26 in Figure 1) in which the HBV DNA had integrated into sites near minisatellite or VNTR (variable number of tandem repeat) sequences.[47,48] Zerial et al.,[49] however, claimed that HBV may integrate into a chromosomal DNA region of particular GC content. The selection of the target sequence for integration may reflect the structure of the chromosome. The elucidation of the actual selection process requires further study.

Do HBV genomes select preferential chromosomal site(s) for integration as has been reported for some retroviruses?[50-55] Mapping of the flanking cellular DNA sequences within human chromosomes has been performed, in particular by Tokino,[55a] who mapped 13 integration sites. His results are shown together with 12 other published mapping results in Figure 2. The filled circle above a chromosome shows that only the chromosome number has been assigned, not the site. The results suggest that the sites for HBV integration on chromosomes are not limited to only a few places, or to the site(s) whose function is directly linked to hepatocarcinogenesis, although the number of cases is too few to draw a definite conclusion. Because most HCCs carry multiple integrants, correlation of the integration site with hepatocarcinogenesis must be approached with caution.

VI. MUTAGENESIS BY INTEGRATION, AND REARRANGEMENT OF HOST CHROMOSOMES FOLLOWED BY CELL SELECTION

As discussed previously, one of the most important effects associated with the integration of HBV DNA may be mutagenesis. Random selection of the target sequence and the chromosomal site for both the insertion and the obligatory loss of chromosomal DNA suggest that this virus acts as a nonselective mutagenic insertional agent.

In addition to acting as insertional mutagens and inducing microdeletions, some integrated HBV genomes appear to provoke secondary rearrangements of chromosomes. Records of rearrangements are seen in several of the integrants shown in Figure 1. One group of integrants (DA2-2, DA2-6, C4) represents chromosome translocations mediated by integrated HBV DNA. In these integrants the right and left arms of the HBV DNA are joined to cell DNA from different chromosomes. Another group of integrants (P4, C15, λhu489, PT7/9.0, and possibly C6) represents chromosome inversions mediated by HBV DNA. The structure of the integrating viral DNA is inverted (both ends having identical sequences) and joined to identical cell DNA sequences, which signifies that they must have been made first by integration, then by amplification followed by recombination within the virus DNA sequences in a head-to-head fashion. Several models for the mechanism of rearrangement are possible, but in the absence of experimental data it would be premature to discuss them.

Integrant HBV DNA also may trigger chromosome deletions, but the process may differ from microdeletion formation. Rogler et al.[56] first recorded the loss of cellular DNA (more than 13.5 kbp) in association with HBV integration. Tamura et al.[56a] analyzed three integrants (C5, C35, and P10-1), each of which carries only a small fragment of the HBV genome. They found that the respective deletions of 12, 11, and 25 kbp of cellular DNA were associated with viral DNA integration. In addition to these chromosome deletions, two examples of the amplification of a region of a chromosome in which an HBV DNA is integrated have been found.[57,57a]

Other DNA viruses (such as the papova and the Epstein–Barr viruses) also cause chromosomal abnormalities along with their integration.[58] It is not clear whether HBV is more potent than these DNA viruses for chromosomal change, but it is noteworthy that of 31 integrants prepared from adult HCC and examined closely, 11 are the products of secondary rearrangements. Thus, HBV acts as a potent nonspecific insertional mutagen on one hand, and as a causative agent for secondary rearrangement of chromosomes on the other.

Hino et al.[59] noted that a transgenic mouse that carries integrant HBV DNA obtained from a human HCC sometimes undergoes changes in its viral DNA. Although the changes have yet to be fully analyzed, their experiment suggests the potential use of animal models to solve the problem of such changes.

The rearrangement of chromosomes distorts the balance of the control circuits of cell metabolism and, in many cases, predisposes the chromosomes of the cell to further amplification, loss, rearrangement, and mutation. The mutagenic effects of HBV often may be combined with cell selection (as caused by hepatitis), the second important effect associated with infection by HBV. In a hepatitis patient's liver, the hepatocytes undergo recurring cell damage and regeneration, which produce average lifetimes markedly shorter than normal, about 400 d in an adult liver. In a population of hepatocytes under constant pressure for cell replacement, cells that grow better than others will be selected and enriched and will eventually predominate in the population, as has been well established using chemostats, a model system to select better growers.[60,61] Because cells do not migrate freely in the liver, the selected cells form a local cluster(s). The clonality of these cells, as identified with bands of integrated HBV DNA in Southern blotting profiles (see Section I) in some biopsy and autopsy samples of liver before HCC develops, supports this idea. The combination of cell selection and insertional mutagenesis along with chromosomal rearrangements renders the affected liver a "chemostat with mutagens". The synergistic effect on the development of HCCs produced by chronic hepatitis and chemical carcinogens such as mycotoxins (taken by eating mold-contaminated food[62]) is in line with this mutagenesis-selection hypothesis.

VII. GENETIC CHANGES ASSOCIATED WITH HEPATOCARCINOGENESIS

Using the broad term "mutation" one cannot specifically identify the cellular change(s) producing the correlation between the integration of HBV and the transformation of hepatocytes, or the subsequent development as cancer. Because the HBV genome does not carry oncogenes, the obvious question is whether HBV DNA integration activates or inactivates cellular genes that have crucial roles in the regulation of cellular functions. Investigations of this question have only recently been started. Given here is a brief overview of the genetic changes in the hepatocytes associated with hepatocarcinogenesis and of the cis- and trans-acting genetic controlling elements in the HBV genome, and a summary of the results of current correlation studies.

The genetic or epigenetic changes that have been correlated with hepatocarcinogenesis, or its subsequent development, can be placed in four categories. The first comprises activated oncogenes that are expressed dominantly. Included in this category are *hst*,[63] *lca*,[64] the *ras* family genes,[65-69,30] *raf*,[68] *hhc*,[70] and some unidentified transforming genes.[71,72,72a] But it must be said that in HCCs no major or predominant transforming oncogene has yet been discovered. A number of efforts have been made to isolate activated transforming oncogenes with the associated integrated HBV genome found in HCCs, but without success.

The second category of change constitutes such structural alterations in genes or genetic regions as amplification, translocation, rearrangement, or all three. Little is known for HCCs, except for scattered observations that include rearrangement of *myc*,[73,74] H-*ras*,[73] *hhc*,[70] and an endogenous retrovirus sequence.[75] Amplification of a DNA region has been described by Hatada et al.[57] in the region covering the *hst-1* and *int-2* genes, close to which integrated HBV DNA was found. Tokino[75a] has seen a similar, but different, cellular DNA amplification in association with integrated HBV in the long arm of chromosome 11. None of these changes seems to predominate in HCCs.

The third category, chromosomal deletions unique to some primary neoplastic tissues, has attracted much interest because such loss might include oncosuppressor genes or genes that are important in the regulation of cellular metabolism and functions. Loss of a chromosome region in HCCs has been observed in chromosomes 11p,[56,76,77] 13q,[77] and 4q.[78,79] Other deletions, such as 16q,[79a] have been noted and undoubtedly more will be added to the list. These findings were made with a relatively small number of polymorphic markers. As none of the reports surveyed loss of heterozygosity in other regions of chromosomes in detail, the question as to whether the recorded loss of heterozygosity represents the major genomic change or whether there are other changes yet to be discovered has not been answered. Moreover, none of these deletions has been found in the majority of the HCCs. Nevertheless, in light of the importance of oncosuppressor genes and the ability of HBV to cause chromosomal deletions, the possible link between deletion formation and HCC formation requires detailed study.

The fourth category of change is reflected by the augmented or repressed expression of genes in HCCs. Studies of their expression have become possible owing to the expanding repertoire of available oncogenes and other genes for the analysis of RNA. Gene c-*myc* and a member(s) of the *ras* family have been shown to have elevated expression in rat hepatomas induced by chemical carcinogens, the most widely used experimental system for carcinogenesis. Several scattered observations also have shown elevated expressions of *fos*, *raf*, *bas*, *erbB*, *p53*, *src*, and the gene encoding γ-glutamyltransferase in human and other mammalian HCCs. As reported hereafter, about one third of the woodchuck HCCs produced by the woodchuck hepatitis virus show activated c-*myc* expression, ranging from 5 to 50 times the normal level. In rodent HCCs, c-*myc* and members of the *ras* family genes are frequently expressed above the normal level.

In human HCCs as well, augmented expressions of c-*myc* and *ras* family member genes are frequent. The expression of genes such as *hap* and insulin-like growth factor II (*IGFII*) also has been noted, but only sporadically. Augmented expression of *IGFII* has been discussed in connection with a possible autocrine effect on cells that bestows selective growth advantages.[80] Overall, in neither human nor rodent HCCs does there seem to be an oncogene that is activated ubiquitously. Exceptions are the activated expression of c-*myc* and some *ras* genes, but what implication this has in HCC formation or development is not clear. Lee et al.[81] claim that within a given section of HCC, some, but not all, cells show elevated expression of c-*myc*. In contrast, elevated expressions of c-*myc*[81,82] and H-*ras*[83] have been found both in the tumorous part and in adjacent cirrhotic or seminodule tissues. Conceivably, such elevated oncogene expression is unnecessary in certain instances, whereas it is insufficient in others for the development of fully malignant HCCs. In the latter instance, elevated expression may contribute to increasing the risk of carcinogenesis, or to the development of HCC.

VIII. CIS- AND TRANS-ACTING GENETIC ELEMENTS IN THE HBV GENOME: THEIR POSSIBLE ROLES IN HEPATOCARCINOGENESIS

The promoters for the transcription of genes in the HBV genome have been well characterized, and the promoters present in the HBV integrants observed in the PLC/PRF/5 (Alexander) hepatoma cell line have been shown to be transcriptionally active. Theoretically, if the viral transcript reads through into flanking cellular sequences, it could have some effect on cellular regulation. Dejean et al.,[84] DeThe et al.,[85] and Brand et al.[86] found in one HCC a cellular gene for the retinoic acid-binding receptor (*RAR*-β) that was fused to an integrated HBV DNA gene. These authors later showed that the *RAR*-β gene is expressed in many HCCs, but not in normal adult liver. Examples other than the production of a fusion protein also can be imagined in connection with promoter insertion. Such "portable" promoter insertions may result in readthrough transcription into a sequence that has a unique controlling element(s), premature termination, change in efficiency of posttranscriptional expression, and transcription of a strand complementary to a normally transcribed strand.

Probably a more important portable cis-acting element for transcription is the enhancer in the HBV genome.[87] This element lies in the coding frame of the DNA polymerase gene, upstream of the promoter for the X gene, and is the major target for the trans-activating X protein that is coded for by the HBV genome (see below). In addition, the same enhancer can be acted upon by cellular trans-acting factors.[88] The various cellular factors that have been identified as binding to specific sites within the enhancer sequence probably affect the function of this enhancer. The efficiency of the activation and cell specificity of the enhancer may be determined by the binding of these proteins and through the factors that affect this binding. In addition to this enhancer, Yee[89] recently found another enhancer in the viral genome in the region close to the core promoter. Although cellular gene activation through viral enhancer insertion has yet to be well characterized (see below), further analyses should contribute much toward that characterization.

Moroy et al.[90] and Hsu et al.[91] studied HCCs in woodchucks infected by woodchuck hepatitis virus. Of the nine HCCs examined, three showed elevated expression of c-*myc* RNAs, some of which were transcribed from novel initiation sites, and two had integrated virus DNA in, or close to, the cellular c-*myc* gene. Whether cellular oncogene expression was activated through the viral enhancer is not known, but a crucial question can be raised: does DNA from the integrants in woodchuck HCCs bring about phenotypic change in normal hepatocytes when a transgenic mouse that carries the integrant DNA is produced?

Trans-acting genetic elements in the HBV genome are also important factors to consider as potential affectors of cellular functions. As noted previously, in the HBV genome the major element that has a trans-activating function is the product of the X gene. This protein activates transcription from a variety of promoters,[92] which includes several HBV promoters under the control of the HBV enhancer.[93-97] Promoters under the control of heterologous enhancers such as simian virus 40 (SV40), human immunodeficiency virus (HIV), herpes simplex virus (HSV), and Rous sarcoma virus (RSV), also are trans-activated.[94,96-100] In addition, the X protein activates several cellular genes. Apparently, the X product is a general transactivator, as is the *tax-1* product of the HTLV-I virus.[101,102] The X product itself is not a DNA-binding protein, and its trans-activating potential differs from one cell type to another, which suggests that its action is mediated through interaction with cellular factors or through the modification of such factors. Thus, the expression of the X gene in the integrated fragment of the HBV genome may affect some, if not all, cellular functions.

We must also take into account the fact that almost without exception, the X gene in the integrant HBV genome is incomplete because part of the X-ORF lies in the cohesive end region of the viral genome that is preferentially used for joining with cellular DNA (see Section IV). Of special interest is the observation that deletion mutants of the X protein that lack some amino acids from the C-terminus still have trans-activating ability.[102a] Therefore, there is the possibility that in some cases the interrupted X ORFs in the integrants might code for truncated X protein fused to an unrelated amino acid sequence coded for by the flanking cellular sequence. Wollersheim et al.[103] detected such a fusion protein in an HCC-carrying integrant HBV DNA. The collection of data on HBV integrants (Figure 1) indicates that similar cases will be forthcoming.[61] Uncharacterized transcripts from the X region present in the integrant HBV sequences also have been observed.[104-106]

Kekulé et al.[107] recently discovered another trans-activator coding sequence in the HBV genome around the preS/S region. Interestingly, this trans-activator also targets the same enhancer sequence. Thus, HBV integration potentially can bring about "portable insertional

changes" that include the production of additional trans-activators, as well as incorporation of the targets for their action, into cellular DNA. Research on these changes has just been started, and systematic studies are awaited.

Do "portable insertional changes" or "mutations and chromosomal rearrangements in general" have a major role in hepatocarcinogenesis? It is too soon to answer this question, but we speculate that there may be diverse paths leading to hepatocarcinogenesis, each involving factors affected by multiple genetic and epigenetic changes. Because of this complexity, there may be diverse causative changes ascribed to the integrating HBV genome. We are far from being able to link findings of change in the genome structure or change in the gene function directly to hepatocarcinogenesis. For this reason, inference of a role for integration or integrated HBV in hepatocarcinogenesis requires critical examination, but this often has been neglected. Discussions of the role of HBV integration also often ignore the fact that there are multiple integrants in the majority of HCCs. Even for HCCs that seemed to have a single integrant, the possibility that undetectable integrants were present, or that rearrangement may have eliminated HBV DNA for a specific cellular site, needs careful consideration. The current information available on individual cases is still too limited for generalization. Organized systematic surveys of the changes found in HCCs, and those found in precancerous cell masses and the nodules occasionally observed in livers, are necessary.

IX. CONCLUSIONS

The liver, which allows vigorous cell regeneration, frequent production of clonal hyperplastic nodules, and HCC development, provides an ideal system with which to study the regulatory elements that act in cell proliferation, differentiation, carcinogenesis, and the acquisition of malignancy. Studies on structural and functional changes in transforming oncogenes, oncosuppressor genes, genes related to growth factors, and receptor genes that might mediate or be altered by these cellular changes are now being performed. Although much more research is needed, we are now approaching the time when there will be a clear, coherent picture of all the processes of change and the interrelation of cellular changes involved in the development of HCC. We shall then be able to propose reasonable ways to counter hepatocarcinogenesis.

ACKNOWLEDGEMENTS

I am grateful to Kyoko Mimura for excellent secretarial assistance.

NOTE ADDED IN PROOF

After this manuscript was prepared, two important papers relevant to Section VIII have appeared. Fourel et al.[108] examined woodchuck HCCs and observed that woodchuck hepatitis virus DNA integrates with high frequency into cellular *myc* genes, causing augmented expression of these oncogenes. Wang et al.[109] observed a human HCC that has integrated HBV DNA within the cyclin A gene and has disrupted the function of this gene, which is important in the control of cell division.

REFERENCES

1. **Szmuness, W.,** Hepatocellular carcinoma and the hepatitis B virus: evidence for causal association, *Progr. Med. Virol.*, 14, 40, 1978.
2. **Beasley, R. P., Hwang, L. Y., Lin, C. C., and Chien, C. S.,** Hepatocellular carcinoma and hepatitis B virus, *Lancet*, ii, 1129, 1981.
3. **Blumberg, B. S. and London, W. T.,** Hepatitis B virus and the prevention of primary hepatocellular carcinoma, *N. Engl. J. Med.*, 304, 782, 1981.
4. **Korba, B. E., Wells, F., Tennant, B. C., Yakum, G. H., Purcell, R. H., and Gerin, J. L.,** Hepadnavirus infection of peripheral blood lymphocytes *in vitro*: woodchuck and chimpanzee models of viral hepatitis, *J. Virol.*, 58, 1986.
5. **Brechot, C., Pourcel, C., Louis, A., Rain, B., and Tiollais, P.,** Presence of integrated hepatitis B virus DNA sequences in cellular DNA in human hepatocellular carcinoma, *Nature (London)*, 286, 533, 1980.
6. **Marion, P. L., Salazar, F. H., Alexander, J. J., and Robinson, W. S.,** State of hepatitis B viral DNA in a human hepatoma cell line, *J. Virol.*, 33, 795, 1980.
7. **Edman, J. C., Gray, P., Valenzuela, P., Rall, L. B., and Rutter, W. J.,** Integration of hepatitis B virus sequences and their expression in a human hepatoma cell, *Nature (London)*, 286, 535, 1980.
8. **Chakraborty, P. R., Ruiz-Opazo, H., Shouval, D., and Shafritz, D. A.,** Identification of integrated hepatitis B virus DNA and expression of viral RNA in an HBsAg-producing human hepatocellular carcinoma cell line, *Nature (London)*, 286, 531, 1980.
9. **Shafritz, D. A., Shouval, D., Sherman, H., Hadziyannis, S. J., and Kew, M. C.,** Integration of hepatitis B virus DNA into the genome of liver cells in chronic liver disease and hepatocellular carcinoma, *N. Engl. J. Med.*, 305, 1067, 1981.
10. **Koshy, R., Maupas, P., Muller, R., and Hofschneider, P. H.,** Detection of hepatitis B virus-specific DNA in the genomes of human hepatocellular carcinoma and liver cirrhosis tissues, *J. Gen. Virol.*, 57, 95, 1981.
11. **Twist, M., Clark, H. F., Aden, D. P., Knowles, B. B., and Plotkin, S. A.,** Integration pattern of hepatitis B virus DNA sequences in human hepatoma cell lines, *J. Virol.*, 37, 239, 1981.
12. **Chen, D. S., Hoyer, B. H., Nelson, J., Purcell, R. H., and Gerin, J. L.,** Detection and properties of hepatitis B viral DNA in liver tissues from patients with hepatocellular carcinoma, *Hepatology*, 2, 42S, 1982.
13. **Tsurimoto, T., Fujiyama, A., and Matsubara, K.,** Stable expression and replication of hepatitis B virus genome in an integrated state in a human hepatoma cell line transfected with the cloned viral DNA, *Proc. Natl. Acad. Sci. U.S.A.*, 84, 444, 1987.
14. **Sureau, C., Romet-Lemonne, J.-L., Mullins, J. I., and Essex, M.,** Production of hepatitis B virus by a differentiated human hepatoma cell line after transfection with cloned circular HBV DNA, *Cell*, 47, 37, 1986.
15. **Yaginuma, K., Shirakata, Y., Kobayashi, M., and Koike, K.,** Hepatitis B virus (HBV) particles are produced in a cell culture system by transient expression of transfected HBV DNA, *Proc. Natl. Acad. Sci. U.S.A.*,84, 2678, 1987.
16. **Alexander, J. J.,** Human hepatoma cell lines, in *Neoplasms of the Liver*, Okuda, K. and Ishak, K. G., Eds., Springer-Verlag, New York, 1987, 47.
17. **Ziemer, M., Garcia, P., Shaul, Y., and Rutter, W. J.,** Sequence of hepatitis B virus DNA incorporated into the genome of a human hepatoma cell line, *J. Virol.*, 53, 885, 1985.
18. **Rogler, C. E. and Summers, J.,** Cloning and structural analysis of integrated woodchuck hepatitis virus sequences from a chronically infected liver, *J. Virol.*, 50, 832, 1984.
19. **Shafritz, D. A. and Lieberman, H. M.,** The molecular biology of hepatitis B virus, *Annu. Rev. Med.*, 35, 219, 1984.
20. **Standring, D. N. and Rutter, W. J.,** The molecular analysis of hepatitis B virus, *Prog. Liver Dis.*, 8, 311, 1986.
21. **Hada, H., Arima, T., Togawa, K., Okada, Y., Morichika, S., and Nagashima, H.,** State of hepatitis B viral DNA in the liver of patients with hepatocellular carcinoma and chronic liver disease, *Liver*, 6, 189, 1986.
22. **Harrison, T. J., Anderson, M. G., Murray-Lyon, I. M., and Zuckerman, A. J.,** Hepatitis B virus DNA in the hepatocyte. A series of 160 biopsies, *J. Hepatol.*, 2, 1, 1986.
23. **Fowler, M. J. F., Greenfield, C., Chu, C. M., Karayiannis, P., Dunk, A., Lok, A. S. F., Lai, C. L., Yeoh, E. K., Monjardino, J. P., Wankya, B. M., and Thomas, H. C.,** Integration of HBV-DNA may not be a prerequisite for the maintenance of the state of malignant transformation. An analysis of 110 liver biopsies, *J. Hepatol.*, 2, 218, 1986.
24. **Yaginuma, K., Kobayashi, H., Kobayashi, M., Morishima, T., Matsuyama, K., and Koike, K.,** Multiple integration sites of hepatitis B virus DNA in hepatocellular carcinoma and chronic active hepatitis tissues from children, *J. Virol.*, 61, 1808, 1987.

25. **Benhamou, J. P., Bismuth, H., Tiollais, P., and Brechot, C.,** Hepatocellular carcinoma with normal adjacent liver. Hepatitis B virus DNA status, *J. Hepatol.*, 8, 249, 1989.
26. **Imazeki, F., Yaginuma, K., Omata, M., Okuda, K., Kobayashi, M., and Koike, K.,** Integrated structures of duck hepatitis B virus DNA in hepatocellular carcinoma, *J. Virol.*, 62, 861, 1988.
27. **Lai, M. Y., Chen, D. S., Chen, P. J., Lee, S. C., Sheu, J. C., Huang, G. T., Wei, T. C., Lee, C. S., Yu, S. C., Hsu, H. C., and Sung, J. L.,** Status of hepatitis B virus DNA in hepatocellular carcinoma: a study based on paired tumor and nontumor liver tissues, *J. Med. Virol.*, 25, 249, 1988.
28. **Esumi, M., Tanaka, Y., Tozuka, S., and Shikata, T.,** Clonal state of human hepatocellular carcinoma and non-tumorous hepatocytes, *Cancer Chemother. Pharmacol.*, 23 (Suppl. S1), 1989.
29. **Yasui, H. and Hino, O.,** Molecular cloning of clonal integrated HBV DNA from a cirrhotic nodule, *Acta Hepatol. Jpn.*, 30, 1549.
30. **Tsuda, H., Hirohashi, S., Shimosato, Y., Yoshida, T., and Terada, M.,** Low incidence of point mutation of c-Ki-ras and N-ras oncogenes in human hepatocellular carcinoma, *Jpn. J. Cancer Res.*, 80, 196, 1989.
31. **Kaneko, S., Oshima, T., Kodama, K., Aoyama, S., Yoshikawa, H., Unoura, M., Fukuoka, K., Matsushita, F., Morimoto, H., Kobayashi, K., Hattori, N., and Murakami, S.,** Stable integration of woodchuck hepatitis virus DNA in transplanted tumors and established tissue culture cells derived from a woodchuck primary hepatocellular carcinoma, *Cancer Res.*, 46, 3608, 1986.
32. **Blum, H. E., Offensperger, W. B., Walter, E., Offensperger, S., Wahl, A., Zeschnigk, C., and Gerok, W.,** Hepatocellular carcinoma and hepatitis B virus infection; molecular evidence for monoclonal origin and expansion of malignantly transformed hepatocytes, *J. Cancer Res. Clin. Oncol.*, 113, 466, 1987.
33. **Okubo, K., Nakamura, T., Tokino, T., and Matsubara, K.,** Different type of hepatitis B virus (HBV) DNA integrants that may reflect the integration process, *Gastroenterologie Japonica*, 25 (Suppl. 2), 23, 1990.
34. **Ochiya, T., Tsurimoto, T., Ueda, K., Okubo, K., Shiozawa, M., and Matsubara, K.,** An *in vitro* system for infection with hepatitis B virus that uses primary human fetal hepatocytes, *Proc. Natl. Acad. Sci. U.S.A.*, 86, 1875, 1989.
35. **Ogston, W., Jonak, G. J., Rogler, C. E., Astrin, S. M., and Summers, J.,** Cloning and structural analysis of integrated woodchuck hepatitis virus sequences from hepatocellular carcinomas of woodchucks, *Cell*, 29, 385, 1982.

35a. **Fujiyama, A., Miyanohara, A., Nozaki, T., Yoneyama, T., Ohtomo, N., and Matsubara, K.,** Cloning and structural analysis of hepatitis B virus DNAs subtype *adr*, *Nucleic Acid Res.*, 11, 4601, 1983.

35b. **Ono, Y., Onda, H., Sasada, R., Igarashi, K., Sugino, Y., and Nishioka, K.,** The complete nucleotide sequences of the cloned hepatitis B virus DNA: subtype *adr* and *adw*, *Nucleic Acid Res.*, 11, 1747, 1983.

36. **Nakamura, T., Tokino, T., Nagaya, T., and Matsubara, K.,** Microdeletion associated with the integration process of hepatitis B virus DNA, *Nucleic Acids Res.*, 16, 4865, 1988.
37. **Koshy, R., Koch, S., Freytag von Loringhoven, A., Kahmann, R., Murray, K., and Hofschneider, P. H.,** Integration of hepatitis B virus DNA: evidence for integration in the single-stranded gap, *Cell*, 34, 215, 1983.
38. **Yaginuma, K., Kobayashi, M., Yoshida, E., and Koike, K.,** Hepatitis B virus integration in hepatocellular carcinoma DNA: duplication of cellular flanking sequences at the integration site, *Proc. Natl. Acad. Sci. U.S.A.*, 82, 4458, 1985.
39. **Shih, C., Burke, K., Chou, M. J., Zeldis, J. B., Yang, C. S., Lee, C. S., Isselbacher, K. J., Wands, J. R., and Goodman, H. M.,** Tight clustering of human hepatitis B virus integration sites in hepatomas near a triple-stranded region, *J. Virol.*, 61, 3491, 1987.
40. **Berger, I. and Shaul, Y.,** Integration of hepatitis B virus: analysis of unoccupied sites, *J. Virol.*, 61, 1180, 1987.
41. **Hino, O., Ohtake, K., and Rogler, C. E.,** Features of two hepatitis B virus (HBV) DNA integrations suggest mechanisms of HBV integration, *J. Virol.*, 63, 2538, 1989.
42. **Ikeda, H.,** Illegitimate recombination mediated by T4 DNA topoisomerase *in vitro*. Recombinants between phage and plasmid DNA molecules, *Mol. Gen. Genet.*, 202, 578, 1986.
43. **Henderson, A. S., Ripley, S., Hino, O., and Rogler, C. E.,** Identification of a chromosomal aberration associated with a hepatitis B DNA integration site in human cells, *Cancer Genet. Cytogenet.*, 30, 269, 1988.
44. **Shaul, Y., Garcia, P. D., Schonberg, S., and Rutter, W. J.,** Integration of hepatitis B virus DNA in chromosome-specific satellite sequences, *J. Virol.*, 59, 731, 1986.
45. **Nagaya, T., Nakamura, T., Tokino, T., Tsurimoto, T., Imai, M., Mayumi, T., Kamino, K., Yamamura, K., and Matsubara, K.,** The mode of hepatitis B virus DNA integration in chromosomes of human hepatocellular carcinoma, *Genes Dev.*, 1, 773, 1987.
46. **Ogata, N.,** Mode of HBV DNA integration into host chromosomes of hepatocellular carcinoma (HCC) and noncancerous liver tissues, *Acta Hepatol. Jpn..*, 29, 60, 1988.
47. **Jeffreys, A. J., Wilson, V., and Their, S. L.,** Hypervariable minisatellite regions in human DNA, *Nature (London)*, 341, 67, 1985.

48. **Nakamura, Y., Leppert, M., O'Connell, P., Wolff, R., Holm, T., Culver, M., Martin, C., Fujimoto, E., Hoff, M., Kumlin, E., and White, R.**, Variable number of tandem repeat (VNTR) markers for human gene mapping, *Science*, 235, 1616, 1987.
49. **Zerial, M., Salinas, J., and Bernardi, G.**, Genomic localization of hepatitis B virus in a human hepatoma cell line, *Nucleic Acids Res.*, 14, 8373, 1986.
50. **Hayward, W. S., Neel, B. G., and Astrin, S. M.**, Activation of a cellular oncogene by promoter insertion in ALV-induced lymphoid leukosis, *Nature (London)*, 290, 475, 1981.
51. **Neel, B. G., Hayward, W. S., Robinson, H. L. Fang, J., and Astrin, S. M.**, Avian leukosis-virus-induced tumors have common proviral integration sites and synthesize discrete new RNAs: oncogenesis by promoter insertion, *Cell*, 23, 323, 1981.
52. **Nusse, R. and Varmus, H. E.**, Many tumors induced by the mouse mammary tumor virus contain a provirus integrated in the same region of the host genome, *Cell*, 31, 99, 1982.
53. **Corcoran, L. M., Adams, J. M., Dunn, A. R., and Cory, S.**, Murine T lymphomas in which the cellular myc oncogene has been activated by retroviral insertion, *Cell*, 37, 113, 1984.
54. **Dickson, C., Smith, R., Brookes, S., and Peters, G.**, Tumorigenesis by mouse mammary tumor virus: proviral activation of a cellular gene in the common integration region int-2, *Cell*, 37, 529, 1984.
55. **Varmus, H. E.**, Do hepatitis viruses make a genetic contribution to primary hepatocellular carcinoma?, in *Viral Hepatitis and Liver Disease*, Vyas, G. N., Hoofnagle, J. H. and, Dienstag, J. L., Eds., Grune & Stratton, Orlando, 1984, 411.
55a. **Tokino, T., et al.**, personal communication, 1990.
56. **Rogler, C. E., Sherman, M., Su, C. Y., Shafritz, D. A., Summers, J., Shows, T. B., Henderson, A., and Kew, M.**, Deletion in chromosome 11p associated with a hepatitis B integration site in hepatocellular carcinoma, *Science*, 230, 319, 1985.
56a. **Tamura, et al.**, personal communication, 1990.
57. **Hatada, I., Tokino, T., Ochiya, T., and Matsubara, K.**, Co-amplification of integrated hepatitis B virus DNA and transforming gene *hst*-1 in a hepatocellular carcinoma, *Oncogene*, 3, 537, 1988.
57a. **Tokino, T.**, personal communication, 1990.
58. **Tooze, J.**, *The Molecular Biology of Tumor Viruses*, 2nd ed., Cold Spring Harbor Laboratory, Cold Spring Harbor, New York, 1980.
59. **Hino, O., Nomura, K., Ohtake, K., Kawaguchi, T., Sugano, H., and Kitagawa, T.**, Instability of integrated hepatitis B virus DNA with inverted repeat structure in a transgenic mouse, *Cancer Genet. Cytogenet.*, 37, 273, 1989.
60. **Novick, A. and Szilard, L.**, Experiments on spontaneous and chemically induced mutations of bacteria growing in the chemostat, *Cold Spring Harbor Symp. Quant. Biol.*, 16, 337, 1951.
61. **Matsubara, K. and Tokino, T.**, Integration of hepatitis B virus DNA and its implications for hepatocarcinogenesis, *J. Mol. Med.*, in press, 1989.
62. **Kew, M. C. and Geddes, E. W.**, Hepatocellular carcinoma in rural South African blacks, *Medicine*, 61, 98, 1982.
63. **Yuasa, Y. and Sudo, K.**, Transforming genes in human hepatomas detected by a tumorigenicity assay, *Jpn. J. Cancer Res.*, 78, 1036, 1987.
64. **Ochiya, T., Fujiyama, A., Fukushige, S., Hatada, I., and Matsubara, K.**, Molecular cloning of an oncogene from a human hepatocellular carcinoma, *Proc. Natl. Acad. Sci. U.S.A.*, 83, 4993, 1986.
65. **Notario, V., Sukumar, S., Sautos, E., and Barbacid, M.**, A common mechanism for the malignant activation of ras oncogenes in human neoplasia and in chemically induced animal tumors, in *Cancer Cells, 2*, Van de Woude, G. F., Ed., Cold Spring Harbor Laboratory, Cold Spring Harbor, New York, 1984, 425.
66. **Gu, J., Tian, P., and Wan, D.**, Expression of N-ras gene in human primary hepatocellular carcinoma, *Scientia Sinica*, 29, 181, 1986.
67. **Sinha, S., Webber, C., Marshall, C. J., Knowles, M. A., Proctor, A., Barrass, N. C., and Neal, G. E.**, Activation of ras oncogene in aflatoxin-induced rat liver carcinogenesis, *Proc. Natl. Acad. Sci. U.S.A.*, 85, 3673, 1988.
68. **Watatani, M., Perantoni, A. O., Reed, C. D., Enomoto, T., Wenk, M. L., and Rice, J. M.**, Infrequent activation of K-ras, H-ras, and other oncogenes in hepatocellular neoplasms initiated by methyl (acetoxymethyl) nitrosamine, a methylating agent, and promoted by phenobarbital in F344 rats, *Cancer Res.*, 49, 1103, 1989.
69. **Takada, S. and Koike, K.**, Activated N-ras gene was found in human hepatoma tissue but only in a small fraction of the tumor cells, *Oncogene*, 4, 189, 1989.
70. **Yang, S. S., Modali, R., Parks, J. B., and Taub, J. V.**, Transforming DNA sequences of human hepatocellular carcinomas, their distribution and relationship with hepatitis B virus sequence in human hepatomas, *Leukemia*, 2 (Suppl. 102s), 1988.

71. **Iwamura, Y., Mitamura, K., Yanagi, K., Hasimoto, T., and Kato, K.,** Transforming potential of DNA of the human PLC/PRF/5 hepatoma cell line, *Intervirology*, 26, 223, 1986.
72. **Morizane, T., Nakamura, T., Saito, H., Watanabe, T., Inagaki, Y., Satoh, I., Tsuchimoto, T., and Tsuchiya, M.,** Transformation of NIH-3T3 cells by DNA from a human hepatoma cell line with integrated hepatitis B virus DNA, *Eur. J. Cancer Clin. Oncol.*, 23, 163, 1987.
72a. **Nagahara et al.,** personal communication, 1990.
73. **Fukuda, K., Ogasawara, S., Maruiwa, M., Yano, H., Murakami, T., and Kojiro, M.,** Structural alteration in c-myc and c-Ha-ras proto-oncogenes in human hepatocellular carcinoma, *Kurume Med. J.*, 35, 77, 1988.
74. **Moroy, T., Etiemble, J., Bougueleret, L., Hadchouel, M., Tiollais, P., and Buendia, M. A.,** Structure and expression of hcr, a locus rearranged with c-myc in a woodchuck hepatocellular carcinoma, *Oncogene*, 4, 59, 1989.
75. **Chandar, N., Lombardi, B., Schultz, W., and Locker, J.,** Analysis of ras genes and linked viral sequences in rat hepatocarcinogenesis, *Am. J. Pathol.*, 129, 232, 1987.
75a. **Tokino, T.,** personal communication, 1990.
76. **Little, M. H., Thomson, D. B., Hayward, N. K., and Smith, P. J.,** Loss of alleles on the short arm of chromosome 11 in a hepatoblastoma from a child with Beckwith-Wiedemann syndrome, *Hum. Genet.*, 79, 186, 1988.
77. **Wang, H. R. and Rogler, C. E.,** Deletions in human chromosome arms 11p and 13q in primary hepatocellular carcinomas, *Cytogenet. Cell Genet.*, 48, 72, 1988.
78. **Pasquinelli, C., Garreau, F., Bougueleret, L., Cariani, E., Grzeschik, K. H., Thiers, V., Croissant, O., Hadchouel, M., Tiollais, P., and Brechot, C.,** Rearrangement of a common cellular DNA domain on chromosome 4 in human primary liver tumors, *J. Virol.*, 62, 629, 1988.
79. **Blanquet, V., Garreau, F., Chenivesse, X., Brechot, C., and Turleau, C.,** Regional mapping to 4q32.1 by *in situ* hybridization of a DNA domain rearranged in human liver cancer, *Hum. Genet.*, 80, 274, 1988.
79a. **Hirohashi,** personal communication, 1990.
80. **Cariani, E., Lasserre, C., Seurin, D., Hamelin, B., Kemeny, F., Franco, D., Czech, M. P., Ullrich, A., and Brechot, C.,** Differential expression of insulin-like growth factor II mRNA in human primary liver cancers, benign liver tumors, and liver cirrhosis, *Cancer Res.*, 48, 6844, 1988.
81. **Lee, H.-S., Rajagopalan, M. S., and Vyas, G. N.,** A lack of direct role of hepatitis B virus in the activation of ras and c-myc oncogenes in human hepatocellular carcinogenesis, *Hepatology,* 8, 116, 1988.
82. **Himeno, Y., Fukuda, Y., Hatanaka, M., and Imura, H.,** Expression of oncogenes in human liver disease, *Liver*, 8, 208, 1988.
83. **Galand, P., Jacobovitz, D., and Alexandre, K.,** Immunohistochemical detection of c-Ha-ras oncogene p21 production pre-neoplastic and neoplastic lesions, *Int. J. Cancer*, 41, 155, 1988,
84. **Dejean, A., Bougueleret, L., Grzeschik, K.-H., and Tiollais, P.,** Hepatitis B virus DNA integration in a sequence homologous to v-erb-A and steroid receptor genes in a hepatocellular carcinoma, *Nature (London)*, 322, 70, 1986.
85. **DeThe, H., Marchio, A., and Tiollais, P.,** A novel steroid thyroid hormone receptor-related gene inappropriately expressed in human hepatocellular carcinoma, *Nature (London)*, 330, 667, 1987.
86. **Brand, N., Petkovich, M., Krust, A., Chambon, P., DeThe, H., Marchio, A., Tiollais, P., and Dejean, A.,** Identification of a second human retinoic acid receptor, *Nature (London)*, 332, 850, 1988.
87. **Shaul, Y., Rutter, W. J., and Laub, O.,** A human hepatitis B viral enhancer element, *EMBO J.*, 4, 427, 1985.
88. **Karpen, S., Banerjee, R., Zelent, A., Price, P., and Acs, G.,** Identification of protein-binding sites in the hepatitis B virus enhancer and core promoter domains, *Mol. Cell. Biol.* 8, 5159, 1988.
89. **Yee, J. K.,** A liver-specific enhancer in the core promoter region of human hepatitis B virus, *Science*, 246, 658, 1989.
90. **Moroy, T., Marchio, A., Etiemble, J., Trepo, C., Tiollais, P., and Buendia, M. A.,** Rearrangement and enhanced expression of c-myc in hepatocellular carcinoma of hepatitis virus-infected woodchucks, *Nature (London)*, 324, 276, 1986.
91. **Hsu, T., Moroy, T., Etiemble, J., Louise, A., Trepo, C., Tiollais, P., and Buendia, M. A.,** Activation of c-myc by woodchuck hepatitis virus infection in hepatocellular carcinoma, *Cell*, 55, 627, 1988.
92. **Koshy, R. and Hofschneider, P. H.,** Transactivation by hepatitis B virus may contribute to hepatocarcinogenesis, *Curr. Top Microbiol. Immunol.*, 144, 265, 1989.
93. **Twu, J.-S. and Schloemer, R. H.,** Transcriptional *trans*-activating function of hepatitis B virus, *J. Virol.*, 61, 3448, 1987.
94. **Spandau, D. F. and Lee, C. H.,** *Trans*-activation of viral enhancers by the hepatitis B virus X protein, *J. Virol.*, 62, 3448, 1987.

95. **Faktor, O., De-Medina, T., and Shaul, Y.,** Regulation of hepatitis B virus S gene promoter in transfected cell lines, *Virology*, 162, 362, 1988.
96. **Colgrove, R., Simon, G., and Ganem, D.,** Transcriptional activation of homologous and heterologous genes by the hepatitis B virus X gene product in cells permissive for viral replication, *J. Virol.*, 63, 4019, 1989.
97. **Siddiqui, A., Gaynor, R., Srinivasan, A., Mapoles, J., and Farr, R. W.,** *Trans*-activation of viral enhancers including long terminal repeat of the human immunodeficiency virus by the hepatitis B virus X protein, *Virology*, 169, 479, 1989.
98. **Zahm, P., Hofschneider, P. H., and Koshy, R.,** The HBV X-ORF encodes a transactivator: a potential factor in viral hepatocarcinogenesis, *Oncogene*, 3, 169, 1988.
99. **Seto, E., Yen, T. S. B., Peterlin, B. M., and Ou, J.-H.,** *Trans*-activation of the human immunodeficiency virus long terminal repeat by the hepatitis B virus X protein, *Proc. Natl. Acad. Sci. U.S.A.*, 85, 8286, 1988.
100. **Twu, J. S., Chu, K., and Robinson, W. S.,** Hepatitis B virus X gene activates kappa B-like enhancer sequences in the long terminal repeat of human immunodeficiency virus 1, *Proc. Natl. Acad. Sci. U.S.A.*, 86, 5168, 1989.
101. **Yoshida, M. and Seiki, M.,** Recent advances in the molecular biology of HTLV-1: *trans*-activation of viral and cellular genes, *Annu. Rev. Immunol.*, 5, 541, 1987.
102. **Taniguchi, T.,** Regulation of cytokine gene expression, *Annu. Rev. Immunol.* 6, 439, 1988.
102a. **Will, H., Koike, K., and Koshy, R.,** personal communication, 1990.
103. **Wollersheim, M., Debelka, U., and Hofschnieder, P. H.,** A transactivating function encoded in the hepatitis B virus X gene is conserved in the integrated state, *Oncogene*, 3, 545, 1988.
104. **Freytag von Loringhoven, A. F., Koch, S., Hofschneider, P. H., and Koshy, R.,** Co-transcribed 3′ host sequences augment expression of integrated hepatitis B virus DNA, *EMBO J.*, 4, 249, 1985.
105. **Ou, J. H. and Rutter, W. J.,** Hybrid hepatitis B virus-host transcripts in a human hepatoma cell, *Proc. Natl. Acad. Sci. U.S.A.*, 82, 83, 1985.
106. **Miyaki, M., Sato, C., Gotanda, T., Matsui, T., Mishiro, H., Imai, M., and Mayumi, M.,** Integration of region X of hepatitis B virus genome in human primary hepatocellular carcinomas propagated in nude mice, *J. Gen. Virol.*, 67, 1449, 1986.
107. **Kekulé, A. S., Lauer, U., Meyer, M., Caselmann, W. H., Hofschneider, P. H., and Koshy, R.,** The preS2/S region of integrated hepatitis B virus DNA encodes a transcriptional transactivator, *Nature (London)*, 343, 457, 1990.
108. **Fourel, G., Treop, C., Bougueleret, L., Henglein, B., Ponzetto, A., Tiollais, P., and Buendia, M.-A.,** Frequent activation of N-myc genes by hepadnavirus insertion in woodchuck liver tumours, *Nature (London)*, 347, 294, 1990.
109. **Wang, J., Chenivesse, X., Henglein, B., and Bréchot, C.,** Hepatitis B virus integration in a cyclin A gene in a hepatocellular carcinoma, *Nature (London)*, 343, 555, 1990.

Chapter 14

THE MOLECULAR BIOLOGY OF CAULIFLOWER MOSAIC VIRUS

Robert J. Shepherd

TABLE OF CONTENTS

I. INTRODUCTION

Cauliflower mosaic virus (CaMV) is the prototype member of the caulimoviruses, a group of double-stranded DNA viruses that infect higher plants. This group has been the subject of a number of recent reviews, especially Hohn et al.,[1] Howell,[2,3] Dixon and Hohn,[4] Covey,[5] and Shepherd.[6-8] In addition, the evidence for reverse transcription of the CaMV genome has been reviewed by Hohn et al.,[9] Mason et al.,[10] and Bonneville et al.[11]

In the caulimovirus group more than a dozen viruses have been described and new members are found each year. The viruses are roughly spherical particles with a diameter of about 50 nm and contain a single relaxed, circular, double-stranded DNA genome about 8000 bp in length.

Two features facilitate the use of these viruses as model systems to study gene expression and pathogenesis in plants. First, their naked DNA is infectious when rubbed on the leaves of healthy plants with a mild abrasive. During mechanical inoculation, spontaneous infections occur at a few sites on a leaf. The viruses then multiply and move cell to cell to establish a systemic infection. Second, the DNA can be cloned in an infectious form in bacteria. This allows genetically restructured genomes to be amplified to sufficient quantity to test on whole plants. The ease with which these viruses can be genetically manipulated has made them favorite subjects for studying viral pathogenesis in plants.

II. BIOLOGICAL PROPERTIES

CaMV infects many *Brassica* spp., other Cruciferae, and some strains infect *Nicotiana* and *Datura* spp. (Solanaceae). The diseases are characterized by mosaic- and mottle-type leaf discolorations that may be accompanied by vein banding, vein clearing, necrosis, puckering, and other forms of leaf distortion and death, particularly of the older leaves, and by varying degrees of stunting. Virus strain, host species, and genotype influence disease severity.

A prominent feature of CaMV-infected plants is the presence of conspicuous inclusion bodies in the cytoplasm. These electron-dense, refractive bodies (Figure 1) occur in virtually every cell of systemically infected plants. They consist of round or elongated bodies of one to several microns that enlarge throughout the course of infection. No external membrane surrounds the body. They appear as a granular matrix interspersed with less dense, vacuole-like regions. Spherical virions pack both the matrix and the vacuoles (Figure 1). Nearly all of the virus in the cell occurs within the inclusion bodies, with only a few virions found free in the cytoplasm or in enlarged plasmodesmata. The abnormal enlargement of the plasmodesmata (protoplasmic channels between cells) is associated with CaMV infections and may be involved in cell-to-cell movement of the virus.[12]

Incipient inclusion bodies appear as small patches of electron-dense matrix material surrounded by clusters of ribosomes.[12] These incipient bodies are observed before the appearance of virions. This suggests that the inclusion bodies are the major site for virus assembly. Indeed, autoradiographic studies indicate that the inclusion bodies are the major site of viral DNA synthesis.[13-15]

In nature CaMV is transmitted by aphids (Aphididae). These soft-bodied sucking insects carry the virus on their mouthparts and transmission occurs as a result of feeding activity. Complementation experiments using transmissible and defective strains of CaMV have shown that an accessory factor, probably a virus-specified protein produced in infected plants, is necessary for insect transmission.[16,17] As mentioned later, this function has been mapped to gene II of CaMV.

III. PROPERTIES OF CaMV AND ITS DNA GENOME

CaMV virions are isometric particles with little substructure apparent in the electron

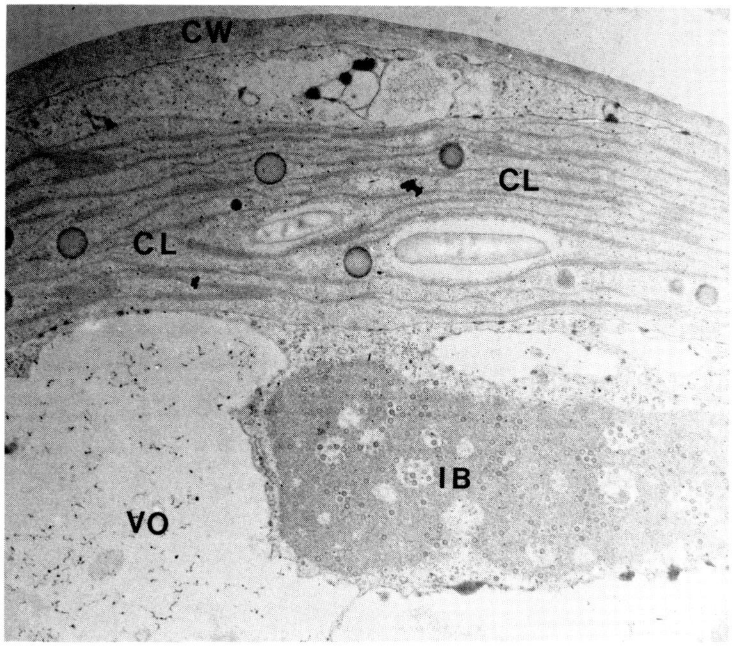

FIGURE 1. A portion of a mesophyll cell of *Brassica campestris* infected with cauliflower mosaic virus. A well-developed virus inclusion (IB) body is shown in the cytoplasm. Note the electron-dense matrix with numerous small, spherical virions embedded in the matrix and filling the vacuole-like areas. CL, chloroplast; CW, cell wall; VO, cell vacuole.

microscope. With neutron small-angle scattering, the molecular weight equivalent for the particle has been estimated to be 19.3 to 20.0 × 10^6.[18] Hull et al.[19] determined that purified CaMV contains 16% DNA based on its phosphorus content (1.63%) and quantitative diphenylamine tests; its sedimentation coefficient is 208S in 0.1 *M* NaCl, 0.01 M phosphate, pH 7.2, at infinite dilution; its partial specific volume is 0.74 g/cm; and its diffusion coefficient is 0.75 (±0.04) × 10^{-7} cm^2/sec. From these measurements the molecular weight equivalent for the virus particle was calculated to be 22.8 × 10^6.

CaMV DNA free of protein can infect turnip (*Brassica campestris*) plants during inoculation with a mild abrasive.[20] At a concentration of about 0.5 µg/ml, about half of the seedlings become infected. The abrasive probably causes breaks in the cuticle and outer cell wall, exposing the plasma membrane through which viral DNA is absorbed. Cloned DNA, when released from the plasmid vector by restriction enzyme digestion at the cloning site, is also infectious.[21,22]

Under the electron microscope, native CaMV DNA appears as linear and relaxed circular molecules of 2.31-µm contour length.[23] Hull and Shepherd[24] found only the circular form to be infectious. The DNA has a T_m of 87.2°C in 0.15 *M* NaCl, 0.015 *M* sodium citrate and a buoyant density in CsCl of 1.70 g/cm^3.[20]

High-resolution electron microscopy reveals that CaMV DNA released from degraded virions takes several topological forms. These appear as relaxed loops, or loops with one or more knots (Figure 2).[25] Electrophoresis in 1% agarose gels separates these forms into a series of closely spaced bands (Figure 2). Knotting probably occurs first in the linear full-length RNA molecule, which then serves as a replicative intermediate in viral replication to produce knotted circular DNA (see Section VII on replication).

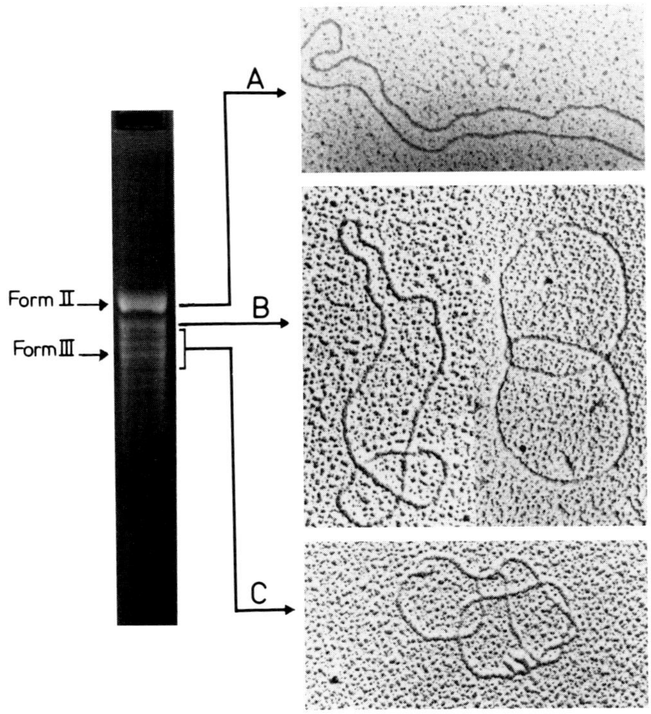

FIGURE 2. Topological forms of encapsidated CaMV DNA. The multiplicity of forms exhibited during gel electrophoresis of viral DNA is related to knotting of the molecules. Supercoiled DNA (form I, not shown) is found in nuclei of infected cells but not in virions. Open circular (form II) is the most slowly migrating form in 1% agarose gels (A). Linear molecules (form III) and knotted molecules migrate more rapidly in these gels. (B and C) Knotted circular molecules taken from 1% agarose gels. (Courtesy of J. Menissier.)

A supercoiled form of the DNA is found in the nucleus probably as a minichromosome similar to simian virus 40 DNA in the nuclei of mammalian cells.[26,27] The minichromosome is active in transcription and contains DNA associated with histones in a nucleosomal structure (Figure 3).[25,28] The repeating subunit of CaMV DNA nucleosomal structures is similar to that of plant chromatin.[29]

There are three single-stranded interruptions at specific sites in the native DNA of CaMV.[30,31] Hence, the circular DNA molecule is relaxed, rather than supercoiled like many other small circular viral double-stranded DNAs. There are two interruptions in the plus DNA strand and one in the minus DNA strand (Figure 4).

Based on DNA sequencing, the interruptions are actually triple-stranded overlaps where a short, redundant, non-base-paired tail protrudes over the double helix. Most likely the 5′-OH terminus is displaced because it is most reactive to labeling with polynucleotide kinase.[32] Richards et al.[33] found ribonucleotides, probably residual RNA primers left from reverse transcription, covalently attached to the 5′ termini at the interruption sites.

IV. GENETIC ORGANIZATION

Our understanding of the genetic organization of CaMV is based largely on the DNA sequences of three strains of the virus. The genome has eight open reading frames (tentative genes) large enough to code for polypeptides greater than 11 kDa. These are indicated by the

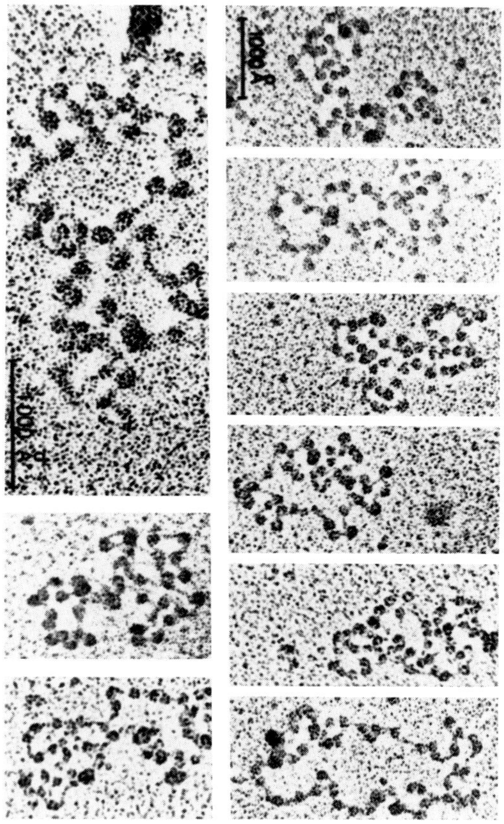

FIGURE 3. Minichromosomal DNA of CaMV isolated from nuclei of infected *Brassica campestris*. Each panel shows a single circular DNA molecule. The latter exhibit an average of 41 globular nucleosomes per DNA molecule. (From Menissier, J., de Murcia, G., Lebeurier, G., and Hirth, L., *EMBO J.*, 2, 1067, 1983. With permission. Courtesy of Dr. J. Menissier, 1990.)

peripheral arrows in Figure 4. Region VIII, which overlaps the C-terminal portion of region IV, and the small open reading region VII in the larger intergenic region (Figure 4), are found in all three strains of CaMV that have been sequenced. However, gene VIII is not required for virus infectivity. Moreover, gene VII has no homology at the amino acid level with similarly located open reading regions of other caulimoviruses that have been sequenced.[34,35] Only the six major genes (I through VI in Figure 4) are conserved in different members of the caulimovirus group.[8]

Except for region VI, the major open reading regions are very closely spaced, usually with only one or two nucleotides between the termination of one gene and the start codon of the next. Between genes IV and V a short overlap occurs. A short intergenic region of about 100 bp separates gene VI from coding region V. A large intergenic region of about 700 bp separates the terminus of gene VI and the start of gene VII (Figure 4). Of the major genes of caulimoviruses, gene V is the most highly conserved and gene VI is the most variable.[35]

Evidence for the biological function of genes I to VI has been provided by analysis of mutants.[36-38] In gene II, large insertions,[39] or deletion of almost the whole region,[40] did not destroy virus infectivity. However, insect transmission is destroyed. Hence, gene II is thought to provide an insect acquisition factor for plant-to-plant transmission in nature. Likewise, in gene VII, large insertions,[41] or deletion of the whole region,[42] do not affect virus activity. This

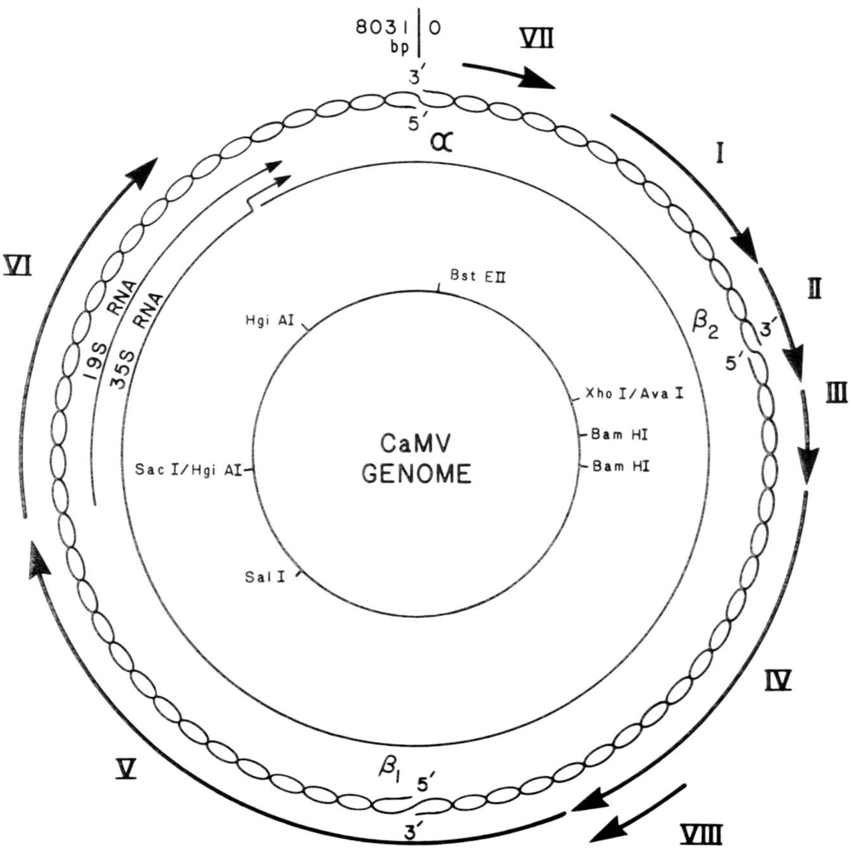

FIGURE 4. A physical map of the CaMV genome. The interwoven line indicates the 8031-bp circular double-stranded DNA. Encapsidated DNA with its three single-stranded interruptions at specific sites is indicated here. One in the transcribed, or minus-strand (designated α), is used as the zero position of the map. Two interruptions (designated β_1 and β_2) occur in the complementary or plus strand. The heavy peripheral arrows indicate the eight translational open reading regions that occur in the full-length (35S) RNA transcript. The latter transcript and the 19S transcript are indicated by the lighter inner lines in the diagram. The innermost circle shows the location of some key restriction sites used for exchanging segments of the genome among strains of CaMV in order to map viral functions.

region may have some regulatory function not related to its coding capacity since all sequenced caulimoviruses have a small open reading region upstream from gene I, the first essential coding region.

V. CaMV MESSENGER RNAs

Only the minus strand of the CaMV genome is transcribed to RNA.[43-45] There are two major polyadenylated virus-specific transcripts found both in infected plants[46,47] and protoplast systems.[43] These RNAs hybridize specifically to the DNA strand with a single interruption. Covey and Hull[48] found that these transcripts make up about 5% of the poly(A) RNA in infected plants.

The 19S poly(A) RNA transcript spans gene VI and part of the large intergenic region.[44,47,48] The 35S transcript spans the entire virus genome. Its synthesis starts in the large intergenic region upstream from the α interruption (Figure 4).[6,49,50] It terminates 180 nucleotides downstream from its initiation site, leaving a direct repeat of 180 nucleotides at each end of the RNA molecule.[46,50] The 19S transcript ends at the same point on the genome as the 35S transcript. Thus, the transcriptional terminator is the same for both RNA species.

A minor 22S RNA that spans region V and is translatable in reticulocyte lysates to produce a 75-kDa polypeptide has been described in CaMV-infected plants by Plant et al.[51] It is a polyadenylated transcript of about 2.5 kb that probably arises from a weak promoter overlapping the sequence of gene IV. The 75-kDa protein is believed to be the reverse transcriptase of CaMV.

The promoter for the 35S RNA of CaMV has been shown to function as a strong promoter in transgenic plants. It is constitutive in nature and has been widely used in plant transformation technology. Some of the upstream regulatory elements of the promoter have recently been characterized (Lam and Chua, 1989, and references therein).[52]

The products of genes I through IV of CaMV probably arise by coupled translation of the 35S polycistronic transcript. Several unique features of the arrangement of these genes plus CaMV mutants, which appear to be polar in nature, were some of the first observations that suggested coupled translation.

Nuclei isolated from CaMV-infected *Brassica* have a high level of virus-specific transcription. Its sensitivity to α-amanitin and ionic strength indicates that transcription is catalyzed by the endogenous DNA-dependent RNA polymerase II.[45] A supercoiled form of CaMV DNA with a nucleosome structure similar to chromatin, found in nuclei of CaMV-infected *Brassica*, has been shown to be transcriptionally active.[25,26]

CaMV-infected plants contain a polypeptide product of about 62 kDa, called P62, which is not found in healthy plants.[44,48,53] This protein has been associated with region VI of the CaMV genome, as demonstrated by hybrid-arrested translation[44,53] and by hybrid selection translation experiments[48] with 19S RNA.

For the three caulimoviruses that have been sequenced, only one or two nucleotides intervene between the stop codon of one gene and the start codon of the next, or in some cases short overlaps occur between successive genes. In addition, the open reading frames alternate in frame and canonical splicing signals are largely lacking in these viral transcripts,[32,34,35,54] except in strains that undergo spontaneous deletion.[55] With one exception (gene V) no subgenomic form of the RNA spanning these genes has been found. Consequently, it appears that these genes, including the coat protein gene, may not be translated from RNAs in which they appear in 5′-proximal positions.

The great majority of known polyadenylated eukaryotic mRNAs are functionally monocistronic, with only the 5′-proximal genes being translated with any degree of efficiency. Internal cistrons, even those with start codons in favorable contexts, are generally poorly translated (see Kozak, 1986, for a review).[56] This is probably a reflection of the scanning behavior of eukaryotic ribosomes that attach to the 5′ ends of mRNA and scan inward to initiate translation on favorable AUG codons. The translation of the four closely spaced genes I through IV of the caulimoviruses may be exceptions to the 5′-proximal rule for eukaryotes. Their positions with respect to one another and the lack of splicing signals suggest this. Another unusual feature is the occurrence of the sequence ATGA at the junction of these closely spaced genes. With this transition sequence the stop codon of the upstream open reading region overlaps by one nucleotide the start codon of the neighboring downstream region. With carnation etched ring virus DNA, the ATGA sequence occurs at every junction of genes I through V.[34] This sequence plus polar effects of nonsense or frameshift mutations in the nonessential region II of CaMV have suggested that these genes are translated from a polycistronic mRNA.[57] Perhaps it is significant that the transition sequence ATGA is common at the junction of successive genes of bacterial operons, which also exhibit a strongly polar effect of nonsense or frameshift mutations. Translation of successive cistrons of some of these operons is known to be coupled.

Observations of this sort have led to the hypothesis that genes I through IV of CaMV are expressed from a polycistronic mRNA, the 35S transcript, by a relay race type of translation.[57] In this proposed model ribosomes do not dissociate completely from the mRNA when reaching a stop codon. Instead, the 40S subunit continues scanning for a short distance. If an AUG codon is encountered nearby, 60S ribosomal subunits are believed to reassociate with the 40S subunits to reinitiate translation.[57]

The protein of gene VI of the caulimoviruses appears to have a regulatory role in the translation of the coupled genes of 35S RNA. This has now been observed with two different caulimoviruses.[58,59] If a polycistronic construct is attached to a transcription unit consisting of the promoter for the full-length transcript and the 5′-untranslated leader sequence of this transcript followed by successive open reading regions, only the 5′-proximal gene is expressed efficiently when these DNA constructs are electroporated into plant protoplasts. However, when gene VI-expressing plasmids are included with these polycistronic transcription units, the interior genes of the latter are expressed at a high level. This effect has been observed with CaMV[58] and the figwort mosaic virus.[59] Mutagenesis of gene VI indicates that it is the protein of this gene that is responsible for enhanced expression.

Deletion analysis of transcription units showing the enhanced expression in response to the gene VI product has revealed that some sequence in the long 5′ leader of the full-length polycistronic transcript of figwort mosaic virus interacts with the gene VI protein during the trans-activation response.[59] When various portions of the leader are removed, only those containing the coding region of gene VII (a small, nonconserved open reading region) show enhancement of reporter gene expression when located in internal downstream positions on the polycistronic transcript. The open reading frame associated with this region probably has no relationship to a sequence element in this position that is active in the gene VI trans-activation response. In fact, frameshift mutations of gene VII retain the ability to respond to gene VI trans-activation in protoplasts, thus demonstrating that there is no requirement for a polypeptide from the gene VII open reading region.[59a]

VI. MAPPING GENE FUNCTIONS OF CaMV

Artificial recombinants between different phenotypic strains of CaMV in addition to insertional mutants have allowed the association of biological functions with particular segments of the CaMV genome. For the six largest open reading regions gene products have been identified in infected plants.

Region IV has been identified as the coat protein gene. Analysis of the DNA sequence of CaMV showed that only region IV codes for a protein with an unusually high basic amino acid content[32] like CaMV coat protein (23% lysine plus arginine).[60] Daubert et al.[61] showed antigenic activity of coat protein occurred in *Escherichia coli* only when the gene IV sequence was expressed.

Although polypeptides of CaMV particles give rise to a number of electrophoretic components in gels,[60,62] the major components of mature virions are proteins of 44 and 37 kDa. They are probably cleaved from a larger precursor protein.[63] The precursor and to a lesser degree the reduced forms are phosphorylated[64] and glycosylated.[65] A kinase is associated with CaMV virions that can phosphorylate endogenous virus capsid proteins.[66,67]

Region VI codes for a 62-kDa protein (P62) that accumulates in infected cells to a far greater degree than other viral proteins. This protein is also produced in conspicuous amounts when total RNA or poly(A) RNA from infected leaves is translated in a cell-free system.[48] A major constituent of inclusion bodies isolated from infected plants was found to be a 62-kDa protein.[68,69] The electron-dense matrix material seen in electron micrographs of viral inclusion bodies may be composed largely of P62, the product of gene VI.

Gene VI appears to have a major role in disease induction by CaMV, as suggested by early observations that disease symptoms are drastically affected by small, noninactivating insertions in this region.[38] In addition, transgenic tobacco plants transformed with gene VI (*Agrobacterium*-mediated Ti plasmid vector) develop chlorotic mottling similar to virus infected plants.[70-72] Experiments using different cloned strains of CaMV to create hybrid genomes have shown that gene VI determines whether the virus elicits a hypersensitive reaction (local lesion) or a compatible reaction (systemic infection) in systemically susceptible hosts.[75] Hence, gene VI plays a major role in determining host range.[75] In fact, it appears to be the major gene mediating the virus–host interaction.

Although the biological function of the gene I product has not been firmly established, its amino acid sequence has been shown to have at least a low level of homology with the 30K cistron of tobacco mosaic virus (TMV).[34] The latter is known to be involved in cell-to-cell movement of the virus.[76-79] Gene I protein is deposited in plant cell walls,[80,81] which is another feature typical of viral proteins involved in cell-to-cell transport.

Hybrid genomes formed between aphid-transmissible and nontransmissible strains of CaMV and insertional mutants have demonstrated that gene II functions in insect transmission.[38,82-84] The product of this gene may be an 18-kDa polypeptide.[84]

The probable protein product of gene III has been detected in CaMV-infected plants. Antibodies to a synthetic 19-amino acid peptide corresponding to the N-terminal sequence of gene III reacted with a 15-kDa protein from infected turnips.[85] A similarly reactive polypeptide of 11 kDa was shown to be localized in virions, in addition to a 15-kDa form found in semipurified inclusion bodies.[86] The latter protein was found to have a strong affinity for double-stranded DNA.[86]

Gene V of CaMV specifies the reverse transcriptase, which participates in replication of viral DNA. Comparisons of the amino acid sequence (inferred from the nucleotide sequence of gene V) with that of known reverse transcriptases show a high degree of homology.[87-89] Furthermore, when the virus is cloned in *Saccharomyces cerevisiae*, gene V expresses a high level of reverse transcriptase activity.[90]

Products of genes VII and VIII have not been observed. It is not known if these regions are expressed in infected plants.

VII. REPLICATION

Although evidence indicates that the DNA of CaMV and the other caulimoviruses replicates by reverse transcription, the details of this process are not well known. Observations of the replicative intermediates indicate that the two DNA strands are assembled by separate processes instead of the more conventional semiconservative mode, in which a replication fork moves around the circular DNA molecule.

Inclusion bodies found in the cytoplasm of CaMV-infected cells have long been thought to function in replication of viral DNA.[13,91] Autoradiographic evidence with isolated cellular components supports this suggestion. Mazzolini et al.[92] found that inclusion bodies prepared essentially free of host cellular DNA retained a high degree of CaMV-specific replication activity, especially minus-strand synthesis. DNA polymerase activity in the isolated inclusion bodies has the characteristic template preference of reverse transcriptase.

Three unusual characteristics of the CaMV genome suggested that it replicates by reverse transcription of its 35S RNA. First, it was observed that the 3′-terminus of the 35S RNA

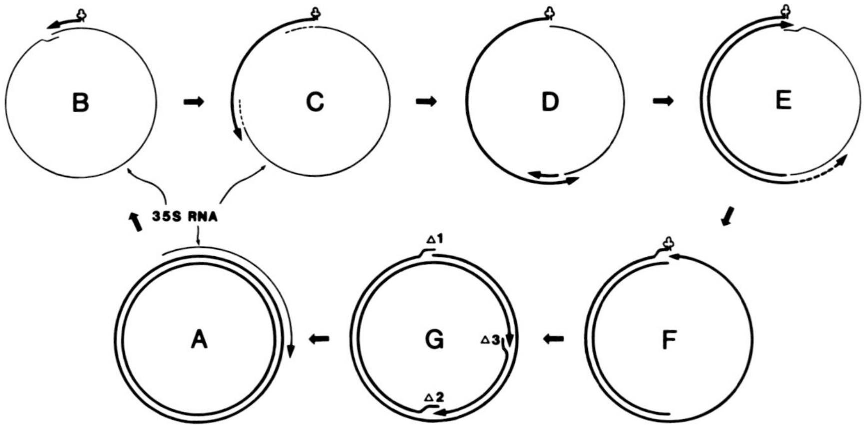

FIGURE 5. Steps in the replication of CaMV DNA by reverse transcription. Replication starts by production of a terminally redundant full-length RNA (35S RNA) transcribed from minichromosomal DNA in the nucleus, as indicated in step A. All subsequent steps probably occur in the cytoplasm. In the second step (B), 35S RNA serves as template for reverse transcription, initiating with a $tRNA^{Met}$ primer that binds to 35S RNA at a point about 600 nucleotides downstream from the 5′ cap site. First-strand DNA synthesis continues past the overlap region where the DNA polymerase is believed to transfer from one end to the other of the RNA template (C). When first-strand synthesis proceeds past a point coinciding with the β_1 interruption (see Figure 4), second-strand DNA synthesis is initiated (D). The completed genome has interruptions (overlaps) at three points, as indicated by Δ1, Δ2, and Δ3 (G). Figure redrawn from that of Marco and Howell.[105]

transcript occurred downstream of its own 5′ initiation site.[46,50] This fact implied that the 35S transcript is more than a full-length copy of the DNA strand, i.e., there is a 180-nucleotide direct repeat at each end of the RNA molecule. Second, a 14-nucleotide sequence adjacent to the α discontinuity in the plus strand (Figure 4) is complementary to the 3′ end of the bean methionine initiator tRNA.[93-97] These observations implied that $tRNA^{Met}$ binds to 35S RNA at this point and primes DNA synthesis by addition of deoxyribonucleotides to its free 3′-OH group. Third, some unusual short single-stranded viral DNA molecules are found with ribonucleotides covalently attached to their 5′ end. A small (725 nucleotide) DNA with minus polarity (sa-DNA), which maps to the large intergenic region upstream of the α interruption, has covalently attached RNA.[98] Its 5′ end (at the α interruption) is the same as that of viral DNA and its 3′ end maps very near to the start site of 35S RNA transcription. This short DNA strand has ribonucleotides at its 5′ end, suggesting that a RNA-priming event originated close to or at the α interruption to initiate viral DNA synthesis.[93,98,99]

The two strands of CaMV DNA are probably synthesized by separate events. Independent initiation of minus-strand synthesis is suggested by the presence of sa-DNA with minus polarity without any corresponding short single strands of plus-polarity DNA. Hypotonic extraction of plant organelle preparations releases CaMV complexes that use 35S RNA as a template for minus-strand DNA synthesis.[100] Characterization of CaMV DNA replication intermediates, and evidence from experiments carried out *in vitro* using purified nuclei, inclusion bodies,[101] or synchronously infected turnip protoplasts,[102] suggest that minus and plus strands are synthesized on different templates. Actinomycin D inhibits plus-strand synthesis but not minus-strand synthesis.[101] Plus-strand DNA of replication complexes occurs as DNA duplex molecules with regions of single-stranded minus polarity; in contrast, the minus-strand DNA is found as single strands in DNA/RNA hybrids.[101,102]

Some evidence suggests that replication may be occurring in partially assembled virions. Partial nuclease resistance of the templates for CaMV DNA synthesis suggests its association with partially completed virions.[101,102] Also, DNA synthesis, associated with sedimenting components with properties similar to assembled virions, is resistant to both DNase and RNase.[101-103] DNA polymerase activity occurs in purified virions. Reverse transcriptase may be encapsidated along with the viral template to which it is bound.[28]

Several laboratories have proposed a model for the replication of CaMV DNA (Figure 5).[93,95-97,104,105] The initial step (A in Figure 5) occurs in the nucleus, while subsequent steps probably take place in association with viral inclusion bodies found in the cytoplasm.

In the first step after infection, the viral minichromosome is transcribed in the nucleus to produce the 35S RNA transcript. This transcript is probably transported to the cytoplasm to begin reverse transcription. There is cytological evidence that the matrix material, believed to consist largely of P62, the product of gene VI, appears in infected cells before mature virions,[12] indicating that production, transport, and expression of the 19S RNA transcript of gene VI may occur before replication. The presence of P62 and partially assembled virions may be necessary to initiate the processes that follow.

In the second step (B, in Figure 5), first (minus)-strand synthesis is primed by $tRNA^{Met}$. The latter anneals to the 35S RNA at a point about 590 nucleotides downstream from its 5′ end (Figure 5), at the α interruption with respect to the CaMV DNA map (Figure 4). DNA polymerase (reverse transcriptase) begins phosphodiester bond formation by adding deoxyribonucleotides to the 3′-OH of the tRNA molecule. Replication proceeds along the RNA template until the 5′ end of the template is reached and DNA synthesis stops. Then the replication complex must switch to the 3′ end of the 35S RNA to continue minus-strand synthesis. As first-strand synthesis proceeds, RNase H activity of the reverse transcriptase probably removes the RNA portion of the hybrid molecule.

During the third step (C, in Figure 5), strand switching by the replication complex is completed and minus-strand synthesis proceeds past the location of the β_1 interruption on the genomic map (Figure 4). Second (plus)-strand synthesis (D in Figure 5) begins when replication passes this priming site. Plus strand is probably synthesized by the same enzyme (reverse transcriptase). Both plus-strand primer-binding sites may consist of purine-rich sequences of RNA left *in situ*. Minor primer-binding regions that initiate plus-strand synthesis can be found elsewhere along the genome.[106]

Specific breaks in the assembled CaMV genome that coincide with priming events during replication are a reflection of the stepwise mode of its synthesis (G, in Figure 5). One to several ribonucleotides remain *in situ* at each priming site. At the interruptions in the DNA, the 5′ ends are at fixed positions, whereas the 3′ ends are variable. This variability may be the result of displacement of 5′ ends for various distances during DNA synthesis near the interruptions.

RNA splicing has been experimentally demonstrated to occur during virus multiplication, giving additional evidence for reverse transcription of the CaMV genome.[107] When cloned chimeric CaMV DNA containing a fragment of the leghemoglobin gene with its intervening sequence (intron) was inoculated into plants, the intron disappeared from progeny virus as if RNA splicing had occurred during the course of replication. A spontaneous deletion at canonical intron–exon borders strongly suggests that RNA splicing occurred during DNA replication via an RNA intermediate. The nature of a naturally occurring deletion mutant that accompanies full-length genomes in CaMV infections provides further evidence for the removal of intervening sequences by RNA splicing during virus multiplication.[55]

Genomic-length 35S RNA may be the form of the viral genome encapsidated rather than viral DNA. Virion assembly may accompany the process by which 35S RNA is trapped in the inclusion bodies prior to reverse transcription.

The coat protein gene contains a highly conserved sequence of cysteine residues, the Cys-X_2-Cys-X_4-His-Cys motif,[108] now referred to as a "zinc finger".[109] The zinc finger may play a role in specific RNA recognition within partially assembled virions. This structure is also a feature of the *gag* gene product of retroviruses, which binds to RNA. Zinc fingers may be

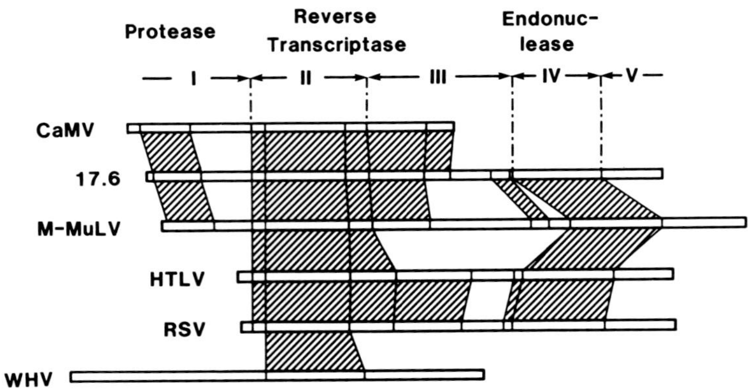

FIGURE 6. Homology profile of the amino acid sequence of various reverse transcriptases. Five regions of highly conserved sequences among the various retroid agents are indicated. Note that the CaMV gene V sequence contains no homologous sequence corresponding to tract IV, the endonuclease believed to bring about the integration of proviruses or retrotransposons into host DNA. 17.6, A *copia*-like transposon of *Drosophila*; RSV, Rous sarcoma virus; M-MuLV, Moloney murine leukemia virus; HTLV, human adult T cell leukemia virus; WHV, woodchuck hepatitis virus. This figure is redrawn from that of Toh et al.[87]

a universal characteristic of the coat proteins, or their precursors, encoded by viruses that utilize tRNA as a primer for reverse transcription. In these cases the structure is probably involved in encapsidation or RNA recognition by immature virions.[110,111]

VIII. REVERSE TRANSCRIPTASE

Evidence indicates that CaMV codes for its own DNA polymerase. The enzyme is an RNA-dependent DNA polymerase (reverse transcriptase). In addition to its polymerase activity it probably has other functions, including an acid protease and RNase H activity.

The most thoroughly studied RNA-dependent DNA polymerase is the reverse transcriptase of Rous sarcoma virus.[112] It consists of dimers of a single protein and utilizes either polyribonucleotides or polydeoxyribonucleotides as templates to direct the synthesis of complementary polydeoxyribonucleotides. As with all DNA polymerases it needs a preformed primer to initiate DNA synthesis.

Gene V, which encodes the reverse transcriptase, is the most highly conserved gene among strains for CaMV.[54] It is related to varying degrees to reverse transcribing elements in other systems.[87,88] Of these, it is most similar to a reverse-transcriptase of a retrotransposon of *Drosophila*, the *copia*-like element designated "17.6" (Figure 6).

The reverse transcriptases of the retroviruses are large, complex proteins with several domains specifying at least three different enzymatic activities, i.e., DNA polymerase and RNase H are located near the N terminus and a DNA endonuclease is located near the C terminus (Figure 6). There is significant amino acid homology between the highly conserved N-terminal region of CaMV gene V and the polymerase genes of hepatitis B virus. Toh et al.[87] examined the relationships among these polymerases and on the basis of conserved clusters of amino acids assigned five domains (Figure 6). The N-terminal domain, which is present in CaMV and the 17.6 transposon, but missing in some viral polymerases, is strongly homologous in amino acid sequence to the *gag*-specific protease of Rous Sarcoma virus and the *gag–pol* spacer of human adult T cell leukemia virus (HTLV). These sequences have a highly conserved series of amino acids very similar to the active site Asp-Thr-Gly of the acid protease family.[87]

Two blocks of sequence (II and III; Figure 6) that are common to the 17.6 transposon and the retrovirus polymerases are found in the central region of the proteins. This central region has a conserved element of Tyr-hydrophobic residues-Asp-Asp flanked by three hydrophobic residues, similar to the element Tyr-Gly-Asp-Asp flanked by hydrophobic residues found in polymerases of picornaviruses and some bacteriophages. This highly conserved element may play a pivotal role in polymerase function.[87] Another conserved 14-residue element common to retrovirus reverse transcriptase, influenza virus, CaMV, and hepatitis B virus was previously pointed out as the probable active site for these viral polymerases.[113]

In the central conserved region, the first portion probably specifies RNase H activity with the DNA polymerase situated near the C-terminal end. There the polymerase of CaMV ends; however, the 17.6 retrotransposon and the retroviruses have a series of amino acids associated with DNA endonuclease activity that is lacking in CaMV polymerase[87] (Figure 6). The latter half of the protein may be active in integrating the retroid elements into the chromosomes of the host organism. In contrast, CaMV has not been found in an integrated state in plant cells.

Toh et al.[87] suggest that CaMV may have evolved from an ancestral *copia*-like element of insects. They base this suggestion on the strong resemblance of the amino acid sequences of CaMV polymerase and the *Drosophila* 17.6 transposon. The homology values between these sequences exceed 30% for regions II and III (Figure 6).

Because of its template-bound nature and the lack of highly purified enzyme, the catalytic properties of CaMV DNA polymerase have not been thoroughly characterized. Most studies have focused on subcellular fractions from infected plants that synthesize virus DNA on an endogenous template.[95,97,100,114] Although high levels of CaMV-specific endogenous activity were found, Pfeiffer et al.[100] were unable to prepare template-dependent activity with fractions from infected plants.

Template-dependent activity with solubilized CaMV polymerase fractions has been obtained by two investigators.[88,102] Volovitch et al.[88] prepared a subcellular fraction specific to infected plants that copied added poly(C) and several natural mRNAs. The poly(C) activity was isolated by phosphocellulose chromatography and eluted as a fraction specific to CaMV-infected plants. It was clearly different from a normal γ-like DNA polymerase found in both healthy and infected plants. A soluble template-dependent DNA polymerase was also prepared by Thomas et al.,[102] using a freeze–thaw cycle of glycerol-containing fractions. The polymerase could copy cowpea mosaic virus RNA to produce extensive runs (greater than 2.5 kb) of cDNA. CaMV-specific DNA synthesis is unaffected by aphidicolin, an inhibitor of plant DNA polymerase.[100]

Both plus- and minus-strand synthesis are similarly sensitive to a variety of inhibitors, implying that a single enzyme probably synthesizes both strands during the replication cycle.[100] Furthermore, Thomas et al.[102] observed that *in vitro*-labeled DNA, produced by an active fraction from infected turnip protoplasts, hybridized to several strand-specific clones, showing that DNA of both polarities is synthesized on the endogenous template.

An 80-kDa protein has been found in infected plants with antibody to a synthetic polypeptide corresponding to the sequence of gene V.[115,116] This antibody inhibited viral DNA polymerase activity, providing further evidence for association of polymerase activity with gene V of CaMV.

IX. INSECT TRANSMISSION

The caulimoviruses are dependent on insects for movement from infected plants to healthy ones, except in cases where the infected host is vegetatively propagated. In the latter case infected individuals may also arise from infected parent plants. During insect transmission, virus is associated with the mouthparts and is acquired through feeding activity of the aphid vectors. The association is a biological one confined to insects of the family Aphididae, order Homoptera. Aphids cannot acquire virus by feeding on purified CaMV through a membrane. However, insects allowed to probe infected plants before feeding through a membrane will acquire and transmit virus.[16,17] A factor active in insect transmission is acquired from infected plants but does not accompany virus during the course of purification.

Insertions and deletions into gene II of CaMV DNA inactivate aphid transmission but not virus infectivity.[38,82,84] Restriction fragment exchanges have also been used to map the insect transmission function to gene II.[83,84] These exchanges between transmissible and nontransmissible strains of CaMV have been used to associate a particular mutation with the defect in transmission. In these mutants amino acid exchanges in gene II (Thr → Asn at amino acid 89 or a Gly → Arg at amino acid 94) can account for loss of the aphid transmission function.[38,114]

Transmission tests with hybrid genomes of a second aphid-nontransmissible isolate of CaMV and a transmissible strain showed that a particular small restriction fragment conferred insect transmissibility on the defective strain (Campbell strain). DNA sequence comparisons of region II showed two changes in the relevant segment of gene II. One of these caused a Gly → Arg exchange at position 94 of the gene II product of the Campbell strain,[118] identical to one of the exchanges previously thought to cause the defect in aphid transmission of strain CM1841.[38] The exchange occurs in a region of the protein that has been reported to show a significant level of amino acid homology with the aphid helper component sequence of an RNA plant virus,[119] suggesting that this region of the gene II protein plays an important role in insect transmission.

Of the three caulimoviruses that have been sequenced, gene II amino acid sequences show a high degree of relatedness (37.4 to 40.9% identity). The extent of similarities, however, varies with different regions of the polypeptide, with regions toward the C-terminal and N-terminal parts showing a higher degree of homology.[34,35] The functional similarity of these genes is illustrated by the complementation for aphid transmission between defective strains of CaMV and two other caulimoviruses.[120]

The 18-kDa protein found in plants infected with aphid-transmissible strains of CaMV,[84,121,122] but not in plants infected by strains with extensive deletions in gene II,[117,121,123] suggests that this polypeptide is probably the accessory factor that brings about insect transmission. The gene II product occurs in isolated inclusion bodies of CaMV[122,123] and appears to influence structural characteristics of these bodies.[120]

REFERENCES

1. **Hohn, T., Richards, K., and Lebeurier, G.**, Cauliflower mosaic virus on its way to becoming a useful plant vector, *Curr. Top. Microbiol. Immunol.*, 96, 193, 1982.
2. **Howell, S. H.**, Plant molecular vehicles: potential vectors for introducing foreign DNA into plants, *Annu. Rev. Plant Physiol.*, 33, 609, 1982.
3. **Howell, S. H.**, The molecular biology of plant DNA viruses, *CRC Crit. Rev. Plant Sci.*, 2, 287, 1985.
4. **Dixon, L. K. and Hohn, T.**, Cloning and manipulating cauliflower mosaic virus, in *Recombinant DNA Research and Virus*, Becker, Y., Ed., Martinius Nijhoff, The Hague, 1985, 247.
5. **Covey, S. N.**, Organization and expression of the cauliflower mosaic virus genome, in *Molecular Biology of Plant Viruses*, Vol. 2, Davies, J., Ed., CRC Press, Boca Raton, Florida, 1985, 121.
6. **Shepherd, R. J.**, DNA plant viruses, *Annu. Rev. Plant Physiol.*, 30, 405, 1979.
7. **Shepherd, R. J.**, Caulimoviruses as potential gene vectors for higher plants, in *Genetic Engineering: Principles and Methods*, Vol. 8, Setlow, J. K. and Hollaender, A., Eds., Plenum Press, New York, 1985, 241.
8. **Shepherd, R. J.**, Biochemistry of DNA plant viruses, in *The Biochemistry of Plants*, Vol. 15, Marcus, A., Ed., Academic Press, New York, 1989, 563.
9. **Hohn, T., Hohn, B., and Pfeiffer, P.**, Reverse transcription in CaMV, *Trends Biochem. Sci.*, 10, 205, 1985.
10. **Mason, W. S., Taylor, J. M., and Hull, R.**, Retroid virus genome replication, *Adv. Virus Res.*, 32, 35, 1987.
11. **Bonneville, J. M., Hohn, T., and Pfeiffer, P.**, Reverse transcription in the plant virus, cauliflower mosaic virus, in *RNA Genetics*, Vol. 2, Domingo, E., Holland, J. J., and Ahlquist, P., Eds., CRC Press, Boca Raton, Florida, 1988, 23.
12. **Kitajima, E. W. and Lauritis, J. A.**, Plant virions in plasmodesmata, *Virology*, 37, 681, 1969.
13. **Kamei, T., Rubio-Huertos, M., and Matsui, C.**, Thymidine-^{3}H uptake by X-bodies associated with cauliflower mosaic virus infection, *Virology*, 37, 506, 1969.
14. **Fujisawa, I., Rubio-Huertos, M., and Matsui, C.**, Incorporation of thymidine -H^3 into carnation etched ring virus, *Phytopathology*, 61, 681, 1971.
15. **Fujisawa, I., Rubio-Huertos, M., and Matsui, C.**, Deoxyribonuclease digestion of the nucleic acid from carnation etched ring virus, *Phytopathology*, 62, 810, 1972.
16. **Lung, M. C. Y. and Pirone, T. P.**, Studies on the reason for differential transmissibility of cauliflower mosaic virus isolates by aphids, *Phytopathology*, 63, 910, 1973.
17. **Lung, M. C. Y. and Pirone, T. P.**, Acquisition factor required for aphid transmission of purified cauliflower mosaic virus, *Virology*, 60, 260, 1974.
18. **Kruse, J., Timmins, P., and Witz, J.**, The spherically averaged structure of a DNA isometric plant virus: cauliflower mosaic virus, *Virology*, 159, 166, 1987.
19. **Hull, R., Shepherd, R. J., and Harvey, J. D.**, *J. Gen. Virol.*, 31, 93, 1976.
20. **Shepherd, R. J., Bruening, G. E., and Wakeman, R. J.**, Double-stranded DNA from cauliflower mosaic virus, *Virology*, 41, 339, 1970.
21. **Howell, S. H., Walker, L. L., and Dudley, R. K.**, Cloned cauliflower mosaic virus DNA infects turnips (*Brassica rapa*), *Science*, 208, 1265, 1980.
22. **Lebeurier, G., Hirth, L., Hohn, T., and Hohn, B.**, Infectivities of native and cloned DNA of cauliflower mosaic virus, *Gene*, 12, 139, 1980.
23. **Shepherd, R. J. and Wakeman, R. J.**, Observation on the size and morphology of cauliflower mosaic virus deoxyribonucleic acid, *Phytopathology*, 61, 188, 1971.
24. **Hull, R. and Shepherd, R. J.**, The structure of cauliflower mosaic virus genome, *Virology*, 79, 216, 1977.
25. **Menissier, J., de Murcia, G., Lebeurier, G., and Hirth, L.**, Electron microscopie studies of the different topological forms of the cauliflower mosaic virus DNA: knotted encapsidated DNA and nuclear minichromosome, *EMBO J.*, 2, 1067, 1983.
26. **Olszewski, N., Hagen, G., and Guilfoyle, T. J.**, A transcriptionally active covalently closed minichromosome of cauliflower mosaic virus DNA isolated from turnip leaves, *Cell*, 29, 395, 1982.
27. **Olszewski, N. E. and Guilfoyle, T. J.**, Nuclei purified from cauliflower mosaic virus-infected turnip leaves contain subgenomic, covalently closed circular cauliflower mosaic virus DNA's, *Nucleic Acids Res.*, 11, 8901, 1983.
28. **Menissier, J., Laquel, P., Lebeurier, G., and Hirth, L.**, A DNA polymerase activity is associated with cauliflower mosaic virus, *Nucleic Acids Res.*, 12, 8769, 1984.
29. **Phillips, G. and Gigot, C.**, *Nucleic Acids Res.*, 4, 3617, 1977.
30. **Volovitch, M., Dumas, J. P., Drugeon, G., and Yot, P.**, Single-stranded interruptions in cauliflower mosaic virus DNA, in *Acids Nucleiques et Synthese des Proteines Chez lex Vegetaux*, Bogorad, L. and Weil, J. H., Eds., Centre National de la Recherche Scientifique (CNRS), Paris, 1976, 635.

31. **Volovitch, M., Modjtahedi, N., Brun, G., Drugeon, G., and Yot, P.,** Studies on the single stranded discontinuities of the cauliflower mosaic virus genome, *Nucleic Acids Res.*, 5, 2913, 1978.
32. **Frank, A., Guilley, H., Jonard, G., Richards, K., and Hirth, L.,** Nucleotide sequence of cauliflower mosaic virus DNA, *Cell*, 21, 285, 1980.
33. **Richards, K. E., Guilley, H., and Jonard, G.,** Further characterization of the discontinuities of cauliflower mosaic virus DNA, *FEBS Lett.*, 134, 67, 1981.
34. **Hull, R., Sadler, J., and Longstaff, M.,** The sequence of carnation etched ring virus DNA: comparison with cauliflower mosaic virus and retroviruses, *EMBO J.*, 5, 3083, 1986.
35. **Richins, R. D., Scholthof, H. B., and Shepherd, R. J.,** Sequence of figwort mosaic virus DNA (caulimovirus group), *Nucleic Acids Res.*, 15, 8451, 1987.
36. **Howell, S. H., Walker, L. L., and Walden, R. M.,** Rescue of *in vitro* generated mutants of cloned cauliflower mosaic virus genome in infected plants, *Nature (London)*, 293, 483, 1982.
37. **Dixon, L. K., Koenig, I., and Hohn, T.,** Mutagenesis of cauliflower mosaic virus, *Gene*, 25, 189, 1983.
38. **Daubert, S., Shepherd, R. J., and Gardner, R. C.,** Insertional mutagenesis of the cauliflower mosaic virus genome, *Gene*, 25, 201, 1983.
39. **Gronenborn, B., Gardner, R. C., Schaefer, S., and Shepherd, R. J.,** Propagation of foreign DNA in plants using cauliflower mosaic virus as vector, *Nature (London)*, 294, 773, 1981.
40. **Howarth, A. J., Gardner, R. C., Messing, J., and Shepherd, R. J.,** Nucleotide sequence of naturally occurring deletion mutants of cauliflower mosaic virus, *Virology*, 112, 678, 1981.
41. **Gardner, R. C.,** CaMV as an experimental tool, in *Genetic Engineering of Plants: An Agricultural Perspective*, Kosuge, T., Meredith, C. P., and Hollaender, A., Eds., Plenum Press, New York, 1982, 121.
42. **Dixon, L. K. and Hohn, T.,** Initiation of translation of the cauliflower mosaic virus genome from a polycistronic mRNA: evidence from deletion mutagenesis, *EMBO J.*, 3, 2731, 1984.
43. **Howell, S. H. and Hull, R.,** Replication of cauliflower mosaic virus and transcription of its genome in turnip leaf protoplasts, *Virology*, 86, 468, 1978.
44. **Odell, J. and Howell, S. H.,** The identification, mapping, and characterization of mRNA for P66, a cauliflower mosaic virus-coded protein, *Virology*, 102, 349, 1980.
45. **Guilfoyle, T. J.,** Transcription of the cauliflower mosaic virus genome in isolated nuclei from turnip leaves, *Virology*, 107, 71, 1980.
46. **Covey, S. N., Lomonossoff, G. P., and Hull, R.,** Characterization of cauliflower mosaic virus DNA sequences which encode major polyadenylated transcripts. Characterization of cauliflower mosaic virus DNa sequences which encode major polyadenylated transcripts, *Nucleic Acids Res.*, 9, 6735, 1981.
47. **Odell, J., Dudley, R. K., and Howell, S. H.,** Structure of the 19S RNA transcript encoded by cauliflower mosaic virus genome, *Virology*, 111, 377, 1981.
48. **Covey, S. N. and Hull, R.,** Transcription of cauliflower mosaic virus DNA: detection of transcripts, properties, and location of the gene encoding the virus inclusion body protein, *Virology*, 111, 463, 1981.
49. **Dudley, R. K., Odell, J. T., and Howell, S. H.,** Structure and 5′ termini of the large and 19S RNA transcripts encoded by cauliflower mosaic virus genome, *Virology*, 117, 19, 1982.
50. **Guilley, H., Dudley, R. K., Jonard, G., Balazs, E., and Richards, K. E.,** Transcription of cauliflower mosaic virus DNA: detection of promoter sequences and characterization of transcripts, *Cell*, 30, 763, 1982.
51. **Plant, A. L., Covey, S. N., and Grierson, P.,** Detection of a subgenomic mRNA for gene V, the putative reverse transcriptase gene of cauliflower mosaic virus, *Nucleic Acids Res.*, 13, 8305, 1985.
52. **Lam, E. and Chua, N.-H.,** A factor that binds to the cauliflower mosaic virus 35S promoter and a conserved gata motif in cab promoters, *Plant Cell*, 1, 1147, 1989.
53. **Al Ani, R., Pfeiffer, P., Whitechurch, 0., Lesot, A., Lebeurier, G., and Hirth, L.,** A virus specified protein produced upon infection by cauliflower moasic virus, *Ann. Virol. (Inst. Pasteur)*, 131E, 33, 1980.
54. **Gardner, R. C., Howarth, A. J., Hahn, P., Brown-Leudi, M., Shepherd, R. J., and Messing, J.,** The complete nucleotide sequence of an infectious clone of cauliflower mosaic virus by M13 mp 7 shotgun sequencing, *Nucleic Acids Res.*, 9, 2871, 1981.
55. **Hirochika, H., Takatsuji, H., Ubasawa, A., and Ikeda, J.-E.,** Site specific deletion in cauliflower mosaic virus DNA: possible involvement of RNA splicing and reverse transcription, *EMBO J.*, 4, 1673, 1985.
56. **Kozak, M.,** Regulation of protein synthesis in virus-infected animal cells, *Adv. Virus Res.*, 31, 229, 1986.
57. **Gronenborn, B.,** The molecular biology of cauliflower mosaic virus and its application as plant gene vector, in *DNA Infectious Agents*, Hohn, T. and Schell, J., Eds., Springer-Verlag, New York, Wein, 1987, 1.
58. **Bonneville, J. M., Sanfacon, H., Futterer, J., and Hohn, T.,** Posttranscriptional *trans*-activation in cauliflower mosaic virus, *Cell*, 59, 1135, 1989.
59. **Gowda, S., Wu, F. C., Scholthof, H. B., and Shepherd, R. J.,** Gene VI of figwort mosaic virus (caulimovirus group) functions in posttranscriptional expression, *Proc. Natl. Acad. Sci. U.S.A.*, 86, 9203, 1989.

59a. **Gowda, S.,** author's laboratory, unpublished observation, 1990.

60. **Brunt, A. A., Barton, R. J., Tremaine, J. H., and Stace-Smith, R.,** The composition of cauliflower mosaic virus protein, *J. Gen. Virol.*, 27, 101, 1975.

61. **Daubert, S., Richins, R., and Shepherd, R. J.,** Mapping of the coat protein gene of cauliflower mosaic virus by its expression in a prokaryotic system, *Virology*, 122, 444, 1982.

62. **Hull, R. and Shepherd, R. J.,** The coat proteins of cauliflower mosaic virus, *Virology*, 70, 217, 1976.

63. **Hahn, P. and Shepherd, R. J.,** Evidence for a 58-kilodalton polypeptide as precursor of the coat protein of cauliflower mosaic virus, *Virology*, 116, 480, 1982.

64. **Hahn, P. and Shepherd, R. J.,** Phosphorylated proteins in cauliflower mosaic virus, *Virology*, 107, 295, 1980.

65. **duPlessis, D. H., and Smith, P.,** Glycosylation of the cauliflower mosaic virus capsid polypeptide, *Virology*, 109, 403, 1981.

66. **Martinez-Izquierdo, J., and Hohn, T.,** Cauliflower mosaic virus coat protein is phosphoryolated *in vitro* by a virion-associated protein kinase, *Proc. Natl. Acad. Sci. U.S.A.*, 84, 1824, 1987.

67. **Menissier, J., deMurcia, G., Geldreich, A., and Lebeurier, G.,** Evidence for a protein kinase activity associated with purified particles of cauliflower mosaic virus, *J. Gen. Virol.*, 67, 1885, 1986.

68. **Shepherd, R. J., Richins, R., and Shalla, T. A.,** Isolation and properties of the inclusion bodies of cauliflower mosaic virus, *Virology*, 102, 389, 1980.

69. **Shockey, M. W., Gardner, C. O., Melcher, U., and Essenberg, R. C.,** Polypeptides associated with inclusion bodies from leaves of turnip infected with cauliflower mosaic virus, *Virology*, 105, 575, 1980.

70. **Baughman, G. A., Jacobs, J. D., and Howell, S. H.,** Cauliflower mosaic virus gene VI produces a symptomatic phenotype in transgenic tobacco plants, *Proc. Natl. Acad. Sci. U.S.A.*, 85, 733, 1988.

71. **Shepherd, R. J., Goldberg, K., Kiernan, J., Gowda, S., Schoelz, J., Young, M., and Richins, R.,** Genomic changes during host adaptation, in *Physiology and Biochemistry of Plant-Microbial Interactions,* Keen, N. T., Kosuge, T., and Walling, L. L., Eds., American Society Plant Physiologists, 1988, 131.

72. **Takahashi, H., Shimamoto, K., and Ehara, Y.,** Cauliflower mosaic virus gene VI causes growth suppression development of necrotic spots and expression of defense-related genes in transgenic tobacco plants, *Mol. Gen. Genet.*, 216, 188, 1989.

73. **Schoelz, J. E., Shepherd, R. J., and Daubert, S.,** Region VI of cauliflower mosaic virus encodes a host range determinant, *Mol. Cell. Biol.*, 6, 2632, 1986.

74. **Schoelz, J. E. and Shepherd, R. J.,** Host range control of cauliflower mosaic virus, *Virology*, 162, 30, 1988.

75. **Schoelz, J. E., Shepherd, R. J., and Daubert, S. D.,** Host response to cauliflower mosaic virus (CaMV) in solanaceous plants is determined by a 496 bp DNA sequence within gene VI, in *Molecular Strategies for Crop Protection*, Arntzen, C. J. and Ryan, C. A., Eds., Alan R. Liss, New York, 1987, 253.

76. **Albrecht, H., Geldreich, A., Menissier, J., de Murcia, G., Kirchherr, D., Mesnard, J.-M., and Lebeurier, G.,** Cauliflower mosaic virus gene I product detected in a cell-wall-enriched fraction, *Virology* , 163, 503, 1988.

77. **Linstead, P. J., Hills, G. J., Plaskitt, K. A., Wilson, I. G., Harker, C. L., and Maule, A. M.,** The subcellular location of the gene I product of cauliflower mosaic virus is consistent with a function associated with virus spread, *J. Gen. Virol.*, 69, 1809, 1988.

78. **Nishiguchi, M., Motoyoshi, F., and Oshima, N.,** Behavior of a temperature sensitive strain of tobacco mosaic virus in tomato leaves and protoplasts, *J. Gen. Virol.*, 39, 53, 1978.

79. **Nishiguchi, M., Motoyoshi, F., and Oshima, N.,** Further investigation of a temperature-sensitive strain of tobacco mosaic virus: its behavior in tomato leaf epidermis, *J. Gen. Virol.*, 46, 497, 1980.

80. **Ohno, T., Tokamatsu, N., Meshi, T., Okada, Y., Nishiguchi, M., and Kiho, Y.,** Single amino acid substitution in 30K protein of TMV and effective in virus transport function, *Virology*, 131, 255, 1983.

81. **Meshi, T., Watanabe, Y., Saito, T., Sugimoto, A., Moeda, T., and Okada, Y.,** Function of the 30-kD protein of tobacco mosaic virus involvement in cell-to-cell movement and dispensability for replication, *EMBO J.*, 6, 2557, 1987.

82. **Armour, S. L., Melcher, U., Pirone, T. P., Lyttle, D. J., and Essenberg, R. C.,** Helper component for aphid transmission encoded by region II of cauliflower mosaic virus DNA, *Virology*, 129, 25, 1983.

83. **Daubert, S. D., Schoelz, J., Li, D., and Shepherd, R. J.,** Expression of disease symptoms in cauliflower mosaic virus genomic hybrids, *J. Mol. Appl. Genet.*, 2, 537, 1984.

84. **Woolston, C., Covey, S., Penswick, J., and Davies, J.,** Aphid transmission and a polypeptide are specified by a defined region of the cauliflower mosaic virus genome, *Gene*, 23, 15, 1983.

85. **Xiong, C., Lebeurier, G., and Hirth, L.,** Detection *in vivo* of a new gene product (gene III) of cauliflower mosaic virus, *Proc. Natl. Acad. Sci. U.S.A.*, 81, 6608, 1984.

86. **Giband, M., Mesnard, J. M., and Lebeurier, G.,** The gene III product (P15) of cauliflower mosaic virus is a DNA-binding protein while an immunologically related P11 polypeptide is associated with virions, *EMBO J.*, 5, 2433, 1986.
87. **Toh, H., Kikuno, R., Hayashida, H., Miyata, T., Kugimiya, W., Inouye, S., Yuki, S., and Saigo, K.,** Close structural resemblance between putative polymerase of a *Drosophila* transposable genetic element 17.6 and pol gene product of Moloney murine leukaemia virus, *EMBO J.*, 4, 1267, 1985.
88. **Volovitch, M., Modjtahedi, N., Yot, P., and Brun, G.,** RNA-dependent DNA polymerase activity in cauliflower mosaic virus-infected plant leaves, *EMBO J.*, 3, 309, 1984.
89. **Saigo, K., Kugimiya, W., Matsuo, Y., Inouye, S., Yoshiaka, K., and Yuki, S.,** Identification of the coding sequence for a reverse transcriptase-like enzyme in a transposable genetic element in *Drosophila melanogaster*, *Nature (London)*, 312, 659, 1984.
90. **Takatsuji, H., Hirochika, H., Fukushi, T., and Ikeda, J.-E.,** Expression of cauliflower mosaic virus reverse transcriptase in yeast, *Nature (London)*, 319, 240, 1986.
91. **Favali, M. A., Bassi, M., and Conti, G. G.,** A quantitative autoradiographic study of intracellular sites for replication of cauliflower mosaic virus, *Virology*, 53, 115, 1973.
92. **Mazzolini, L., Bonneville, J. M., Volovitch, M., Magazin, M., and Yot, P.,** Strand-specific viral DNA synthesis in purified viroplasms isolated from turnip leaves infected with cauliflower mosaic virus, *Virology*, 145, 293, 1985.
93. **Guilley, H., Richards, K. E., and Jonard, G.,** Observations concerning the discontinuous DNA's of cauliflower mosaic virus, *EMBO J.*, 2, 277, 1983.
94. **Hull, R. and Covey, S. N.,** DNA associated with nucleosomes in plants, *Trends Biochem. Sci.*, 8, 119, 1983.
95. **Pfieffer, P. and Hohn, T.,** Involvement of reverse transcription in the replication of plant virus CaMV: a detailed model and test of some aspects, *Cell*, 33, 781, 1983.
96. **Howell, S. H., Walden, R. M., and Marco, Y.,** Recombination and replication of cauliflower mosaic virus DNA, in *Plant Molecular Biology*, Goldberg, R., Ed., Alan R. Liss, New York, 1983, 137.
97. **Guilfoyle, T. J., Olszewski, N., Hagen, G., Kuzj, A., and McClure, B.,** Transcription and replication of the cauliflower mosaic virus genome: studies with isolated nuclei, nuclear lysates, and extranuclear cell-free leaf extracts, in *Plant Molecular Biology*, Goldberg, R., Ed., Alan R. Liss, New York, 1983, 117.
98. **Covey, S. N., Turner, D., and Mulder, G.,** A small DNA molecule containing covalently-linked ribonucleotides originates from the large intergenic region of the cauliflower mosaic virus genome, *Nucleic Acids Res.*, 11, 251, 1983.
99. **Turner, D. S. and Covey, S. N.,** A putative primer for the replication of cauliflower mosaic virus by reverse transcription is virion associated, *FEBS Lett.*, 165, 285, 1984.
100. **Pfieffer, P., Laquel, P., and Hohn, T.,** *Plant Mol. Biol.*, 3, 261, 1984.
101. **Marsh, L., Kuzj, A., and Guilfoyle, T.,** Identification and characterization of cauliflower mosaic virus replication complexes: analogy to hepatitis B viruses, *Virology*, 143, 212, 1985.
102. **Thomas, C. M., Hull, R., Bryant, J. A., and Maule, A. J.,** Isolation of a fraction from cauliflower mosaic virus-infected protoplasts which is active in the synthesis of (+) and (–) strand viral DNa and reverse transcription of primed RNA templates, *Nucleic Acids Res.*, 13, 4557, 1985.
103. **Marsh, L. and Guilfoyle, T.,** Cauliflower mosaic virus replication intermediates are encapsidated into virion-like particles, *Virology*, 161, 129, 1987.
104. **Hull, R. and Covey, S. N.,** Characterization of cauliflower mosaic virus DNA forms isolated from infected turnip leaves, *Nucleic Acids Res.*, 11, 1881, 1983.
105. **Marco, Y. and Howell, S. H.,** Intracellular forms of viral DNA consistent with a model of reverse transcriptional replication of the cauliflower mosaic virus, *Nucleic Acids Res.*, 12, 1517, 1984.
106. **Maule, A. J. and Thomas, C. M.,** Evidence from cauliflower mosaic virus virion DNA for additional discontinuities in the plus strand, *Nucleic Acids Res.*, 13, 7359, 1985.
107. **Hohn, B., Balazs, E., Ruegg, D., and Hohn, T.,** Splicing of an intervening sequence from hybrid cauliflower mosaic viral RNA, *EMBO J.*, 5, 2759, 1986.
108. **Covey, S. N.,** Amino acid sequence homology in the *gag* region of reverse transcribing elements and the coat protein gene of cauliflower mosaic virus, *Nucleic Acids Res.*, 14, 623, 1986.
109. **Miller, J., McLachlan, A. D., and Klug, A.,** Repetitive zinc-binding domains in the protein transcription factor IIIA from *Xenopus* oocytes, *EMBO J.*, 4, 1609, 1985.
110. **Prats, A. C., Sarih, L., Gabus, C., Litvak, S., Keith, G., and Darlix, J. L.,** Small finger protein of avian and murine retroviruses has nucleic acid annealing activity and positions the replication primer tRNA onto genomic RNA, *EMBO J.*, 7, 1777, 1988.
111. **Gorelick, R. J., Henderson, L. E., Hanser, J. P., and Rein, A.,** Point mutants of Moloney murine leukemia virus that fail to package viral RNA: evidence for specific RNA recognition by a "zinc finger-like" protein sequence, *Proc. Natl. Acad. Sci. U.S.A.*, 85, 8420, 1988.

112. **Verma, I. M.,** Reverse transcriptase, in *The Enzymes*, Vol. 14A, Boyer, P. D., Ed., Academic Press, New York, 1981, 87.
113. **Kamer, G. and Argos, P.,** Primary structural comparison of RNA-dependent polymerases from plant, animal, and bacterial viruses, *Nucleic Acids Res.*, 12, 7269, 1984.
114. **Modjtahedi, N., Volovitch, M., Sossountzov, L., Habricot, Y., Bonneville, J. M., and Yot, P.,** Cauliflower mosaic virus-induced viroplasms support viral DNA synthesis in a cell-free system, *Virology*, 133, 289, 1984.
115. **Ziegler, V., Laquel, P., Guilley, H., Richards, K., and Jonard, G.,** Immunological detection of cauliflower mosaic virus gene V protein produced in engineered bacteria or infected plants, *Gene*, 36, 271, 1985.
116. **Laquel, P., Ziegler, V., and Hirth, L.,** The 80K polypeptide associated with the replication complexes of cauliflower mosaic virus is recognized by antibodies to gene V translation product, *J. Gen. Virol.*, 67, 197, 1986.
117. **Modjtahedi, N., Volovitch, M., Mazzolini, L., and Yot, P.,** Comparison of the predicted secondary structure of aphid transmission factor for transmissible and nontransmissible cauliflower mosaic virus trains, *FEBS Lett.*, 181, 223, 1985.
118. **Woolston, C. J., Czoplewski, L. G., Markham, P. G., Good, A. S., Hull, R., and Davies, J. W.,** Location and sequence of a region of cauliflower mosaic virus gene 2 responsible for aphid transmissibility, *Virology*, 160, 246, 1987.
119. **Domier, L. L., Shaw, J. G., and Rhoades, R. E.,** Polyviral proteins share amino acid sequence homology with picorna-, como-, and caulimoviral proteins, *Virology*, 158, 20, 1987.
120. **Markham, P. G. and Hull, R.,** Cauliflower mosaic virus aphid transmission facilitated by transmission factors from other caulimoviruses, *J. Gen. Virol.*, 66, 921, 1985.
121. **Givord, L., Xiong, C., Giband, M., Koenig, I., Hohn, T., Lebeurier, G., and Hirth, L.,** A second cauliflower mosaic virus gene product influences the structure of the viral inclusion body, *EMBO J.*, 3, 1423, 1984.
122. **Rodriguez, D., Lopez-Abella, D., and Diaz-Ruiz, J. R.,** Viroplasms of an aphid-transmissible isolate of cauliflower mosaic virus contain helper component activity, *J. Gen. Virol.*, 68, 2063, 1987.
123. **Harker, C. S., Woolston, C. J., Markham, P. G., and Maule, A. J.,** Cauliflower mosaic virus aphid transmission factor protein is expressed in cells infected with some aphid nontransmissible isolates, *Virology*, 160, 252, 1987.

Chapter 15

USE OF RECOMBINANT HBV PROTEINS IN IMMUNOLOGICAL STUDIES

David R. Milich

TABLE OF CONTENTS

I. INTRODUCTION

Until recently, purified hepatitis B virus (HBV) proteins were obtained only from the serum and liver tissue of infected patients. Therefore, acquiring sufficient quantities of protein to perform antigenic and immunologic analysis has been a limiting factor. For example, hepatitis B core antigen (HBcAg) is not a secreted protein and exists primarily in the liver and within HBV particles. Although HBeAg is secreted from the liver, it is relatively difficult to purify from serum. Furthermore, procedures used to purify HBV proteins from sera have the potential to alter biochemical and structural features of these proteins. For instance, the use of a pepsin digestion step during the purification of hepatitis B surface antign (HBsAg) most likely obscured the detection of larger preS-containing HBsAg polypeptides for a number of years.

The ability to produce recombinant HBV proteins has provided the means to study the immune response to these antigens *in vivo* in animal systems and *in vitro* in human systems, to examine antigen–antibody interactions, to develop alternative vaccines (Chapter 16, this volume) and novel diagnostics, and to produce transgenic animal models (Chapter 4, this volume). This chapter is an attempt to chronicle a number of the advances made possible by the availability of recombinant HBV proteins.

II. DEVELOPMENT OF DIAGNOSTICS FOR THE DETECTION OF PreS ANTIGENS IN SERUM AND LIVER

A. PreS1 ANTIGEN DETECTION

Several investigators have examined the occurrence of preS1 region antigen in sera and/or liver tissue of acutely and chronically HBV-infected patients.[1-8] Sera were evaluated by Western blotting or by enzyme-linked immunosorbent assay (ELISA), and the results of these early studies were consistent: most sera positive for preS1 proteins were also HBV DNA positive, and conversely HBV DNA-negative sera were mostly negative for the preS1 proteins. The presence of preS1 proteins correlated well with HBeAg-positive status. After seroconversion to anti-HBe, neither HBV DNA nor preS1 proteins were detectable. However, as the assays for the detection of the preS1 antigen became more sensitive the correlation with HBV DNA and HBeAg positivity was no longer as apparent.[9-13] However, levels of preS1 are significantly different in viremic vs. nonviremic sera.[14] These findings suggest that the synthesis of the preS1 protein may occur in the absence of active viral replication (i.e., within hepatocytes that have integrated the entire HBsAg open reading frame [ORF]) or occur during low-grade HBV replication in which serum HBV DNA is undetectable. Although preS1 antigens were detected during acute hepatitis B, their titer was relatively low, and they disappeared rapidly from the sera of patients who completely recovered.[8] In patients with chronic active hepatitis B, serum preS1 antigen titers correlated with viral replication; however, preS1 levels were lower than HBsAg/P25 titers by a factor of 10^2 to 10^3.[7] These observations are most likely explained by the preferential distribution of P39/GP42 on intact virions, which are present in serum in smaller amounts than are subviral particles, even during active replication, and disappear more rapidly in resolving disease.[15] Examination of preS1 proteins in liver tissue by Western blotting[2] or immunofluorescence[6,11] revealed that the presence of P39/GP42 polypeptides in hepatocytes during HBV infection did not correlate with the type or severity of liver disease. The preS1 sequence was present in the cytoplasm of hepatocytes in most HBsAg chronic carriers, independent of viral replication, and it was also present in the cytoplasm in the liver of anti-HBe-positive carriers with integrated HBV DNA sequences. PreS1 antigens were not detected in two patients with hepatocellular carcinoma who were negative for HBeAg.[2]

B. PreS2 ANTIGEN DETECTION

The detection of preS2 protein in serum was approached indirectly by measuring serum polymerized human serum albumin (pHSA) receptor activity in early studies. Since Imai et al.[16] first described the existence of a species-specific receptor for pHSA on HBV particles obtained from HBeAg-positive sera, numerous investigators have examined this phenomenon.[17-28] It has been demonstrated that the pHSA receptor is present on both GP33/GP36[17] and P39/GP42[19] HBsAg polypeptides, but not on P25, and is contained within the 55 residues of the preS2 region.[18] More recently, a candidate sequence of seven amino acids within the preS2 region was suggested to be required for pHSA receptor activity.[29] Furthermore, preS2-specific monoclonal antibodies have been shown to inhibit the HBsAg–pHSA interaction.[28]

A number of studies have examined the presence of preS2 antigen in the serum of acutely and chronically HBV-infected patients using pHSA receptor activity[25,26] or preS2-specific antibodies[8,26,30] as probes. In general, the results parallel those reported for the measurement of preS1 antigen in clinical samples. The preS2 antigen was detected concomitantly with HBeAg in patients with active viral replication and was independent of HBsAg titer.[8,25,26] However, the inability to detect preS antigens in the sera of nonviremic, anti-HBe-positive carriers may reflect quantitative rather than qualitative differences between viremic and nonviremic carriers. Hu et al.[31] recently suggested that the presence of preS antigens in serum correlates with HBsAg titer and not HBeAg status, as reported by others. These differences may relate to the sensitivity of the assays used. The preS2 antigen was present at clinical onset of acute hepatitis B infection but was lost rapidly during the acute phase in patients who recovered.[8,25] In chronic HBV infection, high-titer preS2 antigen was continuously expressed in patients seropositive for HBeAg, whereas anti-HBe-positive patients had significantly less or no preS2 antigen in serum.[8,25,26] Levels of preS2 antigen in chronic infection did not correlate with histological or biochemical parameters. The highest preS2 antigen titers correlated with persistence of virus replication in both acute and chronic infection, whereas low or no detectable preS2 antigen correlated with disappearance of viral replication and seroconversion to anti-HBe status.[25] However, in a more recent study of 19 episodes of acute exacerbation of hepatitis occurring in patients with chronic active hepatitis, a significant elevation of preS2 antigen titer was observed, which closely overlapped an HBV DNA increase and preceded alanine aminotransferase (ALT) elevations by 1 to 11 weeks.[32] This finding suggests that quantitative measurements of preS2 antigen during the course of chronic infection may predict acute exacerbation of disease.

Although the existence of virally encoded pHSA receptor within the preS region has been clearly demonstrated, the availability of preS2-specific monoclonal antibodies[18,26,33] and recombinant and synthetic preS2 proteins[26,30,34,35] will allow preS2 antigen and antibody to be measured directly. This is desirable inasmuch as the pHSA-binding assay does not distinguish between preS1 and preS2 antigens, since GP33/GP36 and P39/GP42 polypeptides both exhibit pHSA-binding activity. Furthermore, components of normal human serum[24,36] and anti-pHSA autoantibodies[37] interfere with the HBsAg–pHSA interaction. For example, human C1q can also bind pHSA, and HBsAg-associated Clq may mediate preS2-independent pHSA binding.[38]

III. CLINICAL EVALUATION OF PreS-SPECIFIC ANTIBODY PRODUCTION

A. ANTI-PreS1 ANTIBODIES

Since the synthesis of preS antigens appears maximal during viral replication, it is of interest to examine the immune response to these antigens during infection. As early as 1974, Moodie et al.[39] proposed the existence of a third antigen–antibody system specific for intact virions and not related to anti-HBs or anti-HBc. Subsequently, additional antigenic sites distinct from those on spherical subviral particles of HBsAg were demonstrated by the ability of anti-HBs-negative, HBV-positive sera to aggregate intact virions and filamentous particles but not spherical particles.[40] In 1978, Alberti et al.[41] demonstrated antibodies in the serum of patients with acute hepatitis B that were reactive with radiolabeled virions, but not with HBsAg or HBcAg. These early "pre-DNA sequence" observations most likely chronicled the detection of anti-preS antibodies in HBV-infected patients.

Several groups have initiated studies to determine the occurrence and kinetics of antibody production to the preS1 region during HBV infection.[1,4,42,43] Various methods have been employed, including (1) solid-phase radioimmunoassay (RIA) or ELISA using synthetic peptide analogs of the preS1 region as ligand,[1,42] (2) Western blotting utilizing preS1-containing fusion proteins as ligand,[4] and (3) immunoabsorption utilizing HBsAg particles that either contained or lacked preS1 antigenicity.[43] Neurath et al.[1] demonstrated that the preS1 sequence p12–32 is recognized by antibodies in sera of individuals recovering from HBV infection. It was noted that anti-p12–32 can be detected prior to anti-HBs or anti-HBc seroconversion but declines rapidly after production of anti-HBs and anti-HBc. The fine specificity of anti-preS1 antibody production during acute HBV infection was studied using a panel of preS1 synthetic peptide analogs. It was shown that the human anti-preS1 response can include multiple specificities since peptide sequences represented by p1–21, p12–32, p32–53 and p94–117 were variably recognized by sera from different HBV-infected patients.[42] Analysis of the temporal occurrence of anti-preS1 antibody in an HBV-infected chimpanzee also illustrated the early appearance of this antibody, which was detected shortly after preS1 antigen appeared during acute disease.[4] Similarly, anti-preS1 antibody production was reported during the course of ground squirrel hepatitis virus (GSHV) infection and did not necessarily correlate with anti-S production.[44] One study examined preS1 antibody production in acute and chronic HBV infection and found that antibodies to preS1 appear at the early stage of acute-resolving HBV infection, but the antibodies are absent in the sera of patients with acute HBV infection entering a chronic course and in the sera of chronic carriers.[45] In another study of chronically infected patients an interesting observation was that among asymptomatic carriers anti-preS1 was detected in those seropositive for anti-HBe but not in those positive for HBeAg.[43] The authors suggested that anti-preS1 may have a role in clearing HBV from the circulation since preS1-containing polypeptides are preferentially expressed on virions.

B. ANTI-PreS2 ANTIBODIES

Several groups have examined *in vivo* antibody production to the preS2 region of HBsAg during HBV infection.[1,27,28,46-48] An early study examined acute-phase sera (negative for anti-S) for antibody that would precipitate radiolabeled intact virions and termed this antibody anti-Dane particle (DP).[27] Patients who subsequently seroconvert to anti-HBe status rapidly show high anti-DP activity, whereas patients with low or absent anti-DP activity seroconvert slowly or not at all.[27] The pHSA-binding assay was simultaneously performed, and an inverse correlation was demonstrated. Since pHSA blocked anti-DP activity, the authors suggested that anti-DP reacts with antigens, within or near the preS2 region, that are involved in pHSA binding. In retrospect, the contribution of preS1-specific antibodies cannot be excluded if the anti-DP reactivity was indeed related to the pre-S antigens, as appears likely. Using a synthetic peptide representing residues 120 to 145 within the preS2 region (N-terminal 26 residues) in a solid-phase RIA, it was demonstrated that acute-phase sera from HBV-infected patients contain antibodies that recognize this sequence.[1] Examination of the fine specificity of antibodies reacting with the p120–145 sequence revealed that multiple specificities exist that primarily recognize the C-terminal region p133–143.[46] The preS2-specific antibodies appear early during infection and decline in titer rather rapidly.[1] Several other studies have also suggested a transient nature for anti-preS2 antibodies during HBV infection.[47,48] An examination of antibodies capable of inhibiting the HBsAg–pHSA interaction (i.e., anti-pHSA receptor [rpHSA]) revealed the appearance of anti-rpHSA in the early recovery phase of acute hepatitis co-occurring with HBsAg and elevated transaminase levels and preceding the anti-HBs response. However, in chronic HBV infection, anti-rpHSA was detected in only 1% of 358 asymptomatic carriers seropositive for anti-HBe and none of 67 sera seropositive for HBeAg.[28] Assuming this assay was specific for anti-preS2 antibody (the authors did not state if the HBsAg utilized possessed preS1 protein or if anti-preS1 antibodies inhibit pHSA binding to HBsAg/P39), these results indicate that preS2 antibody is not produced by asymptomatic carriers regardless of HBeAg/anti-HBe status, and in this respect parallels anti-S region antibody production. A caveat to this interpretation is that anti-preS2 antibody may not be detected in a single serum sample due to its apparent transient nature. However, temporal analysis of anti-preS antibody production during acute and chronic HBV infections also indicated an absolute correlation between HBV clearance and anti-preS production. No anti-preS antibody was detected in acute patients who progressed to chronicity or in patients with chronic hepatitis with persistence of HBsAg and preS antigens regardless of HBeAg/anti-HBe status.[36,49]

Cumulatively, these data indicate that the preS1 and preS2 antigens represent new serological markers of HBV infection that provide information relevant to viral replication and infectivity. The corollary immune responses may be important to viral neutralization mechanisms, and analysis of anti-preS antibody production may have diagnostic and prognostic value. Whether these data will provide information beyond that currently derived from HBeAg and HBV DNA determinations remains to be seen.

IV. IMMUNE RESPONSE TO HBV ENVELOPE PROTEINS IN ANIMAL MODEL SYSTEMS

A. T CELL RECOGNITION OF THE PreS REGION CAN CIRCUMVENT NONRESPONSE

In 1982, it was demonstrated that the murine immune response to the S region of HBsAg is regulated by at least two *H-2*-linked immune response (Ir) genes, and S region nonresponder phenotypes ($H\text{-}2^{f,s}$) were identified.[50] This observation has been extended to the human immune response to HBsAg by the reports of an association between HLA phenotype and nonresponsiveness to HBsAg vaccination.[51-53] More recent studies have shown that the T cell response to the preS2 region is regulated by *H-2*-linked Ir genes independently of the S region response, such that immunization of S region nonresponder, preS2 region responder mice ($H\text{-}2^{s}$) with HBsAg/GP33 circumvents nonresponse to the S region.[54,55] Furthermore, the immune response to the preS1 region is influenced by *H-2*-linked Ir genes independently of the S and preS2 region responses.[42] Therefore, HBsAg/P39 immunization in S region and preS2 region nonresponder mice ($H\text{-}2^{f}$) elicits an anti-preS1-specific response as well as anti-S and anti-preS2 antibody production (Table 1). The mechanism responsible for circumvention of nonresponsiveness is mediated at the T cell level, as T cell recognition of a single region is sufficient to provide T cell helper function for multiple B cell specificities present on S, preS2, and preS1 regions of HBsAg.[42,54-56] These studies indicate that the envelope polypeptides of the HBV present an array of T cell determinants to the host's immune response and that the specificity of the recognition process is influenced by major histocompatibility complex-linked genes.

B. T AND B CELL RECOGNITION SITES WITHIN THE PreS REGIONS

Additional studies have been aimed at identifying T cell and B cell (antibody) recognition sites within the preS regions of HBsAg. These studies have been greatly facilitated by the fact

TABLE 1
Influence of *H-2* Genotype on the Humoral Response to HBsAg Particles of Varied Composition

Immunogen	Strain	*H-2*	Specific antibody titer (1/dilution) S	PreS2	PreS1
HBsAg/P25	B10.D2	*d*	81,920	0	0
	B10	*b*	20,480	0	0
	B10.BR	*k*	5,120	0	0
	B10.S	*s*	0	0	0
	B10.M	*f*	0	0	0
HBsAg/P33	B10.D2	*d*	40,960	10,240	0
	B10	*b*	10,240	40,960	0
	B10.BR	*k*	2,280	2,560	0
	B10.S	*s*	640[a]	10,240	0
	B10.M	*f*	0	0	0
HBsAg/P39	B10.D2	*d*	81,920	5,120	640
	B10	*b*	20,480	40,960	10,240
	B10.BR	*k*	5,120	1,280	2,560
	B10.S	*s*	5,120[a]	10,240	1,280
	B10.M	*f*	10,240[a]	1,280[a]	10,240

[a] Represents an antibody response to a specific region of HBsAg, which is not observed when the strain is immunized with that same antigen (i.e., B10.S is nonresponsive to the region when immunized with HBsAg/P25). Adapted from Milich et al.[42]

that unlike the S region, the preS regions of HBsAg contain continuous, as opposed to discontinuous or topographical, determinants.[1,30,42,54,57] Two synthetic peptides derived from the preS2 region sequence elicit antibodies cross reactive with the native preS2 region.[26,30] Denoting the amino terminus of the preS2 region as residue 120, synthetic peptides p120–145[26] and p133–151[30] were shown to bind human antibodies elicited by HBV infection.[26,30] Further analysis using a combination of truncated synthetic peptides and monoclonal antibodies revealed that the murine antibody response to the preS2 region is predominantly focused on residues 133 through 143, and two distinct but overlapping epitopes were identified as p133–139 and p137–143[46] (Figure 1). An independent report of a preS2-specific monoclonal antibody, which binds the p132–145 sequence, confirms the dominance of this region of preS2 as an antibody recognition site.[33] Human HBV-positive sera also bind p133–145 and the constituent overlapping epitopes within this sequence.[46]

Whereas the antibody response to p120–145 is directed toward the C terminus, a predominant T cell recognition site has been localized to the N-terminal residues 120 to 132.[58] It is the presence of this T cell recognition site that explains the immunogenicity of the p120–145 sequence. Immunization with either the isolated T cell site (p120–132) or the B cell site (p133–145) alone induces only a marginal anti-preS2 response, whereas immunization with the composite T and B cell peptide (p120–145) elicits a high-titer, primary anti-preS2 response (i.e., 1:163,840) in strains responsive to p120–132 at the T cell level.[58] This anti-peptide response is termed anti-preS2 since the antibody is totally cross reactive with native HBsAg/GP33 particles.[30,54,58] In contrast to the cross reactivity of anti-p133–145 with HBsAg/GP33, p120–132-primed T cells do not recognize this sequence on the native HBsAg/GP33 particle. Therefore, identification of a peptide sequence that elicits significant T cell activation does not

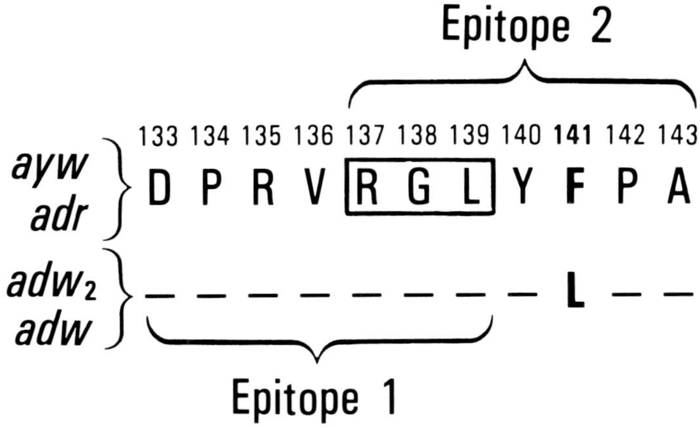

FIGURE 1. Amino acid sequences of two distinct and overlapping antibody-binding sites within the preS2 region. Epitope 1 (P133–139) is conserved in the *ayw*, *adr*, *adw*, and adw_2 subtypes. Epitope 2 (P137–143) contains a subtype-dependent amino acid substitution at residue 141 (phenylalanine [*ayw*, *adr*]; leucine [*adw*, adw_2]). The three-amino acid overlap between epitopes 1 and 2 is depicted. (From Milich, D.R., McLachlan, A., Chisari, F.V., Nakamura, T., and Thornton, G.B., *J. Immunol.*, 137, 2703, 1986. With permission.)

guarantee that the resulting T cell population(s) will recognize the same sequence in the native protein. In support of this, it was demonstrated that the anti-p120–145 response is regulated by distinct *H-2*-linked genes as compared to the anti-HBsAg/GP33 response, indicating that the enhanced immunogenicity of the native preS2 region results from a different mechanism than that of the anti-p120–145 response.[58] This is in disagreement with a report that the p120–145 peptide can prime T cells that are relevant to the native preS2 region.[59]

C. THE MURINE T CELL RESPONSE TO THE PreS2 REGION IS SUBTYPE DEPENDENT

The antibody response to the preS2 region is largely group specific.[42,60,61] Anti-preS2 antibody production elicited by the *d* subtype of HBsAg/P33 is highly cross reactive with the *y* subtype. This is due to the conserved nature of the dominant antibody-binding sites of both subtypes. Within this region (p133–143) the *d/y* subtypes of HBsAg/P33 differ by only a single amino acid substitution at residue 141 (Figure 1). However, the fact that anti-preS2 antibody is primarily group specific does not predict the relevance of subtype in T helper (T_h) cell recognition of the preS2 region of the HBsAg. A recent study examined the influence of viral subtype on T cell recognition of the preS2 region and identified specific T cell recognition sites in a panel of *H-2* congenic strains.[62] Immunization with preS2- containing HBsAg particles of the *d* and *y* subtypes indicated that T cell recognition of the preS2 region is predominantly subtype specific in murine strains of eight different *H-2* haplotypes. Furthermore, the B10.M strain ($H\text{-}2^f$), classified as a T cell nonresponder to the *y* subtype of the preS2 region, responds efficiently to the *d* subtype, indicating that preS2 responder status can be subtype dependent as well as subtype specific.[62]

Studies using a truncated preS2 polypeptide and synthetic peptides illustrated that the C-terminal sequence (p151–174) of the preS2 region is the dominant focus of T cell recognition in multiple murine strains.[62] Specifically, 17 distinct T cell recognition sites were defined within the C-terminal half of the preS2 region. The fine specificity of T cell recognition of the preS2 region was dependent on the *H-2* haplotype of the responding strain. T cell recognition of all 17 sites was subtype specific, which is consistent with the fact that the C-terminal

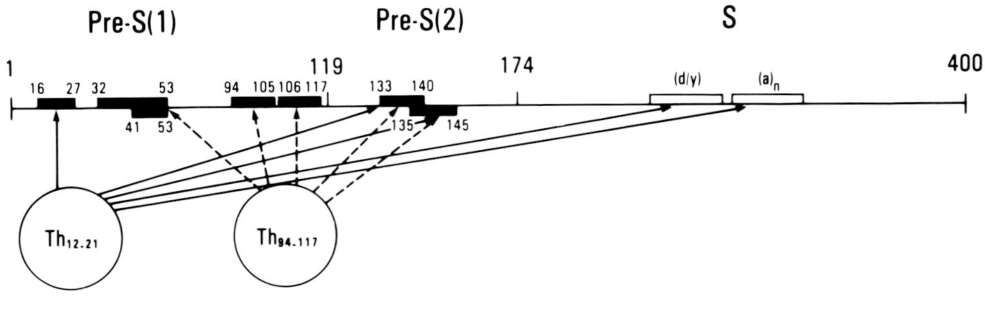

FIGURE 2. Summary of T and B cell recognition of HBsAg/P39. Defined antibody-binding sites are represented by solid boxes, and the N- and C-terminal coordinates are depicted. Serologically determined antibody-binding sites (i.e., S region group [a] and subtype [*d or y* specific]) for which there are no consensus sequences represented by open boxes. T cell helper (T_h) activity and the antibody specificities elicited are represented by the arrows. The specificities of the T_h cells are depicted. Adapted from Milich et al.[56]

sequence is highly polymorphic between the *d* and *y* subtypes of the preS2 region.[62] The preS2-containing HBsAg vaccines currently in clinical trials are composed of a single subtype, either *d* or *y*. The results of this study suggest that both subtypes should be incorporated to increase the frequency of T cell responders to the preS2 region, and to ensure T_h cell memory relevant to infection with hepatitis B virus of either the *d* or *y* subtypes.

The purpose of ongoing studies is to "design" synthetic HBV vaccines. Toward this end a preS2 region immunogen based on combining the dominant B and T cell recognition sites for this region into a single peptide was constructed.[63] A composite peptide consisting of the dominant T cell recognition sequence p151–174 positioned N terminal to the dominant B cell site p133–143 (i.e., p151–174[133–143]) yielded an effective preS2 synthetic immunogen, which was immunogenic in 8 of 8 murine strains of differing major histocompatibility complex (MHC) haplotypes. Furthermore, the antibodies produced and the T cells activated by this synthetic immunogen were highly cross reactive for the native protein. Interestingly, the orientation of the T and B cell determinants and the context of the T cell site within the larger composite peptide influenced both antibody fine specificity and T cell fine specificity.[63]

D. INFLUENCE OF T CELL FINE SPECIFICITY ON ANTIBODY SPECIFICITY

PreS1-specific T cell recognition sites have also been defined that are relevant to both the peptide sequence and native HBsAg/P39. The preS1 sequences p12–21 and p94–117 can induce and elicit HBsAg/P39-specific T cell activation.[56] A number of preS1-specific antibody-binding sites have also been recently elucidated.[56] Identification of distinct T cell and B cell recognition sites within the preS region has allowed examination of the ability of a T cell population specific for a single peptide to provide functional T-cell help for a series of B cell specificities on HBsAg/P39. For this purpose, mice were primed with the T cell determinants p12–21 or p94–117 and challenged 4 weeks later with a suboptimal dose of HBsAg/P39 particles and the fine specificities of the antibodies produced were measured. As illustrated in Figure 2, priming with p12–21 elicited T cell helper function resulting in *in vivo* antibody production to p16–27, p133–140, and p135–145 in the preS region, and group- and subtype-specific determinants in the S region. Similarly, priming with p94–117 elicited antibody production specific for p32–53, p94–105, p106–117, p133–140, and p135–145 in the preS region but did not prime antibody production to the S region determinants (Figure 2). Unprimed mice produced no antibody upon challenge. These results indicate that T cells primed to a single determinant are sufficient to provide functional help to multiple B cell clones, which recognize unique epitopes on a complex HBsAg particle. It is noteworthy that p12–21 and p94–117 primed antibody production to unique as well as common B cell determinants. For example, p94–117 primed an anti-p32–53 response, whereas p12–21 did not. These data provide strong evidence that the fine specificity of the T helper cell can influence the fine specificity of the antibody produced.[56] It is important to note that, due to the limits imposed on T cell antigen recognition by MHC restriction, a given T cell determinant will not be recognized by all *H-2* haplotypes. For example, the p12–21 sequence is recognized only by $H\text{-}2^{s,f}$-bearing strains.[56]

V. *IN VITRO* STUDIES OF THE HUMAN CELLULAR IMMUNE RESPONSE TO HBV ENVELOPE PROTEINS

Extending the information gained from animal models of HBV-specific immune responses to man has been hampered by a limited access to lymphoid compartments, variation and quality of antigen preparations, and difficulty in obtaining antigen induced, antigen-specific antibody and T cell responses *in vitro*. These difficulties are reflected in the numerous and often conflicting results obtained in early studies of human cellular immune responses to HBV envelope antigens.[64] However, recent advances in human T cell cloning techniques as well as the availability of recombinant HBV proteins have made it possible to examine human antigen-specific responses *in vitro*.

Celis et al.[65] first established HBsAg-specific human T cell clones from peripheral blood mononuclear cells of vaccine recipients. Subsequent studies mapped a dominant T cell recognition site to residues 19 to 28 within the N terminus of the S region, and the DPw4 MHC Class II molecule was shown to represent a restriction molecule for this site.[66] The majority of the HBsAg-specific T cells exhibited both proliferative and cytotoxic responses, although all phenotyped as $CD4^+$. More surprisingly, a $CD8^+$, MHC Class I-restricted (HLA-A11) human T cell clone has also been derived from a vaccine recipient.[67] The epitope recognized by the $CD8^+$ clone was mapped to residues 21 to 28 within the preS1 region. Using vaccinia–HBsAg recombinants it was shown that the Class I-restricted as well as two Class II-restricted T cell clones recognized endogenously expressed antigen as well as exogenous antigen.[67] These results suggested that a subunit vaccine can induce Class I-restricted T cells capable of recognizing virally infected cells. In another study peripheral blood mononuclear cells (PBMC) of 14 HBsAg vaccine recipients, who received a plasma-derived vaccine containing trace amounts of preS proteins, were examined for envelope-specific T cell proliferative responses.[68] All subjects displayed S-specific T cell responses, eight subjects responded to the preS1 region, and only one subject responded to the preS2 region. However, the influence of subtype was not examined, and this may have accounted for the low preS2-specific response. Limiting dilution analysis demonstrated that the frequency of preS1-specific T cells in two of the vaccine recipients was higher than that of S-specific T cells. In addition, HBsAg-specific $CD4^+$-cloned T cell lines were generated from two preS-responsive vaccinees. In addition to HBsAg-specific proliferation, the majority of these lines displayed HBsAg-specific, HLA Class II-restricted cytotoxicity for autologous Epstein-Barr virus (EBV)-transformed B cells either preincubated with envelope antigens or infected with recombinant vaccinia virus expressing envelope proteins. It was concluded that the preS1 region may function as a strong T cell immunogen in man.[68] Additionally, it was proposed that Class II-restricted cytotoxicity may contribute to the clearance of HBV-infected hepatocytes inasmuch as HBV infection may induce expression of HLA Class II molecules on hepatocytes. Alternatively, this cytotoxic activity may down regulate T cell activation by lysing antigen-presenting cells.

Other investigators have examined T cell sensitization to HBsAg proteins in the context of chronic HBV infection. T cell clones derived from liver-infiltrating lymphocytes from a patient with chronic active hepatitis were generated by *in vitro* stimulation with preS-containing recombinant HBsAg particles, even though the patients' PBMCs did not respond to the same antigen preparation.[69] MHC Class I- and Class II-restricted T cell clones were identified. Both Class I- and Class II-restricted T cell clones recognized determinants within the preS2 region residues 120 to 134, suggesting that this sequence can bind to human MHC Class I and Class II molecules. Furthermore, it was demonstrated that regardless of CD4 or CD8 phenotype several T cell clones lysed HLA-matched EBV-immortalized lines sensitized with HBsAg particles or p120–134.[69] This study indicates that preS2-specific T cells are present within the liver during chronic infection, and suggests that either Class I- or Class II-restricted T cells may mediate hepatocellular lysis. These investigators have also suggested a mechanism whereby preS2-specific cytotoxic T lymphocytes (CTLs) may lyse HBsAg-specific B cells that present envelope peptides in the context of the MHC Class I pathway.[70] Such a mechanism would result in the suppression of anti-HBs production, a phenomenon that does occur during chronic infection.[70] Additional studies using polyclonal T cells derived from PBMCs have also suggested that T cell immunity specific for the preS2 region may be an important early event in the human immune response to HBV infection.[71]

VI. IMMUNE RESPONSE TO HBV NUCLEOCAPSID ANTIGENS IN ANIMAL MODEL SYSTEMS

A. IMMUNOGENICITY OF THE NUCLEOCAPSID OF HBV

Like HBsAg, the nucleocapsid (HBcAg) is a particulate antigen, although slightly larger (27 nm). A nonparticulate form of HBcAg designated HBeAg is serologically distinct, but the primary amino acid sequences show significant identity. Despite the fact that HBcAg is an internal component of the virus, high-titer anti-HBc antibodies are produced by virtually 100% of HBV-infected patients regardless of clinical status. In contrast, antibodies to HBeAg may not develop or may appear at various times after the appearance of anti-HBc. Envelope- and nucleocapsid-specific cellular immune responses have been suggested to be important in virus elimination and the attendant hepatocellular injury,[72,73] and vaccination with both antigens has been reported to protect against HBV infection.[74,75]

Comparative studies of murine antibody production show that anti-HBc responses are significantly greater (at least 80-fold) than anti-HBs responses in all strains tested.[76] *H-2*-linked genes influence the anti-HBc response, but no nonresponder strains have been identified. The comparative magnitudes of the anti-HBc and anti-HBs responses, and the lack of nonresponsiveness to HBcAg, are consistent with the human immune responses to these HBV antigens. In mice, HBcAg is an extremely efficient immunogen at the T cell level, as measured by the *in vivo* dose required to induce T cell sensitization (1.0 μg), and the minimal *in vitro* concentration required to elicit interleukin 2 (IL-2) production from primed T cells (0.03 ng ml^{-1}).[77] This degree of T cell immunogenicity is approximately 100-fold that observed for HBsAg subviral particles.[78] Similarly, studies of the human T cell response to HBcAg suggest that HBcAg is an efficient T cell antigen.[79]

Examination of the fine specificity of T cell recognition of HBcAg showed that HBcAg-specific T cells from a variety of strains recognized multiple but distinct sites within the HBcAg/HBeAg sequence[77] (HBeAg lacks the carboxyl-terminal 34 residues of HBcAg). T cell recognition of HBcAg and HBeAg is highly cross reactive, which is consistent with the amino acid sequence identity between these two antigens.[76,77] T cell recognition sites have been defined by synthetic peptides of residues 12 to 21. Although several strains recognize multiple T cell sites, each strain recognizes a predominant T cell determinant, and the fine

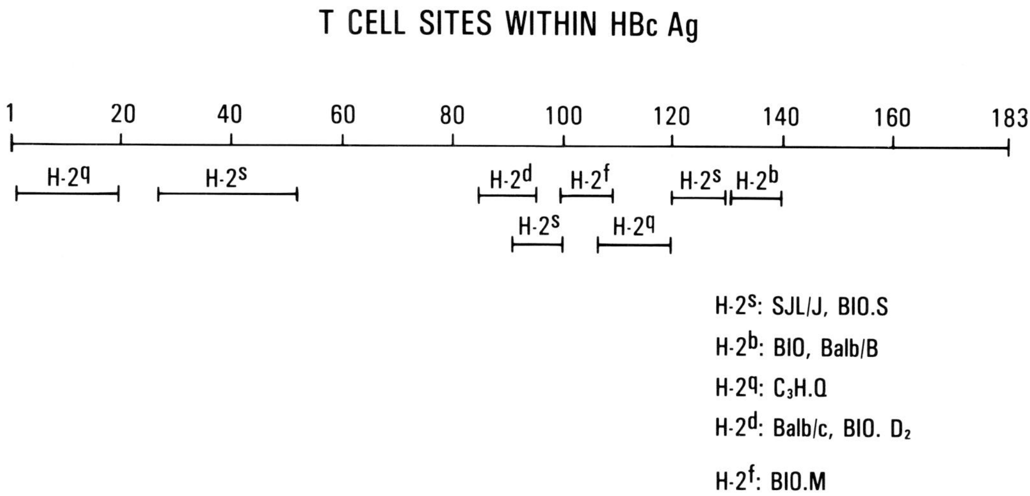

FIGURE 3. Localization of T cell sites within the HBcAg/HBeAg sequence using synthetic peptide analogs. The indicated strains were immunized with HBcAg and T cells cultured with a panel of synthetic peptides from the HBcAg/*ayw* sequence and IL-2 production determined. Peptides that activated HBcAg-primed T cells were then used for immunization in the reciprocal experiment to determine the ability to prime HBcAg-specific IL-2 production. Adapted from Milich.[80a]

specificity of this recognition process is dependent on the *H-2* haplotype of the responding strain. For example, $H\text{-}2^s$ strains recognize P120–131 predominantly, $H\text{-}2^b$ strains recognize p129–140, $H\text{-}2^f$ strains recognized p100–110, $H\text{-}2^q$ strains recognize a sequence within p100–120 distinct from $H\text{-}2^f$ mice, and $H\text{-}2^d$ mice recognize p85–96 predominantly[77,80] (Figure 3).[80a] This murine model predicts that a human outbred population would exhibit similar complexity, and individuals may recognize distinct T cell sites in the context of their HLA genotype.

To determine the ability of synthetic T cell sites to prime T_h cells and induce antibody production *in vivo*, the HBcAg-specific p120–140 and amino-terminal (p120–131) and carboxyl-terminal (p129–140) fragments were studied in the B10.S and B10 strains.[80] Mice were primed with peptide and challenged with a suboptimal dose of HBcAg, and serum anti-HBc measured. In the B10.S strain, p120–140-primed and p120–131-primed mice produced anti-HBc efficiently 7 d after the challenge, whereas p129–140-primed mice did not. Similarly, priming with p120–140 elicited anti-HBc in the B10 strain. In contrast to the B10.S strain, the carboxyl-terminal p129–140 contained the active T cell site. These results are consistent with the T cell proliferation results and indicate a concordance between T cell proliferation and T_h cell fine specificities.[80] Identification of HBcAg-specific T cell recognition sites and characterization of their functional ability to elicit proliferation and to induce anti-HBc antibody *in vivo* have relevance to synthetic vaccine design and possibly to the use of peptide T cell sites as immunotherapeutic reagents in chronic carriers.

B. T CELL-DEPENDENT AND -INDEPENDENT ANTIBODY INDUCTION BY HBcAg

Although HBcAg is an efficient T cell immunogen, this characteristic alone cannot explain the extremely high levels of antibody produced after primary immunization with HBcAg (i.e., anti-HBc titer greater than 1:300,000 in BALB/c mice). HBeAg, the nonparticulate form of HBcAg, is cross reactive with HBcAg at the T cell level, but is a relatively poor immunogen in mice in terms of *in vivo* antibody production.[80,81] Therefore, the ability of HBcAg to activate

B cells directly was examined by immunizing BALB/c euthymic and BALB/c athymic (nude) mice with a mixture of HBcAg and HBsAg. HBcAg could induce antibody production in athymic mice, whereas the comparably sized HBsAg particle was not immunogenic in athymic BALB/c mice.[76] For T cell independence, the HBcAg had to be particulate, since denatured HBcAg and the nonparticulate HBeAg were not immunogenic in athymic mice.[81] The immunogenic effects of an antigen that possesses both T cell-independent and T cell-dependent characteristics may be synergistic in the presence of competent T cells. This property of HBcAg may explain its enhanced immunogenicity in the mouse and during HBV infection in man.

C. INTERMOLECULAR/INTRASTRUCTURAL T CELL HELP

Due to the marked immunogenicity of HBcAg and the observation that in the influenza virus system matrix-specific T cells could elicit hemagglutinin-specific antibody production,[82] the ability of HBcAg-primed T cells to provide help for antibody production to envelope (HBsAg) epitopes was examined.[83] B10.S mice primed with HBcAg and challenged with a mixture of HBcAg and HBsAg/P39 produced no anti-HBs; however, mice challenged with virions produced anti-S-, anti-preS2-, and anti-preS1-specific antibodies. To confirm the T cell nature of this effect, the identical experiment was performed using the synthetic T cell recognition site p120–140 as the priming antigen. The results obtained were similar to those using native HBcAg as the priming antigen. This result indicated that HBcAg-primed T cells could help anti-envelope antibody production to multiple epitopes, and the T_h cell activity did not require that HBcAg and HBsAg be present on the same molecule (intermolecular), but did require them to be within the same particle (intrastructural).[83] This mechanism of T cell help has implications for synthetic vaccine design because multiple T cell sites can be presented within a natural or artificial particle and serve as helper determinants for a B cell epitope(s) exposed on the surface.[81,83]

Production of antibodies specific for the S and preS regions occurs during resolving acute HBV infection, but not during the acute phase or subsequently in chronic asymptomatic infection.[84] However, a subset of chronically infected HBV patients has been reported to produce only anti-preS1 in the absence of anti-preS2 or anti-S antibodies, which correlated with seroconversion from HBeAg to anti-HBe status and viral clearance.[43] It has been proposed that progression to chronic infection required, or is a consequence of, non-responsiveness to the envelope antigens at the T cell level, and furthermore that anti-preS1 production in the absence of anti-preS2 and anti-S production is mediated by T_h cells specific for HBcAg/HBeAg determinants.[83,85] Conceivably, preS antibody is required to neutralize viral infectivity from cell to cell, and HBcAg/HBeAg-specific T_h cell function can substitute for preS-specific T cells in this regard.

Although antibodies to HBcAg do not prevent infection, the fact that HBcAg/HBeAg-specific T_h cells can elicit anti-envelope antibodies[83] that are virus neutralizing[74,86,87] may explain the reported ability of HBcAg vaccination to protect against HBV liver disease.[75] Furthermore, since HBcAg/HBeAg-specific T_h cells induced anti-S antibody production in S region nonresponder mice, this represents another mechanism of circumventing HBsAg nonresponsiveness.[83]

D. ROLE OF IMMUNOLOGIC TOLERANCE IN CHRONIC HBV INFECTION: A TRANSGENIC MODEL

The hepatitis B virus is not directly cytopathic, and the immune response of the host appears to mediate hepatocellular tissue injury and subsequent viral clearance.[88] Women who are chronic carriers of HBV often infect their infants in the perinatal or postnatal periods, whereas intrauterine infection is much less common.[89-92] The vast majority of untreated infants born to HBeAg-positive mothers become infected, and more than 90% of them become chronic carriers.[93] In contrast, approximately 90% of HBV infections occurring in adults are resolved as acute infections, and only 5 to 10% result in chronic infections.[94] This dramatic difference in chronicity rates is believed to reflect the immunologic status of the host at the time of infection. For example, neonates born to HBV carrier mothers may be immunologically tolerant to viral proteins to which they were exposed *in utero*.[95] The tolerogenic potential of the HBc/HBe antigens are of particular interest because there is evidence that these antigens represent important "targets" for immune-mediated viral clearance mechanisms.[73,96] This is an attractive hypothesis; however, the tolerogenic potential of HBV antigens has not been previously investigated. For this purpose, we recently have produced inbred HBeAg-expressing transgenic mice, (B10.S Tg-31e), and also utilized a neonatal tolerance system in order to investigate the tolerogenic characteristics of the HBc/HBe antigens.[97]

HBeAg-expressing transgenic mice represent a model system to examine the consequences of *in utero* exposure to HBeAg on HBcAg/HBeAg-specific immune responses. Characterization of HBe/HBcAg-specific tolerance in B10.S Tg-31e and neonatally tolerized mice indicated (1) T cells but not B cells are tolerized by HBeAg present in the serum at a concentration of 10 ng/ml; (2) T cell tolerance elicited by HBeAg also extends to HBcAg-specific T cells; (3) B10.S Tg-31e mice produce anti-HBc but not anti-HBe antibodies on immunization; (4) the IgG but not the IgM anti-HBc response is diminished in B10.S Tg-31e mice; and (5) the T cell tolerance induced by a single neonatal exposure to HBcAg is reversible and persists for 12 to 16 weeks.[97] It is of interest that many characteristics of immune tolerance demonstrated by B10.S Tg-31e mice parallel the long-term immunologic status of neonates born to HBeAg positive HBV carrier mothers. For example, infants infected perinatally often remain HBeAg positive, and produce anti-HBc but not anti-HBe antibodies. The correlation of the human serological observations with the murine experimental data suggests the hypothesis that the aberrant immunological responses of neonates born to carrier mothers may also result from *in utero* exposure to HBeAg as occurs in the transgenic model.

The finding that *in utero* exposure to HBeAg renders T cells of B10.S Tg-31e mice nonresponsive to HBcAg as well as to HBeAg may be relevant to mechanisms of T cell tolerance in human HBV infection because HBeAg is more likely than HBcAg to transverse the placenta. In support of the possibility that maternal HBeAg may transverse the placenta are the reports that HBeAg has been detected in neonate cord serum of 88,[98] 63,[99] and 54%,[100] of infants born to HBeAg-positive HBV carrier mothers. Placental transit of the intact HBeAg may not be necessary since a peptide fragment of HBeAg (p120–131) injected into neonatal B10.S mice has been shown to elicit T cell tolerance to the entire HBeAg.[101] It seems unlikely that HBsAg functions as a tolerogen *in utero* because a high percentage of infants born to HBeAg-positive mothers vaccinated with HBsAg at birth produce anti-HBs antibodies very efficiently.[102,103] Consistent with the possibility that HBeAg may cross the placenta and induce tolerance *in utero*, the T cell proliferative responses of nontransgenic offspring of B10.S Tg-31e mothers were reduced as compared to control B10.S mice, and were intermediate between transgenic and control mice.[97]

Because HBeAg is a secreted protein it most likely gains access to the thymus through the circulation. It can be predicted that this mode of antigen presentation within the thymus would lead to the functional deletion of MHC Class II-restricted T_h cells. We suggest that the basic immunologic "defect" in neonates born to HBeAg-positive carrier mothers resides in T_h cell tolerance specific for the HBc/HBe antigens.[97] Such tolerance would preclude T_h cell function necessary for anti-HBe and maximal IgG anti-HBc antibody production, and the ability of HBc/HBe-specific T_h cells to elicit anti-envelope antibodies[83] would also be impaired. Addi-

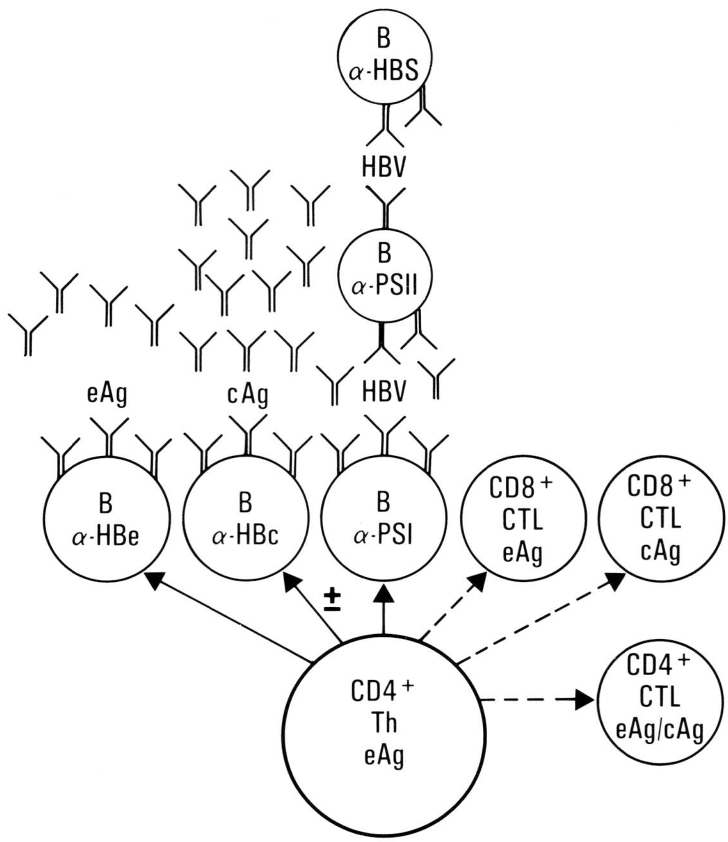

FIGURE 4. Schematic of the known (solid lines) and suggested (broken lines) roles of HBeAg-specific T helper cells in immune responses to HBV antigens. α, Anti-; PSI, preS1 antigen; PSII, preS2 antigen.

tionally, HBc/HBe-specific T_h cell tolerance may diminish development of a CTL response (CD4- and CD8-mediated), which may be required for the elimination of virus-infected hepatocytes. The central importance of T_h cells specific for HBe/HBcAg in terms of the immunoregulation of numerous immune responses during HBV infection is depicted in Figure 4.

Although infants born to HBeAg-positive HBV carrier mothers are at high risk for infection and subsequent viral persistence, the ages of the infants at the time of HBV infection have been shown to be inversely correlated with the rate of chronicity.[90] In the context of an HBc/HBe-specific T_h cell tolerance model, this phenomenon can be explained by the reversibility of T cell tolerance. In the murine system it was demonstrated that a single neonatal dose of HBcAg resulted in T cell tolerance apparent for between 12 and 16 weeks of age, and thereafter waned.[97] Therefore, in order for T cell tolerance to the HBc/HBe antigens to be maintained, the tolerogen must be continually present. This suggests that in the absence of the tolerogen, HBc/HBe-specific thymocytes can emerge from the thymus, and this "repertoire renewal process" requires approximately 12 to 16 weeks in this murine model. Similarly, in the human system the fetus may be exposed *in utero* to tolerogenic HBeAg but not infected at birth. The longer the elapsed time before HBV infection, the greater the probability of renewing the HBc/HBe-specific T cell repertoire because the neonate would no longer be exposed to the tolerogen (HBeAg).[97]

Cumulatively, the human epidemiological and serological observations and the murine experimental data suggest that a function of the HBeAg may be to induce T cell tolerance *in utero*. Indeed, expression of HBeAg may represent a viral strategy to persist in the host subsequent to perinatal infection, which would offer a selective advantage on the conservation of the precore domain.

The experimental murine models offer unique opportunities to study the genetic, cellular, and molecular basis for variable immune responsiveness to HBV-encoded antigens. The mouse has provided a means of studying the immunogenicity of the preS regions of HBsAg and has elucidated the independent *H-2*-linked genes regulating antibody production to preS and S region determinants. The ability to circumvent genetic nonresponsiveness has implications for the design of future HBV vaccines. Mice also afforded the opportunity to examine the ability of HBcAg to activate B cells directly, and to prime T_h cells capable of eliciting anti-envelope antibody production. These observations have potential clinical relevance, and may explain the ability of HBcAg vaccination to protect against HBV infection. The murine system has also facilitated the mapping of T cell and B cell recognition sites within HBV proteins, which at least conceptually enhances the prospects for development of a synthetic HBV vaccine. Antibody-binding sites appear to be similar in humans and mice; the extent to which their T cell repertoires overlap is not known. However, the probable chemical and structural constraints imposed on T cell antigenicity, and the fact that MHC-encoded Class II molecules and the cellular mechanisms mediating T cell recognition are conserved across species, make such overlaps appear likely.

VII. *IN VITRO* STUDIES OF THE HUMAN CELLULAR IMMUNE RESPONSE TO HBV NUCLEOCAPSID PROTEINS

A. HELPER T LYMPHOCYTES

Because of the inability to detect T cell sensitization to the S region of the envelope during both acute and chronic HBV infection, a number of investigators have begun to examine nucleocapsid-specific T cell recognition. Several groups have demonstrated PBMC sensitization to HBcAg in chronically infected patients, using techniques such as T cell proliferation,[79,104] indirect T lymphocyte migration inhibition,[105,106] and γ-interferon (INF) production.[107] In all studies the HBcAg-specific responses were superior to HBsAg-specific responses. These studies are consistent with the murine data from a noninfectious system, which indicated that HBcAg is approximately 100-fold more immunogenic/antigenic compared with HBsAg at both the T cell and B cell levels. This suggests that the differential immunogenicity is due to intrinsic characteristics of these two proteins. It is, therefore, necessary to be cautious in interpreting the inability to detect T cell sensitization to HBsAg in that this may relate more to the sensitivity of the cellular assays than to the presence or absence of HBsAg-specific T cells. It is important to note in this regard that it has been equally difficult to demonstrate T cell sensitization to HBsAg in recovering acute patients and HBsAg vaccine recipients, who nevertheless produce high-titer anti-HBs antibodies indicating the presence of HBsAg-specific T_h cells.

In one report PBMCs of HBeAg-positive and anti-HBe-positive chronic active hepatitis (CAH) patients, and of anti-HBe-positive asymptomatic carriers (ASCs), showed enhanced γ-INF production in response to HBcAg, whereas those of HBeAg-positive ASCs did not respond. Therefore, T cell sensitization to HBcAg appeared to correlate with active disease.[107] Similarly, in a study of patients with acute and chronic hepatitis B, a strong T cell (PBMC) proliferative response to HBcAg or HBeAg was observed in patients with acute self-limited hepatitis, and the response was lower in patients with chronic disease. Moreover, HBcAg-specific T cell sensitization appeared to correlate with the clearance of HBsAg from the serum.[108] These results imply that the T_h cell response to HBcAg/HBeAg is implicated in liver injury and viral clearance in hepatitis B infection. The correlation of T_h cell sensitization to HBcAg with viral clearance may be explained by a requirement of HBcAg-specific cytotoxic T lymphocytes (CTL) for adequate T_h cell function. Second, HBcAg-specific T_h cells have been shown to be capable of eliciting anti-envelope antibody (virus neutralizing) production by virtue of intermolecular/intrastructural T_h cell function.[83] In both studies HBcAg and HBeAg were shown to be highly cross reactive, consistent with the murine data. It is also of interest that HBeAg-positive ASC patients, who demonstrate little or no T cell sensitization to HBcAg, nevertheless produce high-titer anti-HBc.[107] This discrepancy may be accounted for by the observation in the murine system that anti-HBc can be produced by T cell-independent as well as T cell-dependent pathways.[76]

Because antigen-specific T cells may be concentrated at the site of synthesis, several studies have focused on characterization of intrahepatic T cells in patients with chronic active HBV infections.[109,110] These studies revealed the presence of HBcAg-specific, $CD4^+$ and $CD8^+$ T cells in the liver, whereas HBsAg-specific T cells were not detected. HBcAg-specific T cell clones derived from the liver of CAH patients secreted IL-2, γ-INF, and tumor necrosis factor (TNF) on stimulation with HBcAg.[110] Therefore, it has been suggested that $CD4^+$ HBcAg-specific T_h cells may contribute directly to liver cell damage during HBV infection.[110]

B. CYTOTOXIC T LYMPHOCYTES

However, in terms of pathogenicity, most interest has focused on the role of CTLs in the mediation of liver cell necrosis and viral clearance. Mondelli et al.[73] demonstrated CTL activity in patients with CAH, which could be inhibited with anti-HBc but not anti-HBs antibodies. This system employed autologous liver cells as target cells and is technically difficult due to the variability and lability of human hepatocytes in culture. Nevertheless, similar results have been achieved in several laboratories.[96,111] More recently similar experiments showed that HBcAg-specific CTLs were also present in the PBMCs of acute hepatitis B patients as well.[112] However, at least two caveats plague the interpretation of these results. Non-T cell as well as T cell populations express cytotoxic activity in the autologous hepatocyte assay, and more importantly the CTL activity does not appear to correlate with disease activity.[73,112] Therefore, the attractive hypothesis that HBcAg represents the primary target for CTL recognition awaits further confirmation in additional assay systems. It is anticipated that application of molecular technology will provide the necessary systems to address this long-standing problem.

VIII. IMPLICATIONS FOR HBV VACCINE DEVELOPMENT

A number of results from the human and animal studies have indicated that inclusion of preS sequences in future recombinant HBV vaccines should be considered. For example, the native preS2 region is more immunogenic at the T and B cell levels than is the S region in a murine model.[54,55] This difference was reflected in a lower effective immunization dose and a higher quantity of antibody production, and the onset of the preS response preceded the S region-specific response in all murine strains tested.[54] At the T cell level, the effective priming dose of preS2 region protein was lower, the magnitude of the preS2-specific response was greater, and the minimal antigen concentration required to elicit IL-2 production by T cells *in vitro* was lower for the preS2 region than for the S region.[55] Similarly, the preS1 region is an efficient immunogen in the mouse[42] and in man[68] at the T and B cell levels. One study claimed that immunization of outbred mice with preS2-containing HBsAg particles enhanced the S region response as compared to immunization with HBsAg particles lacking the preS2 region.[113] However, preS2-specific antibody was not determined, and the increased anti-HBs titers may have represented preS2-specific and not S-specific antibody. Our studies have not shown that immunization of S region-responder strains with HBsAg/GP33 enhances the S-specific response, but rather it induces a superior preS2-specific response as compared to an S-specific response.[54] However, in the *H-2*^{s} haplotype, an S region-nonresponder mouse strain, immunization with HBsAg/GP33 induces anti-preS2-specific antibody and also elicits anti-S-specific antibody by virtue of the T cell mechanism described previously.[55] The observation that the "enhancement" of the S region response after HBsAg/GP33 immunization is discernable only in S region nonresponders indicates that, not surprisingly, preS2-specific helper T cells are not as efficient at helping S region-specific B cells produce antibody as are S region-specific helper T cells. In the absence of S region-specific T cells, T cell recognition of the preS2 region circumvents S region nonresponsiveness at the level of antibody production. Similarly, immunization of S and preS2 region nonresponder strains with HBsAg/P39 circumvents nonresponsiveness to the S and preS2 regions in terms of antibody production.[42] The independence of *H-2*-linked regulation of the immune responses to the preS1, preS2, and S regions of HBsAg would assure fewer genetic nonresponders to a vaccine containing all three regions.[42]

Since anti-S region antibody protects against HBV infection,[74] is there value in eliciting preS-specific antibodies? Recently, *in vitro* neutralization experiments[114] and *in vivo* immunization with preS-specific synthetic peptides[86,115,116] have shown that preS2- and preS1-specific antibodies in the absence of anti-S region antibody can protect chimpanzees against HBV infection. Presumably, the more neutralizing specificities a vaccine elicits, the more efficacious the vaccine will be.

Cumulatively, these findings are very persuasive that the inclusion of preS sequences in recombinant or synthetic HBV vaccines warrants further study to determine the extent to which the characteristics described for the preS region are applicable to the human immune response after vaccination. Several preliminary vaccine trials utilizing preS-containing HBsAg recombinant particles have yielded rather mixed results.[117-122] One criticism of the proposal to include preS determinants has been that the immunogenicity studies in nonprimate animal species may not pertain to man.[48] In this regard, the preS antigens appear to be immunogenic during HBV infection, and in fact anti-preS antibodies often occur prior to any other HBV-specific antibody.[1,27,28,39-41] The early appearance of preS antibodies may explain their relatively transient nature in infection, since both antigen and antibody occur early during the acute phase and may be cleared rapidly by immune complex formation.

Another objection to the inclusion of the preS sequence in future HBV vaccines relates to the presence of the pHSA receptor within the preS2 region. It has been suggested that on vaccination, pHSA may bind to HBsAg and elicit an anti-albumin autoimmune response or down regulate the anti-HBs response.[48] This seems unlikely since monomeric[17-28] or naturally occurring polymeric human albumin has not been reported to bind to HBsAg.[123] However, a recent study suggested that natural human serum albumin can indeed bind to the preS2 region of HBsAg.[124] Because the pHSA receptor site within the preS2 region[29] overlaps with the dominant antibody binding site,[46] it is possible that albumin binding may mask this important-antibody site. If the presence of the pHSA receptor is a problem, the offending sequences could be deleted from the preS2 region by application of recombinant or synthetic technologies.

A number of characteristics of the immune response to HBcAg suggest that this antigen may have utility in HBV and non-HBV vaccine development. For example, HBcAg has been shown to represent an efficient immunogen at the T and B cell levels both in mice[76] and man;[79]

HBcAg can elicit T cell-independent antibody production (i.e., can activate B cells directly[76]); HBcAg can function as a carrier moiety for envelope antigens of HBV;[83] HBcAg has been shown to function as a carrier moiety for non-HBV-related antigens (i.e., 2,4-dinitrophenol [DNP],[81] FMDV peptide,[125] poliovirus type 1 C3 epitope,[126] human rhinovirus type 2 epitope,[126] FeLV epitope,[126] and a human immunodeficiency virus [HIV] *gag* sequence[127]); HBcAg-DNP_1 can elicit anti-DNP antibodies in a T cell-independent manner;[81] and synthetic T cell sites derived from the HBcAg sequence can function as carriers for B cell epitopes of the envelope of HBV and HIV.[80] These preliminary results indicate that HBcAg and/or synthetic T cell sites of HBcAg represent an efficient T cell carrier moiety for multivalent vaccine development.

REFERENCES

1. **Neurath, A. R., Kent, S. B. H., Strick, N., Taylor, P., and Stevens, C. E.,** Hepatitis B virus contains pre-S gene-encoded domains, *Nature (London)*, 315, 154, 1985.
2. **Theilmann, L., Klinkert, M., Gmelin, K., Salfeld, J., Schaller, H., and Pfaff, E.,** Detection of pre-S1 proteins in serum and liver of HBsAg-positive patients: a new marker for hepatitis B virus infection, *Hepatology*, 6, 186, 1986.
3. **Pfaff, E., Klinkert, M., Theilmann, L., and Schaller, H.,** Characterization of large surface proteins of hepatitis B virus by antibodies to preS-S encoded amino acids, *Virology*, 148, 15, 1986.
4. **Klinkert, M., Theilmann, L., Pfaff, E., and Schaller, H.,** Pre-S1 antigens and antibodies early in the course of acute hepatitis B virus infection, *J. Virol.*, 58, 522, 1986.
5. **Neurath, A. R., Kent, S. B. H., Strick, N., and Parker, K.,** Identification and chemical synthesis of a host cell receptor binding site on hepatitis B virus, *Cell*, 46, 429, 1986.
6. **Thung, S. N., Gerber, M. A., Kasambalides, E. J., Gilja, B. K., Keh, W., and Gerlich, W. H.,** Demonstration of pre-S polypeptides of hepatitis B virus in infected livers, *Hepatology*, 6, 1315, 1986.
7. **Gerken, G., Gerlich, W. H., Manns, M., Hess, G., Slusarczyk, J., and Meyer zum Buschenfelde, K. H.,** Pre-S-1 viral surface antigen in relation to viral polyalbumin receptor and major surface antigen in acute hepatitis-B virus (HBV) infection, *J. Hepatol.*, Suppl. 2, S239, 1985.
8. **Gerken, G., Manns, M., Gerlich, W. H., Hess, G., and Meyer zum Buschenfelde, K. H.,** Pre-S encoded surface proteins in relation to the major viral surface antigen in acute hepatitis B virus infection, *Gastroenterology*, 92, 1864, 1987.
9. **Kakumu, S., Arao, M., Mizokami, M., Orido, E., Yamamoto, M., and Sakamoto, N.,** Pre-S proteins in chronic hepatitis B virus infection: markers of active viral infection, *Am. J. Gastroenterol.*, 84, 1250, 1989.
10. **Budkowska, Dubreuil, P., Gerlich, W. H., Lazizi, Y., and Pillot, J.,** Occurrence of pre-S1 antigen in viremic and nonviremic carriers of hepatitis B surface antigen, *J. Med. Virol.*, 26, 217, 1988.
11. **Fraiese, A., Pontisso, P., Cavalletto, D., Fattovich, G., and Alberti, A.,** Expression of preS1 and preS2 in the liver of chronic hepatitis B virus carriers, *J. Hepatol.*, 7, 157, 1988.
12. **Napoli, N., Schiraldi, O., Fiore, G., and Modugno, A.,** Expression of preS antigens in serum during the chronic phase of hepatitis B virus infection, *Digestion*, 44, 232, 1989.
13. **Petit, M.-A., Zoulim, F., Capel, F., Dubanchet, S., Dauguet, C., and Trepo, C.,** Variable expression of preS1 antigen in serum during chronic hepatitis B virus infection: an accurate marker for the level of hepatitis B virus replication, *Hepatology*, 11, 809, 1989.
14. **Alberti, A., Gerlich, W. H., Heermann, K.-H., and Pontisso, P.,** Nature and display of hepatitis B virus envelope proteins and the humoral immune response, *Springer Semin. Immunopathol.*, 12, 5, 1990.
15. **Krogsgarrd, K., Kryger, P., Aldershville, J., Andersson, P., Brechot, C., and the Copenhagen Hepatitis Acuta Programme,** Hepatitis B virus DNA in serum from patients with acute hepatitis B, *Hepatology*, 5, 10, 1985.
16. **Imai, M., Yanase, Y., Nojiri, T., Miyskawa, Y., and Mayumi, M.,** A receptor for polymerized human and chimpanzee albumins on hepatitis B virus particles co-occurring with HBeAg, *Gastroenterology*, 76, 242, 1979.

17. **Machida, A., Kishimoto, S., Ohnuma, H., Miyamoto, H., Baba, K., Oda, K., Nakamura, T., Miyakawa, Y., and Mayumi, M.,** A hepatitis B surface antigen polypeptide (P31) with the receptor for polymerized human as well as chimpanzee albumins, *Gastroenterology*, 85, 268, 1983.
18. **Machida, A., Kishimoto, S., Ohnuma, H., Baba, K., Ito, Y., Muyamoto, H., Funatsu, G., Oda, K., Usuda, S., Togami, S., Nakamura, T., Miyakawa, Y., and Mayumi, M.,** A polypeptide containing 55 amino acid residues coded by the pre-S region of hepatitis B virus deoxyribonucleic acid bears the receptor for polymerized human as well as chimpanzee albumins, *Gastroenterology*, 86, 910, 1984.
19. **Ohnuma, H., Takahashi, K., Kishimoto, S., Machida, A., Imai, M., Mishiro, S., Usuda, S., Oda, K., Nakamura, T., Miyakawa, Y., and Mayumi, M.,** Large hepatitis B surface antigen polypeptides of Dane particles with the receptor for polymerized human serum albumin, *Gastroenterology*, 90, 695, 1986.
20. **Hansson, B. G. and Purcell, R. H.,** Sites that bind polymerized albumin on hepatitis B surface antigen particles: detection by radioimmunoassay, *Infect. Immunol.*, 26, 125, 1979.
21. **Hollinger, F. B. and Dreesman, G. R.,** Hepatitis B virus antigens and albumin receptors, *Gastroenterology*, 76, 641, 1979.
22. **Stevens, C. E., Neurath, A. R., Szmuness, W., Beasley, R. P., and Ikram, H.,** Correlations between HBeAg/anti-HBe, HBsAg titer, HBsAg-associated albumin binding sites and infectivity, in *Viral Hepatitis*, Vyas, G. N., Cohen, S. N., and Schmid, R., Eds., The Franklin Institute Press, Philadelphia, 1978, 211.
23. **Thung, S. N. and Gerber, M. A.,** Albumin receptors on hepatitis B virus and antibodies to albumin, *Liver*, 1, 75, 1981.
24. **Milich, D. R., Gottfried, T., and Vyas, G. N.,** Characterization of the interaction between polymerized human albumin and hepatitis B surface antigen, *Gastroenterology*, 81, 218, 1981.
25. **Pontisso, P., Alberti, A., Bortolotti, F., and Realdi, G.,** Virus-associated receptors for polymerized human serum albumin in acute and in chronic hepatitis B virus infection, *Gastroenterology*, 84, 220, 1983.
26. **Okamoto, H., Imai, M., Usuda, S., Tanaka, E., Tachibana, K., Mishiro, S., Machada, A., Miyakawa, Y., and Mayumi, M.,** Hemagglutination assay of polypeptide coded by the pre-S region of hepatitis B virus DNA with monoclonal antibody: correlation of pre-S polypeptide with the receptor for polymerized human serum albumin antigens, *J. Immunol.*, 134, 1212 1985.
27. **Alberti, A., Pontisso, P., Schiavon, E., and Realdi, G.,** An antibody which precipitates Dane particles in acute hepatitis type B: relation to receptor sites which bind polymerized human serum albumin on virus particles, *Hepatology*, 4, 220, 1984.
28. **Okamoto, H., Usuda, S., Imai, M., Tachibana, K., Tanaka, E., Kumakura, T., Itabashi, M., Takai, E., Tsuda, F., Nakamura, T., Miyakawa, Y., and Mayumi, M.,** Antibody to the receptor for polymerized human serum albumin in acute and persistent infection with HBV, *Hepatology*, 6, 354 1986.
29. **Fujisawa, Y., Itoh, Y., Miyazaki, T., Koboyashi, M., and Asano, T.,** Synthesis in yeast and characterization of HBsAg modified P31 particles, in *Molecular Biology of Hepatitis B Viruses*, Cold Spring Harbor Laboratory, Cold Spring Harbor, New York, 1986 (abstract).
30. **Neurath, A. R., Kent, S. B. H., and Strick, N.,** Location and chemical synthesis of pre-S gene coded immunodominant epitope of hepatitis B virus, *Science*, 224, 392, 1984.
31. **Hu, P. S., Carithers, R. L., Fiorenza, V., and Peterson, D. L.,** Quantitative studies of the hepatitis B viral pre-S proteins: lack of correlation with the HBeAg status, *J. Virol. Methods*, 16, 97, 1987.
32. **Kurai, K., Iino, S., Koike, K., Mitamura, K., Endo, Y., and Oka, H.,** Serum titers of pre-S(2) antigen in patients with acute and chronic type B hepatitis: relation to serum aminotransferase activity and other hepatitis B virus markers, *Hepatology*, 9, 175, 1989.
33. **Neurath, A. R., Adamowicz, P., Kent, S. B. H., Riottot, M. M., Strick, N., Parker, K., Offensperger, W., Petit, M. A., Wahl, S., Budkowska, A., Girard, M., and Pillot, J.,** Characterization of monoclonal antibodies specific for the pre-S2 region of the hepatitis B virus envelope protein, *Mol. Immunol.*, 23, 991, 1986.
34. **Michel, M. L., Pontissao, P., Sobczak, E., Malpiece, Y., Streeck, R. E., and Tiollais, P.,** Synthesis in animal cells of hepatitis B surface antigen particles carrying a receptor for polymerized human serum albumin, *Proc. Natl. Acad. Sci. U.S.A.*, 81, 7708, 1984.
35. **Langley, K. E., Egan, K. M., Barendt, J. M., Parker, C. G., and Bitter, G. A.,** Characterization of purified hepatitis B surface antigen containing pre-S(2) epitopes expressed in Saccharomyces cerevisiae, *Gene*, 67, 229, 1988.
36. **Budkowska, A., Dubreiul, P., Capel, F., and Pillot, J.,** Hepatitis B virus pre-S gene-encoded antigenic specificity and anti-pre-S antibody; relationship between anti-pre-S response and recovery, *Hepatology*, 6, 360, 1986.
37. **Lenkei, R. and Chetie, V.,** Methods for detection of anti-albumin auto antibodies in hepatic diseases, *J. Immunol. Methods*, 16, 23, 1977.

38. **Milich, D. R., Bhatnagar, P. K., Papas, E., and Vyas, G. N.,** Interactions between polymerized human albumin, hepatitis B surface antigen and complement. I. Binding of polyalbumin to C1q, *Med. Virol.*, 7, 193, 1981.
39. **Moodie, J., Stannard, L. M., and Kipps, A.,** Dane complexes in hepatitis B antigen, *J. Gen. Virol.*, 24, 375, 1974.
40. **Neurath, A. R., Trepo, C., Chen, M., and Prince, A.,** Absence of pre-S2 antibodies in natural hepatitis B virus infection, *J. Gen. Virol.*, 30, 277, 1976.
41. **Alberti, A., Diana, S., Sculard, G. H., Eddleston, A. L. W. F., and Williams, R.,** Detection of an antibody system reacting with Dane particles in hepatitis B virus infection, *Br. Med. J.*, 2, 1056, 1978.
42. **Milich, D. R., McLachlan, A., Chisari, F. V., Kent, S. B. H., and Thornton, G. B.,** Immune response to the pre-S(1) region of the hepatitis B surface antigen (HBsAg): a pre-S(1)-specific T cell response can bypass nonresponsiveness to the pre-S(2) and S regions of the HBsAg, *J. Immunol.*, 137, 315, 1986.
43. **Takai, E., Machida, A., Ohnuma, H., Miyamoto, H., Tanaka, T., Baba, K., Tsuda, T., Usuada, S., Nakamura, T., Miyakawa, Y., and Mayumi, M.,** A solid-phase enzyme immunoassay for the determination of IgM and IgG. Antibodies against translation products of pre-S1 and pre-S2 regions of hepatitis B virus, *J. Immunol. Methods*, 95, 23, 1986.
44. **Persing, D. H., Varmus, H. E., and Ganem, D.,** Antibodies to pre-S and X determinants arise during natural infection with ground squirrel hepatitis virus, *J. Virol.*, 60, 177, 1986.
45. **Theilmann, L., Klinkert, M.-Q., Gmelin, K., Kommerell, B., and Pfaff, E.,** Detection of antibodies against pre-S1 proteins in sera of patients with hepatitis B virus (HBV) infection, *J. Hepatol.*, 4, 22, 1987.
46. **Milich, D. R., McLachlan, A., Chisari, F. V., Nakamura, T., and Thornton, G. B.,** Two distinct but overlapping antibody binding sites in the pre-S(2) region of HBsAg localized within 11 continuous residues, *J. Immunol.*, 137, 2703, 1986.
47. **Petit, M., Maillard, P., Capel, F., and Pillot, F.,** Immunochemical structure of the hepatitis B surface antigen vaccine II. Analysis of antibody responses in human sera against the envelope proteins, *J. Mol. Immunol.*, 23, 511, 1986.
48. **Hellstrom, U., Sylvan, S., Kuhns, M., and Sarin V.,** Absence of pre-S2 antibodies in natural hepatitis B virus infection, *Lancet*, ii, 889, 1986.
49. **Coursaget P., Adamowicz, P., Bourdil, C., Yvonnet, B., Buisson, Y., Barres, J.-L., Saliou, P., Chiron, J.-P., and Mar, I. D.,** Anti-pre-S2 antibodies in natural hepatitis B virus infection and after immunization, *Vaccine*, 6, 357, 1988.
50. **Milich, D. R. and Chisari, F. V.,** Genetic regulation of the immune response to hepatitis B surface antigen (HBsAg). I. H-2 restriction of the humoral immune response to the *a* and *d* determinants of HBsAg, *J. Immunol.*, 129, 320, 1982.
51. **Walker, M. E., Szmuness, W., Stevens, C. E., and Rubinstein, P.,** Genetics of anti-HBs responsiveness. I. HLA-DR7 and non-responsiveness to hepatitis vaccination, *Transfusion*, 21, 601, 1981.
52. **Craven, D. E., Awdeh, Z. L., Kunches, L. M., Yunis, J., Dienstag, J. L., Werner, B. G., Polk, B., Snydman, D. R., Platt, R., Crumpacker, C. S., Grady, G. F., and Alpre, C. A.,** Nonresponsiveness to hepatitis B vaccine in health care workers, *Ann. Int. Med.*, 105, 356, 1986.
53. **Alper, C. A., Kruskall, M. S., Marcus-Bagley, D., Crave, D. E., Katz, A. J., Brink, S. J., Dienstag, J. L., Awdeh, Z., and Yunis, E. J.,** Genetic prediction of nonresponse to hepatitis B vaccine, *N. Engl. J. Med.*, 321, 708, 1989.
54. **Milich, D. R., Thornton, G. B., Neurath, A. R., Kent, S. B., Michel, M., Tiollois, P., and Chisari, F. V.,** Enhanced immunogenicity of the pre-S region of hepatitis B surface antigen, *Science*, 228, 1195, 1985.
55. **Milich, D. R., McNamara, M. K., McLachlan, A., Thornton, G. B., and Chisari, F. V.,** Distinct H-2-linked regulation of T cell responses to the pre-S and S regions of the same HBsAg polypeptide allows circumvention of S region nonresponsiveness, *Proc. Natl. Acad. Sci. U.S.A.*, 82, 8168, 1985.
56. **Milich, D. R., McLachlan, A., Moriarty, A., and Thornton, G. B.,** A single pre-S(1) synthetic peptide can prime antibody production to multiple epitopes within the S, pre-S(2) and pre-S(1) regions of HBsAg, *J. Immunol.*, 138, 4457, 1987.
57. **Vyas, G. N., Roa, K. R., and Ibrahim, A. B.,** Hepatitis associated Australian antigen (HBsAg): a conformational antigen dependent on disulfide bonds, *Science*, 178, 1300, 1972.
58. **Milich, D. R., McLachlan, A., Chisari, F. V., and Thornton, G. B.,** Nonoverlapping T and B cell determinants on an hepatitis B surface antigen (HBsAg) pre-S(2) region synthetic peptide, *J. Exp. Med.*, 164, 532, 1986.
59. **Neurath, A. R., Kent, S. B. H., Strick, N., Stark, D., and Sproul, P. J.,** Genetic restriction of immune responsiveness to synthetic peptides corresponding to sequences in the pre-S region of the hepatitis B virus (HBV) envelope gene, *Med. Virol.*, 17, 119, 1985.

60. **Neurath, A. R., Kent, S. B. H., Strick, N., Parker, K., Courouce, A. M., Riottot, M. M., Petit, M. A., Budkowska, A., Girard, M., and Pillot, J.**, Antibodies to synthetic peptides from the preS1 and preS2 regions of one subtype of the hepatitis B virus (HBV) envelope protein recognize all HBV subtypes, *Mol. Immunol.*, 24, 975, 1987.
61. **Neurath, A. R., Kent, S. B. H., Adamowicz, P., Riottot, M. M., Price, P., Parker, K., Petit, M. A., Budkowska, A., Girard, M., and Pillot, J.**, Immunological cross-reactivity between preS2 sequences of the hepatitis B virus envelope proteins corresponding to serological subtypes *adw2* and *ayw*, *Mol. Immunol.*, 24, 561, 1987.
62. **Milich, D. R., Hughes, J. L., McLachlan, A., Langley, K. E., Thornton, G. B., and Jones, J. E.**, Importance of subtype in the immune response to the pre-S(2) region of the hepatitis B surface antigen, *J. Immunol.*, 144, 3535, 1990.
63. **Milich, D. R., Jones, J. E., McLachlan, A., Bitter, G., Moriarty, A., and Hughes, J. L.**, Importance of subtype in the immune response to the pre-S(2) region of the hepatitis B surface antigen: synthetic pre-S(2) immunogen, *J. Immunol.*, 144, 3544, 1990.
64. **Dienstag, J. L., Bhan, A. K., and Klingenstein, R. J.**, Immunopathogenesis of liver disease associated with hepatitis B, in *Viral Hepatitis*, 1981 Int. Symp., Szmuness, W., Alter, H. J., and Maynard J. E., Eds., Franklin Institute Press, Philadelphia, 1982, 221.
65. **Celis, E., Kung, P. C., and Change T. W.**, Hepatitis B virus-reactive human T lymphocyte clones: antigen specificity and helper function for antibody synthesis, *J. Immunol.*, 132, 1511, 1984.
66. **Celis, E., Ou, D., and Otvos, L., Jr.**, Recognition of hepatitis B surface antigen by human T lymphocytes. Proliferative and cytotoxic responses to a major antigenic determinant defined by synthetic peptides, *J. Immunol.*, 140, 1808, 1988.
67. **Jin, Y., Shih, J. W.-K., and Berkower, I.**, Human T cell response to the surface antigen of hepatitis B virus (HBsAg). Endosomal and nonendosomal processing pathways are accessible to both endogenous and exogenous antigen, *J. Exp. Med.*, 168, 293, 1988.
68. **Ferrari, C., Penna, A., Bertoletti, A., Cavalli, A., Valli, A., Schianchi, C., and Fiaccadori, F.**, The preS1 antigen of hepatitis B virus is highly immunogenic at the T cell level in man, *J. Clin. Invest.*, 84, 1314, 1989.
69. **Barnaba, V., Franco, A., Alberti, A., Balsano, C., Benvenuto, R., and Balsano, F.**, Recognition of hepatitis B virus envelope proteins by liver-infiltrating T lymphocytes in chronic HBV infection, *J. Immunol.*, 143, 2650, 1989.
70. **Barnaba, V., Franco, A., Alberti, A., Benvenuto, R., and Balsano, F.**, Selective killing of hepatitis B envelope antigen-specific B cells by class I-restricted, exogenous antigen-specific T lymphocytes, *Nature (London)*, 345, 258, 1990.
71. **Cupps, T. R., Hoofnagle, J. H., Ellis, R. W., Miller, W. J., Gerin, J. L., Volkman, D. J., and Smith-Haas, S. A.**, *In vitro* immune responses to hepatitis B surface antigens S and preS2 during acute infection by hepatitis B virus in humans, *J. Infect. Dis.*, 161, 412, 1990.
72. **Eddleston, A. L. W. F. and Williams, R.**, Inadequate antibody response to HBsAg or suppressor T-cell defect in development of active chronic hepatitis, *Lancet*, ii, 1543, 1974.
73. **Mondelli, M., Vergani, G. M., Alberti, A., Vergani, D., Portmann, B., Eddleston, A. L. W. F., and Williams, R.**, Specificity of T lymphocyte cytotoxicity to autologous hepatocytes in chronic hepatitis B virus infection: evidence that T cells are directed against HBV core antigen expressed on hepatocytes, *J. Immunol.*, 129, 2773, 1982.
74. **Gerety, F. J., Tabor, E., Purcell, R. H., and Tyeryar, F. J.**, Summary of an international workshop on hepatitis B vaccines, *J. Infect. Dis.*, 140, 642, 1979.
75. **Murray, K., Bruce, S. A., Hinnen, A., Wingfield, P., van Erd, P. M., de Reus, A., and Schellekens, H.**, Hepatitis B virus antigen made in microbial cells immunize against viral infection, *EMBO J.*, 3, 645, 1984.
76. **Milich, D. R. and McLachlan, A.**, The nucleocapsid of hepatitis B virus is both a T cell-independent and T cell-dependent antigen, *Science*, 234, 1398, 1986.
77. **Milich, D. R., McLachlan, A., Moriarty, A., and Thornton, G. B.**, The immune response to the hepatitis B core antigen (HBcAg): localization of T cell recognition sites within HBcAg/HBeAg, *J. Immunol.*, 139, 1223, 1987.
78. **Milich, D. R., Louie, R. E., and Chisari, F. V.**, Genetic regulation of the immune response to hepatitis B surface antigen (HBsAg). V. T cell proliferative response and cellular interactions, *J. Immunol.*, 134, 4194, 1985.
79. **Ferrari, C., Penna, A., Sansoni, P., Giuberti, T., Neri, T. M., Chisari, F. V., and Fiaccadori, F.**, Selective sensitization of peripheral blood T lymphocytes to hepatitis B core antigen in patients with chronic active hepatitis type B, *Clin. Exp. Immunol.*, 67, 497, 1986.

80. **Milich, D. R., Hughes, J. L., McLachlan, A., Thornton, G. B., and Moriarty, A.,** Synthetic hepatitis B immunogen comprised of nucleocapsid T cell sites and an envelope B cell epitope, *Proc. Natl. Acad. Sci. U.S.A.*, 85, 1610, 1988.

80a. **Milich, D. R.,** T- and B-cell recognition of hepatitis B viral antigens, *Immunol. Today*, 9, 380, 1988.

81. **Milich, D. R., McLachlan, A., Stahl, S., Wingfield, P., Thornton, G. B., Hughes, J. L., and Jones J.,** Comparative immunogenicity of hepatitis B virus core and E antigens, *J. Immunol.*, 141, 3617, 1988.

82. **Russell, S. M. and Liew, F. Y.,** T cells primed by influenza virion internal components can cooperate in the antibody response to haemagglutinin, *Nature (London)*, 147, 280, 1979.

83. **Milich, D. R., McLachlan, A., Thornton, G. B., and Hughes, J. L.,** Antibody production to the nucleocapsid and envelope of the hepatitis B virus primed by a single synthetic T cell site, *Nature (London)*, 329, 547, 1987.

84. **Milich, D. R.,** Immunologic response to pre-S antigens of the hepatitis B virus, *Virol. Immunol.*, 1, 83, 1987.

85. **Milich, D. R.,** Genetic and molecular basis of T and B cell recognition of hepatitis B viral antigens, in *Immunological Reviews*, Vol. 99, Moller, G., Ed., International Publications, Munksgaard, Denmark, 1987, 71.

86. **Itoh, Y., Takai, E., Ohnuma, H., Kitajima, K., Tsuda, F., Machida, A., Mishiro, S., Nakamura, T., Miyakawa, Y., and Mayumi, M.,** A synthetic peptide vaccine involving the product of the pre-S(2) region of hepatitis B virus DNA: protective efficacy in chimpanzees, *Proc. Natl. Acad. Sci. U.S.A.*, 83, 9174, 1986.

87. **Thornton, G. B., Milich, D., Chisari, F., Mitamura, K., Kent, S. B., Neurath, R., Purcell, R., and Gerin, J.,** Immune response in primates to the pre-S2 region of hepatitis-B surface antigen: identification of a protective determinant, in *Vaccines 87 Modern Approaches to New Vaccines*, Chanock, R. M., Lerner, R. A., Brown, F., and Ginsberg, H., Eds., Cold Spring Harbor Laboratory, Cold Spring Harbor, New York, 1987, 77.

88. **Alberti, A., Trevisan, A., Fattovich, G., and Realdi, G.,** The role of hepatitis B virus replication and hepatocyte membrane expression in the pathogenesis of HBV-related hepatic damage, in *Advances in Hepatitis Research*, Chisari, F. V., Ed., Masson Publishing USA, New York, 1984, 134.

89. **Beasley, R. P., Trepo, C., Stevens, C.E., and Szmuness, W.,** The e antigen and vertical transmission of hepatitis B surface antigen, *Am. J. Epidemiol.*, 105, 94, 1977.

90. **Beasley, R. P. and Hwang, L.,** Postnatal infectivity of hepatitis B surface antigen-carrier mothers, *J. Infect. Dis.*, 147, 185, 1983.

91. **Li, L., Sheng, M.-H., Tong, S.-P., Chen, H.-Z., and Wen, Y.-M.,** Transplacental transmission of hepatitis B virus, *Lancet*, i, 872, 1986.

92. **Lee, S.-D., Lo, K.-J., Tsai, Y.-T., and Wu, J.-C.,** Maternal hepatitis B virus DNA in mother-infant transmission, *Lancet*, i, 719, 1989.

93. **Stevens, C. E., Beasley, R. P., Tsu, J., and Lee, W. C.,** Vertical transmission of hepatitis B antigen in Taiwan, *N. Engl. J. Med.*, 292, 771, 1975.

94. **Aldershvile, J., Frosner, G. G., Nielsen, J. O., Hardt, F., Dienhardt, F., and Skinhoj, P.,** Hepatitis B e antigen and antibody measured by radioimmunoassay in acute hepatitis B surface antigen-positive hepatitis, *J. Infect. Dis.*, 141, 293, 1980.

95. **Thomas, H. C., Meron, J., Water, J., and Main, J.,** Virus-host interaction in chronic hepatitis B virus infection, *Semin. Liver Dis.*, 8, 342, 1988.

96. **Pignatelli, M., Waters, J., Lever, A., Iwarson, S., Gerety, R., and Thomas, H. C.,** Cytotoxic T-cell responses to the nucleocapsid proteins of HBV in chronic hepatitis. Evidence that antibody modulation may cause protracted infection, *J. Hepatol.*, 4, 15, 1987.

97. **Milich, D. R., Jones, J. E., Hughes, J. L., Price, J., Raney, A. K., and McLachlan, A.,** Is a function of the secreted hepatitis B E antigen to induce immunologic tolerance *in utero*?, *Proc. Natl. Acad. Sci. U.S.A.*, 87, 6599, 1990.

98. **Arakawa, K., Tsuda, F., Takahashi, K., Ise, I., Naito, S., Kosugi, E., Miyakawa, Y., and Mayumi, M.,** Maternofetal transmission of IgG-bound hepatitis B e antigen, *Pediatr. Res.*, 16, 247, 1982.

99. **Biswas, S. C., Gupta, I., Ganguly, N. K., Chawla, Y., and Dilawari, J. B.,** Prevalence of hepatitis B surface antigen in pregnant mothers and its perinatal transmission, *Trans. Royal Soc. Trop. Med. Hyg.*, 83, 698, 1989.

100. **Goudeau, A., Lesage, G., Denis, F., Chiron, J. P., Yvonnet, B., Barin, F., Coursaget, P., and Diop Mar, I.,** Lack of anti-HBc IgM in neonates with HBsAg carrier mothers argues against transplacental transmission of hepatitis B virus infection, *Lancet*, ii, 1103, 1983.

101. **Milich, D. R., Jones, J. E., McLachlan, A., Houghten, R., Thornton, G. B., and Hughes, J. L.,** Distinction between immunogenicity and tolerogenicity among HBcAg T cell determinants: influence of peptide-MHC interaction, *J. Immunol.*, 143, 3148, 1989.

102. **Reichard, O., von Sydow, M., Dennerstam, L., Norder, H., Magnius, L., Berglund, O., Enocksson, E., Ginsburg, B.E., Grillner, L., Lidman, K., and Forsgren, M.,** Prevention by immunisation of hepatitis B in the newborn infant in Stockholm, Sweden, in *Viral Hepatitis & Liver Disease*, Zuckerman, A. J., Ed., Alan R. Liss., New York, 1988, 989.

103. **Poovorawan, Y., Sanpavat, S., Pongpunlert, W., Chumdermpadetsuk, S., Sentrakul, P., and Safary, A.,** Protective efficacy of a recombinant DNA hepatitis B vaccine in neonates of HBe antigen-positive mothers, *JAMA*, 261, 3278, 1989.
104. **Sylvan, S. P. E., Hellstrom, U. B., and Flehmig, B.,** Characterization of cell mediated immune responses to the hepatitis B core protein in man, *Clin. Exp. Immunol.*, 68, 233, 1987.
105. **Vento, S., Hegarty, J. E., Alberti, A., O'Brien, C. J., Alexander, G. J. M., Eddleston, A. L. W. F., and Williams, R.,** T lymphocyte sensitization to HBcAg and T cell-mediated unresponsiveness to HBsAg in hepatitis B virus-related chronic liver disease, *Hepatology*, 5, 192, 1985.
106. **Vento, S., Chen, S.-H., Giuliani-Piccari, G., O'Brien, C. J., Dal Monte, P. R., Eddleston, A. L. W. F., Howard, C. R., and Williams, R.,** Cellular immunity to nucleocapsid and pre-S determinants in asymptomatic carriers of hepatitis B virus, *Immunology*, 62, 593, 1987.
107. **Inoue, M., Kakumu, S., Yoshioka, K., Tsutsumi, Y., Wakita, T., and Arao, M.,** Hepatitis B core antigen-specific IFN-γ production of peripheral blood mononuclear cells in patients with chronic hepatitis B virus infection, *J. Immunol.*, 142, 4006, 1989.
108. **Ferrari, C., Penna, A., Bertoletti, A., Valli, A., Antoni, A., Giuberti, T., Cavalli, A., Petit, M., and Fisccadori, F.**, Cellular immune response to HBV-encoded antigens in acute and chronic HBV infection, *J. Immunol.*, 145, 3442, 1990.
109. **Ferrari, C., Penna, A., Giuberti, T., Tong, M. J., Ribera, E., Fiaccadori, F., and Chisari, F. V.,** Intrahepatic, nucleocapsid antigen-specific T cells in chronic active hepatitis B, *J. Immunol.*, 139, 2050, 1987.
110. **Ferrari, C., Mondelli, M. U., Penna, A., Fiaccadori, F., Chisari, F. V.,** Functional characterization of cloned intrahepatic, hepatitis B virus nucleoprotein-specific helper T cell lines, *J. Immunol.*, 139, 539, 1987.
111. **Wands, J. R. and Isselbacher, K. J.,** Lymphocyte cytotoxicity to autologous liver cells in chronic active hepatitis, *Proc. Natl. Acad. Sci. U.S.A.*, 72, 1301, 1974.
112. **Mondelli, M. U., Bortolotti, F., Pontisso, P., Rondanelli, E. G., Williams, R., Realdi, G., Alberti, A., and Eddleston, A. L. W. F.,** Definition of hepatitis B virus (HBV)-specific target antigens recognized by cytotoxic T cells in acute HBV infection, *Clin. Exp. Immunol.*, 68, 242, 1987.
113. **Coursaget, P., Barres, J. L., Chiron, J. P., and Adamowicz, P.,** Hepatitis B vaccine with and without polymerised albumin receptors, *Lancet*, ii, 1152, 1985.
114. **Gerety, R. J., Tabor, E., Purcell, R. H., and Tyeryar, F. J.,** Summary of an international workshop on hepatitis B vaccines, *J. Infect. Dis.*, 140, 642, 1979.
115. **Thornton, G. B., Moriarty, A. M., Milich, D. R., Eichberg, J. W., Purcell, and Gerin, J. L.,** Protection of chimpanzees from hepatitis-B virus infection after immunization with synthetic peptides: identification of protective epitopes in the pre-S region, in *Vaccines 89*, Brown, F., Chanock, R.M., Ginsberg, H., and Lerner, R., Eds., Cold Spring Harbor Laboratory, Cold Spring Harbor, New York, 1989, 467.
116. **Emini E., Larson, V., Eichberg, J., Conrad, P., Garsky, V. M., Lee, D. R., Ellis, R. W., Miller, W. J., Anderson, C. A., and Gerety, R. J.,** Protective effect of a synthetic peptide comprising the complete preS2 region of the hepatitis B virus surface protein, *J. Med. Virol.*, 28, 7, 1989.
117. **Tron, F., Degos, F., Brechot, C., Courouce, A.-M., Goudeau, A., Marie, F.-N., Adamowicz, P., Saliou, P., Laplanche, A., and Girard, M.,** Randomized dose range study of a recombinant hepatitis B vaccine produced in mammalian cells and containing the S and preS2 sequences, *J. Infect. Dis.*, 160, 199, 1989.
118. **Ellis, R. W., Kniskern, P. J., Hagopian, A., Schultz, L. D., Montogomery, D. L., Maigetter, R. Z., Wampler, D. E., Emini, E. A., Wolanski, B., McAleer, W. J., Hurni, W. M., and Miller, W. J.,** Preparation and testing of a recombinant-derived hepatitis B vaccine consisting of pre-S2 + S polypeptides, in *Viral Hepatitis and Liver Diseases*, Zuckerman, A. J., Ed., Alan R. Liss, New York, 1079.
119. **Fujisawa, Y., Kuroda, S., Van Eerd, P. M. C. A., Schellekens, H., and Kakinuma, A.,** Protective efficacy of a novel hepatitis B vaccine consisting of M (pre-S2 + S) protein particles (a third generation vaccine), *Vaccine*, 8, 193, 1990.
120. **De Wilde, M., Rutgers, T., Cabezon, T., Hauser, P., Van Opstal, O., Harford, N., Van Wijnendaele, F., Desmons, P., Comberbach, M., Roelants, P., Voet, P., Delem, A., and Petre, J.,** PreS containing HBsAg particles from *Saccharomyces cerevisiae*, in *The 1990 Int. Symp. Viral Hepatitis and Liver Disease*, Scientific Program and Abstract Volume, 1990, 117.
121. **Thoma, H. A., Kapfer, G., Koller, E., and Hemmerling, A. E.,** Does the pre-S2 have the same effects in improving the HBV immune response compared to pre-S1? in *The 1990 Int. Symp. Viral Hepatitis and Liver Disease*, Scientific Program and Abstract Volume, 1990, 118.
122. **Iino, S., Suzuki, H., Akahane, Y., and Mayumi, M.,** Phase II study of recombinant HB vaccine containing the S and preS2 antigens, in *The 1990 Int. Symp. Viral Hepatitis and Liver Disease*, Scientific Program and Abstract Volume, 1990, 119.
123. **Yu, M. W., Finlayson, J. S., and Shih, J. W.-K.,** Interaction between various polymerized human albumins and hepatitis B surface antigen, *J. Virol.*, 55, 736, 1985.

124. **Krone, B., Lenz, A., Heermann, K.-H., Seifer, M., Xuangyong, L., and Gerlich, W. H.,** Interaction between hepatitis B surface proteins and monomeric human serum albumin, *Hepatology*, 11, 1050, 1990.

125. **Clarke, B. E., Newton, S. E., Carroll, A. R., Francis, M. J., Appleyard, G., Syred, A. D., Highfield, P. E., Rowlands, D. J., and Brown, F.,** Improved immunogenicity of a peptide epitope after fusion to hepatitis B core protein, *Nature (London)*, 330, 381, 1987.

126. **Clarke, B. E., Brown, A. L., Grace, K. G., Hastings, G. Z., Brown, F., Rowlands, D. J., and Francis, M. J.,** Presentation and immunogenicity of viral epitopes on the surface of hybrid hepatitis B virus core particles produced in bacteria, *J. Gen. Virol.*, 71, 1109, 1990.

127. **Moriarty, A. M., McGee, J. S., Winslow, B., Inman, D., Leturcq, D. J., Thornton, G. B., Hughes, J. L., and Milich, D. R.,** Expression of HIV gag and env B-cell epitopes on the surface of HBV core particles and analysis of the immune responses generated to those epitopes, in *Vaccines 90*, Brown, F., Channock, T., Ginsberg, H., and Lerner, R., Eds., Cold Spring Harbor Laboratory, Cold Spring Harbor, New York, 1990, 225.

Chapter 16

RECOMBINANT HEPATITIS B VACCINES

Ronald W. Ellis and Peter J. Kniskern

TABLE OF CONTENTS

I. INTRODUCTION

Hepatitis B virus (HBV) causes an infection endemic to many parts of the world. The infection is transmitted most commonly from chronically infected mothers to their newborns, who consequently often develop chronic hepatitis B. In addition, many adults percutaneously exposed to HBV suffer acute infections, a minority of which also progress to chronic hepatitis B. The number of hepatitis B carriers worldwide probably exceeds 300 million.[1] Hundreds of thousands of these carriers die annually from the long-term sequelae of chronic hepatitis B, namely cirrhosis and hepatocellular carcinoma.[2] Since there is no effective treatment available for chronic hepatitis B, control and prevention become indispensable.

The first-generation hepatitis B vaccine was derived from the plasma of chronic carriers in the replicative phase of disease.[3] In such individuals, one of the viral envelope proteins (hepatitis B surface antigen; HBsAg) is synthesized in excess by infected liver cells and, in addition to being incorporated into virions, is assembled with cellular lipids into noninfectious lipoprotein spheres approximately 22 nm in diameter. Following secretion from cells, these particles circulate at levels of >100 mg/l blood. These noninfectious particles were isolated to a high degree of purity and formed the basis for the plasma-derived vaccine licensed in the United States in 1981 (Heptavax-B®; Merck Sharp & Dohme, West Point, PA). For this vaccine, each of the three inactivation steps used during the manufacturing process (pepsin at pH 2, 8 *M* urea, and 0.01% [wt/vol] formaldehyde[3]) were demonstrated to be extremely efficient at inactivating HBV as well as representative agents from all known groups of viruses,[4] including human immunodeficiency virus (HIV).[5] However, despite its efficacy and excellent safety profile in millions of recipients worldwide, acceptance of the vaccine was less than anticipated. Furthermore, suitable donors of HBsAg-containing plasma became limiting for the production of sufficient quantities of vaccine for worldwide use. Therefore, a need developed for a second-generation (recombinant-derived) hepatitis B vaccine.

The plasma-derived hepatitis B vaccine was demonstrated to be extremely effective in eliciting antibodies to HBsAg (anti-HBs). Early experiments in chimpanzees, the only commonly available nonhuman primate known to be susceptible to HBV infection, showed that anti-HBs was protective against challenge with HBV in HBsAg-vaccinated animals.[6] Clinical trials then established a level of anti-HBs (10 milli-international units [mIU]/ml; defined relative to an international reference standard[7]) associated with protection against clinical infection,[8,9] which thus became a target level for immunogenicity in developing a recombinant-derived hepatitis B vaccine based on HBsAg.

This chapter will describe recombinant-derived HBsAg-based subunit vaccines that have been developed and licensed on the basis of their efficiency at eliciting anti-HBs. In addition, several experimental recombinant hepatitis B vaccines are described that may elicit immunity based on mechanisms in addition to or instead of anti-HBs.

II. LICENSED SUBUNIT S POLYPEPTIDE VACCINES

In order to produce an effective recombinant hepatitis B vaccine, it was necessary to insert the HBsAg gene into an expression vector that would direct the synthesis of large quantities of HBsAg in a heterologous host cell. Following the cloning in the late 1970s of the HBV genome, a gene designated S was identified as encoding HBsAg.[10] The S gene is part of a larger open reading frame (ORF) that is divided into three domains (preS1, preS2, and S in 5′ to 3′ order within the ORF), each beginning with an in-frame ATG codon that can function as a translational initiation site. Thus, these domains define three polypeptides known as S or

HBsAg (226 amino acids), preS2+S (281 amino acids), and preS1+preS2+S (389 amino acids). The 681-nucleotide S gene encodes a 24-kilodalton (kDa) polypeptide that is the major protein in 22-nm particles from human plasma. Therefore, the S gene was selected initially for expressing a recombinant hepatitis B vaccine.

A. EXPRESSION SYSTEM

Since *Escherichia coli* was the first heterologous expression system to be developed, early attempts at the expression of the *S* gene utilized *E. coli* as the host cell. Unfortunately, all attempts in *E. coli* were unsuccessful due to the lack of stability of the S polypeptide in *E. coli* and its failure therein to assemble into 22-nm particles. However, satisfactory levels of expression were achieved in bakers' yeast *Saccharomyces cerevisiae,*[11-14] mammalian cells,[15-17] yeast *Pichia pastoris,*[18] and in insect cells with the baculovirus expression system.[19] These host cell systems produce HBsAg that assembles into typical 22-nm lipoprotein particles. In part due to the timing of the availability of the expression systems and experience in culture scale-up, most work toward vaccine development has proceeded with HBsAg expressed by *S. cerevisiae* or mammalian cells, both of which have been evaluated clinically, and in the case of *S. cerevisiae* has been licensed as a product. Both vaccine antigens are highly related antigenically to plasma-derived HBsAg and are highly immunogenic for eliciting anti-HBs. Approximately 20 to 25% of mammalian-derived HBsAg is glycosylated[20] with a single asparagine-linked complex-type oligosaccharide at amino acid residue 146,[21] while yeast-derived HBsAg is nonglycosylated. However, this biochemical difference is not reflected in the apparent antigenicity of the different vaccines or in their ability to elicit virus-neutralizing antibodies.

Three general considerations that are important for choosing the system of expression for most recombinant-derived biological products, including vaccines, are the biological activity (immunogenicity, protective efficacy) of the expressed protein, the concern for safety of the final product relative to its host cell for expression, and the scalability of the production process.[22] As mentioned above, the immunogenicities of the yeast- and mammalian-derived HBsAg are comparable, thus presenting no *a priori* reason for choosing one system over the other. Second, the intended worldwide use of hepatitis B vaccines in hundreds of millions of healthy individuals, mostly for the intended eventual eradication of hepatitis B, is a crucial consideration in selecting a host cell for expression. There has been concern that mammalian cells used for expressing foreign genes generally are continuous cell lines (CCLs) possessing one or more properties associated with *in vivo* tumorigenicity or *in vitro* transformation and harboring endogenous protooncogenes and retroviruses, elements of which can be associated with residual DNA in a product. Even though the theoretical risk of oncogenesis associated with residual DNA is extremely low,[23] the apparent risk associated with hundreds of millions of vaccine doses still is finite.[24] In contrast, yeast cells are not associated with these risk factors, such that the presence of minute quantities of yeast DNA in a vaccine is of no apparent concern.[25] Finally, there has been more experience with the scale-up of yeast than with mammalian cells. The stability of recombinant yeast cells during fermentation can be controlled more precisely than that of recombinant mammalian cells. Recombinant proteins, including HBsAg, often can be produced in yeast cells at higher yields and correspondingly lower costs than in mammalian cells. Therefore, for reasons primarily of perceived safety and secondarily of production, yeast has been considered by regulatory agencies as the preferred host for the expression of recombinant-derived hepatitis B vaccines.

The yeast expression systems for the production of HBsAg for Recombivax HB® (Merck Sharp & Dohme), Engerix-B® (SmithKline Beckman Corp., Philadelphia, PA),[13] and Bim-

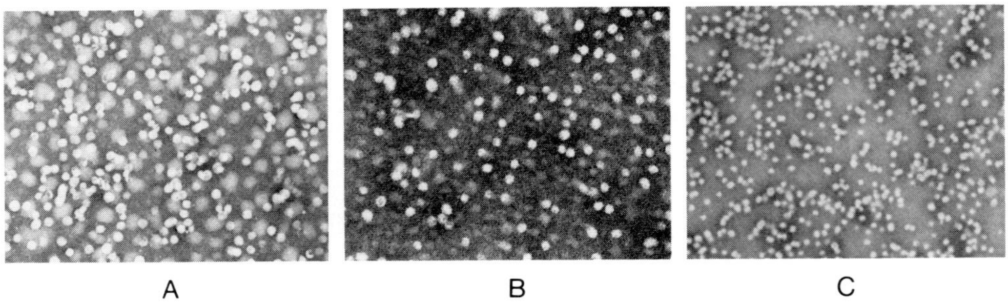

FIGURE 1. Electron micrographs of purified preparations of negatively stained hepatitis B antigens. (A) Plasma-derived HBsAg, (Magnification × 130,000.) (B) Yeast-derived HBsAg, (Magnification × 165,000.) (C) Yeast-derived PreS2+S, (Magnification × 130,000.)

mugen® (Kaketsuken)[14] are similar and typical of systems commonly used for *S. cerevisiae*. For Recombivax HB®, expression of HBsAg is directed by a plasmid containing the yeast 2 μm sequences, which enables amplification to approximately 100 plasmid copies per cell in *cir*° yeast hosts.[26] The *LEU2* gene enables selection for plasmid retention in leu$^-$ host strains by fermenting the recombinant strain in medium lacking leucine. The *GAP491* gene, the product of which is glyceraldehyde-3-phosphate dehydrogenase, represents approximately 5% of total yeast protein[27] and provides one of the strongest known promoters for the constitutive expression of HBsAg in yeast. Typical mammalian cell expression systems have utilized strong mammalian cell promoters and DNA amplification sequences in various host cells[15,16] or the bovine papillomavirus expression vector in susceptible mouse cells.[17]

B. PRECLINICAL EVALUATION

Recombinant yeast cells expressing HBsAg are grown in stirred-tank fermentors inoculated from frozen vials of a stable-production Master Seed. Unlike mammalian cells, which secrete HBsAg, the HBsAg expressed in yeast accumulates intracellularly. Therefore, the yeast cells are harvested, washed, and disrupted by high pressure. For Recombivax HB®, the antigen is purified by conventional biochemical techniques[28,29] to >99% purity with respect to protein content and formulated into a vaccine by adsorption to aluminum hydroxide, a common adjuvant for human use. The final vaccine antigen is a particle approximately 22 nm in diameter, that is morphologically similar to plasma-derived HBsAg (Figure 1). Numerous immunological, chemical, biochemical, and physical assays are performed as quality controls both in process as well as on the final product to assure thorough characterization and lot-to-lot consistency for the product.[28]

The appropriate display of epitopes that elicit the production of anti-HBs is a key property associated with the excellent immunogenicity of yeast-derived hepatitis B vaccines. Most important among these are the *a* epitopes, which are conformational determinants present on all subtypes of HBV (*ayr, ayw, adr, adw*) and which map to amino acids 124 to 137 of the S polypeptide.[30] A monoclonal antibody (MAb) directed against one specific *a* epitope was shown capable of neutralizing the infectivity of HBV in chimpanzees.[33] Anti-*a* antibodies are the most abundant elicited by HBsAg[29,32] and their induction is associated with protection from disease caused by all subtypes.[6,29,32] Appropriate quantitative immunoassays using both anti-*a* MAbs as well as MAbs directed to other epitopes on HBsAg assure the consistent presence of high levels of key epitopes on yeast-derived HBsAg.[29,32]

The immunogenicity of yeast-derived hepatitis B vaccines can be demonstrated in several experimental species, including mice and subhuman primates. However, protective efficacy in chimpanzees is the most important preclinical immunological test. Acute infection in this species results in the appearance of HBsAg, antibodies to hepatitis B core antigen (anti-HBc), elevated levels of liver enzymes in serum, and histopathological changes typical of hepatitis in liver biopsies 6 to 18 weeks postinoculation with standardized HBV inocula.[33] Thus, this model offers an opportunity to test the efficacy of a hepatitis B vaccine by immunization, then challenge with a preparation of HBV known to cause such signs of infection consistently. If the animal remains free of such symptoms for 6 months following challenge, the vaccine is considered effective in preventing infection. This principle first was demonstrated for the plasma-derived hepatitis B vaccine of subtype *ad*, which was shown to protect chimpanzees against challenge with HBV of subtypes *ay* or *ad*. Chimpanzees vaccinated with three doses of yeast-derived hepatitis B vaccine at monthly intervals produced anti-HBs titers of 800 to 3000 mIU/ml.[28,29] Following challenge with HBV, no vaccinated animal showed any signs of viral infection throughout the period of observation, while all control animals developed signs of typical infection. As in the case of the plasma-derived vaccine, the yeast-derived vaccine (subtype *ad*) protected against challenges with HBV subtypes *ad* or *ay*, thus confirming the protective effect of anti-*a* antibodies. This test established yeast-derived HBsAg as a protective vaccine and, thus, a candidate for testing in clinical trials.

C. CLINICAL EVALUATION

The yeast-derived hepatitis B vaccines have been tested in more than 10,000 healthy volunteers of all ages who were seronegative for HBsAg, anti-HBs, and anti-HBc at the time of vaccination. Following the protocol developed for the plasma-derived vaccine, individuals were vaccinated at 0, 1, and 6 months, and sera drawn at 0, 1, 3, 6, and 7 to 9 months were tested for anti-HBs using the AUSAB® (Abbott Laboratories, North Chicago, IL) radioimmunoassay. The first stage of clinical testing for a vaccine is Phase I, in which safety and tolerability are assured in healthy volunteers and initial data provided on immunogenicity. Subsequently, Phase II trials are conducted, during which an optimal dose for each age group is selected and the immunogenicity of the vaccine firmly established. Finally, Phase III trials are conducted to demonstrate that immunization protects against clinical hepatitis B infection. The yeast-derived hepatitis vaccines generally have been tolerated very well.[34-38] Clinical complaints, such as fatigue, headache, malaise, and nausea, or those related to the injection site, have been mild and transient. Initial dose-ranging studies for Recombivax HB® evaluated 2.5-, 5-, 10-, and 20-μg doses of the vaccine in adults.[34,36,39] The 10- and 20-μg formulations elicited much higher geometric mean titers (GMTs) than the lower doses, even though the kinetics of the immune response and the percentage of seroconverting individuals were similar for all doses. The 10-μg dose chosen for most of the studies in adults induced higher GMTs of anti-HBs among younger adults (20 to 29 years) than among older individuals (>50 years). By comparative testing, the 5-μg dose initially was chosen for Recombivax HB® for children 10 years of age or younger;[34] this indication since has been revised to a 2.5-μg dose on the basis of comparable immunogenicity. A 5-μg dose is indicated for adolescents 11 to 19 years of age. In children as compared to adults, seroconversion is more rapid and frequent, and GMTs are invariably much higher even at lower dosage levels of vaccine. For Engerix B®, a 20-μg dose of vaccine has been licensed for adults and a 10-μg dose for children based on dosages needed for consistent immune responses.

The introduction of a second, more productive Master Seed and improvements made to the original purification process for Recombivax HB® resulted in a higher yielding production process and a more highly purified vaccine. As a consequence of these changes, the vaccine was shown to elicit peak GMTs (approximately 1850 mIU/ml in young adults) that were significantly higher than those elicited by the original product.[29,35]

Since the yeast-derived and plasma-derived vaccines have some biochemical differences, it was important to compare the properties of vaccine-induced antibodies to assure immunological equivalence.[29,32] Both vaccines elicit antibodies that (1) have similar kinetics of appearance following inoculation, (2) can be absorbed totally by either plasma or yeast-derived vaccine antigens, (3) are capable of binding with equivalent titers and affinities to a synthetic peptide containing an *a* epitope, (4) are capable of binding to plasma-derived HBsAg with equivalent avidities, (5) have similar avidity and immunoglobulin isotype, and (6) protect chimpanzees against challenge with HBV (subtype *ad* or *ay*). By these criteria, the yeast-derived vaccines are immunologically equivalent to the plasma-derived vaccine.

The ultimate proof of the protective efficacy of the yeast-derived vaccine lay in demonstrating its effectiveness in preventing clinical infection. Efficacy rates of approximately 85 to 90% were demonstrated in clinical trials of the plasma-derived vaccine.[8,9] Even though the yeast-derived vaccine is immunologically comparable to the plasma-derived vaccine, it was deemed critical, in light of the intended worldwide use of this novel vaccine, to demonstrate directly the protective effect in man of anti-HBs elicited by the yeast-derived vaccine in controlled clinical trials. Due to the changing epidemiology of hepatitis B in various high-risk adult populations, it was decided to conduct efficacy studies in newborn children born to chronic hepatitis B carrier mothers.[40] It had been shown previously that infants born to viremic carrier mothers positive for HBsAg and HBeAg (a marker of HBV replication) had a 70 to 90% rate of infection and a very high risk of becoming chronic carriers themselves.[41] In these infants, passive immunization at birth with one dose of hepatitis B immune globulin (HBIG) reduced the rate of infection by half.[42] Combined active immunization with the plasma-derived hepatitis B vaccine plus passive immunization with HBIG reduced the rate of infection even further.[43,44] In lieu of a placebo control group, which would have been unethical, it was appropriate to calculate the efficacy rate of the recombinant-derived vaccine on the basis of historical control data.

Among infants receiving HBIG at birth and 5-μg doses of Recombivax HB® at 0, 1, and 6 months, only 5% became chronic carriers. This 95% rate of protection from the chronic carrier state was comparable to a control group of infants in the same study receiving HBIG and three doses of the plasma-derived vaccine.[40] This study demonstrated that anti-HBs elicited by Recombivax HB® is as effective in preventing viral infection as anti-HBs elicited by the plasma-derived vaccine and is consistent with the immunological equivalence of the vaccines in terms of the *in vivo* properties of the anti-HBs antibodies elicited by each vaccine. On the basis of this study, in concert with many other studies demonstrating safety/immunogenicity, Recombivax HB® was licensed in 1986 in West Germany and the U.S. as the first recombinant-derived vaccine of any type for humans. Subsequently, both Recombivax HB® and Engerix B® have been licensed in numerous countries worldwide, and Bimmugen® has been licensed in Japan.

III. POTENTIAL THIRD-GENERATION VACCINES

A. CLINICAL TARGETS

It has been shown clearly that subunit S polypeptide vaccines (plasma or recombinant derived) are well tolerated, highly immunogenic, and capable of inducing a protective immune response in the vast majority of individuals.[45] However, in a few percent of the general population and in certain specific populations, the antibody response to these vaccines is either absent or lower than protective levels (<10 mIU/ml). These populations, which represent targets for improved third-generation hepatitis B vaccines, include those with certain HLA types,[46-50] older individuals,[51-53] hemodialysis and renal allograft recipients,[9,54-56] institutionalized patients,[57] and those with certain blood disorders.[58-62] These reflect diverse genetic, immunological, and pathological factors that result in suboptimal responses to currently available vaccines. Thus, a third-generation vaccine candidate may result in enhanced immunogenicity in some but not all of the target populations. This possibility has widened the search for improved vaccines to include not only different polypeptide sequences but also vaccines produced in a variety of host recombinant expression systems that might alter immunogenicity due to unknown parameters of their posttranslational modification and protein folding. In addition, the possibly unrelated nature of the immune impairment in such target populations necessitates the simultaneous evaluation of product candidates in multiple clinical trials. In the past, the complexity of vaccine development generally has restricted the search for modifications in existing vaccines to goals of an improved safety profile that is apparently unnecessary for the licensed hepatitis B vaccines. On the other hand, many therapeutic agents have been improved continually with respect to their efficacy, breadth of action, and safety profile. The advent of recombinant genetic engineering has put vaccine development on a potentially even footing with synthetic chemistry, in that molecular biologists now may pursue possible improvements of even highly efficacious vaccines, such as these for hepatitis B. The following sections review the current status of this research.

B. HBcAg VACCINES

Antibodies to HBcAg (anti-HBc) are elicited during acute hepatitis B and persist during chronic hepatitis B without resolution of the disease. Anti-HBc is transferred to the fetus via the placenta but does not prevent perinatal infection. However, HBcAg can induce cell-mediated immune (CMI) responses in laboratory animals and humans, whereas HBsAg is less effective in eliciting CMI responses.[63] Furthermore, immunization with HBcAg has been demonstrated to protect or modify HBV infectivity in chimpanzees.[64,65] A possible explanation for this observed protection has been provided by the recent observation that immunization with HBcAg or a synthetic peptide fragment of HBcAg can prime mice to make a secondary antibody response to HBsAg (Chapter 15, this volume). Recombinant HBcAg has been expressed successfully and purified by several groups.[66-68] However, HBcAg is a nucleic acid-binding protein and, as is the case for core particles isolated from the livers of infected individuals, the recombinant particles also contain nucleic acids.[69] This nucleic acid level is well above that deemed acceptable according to guidelines for recombinant vaccines[70] and thus would represent a potential technical obstacle to be overcome before the initiation of clinical trials. In addition, current work in vaccine development has been deemphasized, apparently due to the unwillingness of the medical community to lose the anti-HBc marker for viral infection, the lack of a clear immunological readout of HBcAg-induced efficacy, and the hoped-for success of other approaches to improved vaccines (such as the inclusion of preS sequences). Depending on the results of these clinical evaluations, inclusion of HBcAg along with HBsAg may be reconsidered in vaccines for nonresponders.

C. PreS-CONTAINING POLYPEPTIDE VACCINES

Numerous lines of clinical and experimental evidence have been accumulating that implicate the preS domains in the infectious cycle of HBV and of anti-preS in successful resolution of infection. Several observations offer hope that the immune response to a preS-containing vaccine formulation would abrogate or prevent infectivity, especially in those individuals who are refractory to immunization with S polypeptide vaccines. These observations are as follows: (1) PreS-specific helper T cells from infected humans stimulate *in vitro* production of anti-HBs,[71] and immunization with preS+S of some inbred mouse strains, which show poor immune responses to the S-domain alone, overcomes this block (Chapter 15, this volume); (2)

during infection with HBV, CMI responses to preS antigens[72-74] and anti-preS[75-79] arise early. This early response to preS may serve a role in elimination of the virus, as retrospective analysis shows that anti-preS is generally absent from sera of patients with acute HBV infection who are entering the chronic phase or in the sera of chronic carriers;[80-82] (3) anti-preS responses either during infection[80] or following vaccination of humans[83] are independent and dissociated from anti-HBs responses; and (4) the theoretical hepatocyte binding site of HBV is located in the preS region,[84,85] anti-preS is virus neutralizing,[87] and immunization with preS peptide antigens protects chimpanzees against challenge with HBV in the absence of anti-HBs.[88,89] Therefore, preS-containing vaccines are being evaluated or developed as either recombinant live viruses or bacteria or as recombinant subunits expressed and purified from mammalian or insect cells or *S. cerevisiae*.

1. Live Recombinant Viruses or Bacteria

Several recombinant vaccinia viruses have been constructed that direct the synthesis of S,[90,91] preS2+S,[92] or preS1+preS2+S[93] polypeptides in mammalian cells. The S and preS2+S polypeptides are processed and secreted as 22-nm particles, whereas preS1+preS2+S polypeptides are not secreted or processed into 22-nm particles (as is the case for the other mammalian, insect, and yeast expression systems discussed below). However, as for insect- or yeast-derived preS1+preS2+S, immunization of rabbits with recombinant vaccinia virus expressing preS1+preS2+S did elicit anti-HBs.[93] Immunization of chimpanzees with S-expressing vaccinia virus did not induce high titers of anti-HBs but still resulted in active protection against HBV challenge, perhaps through an anamnestic antibody or CMI response. However, these recombinant vaccinia constructions generally use vaccinia strain WR, which presumably would not be acceptable as a live human vaccine because of its high neurovirulence;[94] studies employing a better tolerated recombinant vaccinia virus strain are underway.[95] Even if a safe and efficacious live recombinant vaccinia virus expressing HBV antigens is developed as a potential hepatitis B vaccine, the question of benefit to risk will always remain. This must be evaluated by weighing the yet to be determined advantages of such live virus vaccines against the availability of highly efficacious and very well-tolerated S polypeptide vaccines, the potential of preS-containing vaccines, and the lack of an active vaccinia immunization program following the worldwide irradication of smallpox.

Live adenovirus strains 4 and 7 have been administered to more than 1 million military recruits without significant adverse reactions,[96] and recombinant vectors based on them have been developed that express either S[97] or preS2+S[98] polypeptides. Such recombinant viruses elicited appropriate antibodies in laboratory animals, and one of two chimpanzees administered this candidate vaccine was completely protected from disease.[99] Safe, adenovirus vaccines are not employed currently in immunization programs for the general public, and it is likely that worldwide use and acceptance of a recombinant adenovirus would be driven primarily by the need for a vaccine for the other pathogen, e.g., HBV. Thus, as for recombinant live vaccinia viruses, the potential risks and benefits must be considered against available hepatitis B vaccines.

Recombinant vectors from the herpesvirus family also are being developed for the expression of foreign antigens as live vaccine candidates. A live recombinant herpes simplex virus type-1 (HSV-1) vector secretes HBsAg (in Vero cells) that is highly similar to "authentic" HBsAg isolated from the plasma or livers of infected individuals in both its buoyant density in CsCl and electron microscopic appearance.[100] The suitability of such an HSV-1 vector for human administration must await the accumulation of more preclinical safety data, since live attenuated HSV-1 strains have not been administered to man. Although not yet evaluated for expressing HBV polypeptides, a second herpesvirus (the Oka strain of varicella-zoster virus

[VZV]) vector has been developed and demonstrated successfully to express other viral glycoproteins.[101] As VZV(Oka) currently is approved or in extensive clinical trials for pediatric immunization, a recombinant VZV(Oka), if developed to express HBV antigens, might be useful for vaccination.

Genetically engineered bacteria represent yet another live vaccine technology. Nucleotide sequences encoding peptides derived from the S or preS2 domains have been inserted into the flagellin gene of live attenuated *Salmonella dublin* and the resultant recombinant bacteria demonstrated to express appropriate fusion proteins.[102] Laboratory animals infected with such recombinant *Salmonella dublin* produced anti-HBs and anti-preS2. If such recombinant bacteria are sufficiently immunogenic and if they elicit CMI responses or mucosal immunities more efficiently than recombinant subunit vaccines, they may evolve into attractive alternative vaccine strategies for HBV.

2. Subunit Polypeptides Containing PreS Domains

Subunit polypeptides that contain preS together with S domains are being evaluated from eukaryotic expression systems including mammalian cells, insect cells, and *S. cerevisiae*. As discussed above, live vaccinia, adenovirus, and herpesvirus vectors have been employed successfully to infect tissue culture cells and direct the synthesis and secretion of 22-nm particles containing the preS2+S polypeptide.[93,98] PreS1+preS2+S polypeptides expressed by these vectors have not been secreted into the culture medium of infected cells.[94] Therefore, the media of live recombinant virus-infected tissue culture cells can serve as a source for a purified subunit polypeptide vaccine, and the cells themselves could be processed to purify preS1+preS2+S. However, these are transient expression systems with inherent commercial and safety concerns that make them less attractive than nonmalignant CCLs, baculovirus-infected insect cells, or recombinant yeast as a source for a subunit vaccine antigen. Although CCLs derived from hepatocellular carcinomas have been established that secrete the subunit S polypeptide into the media[103,104] and that have been considered for vaccine production, these cells (due presumably to the random and fragmented nature of the virus genome integration) have not been reported to express preS-containing polypeptides in appreciable quantities. In addition, there are safety concerns regarding the use of a frankly malignant cell substrate for producing a vaccine.

Among CCLs, Chinese hamster ovary (CHO) cells have been used most widely as cell substrates for candidate hepatitis B vaccines. These cell lines express and secrete particles containing preS2+S polypeptides[105-108] and, in one case, small amounts of preS1+preS2+S.[109] The lines producing preS2+S particles can be grown at commercial scale, and the antigen has been purified and evaluated in human clinical trials. Such vaccines were capable of eliciting anti-HBs titers comparable to those achieved by the currently licensed plasma or recombinant yeast vaccines. In addition, anti-preS2 was detected in approximately 90% of the recipients after the third 20-μg immunization.[107,108] One such product (GenHevac-B; Pasteur) was licensed for use in France in 1989. Whether such a vaccine will confer an advantage over the licensed S polypeptide vaccines or over the subunit preS2+S or preS1+preS2+S polypeptide vaccines expressed in insect cells or in yeast awaits the results of comparative clinical trials in appropriate target populations.

The baculovirus expression vector system has been used to direct the expression of the S, preS2+S, and the preS1+preS2+S polypeptides. The S polypeptides were secreted as typical 22-nm particles highly similar to those prepared from plasma or other recombinant systems.[19] The preS2+S polypeptides also were secreted in particulate form, but, interestingly, S polypeptides were not secreted concomitantly.[110] As in mammalian systems, preS1+preS2+S polypeptides also were not secreted when expressed in insect cells, nor did they form particles.[111] However, the preS1+preS2+S polypeptides could be purified as detergent micelles capable of eliciting anti-HBs in mice.[111] These vaccine candidates await further development.

Saccharomyces cerevisiae is a versatile choice for the expression of potential third-generation vaccines composed of polypeptides containing preS as well as S domains. The *GAP* promoter system successfully used in development of S polypeptide vaccines also has been exploited successfully to direct the expression of preS2+S polypeptides.[112-119] However, unlike S polypeptides, the preS2+S polypeptides have an apparent deleterious effect on yeast. Constitutive expression results in decreased growth rates and expression plasmid loss, especially for those cells in the culture making the highest levels of antigen. For this reason, some investigators have sought to develop physiologically regulated expression systems for the synthesis of preS2+S in yeast that would be established more easily as stable production seed stocks and fermented more consistently at commercial scale.[118,119] The preS2+S antigens expressed by either the constitutive or regulated systems described above are an approximately 30-kDa polypeptide and its approximately 34-kDa glycosylated derivative, which assemble into 22-nm particles similar to the plasma- or yeast-derived subunit S polypeptides (Figure 1). These preS2+S particles display surface epitopes specific for preS2 as well as the *a* epitope of the S domain and are highly immunogenic in primates, in which they elicit high titers of anti-HBs as well as anti-preS2.[114] The preS1+preS2+S polypeptide is even more toxic for yeast cells than is the preS2+S polypeptide and therefore also requires regulated expression.[120] The purified preS1+preS2+S antigens are composed of an approximately 39-kDa polypeptide and its glycosylated approximately 45-kDa derivative.[117,120] As has been previously discussed for the preS1+preS2+S antigens expressed in either mammalian or insect cells, the purified yeast-expressed antigen consists predominantly of small particles (2 to 3 nm) and a polydisperse population of loosely assembled aggregates (15 to 30 nm), with only small numbers of 22-nm particles.[120] The purified product was immunogenic in laboratory animals, and a significant antibody response occurred after only two immunizations; the response was directed not only to the preS1 and preS2 domains but also to the S domain.[120] However, the anti-HBs response was lower than that typically elicited by either recombinant- or plasma-derived antigens, which contain the S polypeptide alone, or to recombinant preS2+S expressed in yeast.

IV. CONCLUSIONS

The fact that recombinant hepatitis B vaccines were the first recombinant vaccines approved for human use derives primarily from the prior experience with a natural product vaccine (Heptavax-B®) purified from the plasma of chronic carriers. However, it is obvious in retrospect, due to the relative structural simplicity of HBV and the propensity of its major envelope proteins to assume conformations reflective of protective epitopes on the virion, that this virus could serve as the archetype with which the scientific and commercial communities could develop their recombinant vaccine technologies. The subsequent approval of the yeast-derived subunit vaccines added a recombinant prototype, with a clearly proven efficacy and a defined analytical, antigenic, and safety profile, against which subsequent hepatitis B vaccines can and would be measured. In addition, the research on all the hepatitis B vaccines and vaccine candidates, which involves dozens of laboratories and more than two decades of effort, has begun to erect a developmental scaffolding on which other vaccines of the future can be assembled.

REFERENCES

1. **Beasley, R. P. and Hwang, L.-Y.,** Epidemiology of hepatocellular carcinoma, in *Viral Hepatitis and Liver Disease*, Vyas, G. N., Dienstag, J. L., and Hoofnagle, J. H., Eds., Grune & Stratton, Orlando, Florida, 1984, 209.
2. **Szmuness, W.,** Hepatocellular carcinoma and the hepatitis B virus: evidence for a causal association. *Prog. Med. Vir.*, 24, 40, 1978.
3. **Hilleman, M. R., McAleer, W. J., Buynak, E. B., and McLean, A. A.,** The preparation and safety of hepatitis B vaccine. *J. Infect.*, 7, 3, 1983.
4. **Tabor, E., Buynak, E. B., Smallwood, L. A., Snoy, P., Hilleman, M. R., and Gerety, R. J.,** Inactivation of hepatitis B virus by three methods: treatment with pepsin, urea, or formalin. *J. Med. Virol.*, 11, 1, 1983.
5. **Francis, D. P., Feorino, P. M., McDougal, S., Warfield, D., Getchell, J., Cabradilla, C., Tong, M., Miller, W. J., Schultz, L. D., Bailey, F. J., McAleer, W. J., Scolnick, E. M., and Ellis, R. W.,** The safety of the hepatitis B vaccine, *JAMA*, 256, 869, 1986.
6. **Hollinger, F. B., Dressman, G. R., Sanchez, Y., Cabral, G. A., and Melnick, J. L.,** Experimental hepatitis B polypeptide vaccine in chimpanzees, in *Viral Hepatitis*, Vyas, G. N., Cohen, S. N., and Schmid, R., Eds., Franklin Institute Press, Philadelphia, 1978, 557.
7. **Centers for Disease Control,** Update on hepatitis B prevention, *Morbid. Mortal. Wkly. Rep.*, 36, 353, 1987.
8. **Francis, D. P., Hadler, S. C., Thompson, S. E., Maynard, J. E., Ostrow, D. G., Altman, N., Braff, E. H., O'Malley, P., Hawkins, D., Judson, F. N., Penley, K., Nylund, T., Christie, G., Mayers, F., Moore, J. N., Gardner, A., Doto, I. L., Miller, J. H., Reynolds, G. H., Murphy, B. L., Schable, C. A., Clark, B. T., Curran, J. W., and Redeker, A. G.,** The prevention of hepatitis B with vaccine, *Ann. Intern. Med.*, 97, 362, 1984.
9. **Stevens, C. E., Alter, H. J., Taylor, P. E., Zaug, E. A., Harley, E. J., and Szmuness, W.,** Hepatitis B vaccine in patients receiving hemodialysis: immunogenicity and efficacy, *N. Engl. J. Med.*, 311, 496, 1984.
10. **Valenzuela, P., Gray, P., Quiroga, M., Zaldivar, J., Goodman, H. M., and Rutter, W. J.,** Nucleotide sequence of the gene coding for the major protein of hepatitis B virus surface antigen, *Nature (London)*, 280, 815, 1979.
11. **Valenzuela, P., Medina, A., Rutter, W. J., Ammerer, G., and Hall, B. D.,** Synthesis and assembly of hepatitis B virus surface antigen particles in yeast, *Nature (London)*, 298, 347, 1982.
12. **Hitzemann, R. A., Chen, C. Y., Hagie, F. E., Patzer, E. J., Liu, C. C., Estell, A. A., Miller, J. V., Yaffee, A., Kleid, D. G., Levinson, A. D., and Oppermann, H.,** Expression of hepatitis B surface antigen in yeast, *Nucleic Acids Res.*, 11, 2475, 1983.
13. **Miyatsu, Y., Adachi, S., Shiosaki, K., Sugawara, T., and Mizogani, H.,** Development of recombinant hepatitis B vaccine: seed lot system for production of recombinant hepatitis vaccine, *Clin. Rep. Jpn.*, 21, 145, 1987.
14. **Harford, N., Cabezon, T., Colau, B., Delisse, A.-M., Rutgers, T., and DeWilde, M.,** Construction and characterization of a *Saccharomyces cerevisiae* strain (RIT4376) expressing hepatitis B surface antigen, *Postgrad. Med. J.*, 63, 65, 1987.
15. **Moriarty, A. M., Hoyer, B. H., Shih, W.-K., Gerin, J. L., and Hamer, D. H.,** Expression of the hepatitis B virus surface antigen gene in cell culture by using a simian virus 40 vector, *Proc. Natl. Acad. Sci. U.S.A.*, 78, 2606, 1981.
16. **Liu, C.-C., Yansura, D., and Levinson, A.,** Direct expression of hepatitis B surface antigen in monkey cells from an SV40 vector, *DNA*, 1, 213, 1982.
17. **Hsiung, N., Fitts, R., Wilson, S., Milne, A., and Hamer, D.,** Efficient production of hepatitis B surface antigen using a bovine papilloma virus-metallothionein vector, *J. Mol. Appl. Genet.*, 2, 497, 1984.
18. **Gregg, J. M., Tschopp, J. F., Stillman, C., Siegel, R., Akong, M., Craig, W. S., Buckholz, R. G., Madden, K. R., Kellaris, P. A., Davis, G. R., Smiley, B. L., Cruze, J., Torregrossa, R., Velicelebi, R., and Thill, G. P.,** High-level expression and efficient assembly of hepatitis B surface antigen in the methylotropic yeast, *Pichia pastoris*, *Biotechnology*, 5, 479, 1987.
19. **Kang, C. Y., Bishop, D. H. L., Seo, J.-S., Matsuura, Y., and Choe, M.,** Secretion of particles of hepatitis B surface antigen from insect cells using a baculovirus vector, *J. Gen. Virol.*, 68, 2607, 1987.
20. **Heerman, K. H., Goldman, V., Schwartz, W., Seyffarth, T., Baumgarten, H., and Gerlich, W. H.,** Large surface proteins of hepatitis B virus containing the pre-S sequence, *J. Virol.*, 52, 396, 1984.
21. **Peterson, D. L.,** Isolation and characterization of the major protein and glycoprotein of hepatitis B surface antigen, *J. Biol. Chem.*, 256, 6975, 1981.
22. **Ellis, R. W.,** Vaccines, diagnostic proteins and hormones, in *Biotechnology, A Comprehensive Treatise: Gene Technology*, Vol. 76, Rehm, H. J. and Reed, G., Eds., VCH Publishers, Weinheim, Germany, 1989, 167.

23. **Temin, H.,** Overview of biologist effects of addition of DNA molecules to cells, *J. Med. Virol.*, 31, 13, 1990.
24. **Ellis, R. W. and Gerety, R. J.,** Key issues in the selection of an expression system for vaccine antigens, *J. Med. Virol.*, 31, 54, 1990.
25. **Hilleman, M. R. and Ellis, R. W.,** Vaccines made from recombinant yeast cells, *Vaccine*, 4, 75, 1986.
26. **Jayaram, M., Li, Y.-Y., and Broach, J. R.,** The yeast plasmid in 2μ circle encodes components required for its high-copy propagation, *Cell*, 34, 95, 1983.
27. **Holland, J. P. and Holland, M. J.,** Structural expression of two nontandemly repeated yeast glyceraldehyde-3-phosphate dehydrogenase genes, *J. Biol. Chem.*, 255, 2596, 1980.
28. **Ellis, R. W.,** Recombinant-derived hepatitis B vaccine — a paradigm for other subunit vaccines, in *AIDS Vaccine: Research and Clinical Trials*, Putney, S. and Bolognesi, D., Eds., Marcel Dekker, New York, 1990, 381.
29. **Emini, E. A., Ellis, R. W., Miller, W. J., McAleer, W. J., Scolnick, E. M., and Gerety, R. J.,** Production and immunological analysis of recombinant hepatitis B vaccine, *J. Infect.*, 13, 3, 1986.
30. **Waters, J. A., O'Rourke, S. M., Richardson, S. C., Papaevangelou, G., and Thomas, H. C.,** Qualitative analysis of the humoral immune response to the *a* determinant of HBs antigen after inoculation with plasma-derived or recombinant vaccine, *J. Med. Virol.*, 21, 155, 1987.
31. **Iwarson, S., Tabor, E., Thomas, H. C., Godall, A., Waters, J., Snoy, P., Shih, J. W.-K., and Gerety, R.,** Neutralization of hepatitis B virus infectivity by a murine monoclonal antibody: an experimental study in the chimpanzee, *J. Med. Virol.*, 16, 89, 1985.
32. **Hauser, P., Voet, P., Simoen, E., Thomas, H. C., Petre, J., DeWilde, M., and Stephenne, J.,** Immunological properties of recombinant HBsAg produced in yeast, *Postgrad. Med. J.*, 63, 83, 1987.
33. **Tabor, E., Purcell, R. H., London, W. T., and Gerety, R. J.,** Use of and interpretation of results using inocula of hepatitis B virus with known infectivity titers, *J. Infect. Dis.*, 147, 531, 1983.
34. **Zajac, B. A., West, D. J., McAleer, W. J., and Scolnick, E. M.,** Overview of clinical studies with hepatitis B vaccine made by recombinant DNA, *J. Infect.*, 13, 39, 1986.
35. **West, D. J., Zajac, B. A., Ellis, R. W., and Gerety, R. J.,** Improved immunogenicity of a yeast-derived hepatitis B vaccine, *Infektionsklinik*, 1, 1, 1988.
36. **Hollinger, F. B., Troisi, C. L., and Pepe, P. E.,** Anti-HBs responses to vaccination with a human hepatitis B vaccine made by recombinant DNA technology in yeast, *J. Infect. Dis.*, 153, 156, 1986.
37. **André, F. E. and Safary, A.,** Summary of clinical findings on Engerix-B, a genetically engineered yeast-derived hepatitis B vaccine, *Postgrad. Med. J.*, 63, 169, 1987.
38. **Yano, M.,** Recombinant yeast-derived hepatitis B vaccine: phase III clinical study, *Clin. Rep.*, 21, 259, 1987.
39. **Morton, D. and Krugman, S.,** Recombinant yeast hepatitis B vaccine compared with plasma-derived vaccine: immunogenicity and effect of a booster dose, *J. Infect.*, 13, 31, 1986.
40. **Stevens, C. E., Taylor, P. E., Tong, M. J., Toy, P. T., Vyas, G. N., and Krugman, S.,** Yeast-recombinant hepatitis B vaccine: efficacy with hepatitis B immune globulin in prevention of perinatal hepatitis B virus transmission, *JAMA*, 257, 2612, 1987.
41. **Okada, K., Kamiyama, I., Inomata, M., Imai, M., Miyakura, Y., and Mayumi, M.,** *e* antigen and anti-*e* in the serum of asymptomatic carrier mothers as indicators of positive and negative transmission of hepatitis B virus to their infants, *N. Engl. J. Med.*, 294, 746, 1976.
42. **Beasley, R. P., Hwang, L.-Y., Stevens, C. E., Lin, C.-C., Hsieh, F.-J., Wang, K.-Y., Sun, T.-S., and Szmuness, W.,** Efficacy of hepatitis B immune globulin for prevention of perinatal transmission of the hepatitis B carrier state: final report of a randomized double-blind, placebo-controlled trial, *Hepatology*, 3, 135, 1983.
43. **Beasley, R. P., Hwang, L. Y., Lee, G. C.-Y., Lau, C. C., Roan, C. H., Huang, F., and Chen, C. L.,** Prevention of perinatally transmitted hepatitis B virus infections with hepatitis B immune globulin and hepatitis B vaccine, *Lancet*, ii, 1099, 1983.
44. **Stevens, C. E., Toy, P. T., Tong, M. J., Taylor, P. E., Vyas, G. N., Nair, P. V., Gudavalo, M., and Krugman, S.,** Perinatal hepatitis B virus transmission in the United States: prevention by passive-active immunization, *JAMA*, 253, 1740, 1985.
45. **West, D. J.,** Clinical experience with hepatitis B vaccines, *Am. J. Infect. Control*, 17, 172, 1989.
46. **Craven, D. E., Awdel, Z. L., Kunches, L. M. et al.,** Non-responsiveness to hepatitis B vaccine in health care workers, *Ann. Intern. Med.*, 105, 356, 1986.
47. **Jilg, W., Schmidt, M., and Deinhardt, F.,** Immune response to hepatitis B revaccination, *J. Med. Virol.*, 24, 377, 1988.
48. **Weissman, J. Y., Tsuchiyose, M. M., Tong, M. G., Co, R., Chin, K., and Ettenger, R. B.,** Lack of response to recombinant hepatitis B vaccine in nonresponders to the plasma vaccine, *JAMA*, 260, 1734, 1988.
49. **Krämer, A., Herth, D., von Keyserlingk, H.-J., Ludwig, W.-D., Hampl, H., Sommer, D., Hahn, E. G., and Riecken, E.-O.,** Non-responsiveness to hepatitis B vaccination: revaccination and immunogenetic typing, *Klin. Wochenschr.*, 66, 670, 1988.

50. **Alper, C. A., Kruskall, M. S., Marcus-Bagley, D., Craven, D. E., Katz, A. J., Brink, S. J., Dienstag, J. L., Awdeh, Z., and Yunis, E. J.,** Genetic preduction of nonresponse to hepatitis B vaccine, *N. Engl. J. Med.*, 231, 708, 1989.
51. **Denis, F., Mounier, M., Hessel, L., Michel, J. P., Gualde, N., Dubois, F., Barin, F., and Goudeau, A.,** Hepatitis B vaccination in the elderly, *J. Infect. Dis.*, 149, 1019, 1984.
52. **Cook, J. M., Gualde, N., Hessel, L., Mounier, M., Michel, J. P., Denis, F., and Ratinaud, M. H.,** Alterations in the human immune response to the hepatitis B vaccine among the elderly, *Cell. Immunol.*, 109, 89, 1987.
53. **Chiaramonte, M., Floreani, A., and Naccarato, R.,** Dossier: gerontology, *Biomed. Pharmacol.*, 41, 121, 1987.
54. **Grob, P. J., Binswanger, U., Zaruba, K., Joller-Jemelka, H. I., Schmid, M., Hacki, W., Blumburg, A., Abplanalp, A., Herwig, W., Iselin, H., and Descoeudres, C.,** Immunogenicity of a hepatitis subunit vaccine in hemodialysis and in renal transplant recipients, *Antiviral Res.*, 3, 43, 1983.
55. **Jilg, W., Schmidt, M., Weinel, B., Küttler, T., Brass, H., Bommer, J., Müller, R., Schulte, B., Schwarzbeck, A., and Deinhardt, F.,** Immunogenicity of recombinant hepatitis B vaccine in dialysis patients, *J. Hepatol.*, 3, 190, 1986.
56. **Bruguera, M., Cremades, M., Mayor, A., Sánchez Tapias, J. M., and Rodés, J.,** Immunogenicity of a recombinant hepatitis B vaccine in haemodialysis patients, *Postgrad. Med. J.*, 63, 155, 1987.
57. **Heijtink, R. A., De Jong, P., Schalm, S. W., and Masurel, N.,** Hepatitis B vaccination in Down's syndrome and other mentally retarded patients, *Hepatology*, 4, 611, 1984.
58. **Mannucci, P. M., Gringeri, A., Morfini, M., De Biasi, Tirindellim, M. C., Tamponi, G., Baudo, F., Carnelli, V., Ciavarella, N., Colombo, M., and Zanetti, A.,** Immunogenicity of a recombinant hepatitis B vaccine in hemophiliacs, *Am. J. Hepatol.*, 29, 211, 1988.
59. **Sarnaik, S. A., Merline, J. R., and Bond, S.,** Immunogenicity of hepatitis vaccine in children with sickle cell anemia, *Clin. Lab. Obs.*, 112, 429, 1988.
60. **Miller, S. T., Jensen, D., and Sreedhar, P. R.,** Hepatitis vaccine in sickle-cell anemia, *J. Pediatr.*, 113, 955, 1988.
61. **Onwubalili, J. K.,** Sickle cell disease and infection, *J. Infect.*, 7, 2, 1983.
62. **Giammanco, G., De Grandi, V., Pignato, S., Dolci, C., Congia, M., Pisu, C., and De Virgiliis, S.,** Yeast-derived hepatitis B vaccine in thalassaemic patients: a preliminary report, *Postgrad. Med. J.*, 63, 151, 1987.
63. **Gerety, R. J., Hoofnagle, J. H., and Barker, L. F.,** Humoral and cell-mediated immune responses to two hepatitis B virus antigens in guinea pigs, *J. Immunol.*, 113, 1223, 1974.
64. **Murray, K., Bruce, S. A., Hinnen, A., Wingfield, P., van Erd, P. M. C. A., de Reus, A., and Schellekens, H.,** Hepatitis B virus antigens made in microbial cells immunise against viral infection, *EMBO J.*, 3, 645, 1984.
65. **Iwarson, S., Tabor, E., Thomas, H. C., Snoy, P., and Gerety, R. J.,** Protection against hepatitis B virus infection by immunization with hepatitis core antigen, *Gastroenterology*, 88, 763, 1985.
66. **Pasek, M., Goto, T., Gilbert, W., MacKay, P., Leadbetter, G., and Murray, K.,** Hepatitis B virus genes and their expression in *E. coli*, *Nature* (London), 282, 575, 1979.
67. **Kniskern, P. J., Hagopian, A., Montgomery, D. L., Burke, P., Dunn, N. R., Hofmann, K. J., Miller, W. J., and Ellis, R. W.,** Unusually high-level expression of a foreign gene (hepatitis B virus core antigen) in *Saccharomyces cerevisiae*, *Gene*, 46, 135, 1986.
68. **Roossinck, M. J., Jameel, S., Loukin, S. H., and Siddiqui, A.,** Expression of hepatitis B viral core region in mammalian cells, *Mol. Cell. Biol.*, 6, 1393, 1986.
69. **Petit, M.-A. and Pillot, J.,** HBc and HBe antigenicity and DNA-binding activity of major core protein P22 in hepatitis B virus core particles isolated from the cytoplasm of human liver cells, *J. Virol.*, 53, 543, 1985.
70. **Anon.,** Points to Consider in the Production and Testing of New Drugs and Biologicals Produced by Recombinant DNA Technology, Office of Biologics Research and Review, Center for Drugs and Biologics, Food and Drug Administration, Washington, D.C., April 10, 1985.
71. **Cupps, T. R., Hoofnagle, J. H., Ellis, R. W., Miller, W. J., Seeff, L., Guerrera, A., Gerin, J. L., and Haas-Smith, S. A.,** *In vitro* immune responses to hepatitis B surface antigen (pre-S2 and S) following remote infection by hepatitis B virus in humans, *J. Clin. Immunol.*, 9, 229, 1989.
72. **Vento, S., Chen, S.-H., Giuliani-Piccardi, G., O'Brien, C. J., Dal Monte, P. R., Eddleston, A. L. W. F., Howard, C. R., and Williams, R.,** Cellular immunity to nucleocapsid and pre-S determinants in asymptomatic carriers of hepatitis B virus, *Immunology*, 62, 593, 1987.
73. **Ferrari, C., Penna, A., DegliAntoni, A., and Ficcadori, F.,** Cellular immune response to hepatitis B virus antigens, *J. Hepatol.*, 7, 21, 1988.
74. **Vento, S., Ranier, S., Williams, R., Rondanelli, E. G., O'Brien, C. J., and Eddleston, A. L. W. F.,** Prospective study of cellular immunity to hepatitis B virus antigens from the early incubation phase of acute hepatitis B, *Lancet*, 21, 119, 1987.

75. **Alberti, A., Pontisso, P., and Schiavon, E.,** An antibody which precipitates Dane particles in acute hepatitis type B: relation to receptor sites which bind polymerized human serum albumin on virus particles, *Hepatology*, 4, 220, 1984.
76. **Budkowska, A., Dubreuil, P., Capel, F., and Pillot, J.,** Hepatitis B virus pre-S gene-encoded antigenic specificity and anti-pre-S antibody: relationship between anti-pre-S response and recovery, *Hepatology*, 6, 360, 1986.
77. **Okamoto, H., Usuda, S., Imai, M., Tachibana, K., Tanaka, E., Kumakura, T., Itabashi, M., Takai, E., Tsuda, F., Nakamura, T., Miyakawa, Y., and Mayumi, M.,** Antibody to the receptor for polymerized human serum albumin in acute and persistent infection with hepatitis B virus, *Hepatology*, 6, 354, 1986.
78. **Coursegaet, P., Adamowicz, P., Bourdil, C., Yvonnet, B., Buisson, Y., Barrès, Saliou, P., Chiron, J.-P., and Diop Mar, I.,** Anti-pre-S2 antibodies in natural hepatitis B virus infection and after immunization, *Vaccine*, 6, 357, 1988.
79. **Heijtink, R. A., de Wilde, G. A., van Hattum, J., and Schalm, S. W.,** Long-term immune reactivity to pre-S(2)-antigen after acute hepatitis B infection, *J. Med. Virol.*, 27, 95, 1989.
80. **Tsuda, F., Masuko, K., Mitsui, T., Akahane, H., Machida, A., Miyakawa, Y., and Mayumi, M.,** Antibody against the translation product of the pre-S2 region of hepatitis B virus and the resolution of infection after accidental needlestick exposure, *Am. J. Gastroenterol.*, 83, 943, 1988.
81. **Delfini, C., Colloca, S., Taliani, G., Mazzotta, F., D'Agata, A., Buonamici, C., Stroffolini, T., and Carloni, G.,** Clearance of hepatitis B virus DNA and pre-S surface antigens in patients with markers of acute viral replication, *J. Med. Virol.*, 28, 169, 1989.
82. **Theilmann, L., Klinkert, M.-Q., Gmelin, K., Kommerell, B., and Pfaff, E.,** Detection of antibodies against pre-S1 proteins in sera of patients with hepatitis B virus (HBV) infection, *J. Hepatol.*, 4, 22, 1987.
83. **Zanetti, A., Tanzi, E., and Mannucci, P. M.,** Dissociated antibody responses to the S and pre-S2 regions of the hepatitis B virus after vaccination in hemophiliacs, *J. Med. Virol.*, 28, 156, 1989.
84. **Neurath, A. R., Kent, S. B. H., Strick, N., and Parker, K.,** Identification and chemical synthesis of a host cell receptor binding site on hepatitis B virus, *Cell*, 4, 429, 1986.
85. **Petit, M.-A., Dubanchet, S., and Capel, F.,** A monoclonal antibody specific for the hepatocyte receptor binding site on hepatitis B virus, *Mol. Immunol.*, 26, 531, 1989.
86. **Pontisso, P., Petit, M.-A., Bankowski, M. J., and Peeples, M. E.,** Human liver plasma membranes contain receptors for the hepatitis B virus pre-S1 region and, via polymerized human serum albumin, for the pre-S2 region, *J. Virol.*, 63, 1981, 1989.
87. **Neurath, A. R., Seto, B., and Strick, N.,** Antibodies to synthetic peptides from the preS1 region of the hepatitis B virus (HBV) envelope (env) protein are virus-neutralizing and protective, *Vaccine*, 7, 234, 1989.
88. **Itoh, Y., Takai, E., Ohnuma, H., Kitajima, Tsuda, F., Machida, A., Mishiro, S., Nakamura, T., Miyakawa, Y., and Mayumi, M.,** A synthetic peptide vaccine involving the product of the pre-S(2) region of hepatitis B virus DNA: protective efficacy in chimpanzees, *Proc. Natl. Acad. Sci. U.S.A.*, 83, 9174, 1986.
89. **Emini, E. A., Larson, V., Eichberg, J., Conard, P., Garsky, V. M., Lee, D. R., Ellis, R. W., Miller, W. J., Anderson, C. A., and Gerety, R. J.,** Protective effect of a synthetic peptide comprising the complete PreS2 region of the hepatitis B virus surface protein, *J. Med. Virol.*, 28, 7, 1989.
90. **Perkus, M. E., Piccini, A., Lipinskas, B. R., and Paoletti, E.,** Recombinant vaccinia virus: immunization against multiple pathogens, *Science*, 229, 981, 1985.
91. **Chelyapov, N. V., Antonova, T. P., Yanova, N. N., and Chernos, V. I.,** Antigenic properties of vaccinia virus and of the virus recombinant strains expressing heterologous genes, *Acta Virol.*, 32, 409, 1988.
92. **Cheng, K.-C. and Moss, B.,** Selective synthesis and secretion of particles composed of the hepatitis B virus middle surface protein directed by a recombinant vaccinia virus: induction of antibodies to pre-S and S epitopes, *J. Virol.*, 61, 1286, 1987.
93. **Cheng, K.-C., Smith, G. L., and Moss, B.,** Hepatitis B virus large surface protein is not secreted but is immunogenic when selectively expressed by recombinant vaccinia virus, *J. Virol.*, 60, 337, 1986.
94. **Moss, B., Fuerst, T. R., Flexner, C., and Hugin, A.,** Roles of vaccinia virus in the development of new vaccines, *Vaccine*, 6, 161, 1988.
95. **Watanabe, K., Kobayashi, H., Kajiyama, K., Morita, M., Yasuda, A., Gotoh, H., Saeki, S., Sugimoto, M., Saito, H., and Kojima, A.,** Improved recombinant LC16m0 or LC16m8 vaccinia virus successfully expressing hepatitis B surface antigen, *Vaccine*, 7, 53, 1989.
96. **Takafuji, E. T., Gaydos, J. C., Allan, R. G., and Top, F. H., Jr.,** Simultaneous administration of live enteric-coated adenovirus types 4, 7 and 21 vaccines; safety and immunogenicity, *J. Infect. Dis.*, 140, 48, 1983.
97. **Morin, J. E., Lubeck, M. D., Barton, J. E., Conley, A. J., Davis, A. R., and Hung, P. P.,** Recombinant adenovirus induces antibody response to hepatitis B virus surface antigen in hamsters, *Proc. Natl. Acad. Sci. U.S.A.*, 84, 4626, 1987.

98. **Ballay, A., Levrero, M., Buendia, M.-A., Tiollais, P., and Perricaudet, M.,** *In vitro* and *in vivo* synthesis of the hepatitis B virus surface antigen and of the receptor for polymerized human serum albumin from recombinant human adenoviruses, *EMBO J.*, 4, 3861, 1985.
99. **Lubek, M. D., Davis, A. R., Chengalvala, M., Natuk, R. J., Morin, J. E., Molnar-Kimber, C., Mason, B. B., Bhat, B. M., Mizutani, S., Hung, P. P., and Purcell, R. H.,** Immunogenicity and efficacy testing in chimpanzees of an oral hepatitis B vaccine based on live recombinant adenovirus, *Proc. Natl. Acad. Sci. U.S.A*, 86, 6763, 1989.
100. **Shih, M.-F., Arsenakis, M., Tiollais, P., and Roizman, B.,** Expression of hepatitis B virus *S* gene by herpes simplex virus type-1 vectors carrying α and β-regulated gene chimeras, *Proc. Natl. Acad. Sci. U.S.A.*, 81, 5867, 1984.
101. **Lowe, R. S., Keller, P. M., Keech, B. J., Davison, A. J., Whang, Y., Morgan, A. J., Kieff, E., and Ellis, R. W.,** Varicella-zoster virus as a live vector for the expression of foreign genes, *Proc. Natl. Acad. Sci. U.S.A.*, 84, 3896, 1987.
102. **Wu, J. Y., Newton, S., Judd, A., Stocker, B., and Robinson, W. S.,** Expression of immunogenic epitopes of hepatitis B surface antigen with hybrid flagellin proteins by a vaccine strain of *Salmonella, Proc. Natl. Acad. Sci. U.S.A.*, 86, 4726, 1989.
103. **Alexander, J. J., Bey, E. M., Macnab, G., and Saunders, R.,** Studies on *in vitro* production of hepatitis B surface antigen by a human hepatoma cell line, in *Perspectives in Virology*, Vol. 10, Pollard, M., Ed., Raven Press, New York, 1978, 103.
104. **Aspinall, S., Alexander, J., and Bos, P.,** Comparative expression of hepatitis B virus antigens in several cell model systems, *J. Gen. Virol.*, 67, 2322, 1986.
105. **Michel, M.-L., Pontisso, P., Sobczak, E., Malpièce, Y., Sstreeck, R. E., and Tiollais, P.,** Synthesis in animal cells of hepatitis B surface antigen particles carrying a receptor for polymerized human serum albumin, *Proc. Natl. Acad. Sci. U.S.A.*, 81, 7708, 1984.
106. **Coursaget, P., Yvonnet, B., Anthonioz, P., Chotard, J., Bourdil, C., Adamowicz, P., Tron, F., and Girard, M.,** Immunogénicité d'un vaccine contre l'hépatite B obtenu par recombinaison génétique et contenant les produits des génes S et prè S_2, *Presse Médicale*, 17, 1150, 1988.
107. **Michel, M.-L., Sobczak, E., Malpièce, Y., Tiollais, P., and Streeck, R. E.,** Expression of amplified hepatitis B virus surface antigen genes in Chinese hamster ovary cells, *Biotechnology*, 3, 56, 1988.
108. **Tron, F., Degos, F., Bréchot, C., Courouсé, A.-M., Goudeau, A., Marie, F.-N., Adamowicz, P., Saliou, P., Laplanche, A., Benhamou, J.-P., and Girard, M.,** Randomized dose range study of a recombinant hepatitis B vaccine produced in mammalian cells and containing the S and PreS2 sequences, *J. Infect. Dis.*, 160, 199, 1989.
109. **Ou, J.-H. and Rutter, W. J.,** Regulation of secretion of the hepatitis B virus major surface antigen by the preS- protein, *J. Virol.*, 61, 782, 1987.
110. **Price, P. M., Mohamad, A., Zelent, A., Neurath, A. R., and Acs, G.,** Translational selection in the expression of the hepatitis B virus envelope proteins, *DNA*, 7, 417, 1988.
111. **Lanford, R. E., Luckow, V., Kennedy, R. C., Dreesman, G. R., Notvall, L., and Summers, M. D.,** Expression and characterization of hepatitis B virus surface antigen polypeptides in insect cells with a baculovirus expression system, *J. Virol.*, 63, 1549, 1989.
112. **Valenzuela, P., Coit, D., and Kuo, C. H.,** Synthesis and assembly in yeast of hepatitis B surface antigen particles containing the polyalbumin receptor, *Biotechnology*, 3, 317, 1985.
113. **Itoh, Y. and Fujisawa, Y.,** Synthesis in yeast of hepatitis B virus surface antigen modified P31 particles by gene modification, *Biochem. Biophys. Res. Commun.*, 141, 942, 1988.
114. **Ellis, R. W., Kniskern, P. J., Hagopian, A., Schultz, L. D., Montgomery, D. L., Maigetter, R. Z., Wampler, D. E., Emini, E. A., Wolanski, B., McAleer, W. J., Hurni, W. M., and Miller, W. J.,** Preparation and testing of a recombinant-derived hepatitis B vaccine consisting of pre-S2+S polypeptides, in *Viral Hepatitis and Liver Disease*, Zuckerman, A. J., Ed., Alan R. Liss, New York, 1988, 1079.
115. **Rutgers, T., Cabezon, T., Harford, N., Van der Brugge, D., Descurieux, M., Van Opstal, O., Van Wijnendaele, F., Hauser, P., Voet, P., and DeWilde, M.,** Expression of different forms of hepatitis B virus envelope proteins in yeast, in *Viral Hepatitis and Liver Disease*, Zuckerman, A. J., Ed., Alan R. Liss, New York, 1988, 304.
116. **Langley, K. E., Egan, K. M., Barendt, J. M., Parker, C. G., and Bitter, G. A.,** Characterization of purified hepatitis B surface antigen containing pre-S(2) epitopes expressed in *Saccharomyces cerevisiae*, *Gene*, 67, 229, 1988.
117. **Imamura, T., Araki, M., Miyanohara, A., Nakao, J., Yonemura, H., Ohtomo, N., and Matsubara, K.,** Expression of hepatitis B virus middle and large surface antigen genes in *Saccharomyces cerevisiae*, *J. Virol.*, 61, 3543, 1987.

118. **Kniskern, P. J., Hagopian, A., Burke, P., Dunn, N., Montgomery, D. L., Schultz, L. D., Schulman, C. A., Carty, C. E., Maigetter, R. Z., Wampler, D. E., Lehman, E. D., Yamazaki, S., Kubek, D. J., Emini, E. A., Miller, W. J., Hurni, W. M., and Ellis, R. W.,** The application of molecular biology to the development of novel vaccines, *Adv. Exp. Biol. Med.*, 251, 83, 1989.
119. **Kniskern, P. J., Hagopian, A., Montgomery, D. L., Carty, C. E., Burke, P., Schulman, C. A., Hofmann, K. J., Bailey, F. J., Dunn, N. R., Schultz, L. D.,Hurni, W. M., Miller, W. J., Ellis, R. W., and Maigetter, R. Z.,** Constitutive and regulated expression of hepatitis B virus (HBV) preS2+S protein in yeast, in *ACS Symp. Microbiological and Biochemical Technology*, American Chemical Society, Washington, D.C., 1991, in press.
120. **Kniskern, P. J., Hagopian, A., Burke, P., Dunn, N., Emini, E. A., Miller, W. J., Yamazaki, S., and Ellis, R. W.,** A candidate vaccine for hepatitis B containing the complete viral surface protein, *Hepatology*, 8, 82, 1988.

INDEX

A

B

C

D

E

I

J

K

L

M

N

O

P

Q

R

S

T

U

V

W